Computational Methods for Fracture

Computational Methods for Fracture

Special Issue Editor
Timon Rabczuk

MDPI • Basel • Beijing • Wuhan • Barcelona • Belgrade

MDPI

Special Issue Editor
Timon Rabczuk
Bauhaus University Weimar
Germany

Editorial Office
MDPI
St. Alban-Anlage 66
4052 Basel, Switzerland

This is a reprint of articles from the Special Issue published online in the open access journal *Applied Sciences* (ISSN 2076-3417) from 2018 to 2019 (available at: https://www.mdpi.com/journal/applsci/special_issues/Computational_Methods_for_Fracture).

For citation purposes, cite each article independently as indicated on the article page online and as indicated below:

LastName, A.A.; LastName, B.B.; LastName, C.C. Article Title. *Journal Name* **Year**, *Article Number*, Page Range.

ISBN 978-3-03921-686-4 (Pbk)
ISBN 978-3-03921-687-1 (PDF)

Contents

About the Special Issue Editor

Timon Rabczuk is currently Chair of Computational Mechanics, Bauhaus Universität-Weimar since 2009. Prof. Rabczuk's research focus is Computational Solid Mechanics with an emphasis on method development for problems involving fracture and failure of solids and fluid-structure interaction. Prof. Rabczuk is particularly interested in developing multiscale methods and in their application to computational materials design: Constitutive Modeling; Material Instabilities; Fracture, Strain Localization; Numerical Methods (Extended Finite Element and Meshfree Methods); Isogeometric Analysis; Computational Fluid-Structure Interaction; and Biomechanical Engineering. Since 2014, he has been highly cited in engineering and computer science journals. He has published more than 400 papers in peer-reviewed international journals. His citation rate is about 20,000 and he has an h-index of 79 in Google Scolar. Also, he is an editorial board member of the international journal *Applied Sciences*.

Preface to "Computational Methods for Fracture"

The prediction of fracture and material failure is of major importance for the safety and reliability of engineering structures and the efficient design of novel materials. Experimental testing is often cumbersome, expensive, and in certain cases unfeasible (as in civil engineering, when it is not possible to test the structures in the laboratory). Therefore, computational modeling of fracture and failure of engineering systems and materials has been the focus of research for many years, and there has been tremendous advancement in the past two decades with methods such as the Extended Finite Element Method (XFEM) developed in 1999, peridynamics (2000), the cracking particles method (2004) or phase field models (2009). There has been also a great deal of effort in developing multiscale methods for the design of new materials, such as the Extended Bridging Domain Method or the MAD method. The main focus of this book is on computational methods for fracture. However, research related to validation, uncertainty quantification, large-scale engineering applications, and constitutive modeling are also addressed.

Timon Rabczuk
Special Issue Editor

applied
sciences

MDPI

Editorial

Special Issue "Computational Methods for Fracture"

Timon Rabczuk

Institut für Strukturmechanik, Bauhaus University Weimar, Marienstrasse 15, 99423 Weimar, Germany; timon.rabczuk@uni-weimar.de

Received: 19 August 2019; Accepted: 20 August 2019; Published: 21 August 2019

The prediction of fracture and material failure is of major importance for the safety and reliability of engineering structures and the efficient design of novel materials. Experimental testing is often cumbersome, expensive and, in certain cases, unfeasible, for instance in civil engineering when it is not possible to test the structures in the laboratory. Therefore, computational modeling of fracture and failure of engineering systems and materials has been the focus of research for many years, and there has been tremendous advancements in the past two decades with methods such as the Extended Finite Element Method (XFEM) developed in 1999, peridynamics (2000), the cracking particles method (2004) and phase field models (2009). There has also been a great deal of effort made in developing multiscale methods for the design of new materials, such as the Extended Bridging Domain Method or the MAD method. The main focus of this book is computational methods for fracture. However, articles concerning issues related to validation, uncertainty quantification, large-scale engineering applications and constitutive modeling are also addressed.

This book offers a collection of 17 scientific papers about computational modeling of fracture [1–17]. Some manuscripts propose new computational methods or the improvement of existing cutting-edge methods for fracture. Other manuscripts apply state-of-the-art methods to challenging problems in engineering and materials science.

These contributions can be classified into two categories:

1. Methods which treat the crack as strong discontinuity, such as peridynamics, scaled boundary elements or specific versions of the smoothed finite element methods applied to fracture;
2. Continuous approaches to fracture based on, for instance, phase field models or continuum damage mechanics. On the other hand, this book also offers a wide application range where state-of-the-art techniques are employed to solve challenging engineering problems including fractures in rock, glass, and concrete. Larger systems are also studied, including subway stations due to fire, arch dams and concrete decks.

References

1. Gou, Y.; Cai, Y.; Zhu, H. A Simple High-Order Shear Deformation Triangular Plate Element with Incompatible Polynomial Approximation. *Appl. Sci.* **2018**, *8*, 975. [CrossRef]
2. Schreter, M.; Neuner, M.; Hofstetter, G. Evaluation of the Implicit Gradient-Enhanced Regularization of a Damage-Plasticity Rock Model. *Appl. Sci.* **2018**, *8*, 1004. [CrossRef]
3. Li, J.; Gao, X.; Fu, X.; Wu, C.; Lin, G. A Nonlinear Crack Model for Concrete Structure Based on an Extended Scaled Boundary Finite Element Method. *Appl. Sci.* **2018**, *8*, 1067. [CrossRef]
4. Díaz, R.; Wang, H.; Mang, H.; Yuan, Y.; Pichler, B. Numerical Analysis of a Moderate Fire inside a Segment of a Subway Station. *Appl. Sci.* **2018**, *8*, 2116. [CrossRef]
5. Bian, P.; Liu, T.; Qing, H.; Gao, C. 2D Micromechanical Modeling and Simulation of Ta-Particles Reinforced Bulk Metallic Glass Matrix Composite. *Appl. Sci.* **2018**, *8*, 2192. [CrossRef]
6. Freimanis, A.; Kaewunruen, S. Peridynamic Analysis of Rail Squats. *Appl. Sci.* **2018**, *8*, 2299. [CrossRef]

7. Chen, X.; Xie, W.; Xiao, Y.; Chen, Y.; Li, X. Progressive Collapse Analysis of SRC Frame-RC Core Tube Hybrid Structure. *Appl. Sci.* **2018**, *8*, 2316. [CrossRef]
8. Oucif, C.; Mauludin, L. Continuum Damage-Healing and Super Healing Mechanics in Brittle Materials: A State-of-the-Art Review. *Appl. Sci.* **2018**, *8*, 2350. [CrossRef]
9. Bhowmick, S.; Liu, G. Three Dimensional CS-FEM Phase-Field Modeling Technique for Brittle Fracture in Elastic Solids. *Appl. Sci.* **2018**, *8*, 2488. [CrossRef]
10. Lin, P.; Wei, P.; Wang, W.; Huang, H. Cracking Risk and Overall Stability Analysis of Xulong High Arch Dam: A Case Study. *Appl. Sci.* **2018**, *8*, 2555. [CrossRef]
11. Ma, H.; Shi, X.; Zhang, Y. Long-Term Behaviour of Precast Concrete Deck Using Longitudinal Prestressed Tendons in Composite I-Girder Bridges. *Appl. Sci.* **2018**, *8*, 2598. [CrossRef]
12. Zhu, Y.; Wang, X.; Deng, S.; Chen, W.; Shi, Z.; Xue, L.; Lv, M. Grouting Process Simulation Based on 3D Fracture Network Considering Fluid–Structure Interaction. *Appl. Sci.* **2019**, *9*, 667. [CrossRef]
13. Bahmani, B.; Abedi, R.; Clarke, P. A Stochastic Bulk Damage Model Based on Mohr-Coulomb Failure Criterion for Dynamic Rock Fracture. *Appl. Sci.* **2019**, *9*, 830. [CrossRef]
14. Cheng, P.; Zhuang, X.; Zhu, H.; Li, Y. The Construction of Equivalent Particle Element Models for Conditioned Sandy Pebble. *Appl. Sci.* **2019**, *9*, 1137. [CrossRef]
15. Onoda, M. Topological Photonic Media and the Possibility of Toroidal Electromagnetic Wavepackets. *Appl. Sci.* **2019**, *9*, 1468. [CrossRef]
16. Yang, Y.; Chu, S.; Chen, H. Prediction of Shape Change for Fatigue Crack in a Round Bar Using Three-Parameter Growth Circles. *Appl. Sci.* **2019**, *9*, 1751. [CrossRef]
17. Egger, A.; Pillai, U.; Agathos, K.; Kakouris, E.; Chatzi, E.; Aschroft, I.; Triantafyllou, S. Discrete and Phase Field Methods for Linear Elastic Fracture Mechanics: A Comparative Study and State-of-the-Art Review. *Appl. Sci.* **2019**, *9*, 2436. [CrossRef]

applied
sciences

MDPI

Article

A Simple High-Order Shear Deformation Triangular Plate Element with Incompatible Polynomial Approximation

Yudan Gou *, Yongchang Cai and Hehua Zhu

College of Civil Engineering, Tongji University, Shanghai 200092, China; yccai@tongji.edu.cn (Y.C.);
zhuhehua@tongji.edu.cn (H.Z.)
*   Correspondence: 1610257@tongji.edu.cn; Tel.: +86-021-6598-3809

Received: 11 May 2018; Accepted: 12 June 2018; Published: 14 June 2018

Featured Application: Due to the mathematical complexity raised by a high continuity requirement, developing simple/efficient standard finite elements with general polynomial approximations applicable for arbitrary HSDTs seems to be a difficult task at the present theoretical level. In this article, a series of High-order Shear Deformation Triangular Plate Elements (HSDTPEs) are developed using polynomial approximation for the analysis of isotropic thick-thin plates, through-thickness functionally graded plates, and cracked plates. The HSDTPEs have the advantage of simplicity in formulation, are free from shear locking, avoid using a shear correction factor and reduced integration, and provide stable solutions for thick and thin plates. The work can be further applied to plates and shells analysis with arbitrary shapes of elements, as well as more general problems related to the shear deformable effect, such as fracture and functionally graded plates.

Abstract: The High-order Shear Deformation Theories (HSDTs) which can avoid the use of a shear correction factor and better predict the shear behavior of plates have gained extensive recognition and made quite great progress in recent years, but the general requirement of C^1 continuity in approximation fields in HSDTs brings difficulties for the numerical implementation of the standard finite element method which is similar to that of the classic Kirchhoff-Love plate theory. As a strong complement to HSDTs, in this work, a series of simple High-order Shear Deformation Triangular Plate Elements (HSDTPEs) using incompatible polynomial approximation are developed for the analysis of isotropic thick-thin plates, cracked plates, and through-thickness functionally graded plates. The elements employ incompatible polynomials to define the element approximation functions $u/v/w$, and a fictitious thin layer to enforce the displacement continuity among the adjacent plate elements. The HSDTPEs are free from shear-locking, avoid the use of a shear correction factor, and provide stable solutions for thick and thin plates. A variety of numerical examples are solved to demonstrate the convergence, accuracy, and robustness of the present HSDTPEs.

Keywords: plate; FSDT; HSDT; Mindlin; incompatible approximation; fracture

---

1. Introduction

The classic Kirchhoff-Love plate theory based on the assumption that a plane section perpendicular to the mid-plane of the plate before deformation remains plane and perpendicular to the deformed mid-plane after deformation is the simplest plate theory in engineering analysis. However, the Kirchhoff-Love plate theory is only applicable for thin plates due to the neglecting of the shear deformation effects. The most well-known and earliest plate theories that take into account the shear deformation effects were proposed by Reissner [1] and Mindlin [2], in which the Mindlin plate theory was based on an assumption of a linear variation of in-plane displacements through the thickness of the plate, referred to as the First-order Shear Deformation Theory (FSDT). The plate elements derived from the FSDT only require C^0 continuity in approximation fields, have the advantages of physical clarity and simplicity of application [3], and hence were widely accepted and used to model thick-thin plates by scientists and engineers. Unfortunately, the FSDT elements suffer from the shear-locking problem when the thickness to length ration of the plate becomes very small, due to inadequate dependence among transverse deflection and rotations using an ordinary low-order finite element [4]. Quite a large number of techniques have been developed to overcome this problem, such as the assumed shear strain approach, the discrete Kirchhoff/Mindlin representation, the mixed/hybrid formulation, and the reduced/selected integration [5–15]. These formulations are free from shear locking and are applicable to a wide range of practical engineering problems, but in general, it is rather complex and time consuming to include the transverse shear effects for thick plates, which would also lead to complexity and difficulty in the programming. Moreover, the assumption of FSDT causes constant transverse shear strains and stresses across the thickness, which violates the conditions of zero transverse shear stresses on the top and bottom surfaces of plates. A shear correction factor is therefore required to properly compute the transverse shear stiffness. The finding of such a shear correction factor in FSDT is difficult since it depends on geometric parameters, material, loading and boundary conditions, etc. [16].

In recent years, High-order Shear Deformation Theories (HSDTs) have gained extensive recognition and made quite great progress [4,16–43]. Based on polynomial or non-polynomial transverse shear functions, various HSDTs have been proposed to avoid the use of a shear correction factor, and to better predict the shear behavior of the plate, for instance, the third-order shear deformation theory [17,18], the fifth-order shear deformation theory [19], the exponential shear deformation theory [20], the hyperbolic shear deformation theory [21], and the combined or mixed HSDTs [22,23]. Please see Thai and Kim [16] and Caliri et al. [24] for a comprehensive review of HSDTs. In HSDTs, the bending angles of rotation and shear angles can be treated as independent variables, and the shear-locking problem encountered in FSDT can be well-solved [4]. In [30–43], two well-know HSDTs named as equivalent single layer (ESL) and layer-wise (LW) models are developed to evaluate the effective mechanical behavior of composite structures correctly. The accuracy and reliability of HSDTs have been illustrated by numerous examples in the literature [4,17,25–43]. However, the general requirement of C^1 continuity in approximation fields in HSDTs brings difficulties for the numerical implementation of the standard Finite Element Method (FEM), which is similar to that of the classic Kirchhoff-Love plate theory. Most examples in the literature are focused on the analytical/numerical solutions of simple Navier-type or Levy-type square plates. The numerical examples reported for the C^1 rectangular finite element using Lagrange interpolation and Hermite interpolation proposed by Reddy [25] and the C^0 continuous isoparametric Lagrangian finite element with 63 Degrees Of Freedom (DOFs) per element proposed by Gulshan et al. [44] are also limited to the rectangular plate or skew plate. Owing to the striking feature of capturing the high-order continuity well, the Meshless Methods (MM) and IsoGeometric Analysis methods (IGA) appear to be suitable potential methods to construct the numerical formulations for the plate based on HSDTs. The successful implementation of MM [45–48] and IGA [19,23,49–54] in a number of thick-thin plates with arbitrary geometries can be found in the literature.

From the above literature review, it is observed that, due to the mathematical complexity raised by the high continuity requirement, developing simple/efficient standard finite elements with general polynomial approximations applicable for arbitrary HSDTs seems to be a difficult and unreachable task at the present theoretical level. In Cai and Zhu [55], a locking-free MTP9 (Mindlin type Triangular Plate element with nine degrees of freedom) using incompatible polynomial approximation is proposed. It also provides a new way and methodology to develop simple and efficient plate/shell elements based on HSDTs. In this work, with a similar procedure as the MTP9, a series of simple High-order Shear Deformation Triangular Plate Elements (HSDTPEs) using incompatible polynomial approximation are developed for the analysis of isotropic thick-thin plates and through-thickness functionally graded plates. In the HSDTPEs, different orders of general polynomials can be easily employed as element approximation functions, the displacement continuity among the adjacent plate elements can be equivalently enforced by a fictitious thin layer which has a definite physical meaning, and consequently, there are no extra continuity requirements under the theoretical framework of the present HSDTPEs. The HSDTPEs avoid the shear-locking problem and the use of a shear correction factor, and have a good convergence rate and high accuracy for both thick and thin plates. Several representative numerical examples are solved and compared to validate the performance of the present HSDTPEs.

2. Basic Theory of HSDTPEs

2.1. Incompatible Polynomial Approximation over Each Triangular Element

Consider a linear elastic plate with a length a, width b, and thickness h undergoing infinitesimal deformation, as illustrated in Figure 1. The mid-plane of the plate is divided into arbitrary triangular elements, as shown in Figure 2. The displacement function of the most well-known HSDTs [17] for each triangular element e_i is generally defined by:

$$
\begin{cases}
u(x,y,z) = u_0(x,y) - z\theta_x + g(z)\left(\frac{\partial w_0}{\partial x} - \theta_x\right) \\
v(x,y,z) = v_0(x,y) - z\theta_y + g(z)\left(\frac{\partial w_0}{\partial y} - \theta_y\right) \\
w(x,y,z) = w_0(x,y)
\end{cases}
\tag{1}
$$

where u_0, v_0 and w_0 are the in-plane and transverse displacements at the mid-plane, respectively; $\mathbf{u} = [u,v,w]^\mathrm{T}$ denotes the displacements of a point $\mathbf{x}$ on the plate; θ_x and θ_y are the rotations of the normal to the cross section; z is the coordinate in the transverse direction; and $g(z)$ describes the distribution of shear effect in the thickness direction. A review of transverse shear functions $g(z)$ can be found in Nguyen et al. [56]. For isotropic plates with infinitesimal strains, the in-plane displacements $u_0(x,y)$ and $v_0(x,y)$ can be neglected because the thickness h is much smaller than the characteristic length a and b, and the transverse displacement is much smaller than the thickness h, which leads to $u_0(x,y) \approx 0$ and $v_0(x,y) \approx 0$ at the mid-plane. The transverse normal displacement w can also be assumed as $w = w(x,y,z)$, which is not a constant along the z axis, and can be defined by the ESL or LW models [30–43] to capture the effective mechanical behavior along the thickness of composite structures well.

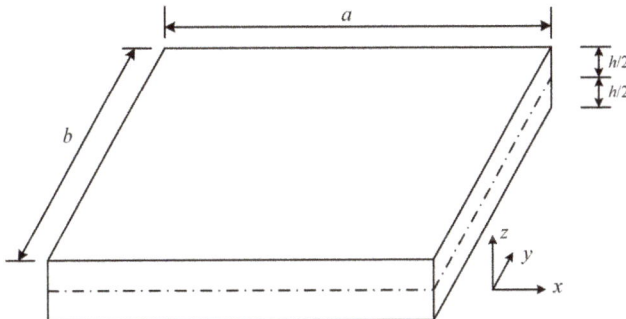

Figure 1. A linear elastic plate.

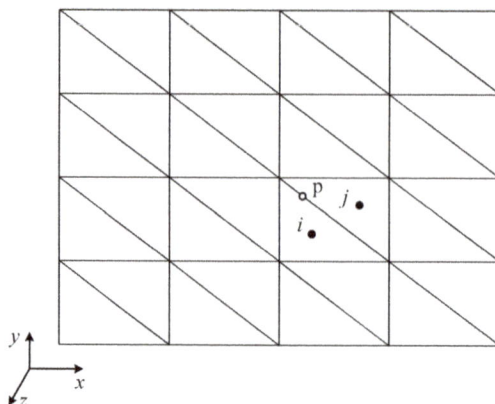

Figure 2. Triangular elements for the mid-plane of the plate.

To demonstrate the performance of the present theory for various transverse shear functions, the third-order shear function $g(z) = -\frac{4z^3}{3h^2}$ [17] and the fifth-order shear function $g(z) = -\frac{z}{8} - \frac{2z^3}{h^2} + \frac{2z^5}{h^4}$ [19] are used to develop the Third-order Shear Deformation Triangular Plate Element (TrSDTPE) and Fifth-order Shear Deformation Triangular Plate Element (FfSDTPE), respectively. For the special case $g(z) = 0$, Equation (1) is actually the expression of Mindlin plate theory (or FSDT). The corresponding plate element is referred to as FiSDTPE (First-order Shear Deformation Triangular Plate Element) for comparison in the paper.

We assume that:

$$\begin{cases} u_0 = \mathbf{P}^2 \mathbf{a}^{u_0} \\ v_0 = \mathbf{P}^2 \mathbf{a}^{v_0} \\ \theta_x = \mathbf{P}^2 \mathbf{a}^{\theta_x} \\ \theta_y = \mathbf{P}^2 \mathbf{a}^{\theta_y} \\ w_0 = \mathbf{P}^3 \mathbf{a}^w \end{cases} \tag{2}$$

where $\mathbf{a}^{u_0} = \begin{bmatrix} a_1 & a_2 & \cdots & a_6 \end{bmatrix}^{\mathrm{T}}$, $\mathbf{a}^{v_0} = \begin{bmatrix} a_7 & a_8 & \cdots & a_{12} \end{bmatrix}^{\mathrm{T}}$, $\mathbf{a}^{\theta_x} = \begin{bmatrix} a_{13} & a_{14} & \cdots & a_{18} \end{bmatrix}^{\mathrm{T}}$, $\mathbf{a}^{\theta_y} = \begin{bmatrix} a_{19} & a_{20} & \cdots & a_{24} \end{bmatrix}^{\mathrm{T}}$, $\mathbf{a}^w = \begin{bmatrix} a_{25} & a_{26} & \cdots & a_{34} \end{bmatrix}^{\mathrm{T}}$ are the vector of generalized approximation DOFs (degrees of freedom) of the triangular element e_i, $\mathbf{P}^2$ is the second-order polynomial basis function, and $\mathbf{P}^3$ is the third-order polynomial basis function in which:

$$\mathbf{P}^2(\mathbf{x}) = \begin{bmatrix} 1 & x_0 & y_0 & x_0^2 & x_0 y_0 & y_0^2 \end{bmatrix} \tag{3}$$

$$\mathbf{P}^3(\mathbf{x}) = \begin{bmatrix} 1 & x_0 & y_0 & x_0^2 & x_0 y_0 & y_0^2 & x_0^3 & x_0^2 y_0 & y_0^2 x_0 & y_0^3 \end{bmatrix} \tag{4}$$

where $x_0 = x - x_i$, $y_0 = y - y_i$, (x_i, y_i) are the coordinates of the central point of element e_i. It should be noted that only triangular elements, as well as second-order and third-order polynomial functions, are implemented in the paper, but actually, arbitrary shape of elements and arbitrary orders of polynomials can also be easily employed to derive high-order shear deformation plate elements in the present work.

Substituting Equation (2) into Equation (1), the displacement approximation over element e_i can be further expressed as:

$$\mathbf{u}^e = \left\{ \begin{array}{c} u \\ v \\ w \end{array} \right\} = \begin{bmatrix} \mathbf{P}^2 & 0 & \alpha_z \mathbf{P}^2 & 0 & g(z)\mathbf{P}^3_{,x} \\ 0 & \mathbf{P}^2 & 0 & \alpha_z \mathbf{P}^2 & g(z)\mathbf{P}^3_{,y} \\ 0 & 0 & 0 & 0 & \mathbf{P}^3 \end{bmatrix} \left\{ \begin{array}{c} \mathbf{a}^{u_0} \\ \mathbf{a}^{v_0} \\ \mathbf{a}^{\theta_x} \\ \mathbf{a}^{\theta_y} \\ \mathbf{a}^{w} \end{array} \right\} = \mathbf{N}^e \mathbf{a}^e \tag{5}$$

where $\alpha_z = -z - g(z)$, $\mathbf{P}^3_{,x} = \frac{\partial \mathbf{P}^3}{\partial x}$, $\mathbf{P}^3_{,y} = \frac{\partial \mathbf{P}^3}{\partial y}$,

$$\mathbf{N}^e = \left\{ \begin{array}{c} \mathbf{N}^u \\ \mathbf{N}^v \\ \mathbf{N}^w \end{array} \right\} = \begin{bmatrix} \mathbf{P}^2 & 0 & \alpha_z \mathbf{P}^2 & 0 & g(z)\mathbf{P}^3_{,x} \\ 0 & \mathbf{P}^2 & 0 & \alpha_z \mathbf{P}^2 & g(z)\mathbf{P}^3_{,y} \\ 0 & 0 & 0 & 0 & \mathbf{P}^3 \end{bmatrix} \tag{6}$$

$$\mathbf{a}^e = \begin{bmatrix} a_1 & a_2 & \cdots & a_{34} \end{bmatrix}^{\mathrm{T}} \tag{7}$$

The strain–displacement relations of the linear elastic problem are given by:

$$\varepsilon_x = \frac{\partial u}{\partial x}, \varepsilon_y = \frac{\partial v}{\partial y}, \varepsilon_z = \frac{\partial w}{\partial z} \approx 0, \gamma_{xy} = \frac{\partial u}{\partial y} + \frac{\partial v}{\partial x}, \gamma_{yz} = \frac{\partial w}{\partial y} + \frac{\partial v}{\partial z}, \gamma_{xz} = \frac{\partial u}{\partial z} + \frac{\partial w}{\partial x} \tag{8}$$

Substituting Equation (5) into Equation (8), we have:

$$\varepsilon = \mathbf{L}\mathbf{u}^e = \mathbf{L}\mathbf{N}^e \mathbf{a}^e = \mathbf{B}\mathbf{a}^e \tag{9}$$

where $\varepsilon = \begin{bmatrix} \varepsilon_x, \varepsilon_y, \gamma_{xy}, \gamma_{yz}, \gamma_{xz} \end{bmatrix}^{\mathrm{T}}$ is the strain vector and B is the strain matrix, where:

$$\mathbf{B} = \mathbf{L}\mathbf{N}^e \tag{10}$$

L is a differential operator where:

$$\mathbf{L} = \begin{bmatrix} \frac{\partial}{\partial x} & 0 & 0 \\ 0 & \frac{\partial}{\partial y} & 0 \\ \frac{\partial}{\partial y} & \frac{\partial}{\partial x} & 0 \\ 0 & \frac{\partial}{\partial z} & \frac{\partial}{\partial y} \\ \frac{\partial}{\partial z} & 0 & \frac{\partial}{\partial x} \end{bmatrix} \tag{11}$$

For an isotropic linear elastic material, the stress–strain relations in element e_i are given by:

$$\sigma = \mathbf{D}\mathbf{B}\mathbf{a}^e \tag{12}$$

where $\sigma = \begin{bmatrix} \sigma_x, \sigma_y, \tau_{xy}, \tau_{yz}, \tau_{xz} \end{bmatrix}^{\mathrm{T}}$, the transverse stress σ_z is assumed to be ignored for plate structures, and the elasticity matrix is:

$$\mathbf{D} = D_0 \begin{bmatrix} 1 & v & 0 & 0 & 0 \\ v & 1 & 0 & 0 & 0 \\ 0 & 0 & \frac{1-v}{2} & 0 & 0 \\ 0 & 0 & 0 & \frac{1-v}{2k} & 0 \\ 0 & 0 & 0 & 0 & \frac{1-v}{2k} \end{bmatrix} \tag{13}$$

where $D_0 = \frac{E}{1-v^2}$, E is the elastic modulus and v is the Poisson ratio. As mentioned above, a shear correction factor k is required to properly compute the transverse shear stiffness in the FiSDTPE with the assumption of FSDT, which causes constant transverse shear strains and stresses across the thickness and violates the conditions of zero transverse shear stresses on the top and bottom surfaces of plates [16]. Usually, k is taken as $k = 1.2$ for the special case of FiSDTPE in Equation (13) according to the principle of the equivalence of strain energy. However, the high-order shear deformation theory gives a parabolic distribution of the transverse stresses/strains directly and avoids the use of a shear correction factor, and thus k is taken as $k = 1.0$ in Equation (13) for the rest of the HSDTPEs.

Therefore, the strain energy of element e_i can be derived as:

$$\Pi^e = \frac{1}{2}(\mathbf{a}^e)^\mathrm{T} \int_{-h/2}^{h/2} \left(\iint_{\Delta e_i} \mathbf{B}^\mathrm{T} \mathbf{D} \mathbf{B} dx dy \right) dz \, \mathbf{a}^e \tag{14}$$

For the plate made of Functionally Graded (FG) materials which is created by mixing two distinct material phases, the composition of the FG materials is in general assumed to be varied continuously from the top to the bottom surface. There are many kinds of FG materials made from all classes of solids. But for the sake of simplicity and convenience, only a ceramic-metal composite is considered and implemented to test the performance of the HSDTPEs in the present study, and the power-law [25,32,45] is used to describe the through-the-thickness distribution of FG materials, which is expressed as:

$$V_c(z) = \left(\frac{1}{2} + \frac{z}{2} \right)^n \quad (0 \leq n \leq \infty) \tag{15}$$

$$P(z) = (P_c - P_m)V_c + P_m \tag{16}$$

where n is the volume fraction exponent, V_c is the volume fraction of the ceramic, P_m represents the material property of the metal, P_c represents the material property of the ceramic, and P denotes the effective material property. In this work, the Young's modulus E in Equation (13) varies according to Equation (16) and the Poisson ratio v is assumed to be constant for the analysis of functionally graded plates.

2.2. Fictitious Thin Layer between Adjacent Triangular Elements

According to the definition of the displacement approximation in Equation (5), the deformation along the share boundary of the adjacent elements e_i and e_j is discontinuous, which means that $\mathbf{u}^{e_i}(x_p, y_p, z_p) \neq \mathbf{u}^{e_j}(x_p, y_p, z_p)$ for an arbitrary point p along the share boundary shown in Figure 2, where point p has local coordinates (s_p, n_p, z_p) and global coordinates (x_p, y_p, z_p). Here, we introduce a fictitious thin layer e_l shown in Figure 3 to enforce the continuous condition over the share boundary of the elements. The geometry dimensions of e_l are the length l, width d, and height h, where $d \ll l$, $d \ll h$, and h is the thickness of the plate in the transverse direction.

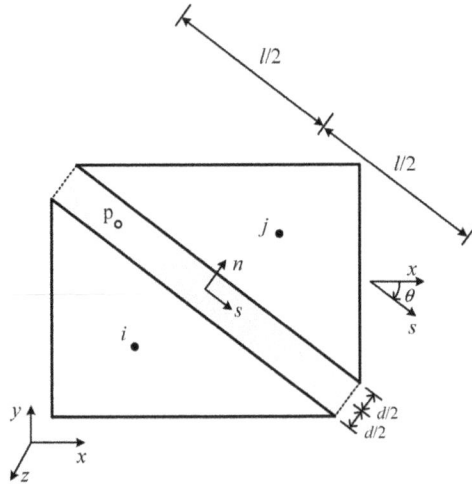

Figure 3. A fictitious thin layer between adjacent triangular elements.

Because $d \ll l$ and $d \ll h$, the strain–displacement relations $\varepsilon^l = [\gamma_{ns}, \varepsilon_n, \gamma_{nz}]^{\mathrm{T}}$ in thin layer e_l can be simplified as:

$$\gamma_{ns} = \frac{\partial \bar{u}}{\partial n} \approx \frac{\bar{u}^{p^j} - \bar{u}^{p^i}}{d}, \varepsilon_n = \frac{\partial \bar{v}}{\partial n} \approx \frac{\bar{v}^{p^j} - \bar{v}^{p^i}}{d}, \gamma_{nz} = \frac{\partial \bar{w}}{\partial n} \approx \frac{\bar{w}^{p^j} - \bar{w}^{p^i}}{d} \tag{17}$$

where $\bar{\mathbf{u}}^{p^i} = \left[\bar{u}^{p^i}, \bar{v}^{p^i}, \bar{w}^{p^i}\right]^{\mathrm{T}}$ is the displacement of point p in the local coordinate (s, n, z) computed by the approximation of triangular element e_i, and $\bar{\mathbf{u}}^{p^j} = \left[\bar{u}^{p^j}, \bar{v}^{p^j}, \bar{w}^{p^j}\right]^{\mathrm{T}}$ is the displacement of point p in local coordinate (s, n, z) computed by the approximation of triangular element e_j. $\bar{\mathbf{u}}^{p^j}$ and $\bar{\mathbf{u}}^{p^i}$ can be calculated using Equation (5).

Substitution of Equation (5) into Equation (17) yields:

$$\varepsilon^l = \frac{1}{d} \mathbf{N}^l \mathbf{a}^l \tag{18}$$

where

$$\mathbf{N}^l = \lambda^l \left[-\mathbf{N}^{e_i}(x_p, y_p, z_p) \quad \mathbf{N}^{e_j}(x_p, y_p, z_p) \right] \tag{19}$$

where $\mathbf{N}^{e_i}(x_p, y_p, z_p)$ is the shape function of point p in triangular element e_i and $\mathbf{N}^{e_j}(x_p, y_p, z_p)$ is the shape function of point p in triangular element e_j. λ^l is the transformation matrix of point p from the global coordinate (x_p, y_p, z_p) to the local coordinate (s_p, n_p, z_p), where:

$$\lambda^l = \begin{bmatrix} \cos\theta & \sin\theta & 0 \\ -\sin\theta & \cos\theta & 0 \\ 0 & 0 & 1 \end{bmatrix} \tag{20}$$

and

$$\mathbf{a}^l = \begin{bmatrix} \mathbf{a}^{e_i} \\ \mathbf{a}^{e_j} \end{bmatrix} \tag{21}$$

where $\mathbf{a}^{e_i}$ is the DOFs of element e_i and $\mathbf{a}^{e_j}$ is the DOFs of element e_j.

9

The stress–strain relations in fictitious thin layer e_l are then given by:

$$\sigma^l = \mathbf{D}^l \varepsilon^l \tag{22}$$

where $\sigma^l = [\tau_{ns}, \sigma_n, \tau_{nz}]^T$ and

$$\mathbf{D}^l = \begin{bmatrix} G_0 & 0 & 0 \\ 0 & E_0 & 0 \\ 0 & 0 & G_0/k \end{bmatrix} \tag{23}$$

where $G_0 = \frac{E}{2(1+v)}$ and $E_0 = \frac{E}{1-v^2}$. Similar to Equation (13), the shear correction factor k is taken as $k = 1.2$ for the special case of FiSDTPE, and $k = 1.0$ for the rest of the high-order shear deformation plate elements, including the TrSDTPE and FfSDTPE, without the need to use a shear correction factor.

The width d of fictitious thin layer e_l is an important artificial parameter for the present HSDTPEs, but it is easy to select a reasonable d to satisfy $d \ll l$ and $d \ll h$ for simplifying the strain–displacement relations of thin layer e_l in Equation (17). Numerical studies show that the variation of d in a large range has little effect on the accuracy of the calculation results. In this paper, width d is taken as $d = 0.0001l$.

Thus, the strain energy of thin layer e_l can be derived as:

$$\Pi^l = \frac{1}{2d} \left(\mathbf{a}^l\right)^T \int_{-h/2}^{h/2} \left(\int_{-l/2}^{l/2} \left(\mathbf{N}^l\right)^T \mathbf{D}^l \mathbf{N}^l ds \right) dz \, \mathbf{a}^l \tag{24}$$

2.3. Imposing Displacement Boundary Condition

As illustrated in Figure 4, along boundary 1–2 of element e_k, rotations $(\bar{\theta}_s, \bar{\theta}_n)$ in the local coordinate (s, n, z) or the displacements $(\bar{u}_0, \bar{v}_0, w_0)$ in the local coordinate (s, n, z) are fixed, where $(\bar{u}_0, \bar{v}_0, w_0)$ represents the in-plane and transverse displacements at the mid-plane. A fictitious thin layer e_b over boundary 1–2 shown in Figure 4 is also introduced to enforce the displacement boundary condition. We divide the displacement approximation in Equation (5) into two parts to the follow to separately enforce the rotation and mid-plane displacement boundaries.

$$\bar{u}^{Pr}\left(s_p, n_p, z_p\right) \approx \lambda^b \mathbf{N}^r\left(x_p, y_p, z_p\right)\mathbf{a}^r \tag{25}$$

$$\bar{u}^{Pt}\left(s_p, n_p, z_p\right) \approx \lambda^b \mathbf{N}^t\left(x_p, y_p, z_p\right)\mathbf{a}^t \tag{26}$$

where $\bar{u}^{Pr}$ represents the displacement function of point p for rotations in triangular element e_k, $\bar{u}^{Pt}$ represents the displacement function of point p for mid-plane displacements in triangular element e_k, and $\mathbf{a}^r$ and $\mathbf{a}^t$ are the corresponding DOFs of element e_k.

$$\mathbf{N}^r = \begin{bmatrix} 0 & 0 & \alpha_z \mathbf{P}^2 & 0 & g(z)\mathbf{P}^3_{,x} \\ 0 & 0 & 0 & \alpha_z \mathbf{P}^2 & g(z)\mathbf{P}^3_{,y} \\ 0 & 0 & 0 & 0 & 0 \end{bmatrix} \tag{27}$$

$$\mathbf{N}^t = \begin{bmatrix} \mathbf{P}^2 & 0 & 0 & 0 & 0 \\ 0 & \mathbf{P}^2 & 0 & 0 & 0 \\ 0 & 0 & 0 & 0 & \mathbf{P}^3 \end{bmatrix} \tag{28}$$

where "**0**" in bold in Equations (27) and (28) represents zero matrix , and λ^b is the transformation matrix similar to Equation (20) where

$$\lambda^b = \begin{bmatrix} \cos\omega & \sin\omega & 0 \\ -\sin\omega & \cos\omega & 0 \\ 0 & 0 & 1 \end{bmatrix} \tag{29}$$

Using the same derivation process as in Equation (24), the strain energy of the thin layer e_b is derived as:

$$\Pi^b = \frac{1}{2d} \int_{-h/2}^{h/2} \int_{-l/2}^{l/2} [(\mathbf{a}^r)^T (\overline{\mathbf{N}}^r)^T \mathbf{D}^r \overline{\mathbf{N}}^r \mathbf{a}^r + (\mathbf{a}^t)^T (\overline{\mathbf{N}}^t)^T \mathbf{D}^t \overline{\mathbf{N}}^t \mathbf{a}^t] ds dz \tag{30}$$

where $\mathbf{D}^b$ is calculated using Equation (23), $\overline{\mathbf{N}}^r = \lambda^b \mathbf{N}^r$, and $\overline{\mathbf{N}}^t = \lambda^b \mathbf{N}^t$.

Please refer to Cai and Zhu [55] for the detailed derivation of the displacement boundary condition fixed at a point or the given displacement boundary condition.

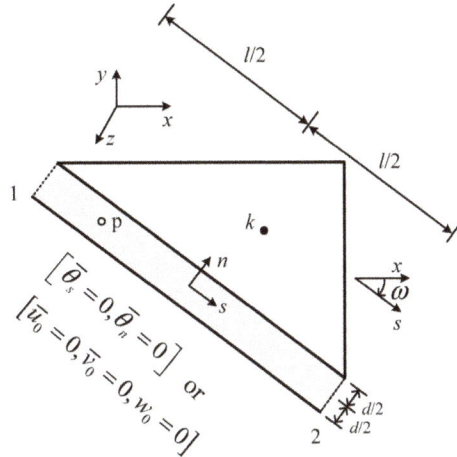

Figure 4. Fixed displacement boundary condition.

2.4. Load Boundary Condition

A distributed force $\mathbf{f}_0 = \begin{bmatrix} 0 & 0 & f_z\ (x,y) \end{bmatrix}^T$ along the transverse direction z is applied at element e_d, as illustrated in Figure 5. By using Equation (5), the external force potential energy of element e_d is written as:

$$\Pi^f = -(\mathbf{a}^{e_d})^T \iint_{\Delta e_d} (\mathbf{N}^{e_d})^T \mathbf{f}_0\ dx\ dy \tag{31}$$

where $\mathbf{N}^{e_d}$ is the shape function of element e_d calculated by Equation (5), and $\mathbf{a}^{e_d}$ is the DOFs of element e_d.

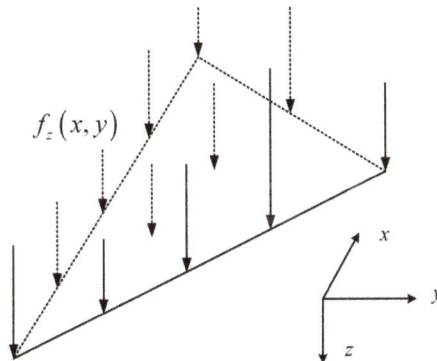

Figure 5. Distributed transverse force.

Similarly, a distributed resultant moment $\mathbf{M}_0 = \begin{bmatrix} M_s & M_{sn} & 0 \end{bmatrix}^{\mathrm{T}}$ is applied to the edge of element e_m, as illustrated in Figure 6. By using Equation (5), the external force potential energy of element e_m is written as:

$$\Pi^m = -(\mathbf{a}^{e_m})^{\mathrm{T}} \int_{-l/2}^{l/2} \left(\boldsymbol{\lambda}^m \tilde{\mathbf{N}}^{e_m} \right)^{\mathrm{T}} \mathbf{M}_0 \, ds \tag{32}$$

where $\mathbf{a}^{e_m}$ is the DOFs of element e_m, and $\tilde{\mathbf{N}}^{e_m}$ is the shape function of element e_m corresponding to the moment, where:

$$\tilde{\mathbf{N}}^{e_m} = \begin{bmatrix} 0 & 0 & \mathbf{P}^2 & 0 & 0 \\ 0 & 0 & 0 & \mathbf{P}^2 & 0 \\ 0 & 0 & 0 & 0 & 0 \end{bmatrix} \tag{33}$$

and the transformation matrix

$$\boldsymbol{\lambda}^m = \begin{bmatrix} \cos\beta & \sin\beta & 0 \\ -\sin\beta & \cos\beta & 0 \\ 0 & 0 & 1 \end{bmatrix} \tag{34}$$

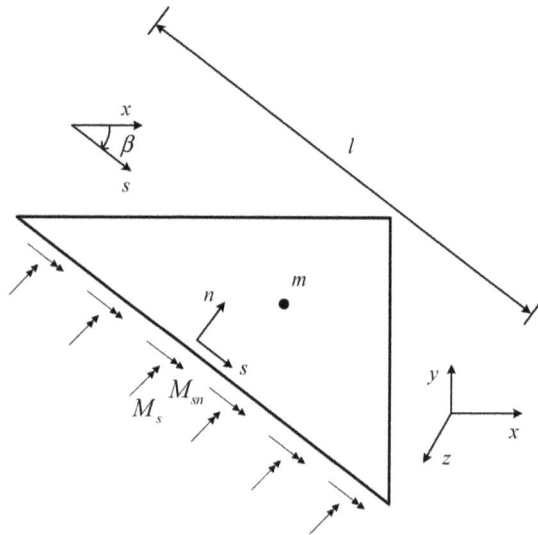

Figure 6. Moment boundary condition.

2.5. Equilibrium Equation

From Equations (14), (24), (30), (31), and (32), the total potential energy of a plate is obtained as:

$$\Pi = \sum \left(\Pi^e + \Pi^l + \Pi^b + \Pi^f + \Pi^m \right) \tag{35}$$

The variation of total potential energy Π results in the following discrete equation:

$$\frac{\partial \Pi}{\partial \mathbf{a}} = \sum \left(\mathbf{K}^e \mathbf{a}^e + \mathbf{K}^l \mathbf{a}^l + \mathbf{K}^b \mathbf{a}^b - \mathbf{F}^{ed} - \mathbf{F}^{em} \right) = 0 \tag{36}$$

where

$$\mathbf{K}^e = \int_{-h/2}^{h/2} \left(\iint_{\Delta e_i} \mathbf{B}^{\mathrm{T}} \mathbf{D} \mathbf{B} dx \, dy \right) dz \tag{37}$$

$$\mathbf{K}^l = \frac{1}{d} \int_{-h/2}^{h/2} \left(\int_{-l/2}^{l/2} \left(\mathbf{N}^l \right)^T \mathbf{D}^l \mathbf{N}^l \, ds \right) dz \tag{38}$$

$$\mathbf{K}^b = \frac{1}{d} \int_{-h/2}^{h/2} \int_{-l/2}^{l/2} \left[\left(\overline{\mathbf{N}}^r \right)^T \mathbf{D}^b \overline{\mathbf{N}}^r + \left(\overline{\mathbf{N}}^t \right)^T \mathbf{D}^b \overline{\mathbf{N}}^t \right] ds \, dz \tag{39}$$

$$\mathbf{F}^{ed} = \iint_{\Delta e_m} \left(\hat{\mathbf{N}}^{e_d} \right)^T \mathbf{f}_0 \, dx \, dy \tag{40}$$

$$\mathbf{F}^{em} = \int_{-l/2}^{l/2} \left(\boldsymbol{\lambda}^m \tilde{\mathbf{N}}^{em} \right)^T \mathbf{M}_0 \, ds \tag{41}$$

Assembling the above stiffness matrix and force vector, the equilibrium equation for a plate is then obtained as:

$$\mathbf{K} \cdot \mathbf{U} = \mathbf{F} \tag{42}$$

where $\mathbf{K}$ is the global stiffness matrix, $\mathbf{F}$ is the force vector, and $\mathbf{U}$ is the vector of DOFs to be solved.

As described in Senjanović et al. [4], the shear-locking problem could be well and naturally solved because the bending angles of rotation and shear angles are treated as independent variables in HSDTs. The regular full integration can be applied to make HSDTPEs valid for the thick-thin plates for the computation of Equation (42), for instance, seven quadrature points for each triangular element [57], four Gauss quadrature points for transverse direction z (where the analytical integration can also be applied for the direction z), and four Gauss quadrature points for the local direction s of each fictitious layer are used for the integration of the TrSDTPE using the third-order shear function $g(z)$ [58–60].

3. Analysis of Cracked Plates

The present HSDTPEs are also applied to the calculation of Stress Intensity Factors (SIFs) of cracked thick-thin plates. As illustrated in Figure 7, the mid plane of a cracked plate is taken as the x-y plane and is divided into arbitrary triangular elements. Accurate computation of SIFs remains challenging in the field of fracture mechanics. For plates loaded by a combination of bending and tension, the SIFs can also be computed by the Virtual Crack Closure Technique (VCCT) [61–63], the path-independent J-integral technique or interaction integral [64,65], and the stiffness derivative method [66]. In this paper, the Virtual Crack Closure Technique (VCCT) [61–63] is employed to calculate the SIFs of the cracked plate. For the convenience of implementing the VCCT, point T_2 shown in Figures 7 and 8 is temporarily moved to T_3 along the extended line direction of $T_1 - T$. In the local coordinates (s, n, z) shown in Figure 9, the relative displacements $[\Delta \overline{u}(s, z), \Delta \overline{v}(s, z), \Delta \overline{w}(s, z)]$ of $T_1 - T$ and the stresses $[\tau_{ns}(s, z), \sigma_n(s, z), \tau_{nz}(s, z)]$ of $T - T_1$ can be easily calculated using Equations (18) and (22) for the fictitious thin layer e_l. For example, assuming that the $T_1 - T$ is simulated by a fictitious thin layer with width d shown in Figure 3, the relative displacements can be evaluated by $\Delta \overline{u}(s, z) = \gamma_{ns} d$, $\Delta \overline{v}(s, z) = \varepsilon_n d$ and $\Delta \overline{w}(s, z) = \gamma_{nz} d$. Then, the energy release rate at crack tip T is obtained by the VCCT as:

$$\begin{cases} G_{\mathrm{I}} \cong \frac{1}{2hr_0} \int_{-h/2}^{h/2} \int_0^{r_0} \sigma_n(s, z) \Delta \overline{v}(s - r_0, z) \mathrm{d}s \, \mathrm{d}z \\ G_{\mathrm{II}} \cong \frac{1}{2hr_0} \int_{-h/2}^{h/2} \int_0^{r_0} \tau_{ns}(s, z) \Delta \overline{u}(s - r_0, z) \mathrm{d}s \, \mathrm{d}z \\ G_{\mathrm{III}} \cong \frac{1}{2hr_0} \int_{-h/2}^{h/2} \int_0^{r_0} \tau_{nz}(s, z) \Delta \overline{w}(s - r_0, z) \mathrm{d}s \, \mathrm{d}z \end{cases} \tag{43}$$

where G_{I} is the energy release rate of crack mode I, G_{II} is that of crack mode II, and G_{III} is that of crack mode III. Then, the SIFs of the crack tip can be computed by means of the relations between the energy release rate and SIFs for the plate theory, for instance, $K_1 = \sqrt{3EG_{\mathrm{I}}}$ [63].

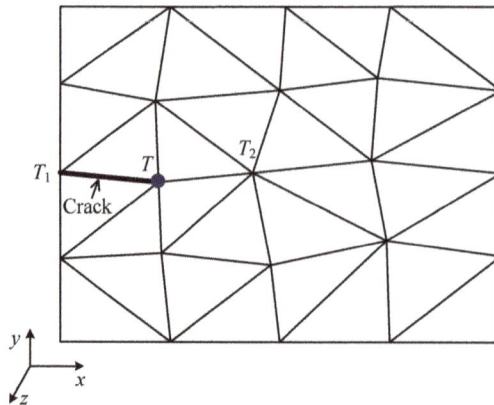

Figure 7. Triangular elements for a cracked plate.

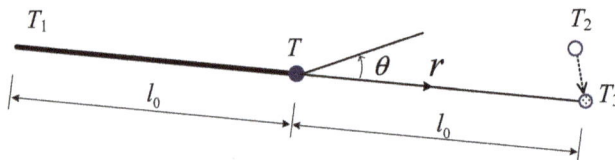

Figure 8. Minor movement for the implementation of VCCT.

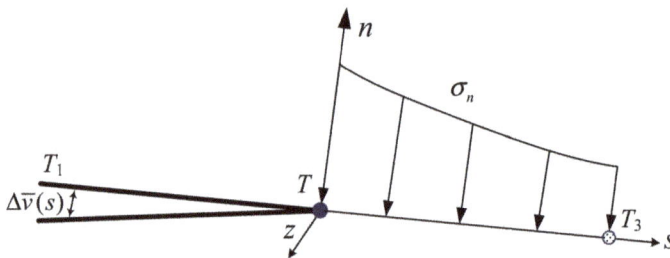

Figure 9. Calculating SIFs by VCCT.

4. Numerical Examples

4.1. Simply Supported Square Plate Subjected to Uniform Load

A simply supported square plate subjected to a uniform load q is tested to show the reliability and convergence of the present elements. The side length of the plate is L, and the thickness of the plate is h. A quarter of the plate is modeled as a result of symmetry, as illustrated in Figure 10. For the isotropic plates, the in-plane displacements (u_0, v_0) and their DOFs in Equations (1), (2), and (5) are neglected in the following analyses. The displacement boundary conditions of the present theory along the simply supported edges in local coordinates are $\bar{\theta}_s = 0$ and $w = 0$. The $n \times n$ regular mesh and irregular mesh illustrated in Figures 11 and 12 are employed for convergence studies.

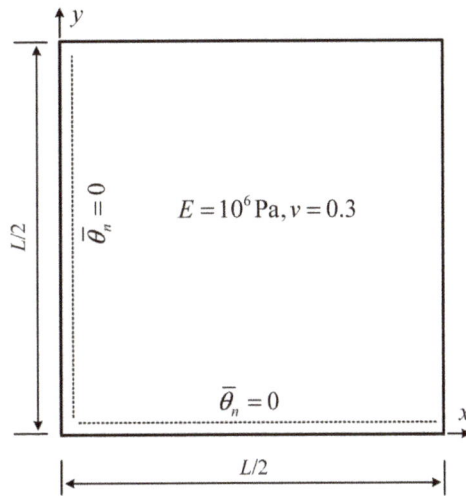

Figure 10. A quarter model of the square plate.

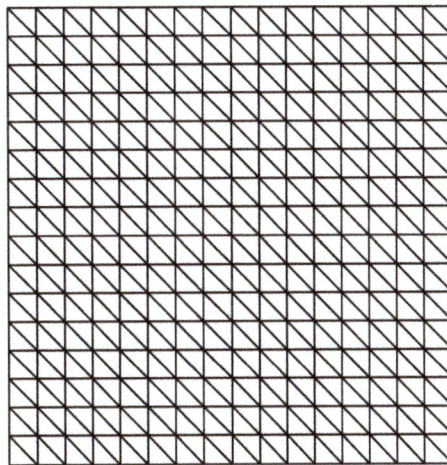

Figure 11. Regular 16 × 16 mesh for the square plate.

The elements DST-BL (Discrete Mindlin triangular plate element) [7] and RDKTM (Re-constituting discrete Kirchhoff triangular plate element) [14] have been selected for comparison with the present elements based on HSDTs. The reference solutions in the following Tables 1–6 are taken from Long et al. [67], which are also labeled as analytical solutions in [67]. Table 1 lists the normalized defection $W_0 = W_c / \frac{qL^4}{100D_b}$ of the simply supported square plate, where W_c is the central deflection of the plate and $D_b = \frac{Eh^3}{12(1-v^2)}$. Table 2 reports the normalized bending moment $M_0 = M_c / \frac{qL^2}{10}$ of the simply supported square plate, where M_c is the central bending moment of the plate. The convergence of the deflection for the simply supported square plate using different elements when the aspect ratio $h/L = 0.1$ is shown in Figure 13. It is observed that all the present TrSDTPE, FfSDTPE, and FiSDTPE shows a good convergence rate and high accuracy, and avoids the shear-locking problem. The results also indicate that the present elements are insensitive to element distortions of the irregular mesh shown in Figure 12.

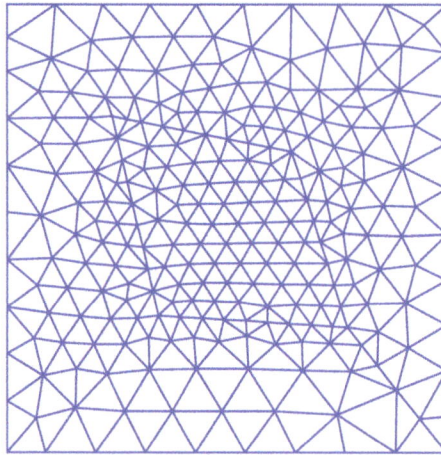

Figure 12. Irregular mesh for the square plate.

Table 1. Normalized deflection for the simply supported square plate.

h/L	0.001	0.01	0.10	0.15	0.20	0.25	0.30
TrSDTPE (4 × 4)	0.4027	0.4050	0.4266	0.4529	0.4897	0.5367	0.5941
TrSDTPE (8 × 8)	0.4058	0.4064	0.4273	0.4536	0.4903	0.5375	0.5949
TrSDTPE (16 × 16)	0.4063	0.4065	0.4274	0.4536	0.4904	0.5375	0.5950
TrSDTPE (Figure 12)	0.4063	0.4066	0.4274	0.4537	0.4904	0.5375	0.5950
FfSDTPE (16 × 16)	0.4063	0.4065	0.427	0.4528	0.4888	0.5350	0.5911
FiSDTPE (16 × 16)	0.4063	0.4065	0.4274	0.4537	0.4905	0.5379	0.5958
DST-BL (16 × 16)	0.4057	0.4059	0.4267	0.4529	0.4896	0.5367	0.5944
RDKTM (16 × 16)	0.4057	0.4059	0.4270	0.4532	0.4899	0.5371	0.5847
Ref. [67]	0.4064	0.4064	0.4273	0.4536	0.4906	0.5379	0.5956

Table 2. Normalized bending moment for the simply supported square plate.

h/L	0.001	0.01	0.10	0.15	0.20	0.25	0.30
TrSDTPE (4 × 4)	0.4266	0.4531	0.4741	0.4771	0.4785	0.4791	0.4794
TrSDTPE (8 × 8)	0.4633	0.4734	0.4786	0.4789	0.4791	0.4791	0.4791
TrSDTPE(16 × 16)	0.4759	0.4778	0.4788	0.4789	0.4789	0.4789	0.4789
TrSDTPE (Figure 12)	0.4807	0.4806	0.4807	0.4807	0.4807	0.4807	0.4807
FfSDTPE (16 × 16)	0.4757	0.4775	0.4788	0.4788	0.4788	0.4788	0.4788
FiSDTPE (16 × 16)	0.4762	0.4789	0.479	0.4790	0.4790	0.4790	0.4790
DST-BL (16 × 16)	0.4792	0.4788	0.4773	0.4770	0.4768	0.4767	0.4767
RDKTM (16 × 16)	0.4792	0.4790	0.4789	0.4790	0. 4790	0.4790	0.4790
Ref. [67]				0.4789			

Table 3. Convergence of normalized deflection with different width-to-length ratios.

d/l	0.1	0.01	0.001	0.0001	0.00001	0.000001	0.0000001
TrSDTPE	0.5235	0.4371	0.4283	0.4274	0.4273	0.4273	0.4273
FfSDTPE	0.5229	0.4367	0.4279	0.4270	0.4269	0.4269	0.4269
FiSDTPE	0.5247	0.4372	0.4283	0.4274	0.4273	0.4273	0.4273
Ref. [67]				0.4273			

Table 4. Normalized deflection for the clamped square plate.

h/L	0.001	0.01	0.10	0.15	0.20	0.25	0.30
TrSDTPE (4 × 4)	0.1171	0.1241	0.1459	0.1708	0.2044	0.2463	0.2962
TrSDTPE (8 × 8)	0.1255	0.1266	0.1491	0.1757	0.2112	0.2553	0.3077
TrSDTPE (16 × 16)	0.1265	0.1268	0.1497	0.1766	0.2124	0.2569	0.3097
TrSDTPE (Figure 12)	0.1266	0.1268	0.1496	0.1764	0.2122	0.2567	0.3095
FfSDTPE (16 × 16)	0.1265	0.1268	0.1491	0.1750	0.2093	0.2516	0.3013
FiSDTPE (16 × 16)	0.1265	0.1268	0.1505	0.1788	0.2173	0.2659	0.3247
DST-BL (16 × 16)	0.1265	0.1267	0.1488	0.1756	0.2127	0.2601	0.3179
RDKTM (16 × 16)	0.1265	0.1267	0.1502	0.1784	0.2167	0.2650	0.3236
Ref. [67]	0.1265	0.1265	0.1499	0.1798	0.2167	0.2675	0.3227

Table 5. Comparisons of the normalized deflections with 3D FEM solutions for thick plates.

h/L	0.20	0.25	0.30
TrSDTPE (16 × 16)	0.2124(0.19%)	0.2569(−0.42%)	0.3097(−1.02%)
FfSDTPE (16 × 16)	0.2093(−1.27%)	0.2516(−2.48%)	0.3013(−3.71%)
FiSDTPE (16 × 16)	0.2173(2.50%)	0.2659(3.06%)	0.3247(3.77%)
DST-BL (16 × 16)	0.2127(0.33%)	0.2601(0.81%)	0.3179(1.60%)
RDKTM (16 × 16)	0.2167(2.22%)	0.2650(2.71%)	0.3236(3.42%)
3D FEM	0.2120	0.2580	0.3129

Note: Value in parentheses is the relative error with respect to 3D FEM.

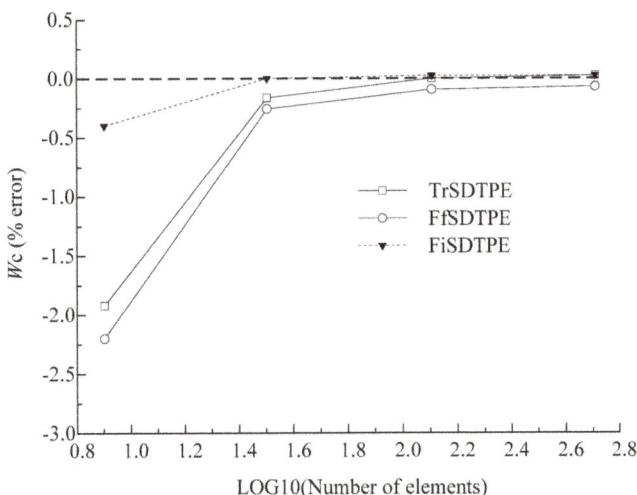

Figure 13. Convergence of normalized deflection for the simply supported square plate.

The width-to-length ratio d/l of the fictitious thin layer e_l plays an important role in the present formulations. Table 3 reports the effect of the ratio d/l on the normalized defection W_0 for the simply supported square plate, where 16 × 16 regular mesh and an aspect ratio of $h/L = 0.1$ are employed. The results in Table 3 indicate that the artificial parameter d/l has little effect on the solution accuracy when $d/l \leq 0.001$, and it is easy to select a reasonable d in the current formulation. In this work, width d is taken as $d = 0.0001l$. The condition numbers of the global stiffness matrices of the simply supported square plate using the present elements are also computed and reported in Figure 14. As seen, the variation of the condition number in Figure 14 reflects that the present elements show a good conditioning and stability in the case of mesh refinement.

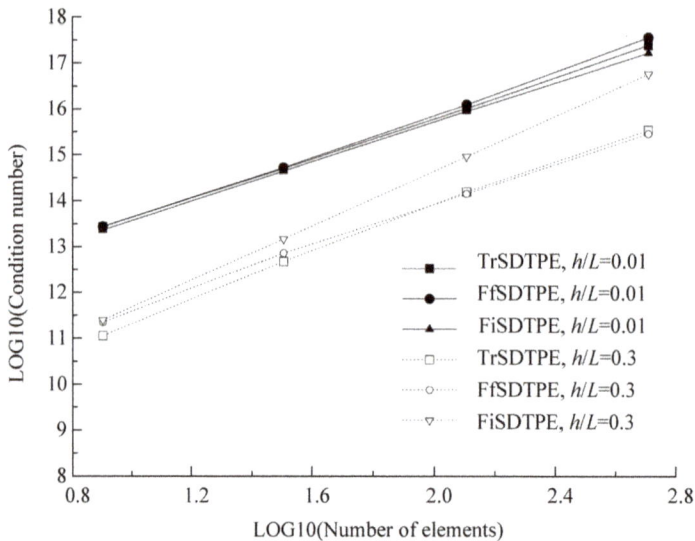

Figure 14. Variation of condition number versus number of elements.

Table 6. Normalized deflection for the clamped circular plate.

h/L	0.001	0.01	0.10	0.15	0.20	0.25	0.30
TrSDTPE (32)	1.3964	1.5225	1.6029	1.6846	1.798	1.9428	2.1187
TrSDTPE (128)	1.5441	1.5572	1.6262	1.7119	1.8312	1.9835	2.1682
TrSDTPE (512)	1.5607	1.5622	1.6316	1.7186	1.8394	1.9935	2.1801
TrSDTPE (2048)	1.5626	1.5632	1.6330	1.7203	1.8414	1.9957	2.1825
FfSDTPE (2048)	1.5620	1.5632	1.6315	1.7166	1.8343	1.9836	2.1637
FiSDTPE (2048)	1.5623	1.5633	1.6340	1.7233	1.8483	2.0090	2.2054
DST-BL (2048)	1.5634	1.5642	1.6452	1.7385	1.8665	2.0293	2.2273
RDKTM (2048)	1.5634	1.5640	1.6346	1.7239	1.8490	2.0098	2.2063
Ref. [67]	1.5625	1.5632	1.6339	1.7232	1.8482	2.0089	2.2054

4.2. Clamped Square Plate Subjected to Uniform Load

A clamped square plate subjected to a uniformly distributed load q is further investigated to test the performance of the present elements for clamp boundary conditions. The geometry and material parameters of the clamped plate are the same as those of the above simply supported plate. The displacement boundary conditions of the present theory along the clamped edges in Figure 10 in local coordinates are $\bar{\theta}_s = 0$, $\bar{\theta}_n = 0$ and $w = 0$.

The results for the normalized central deflection W_0 of the clamped square plate are compared in Table 4. It is seen that, for the plate with the clamped boundary conditions, the predictions of FiSDTPE, DST-BL, and RDKTM based on FSDT agree well with the reference solutions [67] for plates, but the TrSDTPE and FfSDTPE based on HSDTs seem to underestimate the deflections compared with the reference solutions [67] for thick plates of $h/L \geq 0.2$. To further illustrate the accuracy of the present shear elements, the comparisons of the predictions by different elements and the solutions by 3D elasticity FEM software ANSYS using 20-nodes hexahedron isoparametric element and an element side length of 0.05 are listed in Table 5. By taking the 3D FEM solutions as the benchmark, Table 5 indicates that the present TrSDTPE and FfSDTPE show better solution accuracy than the elements DST-BL, RDKTM, and FiSDTPE based on FSDT for the plates involving clamp boundaries. Moreover, the TrSDTPE with a third-order shear function $g(z)$ shows the best solution accuracy among all the elements for the clamped plate.

The DOFs of the different methods employed in Table 5 are compared by taking the case of $h/L = 0.3$ for the clamped plate. Assuming that a 16×16 regular mesh is employed for the plate element discretization and a $16 \times 16 \times 9$ regular mesh for the 20-nodes hexahedron 3D element discretization, we can see that the total DOFs of the DST-BL/RDKTM element, HSDTs, and 3D 20-nodes hexahedron element are 867, 17408, and 41337, respectively. It is seen that the total DOFs and the efficiency of HSDTs are between the DST-BL/RDKTM and the 3D FEM. Although the number of DOFs only decreases to 42% of the 3D FEM method, the present 2D HSDTs have the advantage of simplicity and flexibility in the mesh generation compared with 3D FEM for the plates with different thicknesses. Compared with other 2D plate elements such as DST-BL and RDKTM, the computational DOFs of the present 2D HSDTs seem to be relatively higher, but the formulation and the numerical implementation of the high-order shear deformation theory in the present HSDTs are much simpler than those of the DST-BL/RDKTM. From the point of view of the 2D analysis, the total computational cost of the present elements is bearable and worthy in terms of its advantages in formulation and implementation. HSDTs which have almost the same computational efficiency of DST-BL/RDKTM could also be constructed using the reduced integral method similar to our previous work [55], but the present HSDTs avoiding the reduced integration by paying a certain computational cost are more practicable in engineering analysis.

4.3. Clamped Circular Plate Subjected to Uniform Load

A clamped circular plate subjected to a uniformly distributed load q is taken into consideration in this section. The thickness of the plate is h. The radius of the plate is $r = 100$. A quarter of the plate with symmetry conditions on axes x and y is modeled in Figure 15. The displacement boundary conditions of the present theory along the clamped edges in local coordinates are $\bar{\theta}_s = 0$, $\bar{\theta}_n = 0$, and $w = 0$. Divisions of 32, 128, 512, and 2048 triangular elements are employed for the convergence studies. Typical meshes of 512 and 2048 triangular elements for the circular plate are shown in Figure 16. The results for normalized deflection $W_0 = W_c / \frac{qr^4}{100D_b}$ of the clamped circular plate are listed in Table 6, where W_c is the central deflection of the circular plate. Again, an excellent agreement between the present solutions and the reference solutions is observed for this problem.

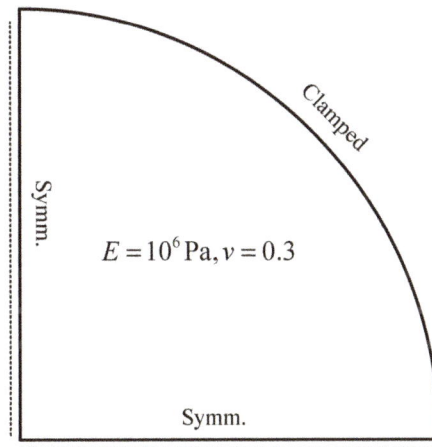

Figure 15. Model of the circular plate.

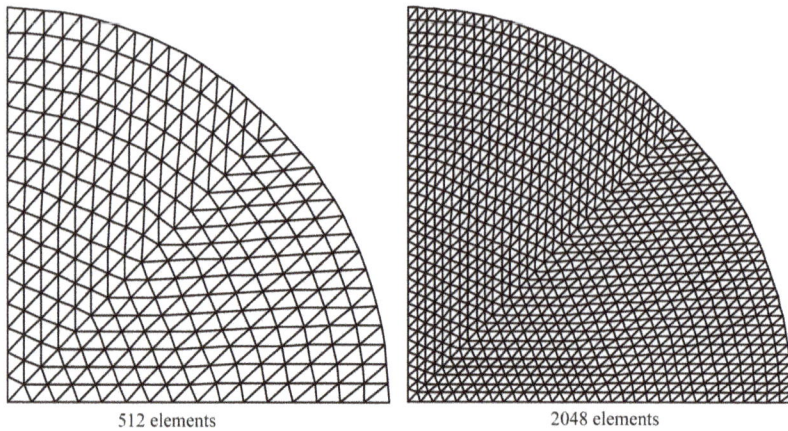

512 elements   2048 elements

Figure 16. Typical meshes for the circular plate.

4.4. Rectangular Plate Involving a Center Crack

Consider a rectangular plate involving a center crack as shown in Figure 17. The material properties are $E = 1.0 \times 10^6$ Pa and $v = 0.3$. The width and length are $2b = 1$m and $2c = 2$m, respectively. The crack length is $2a$ and the plate thickness is h. Divisions of 2728 and 7338 triangular elements are employed for the calculation of SIFs of the center crack plate. The displacement and moment boundary conditions are also illustrated in Figure 17. The numerical results obtained by TrSDTPE, FfSDTPE, and FiSDTPE for different a/h values are reported in Table 7, along with the reference solutions by Tanaka et al. [68] and Boduroglu et al. [69] based on FSDT for comparison. In Tanaka et al. [68], a cracked plate is analyzed by employing the mesh-free reproducing kernel approximation formulated by Mindlin-Reissner plate theory, and the moment intensity factor is evaluated by the *J*-integral with the aid of nodal integration. In Boduroglu et al. [69], the crack problem is solved by the dual boundary element method based on Reissner plate formulation, and the stress resultant intensity factor is calculated by employing the *J*-integral techniques. The SIFs in Table 7 are normalized by $F_1 = \frac{h^2 K_1}{6M\sqrt{\pi a}}$. It is observed that the present elements show a high solution accuracy for the calculation of the SIFs.

Table 7. Normalized SIFs F_1 for the center cracked plate.

a/h	0.8 (0.2/0.25)	1.0 (0.25/0.25)	4.0 (0.2/0.05)	5.0 (0.25/0.05)
TrSDTPE (2728)	0.8577	0.8981	0.7235	0.7602
TrSDTPE (7338)	0.8589	0.9005	0.7243	0.7622
FfSDTPE (7338)	0.8548	0.8968	0.7224	0.7610
FiSDTPE (7338)	0.8627	0.9036	0.7266	0.7633
Tanaka [68]	0.8683	0.9096	0.7287	0.7663
Boduroglu [69]	0.8694	0.9094	0.7347	0.7702

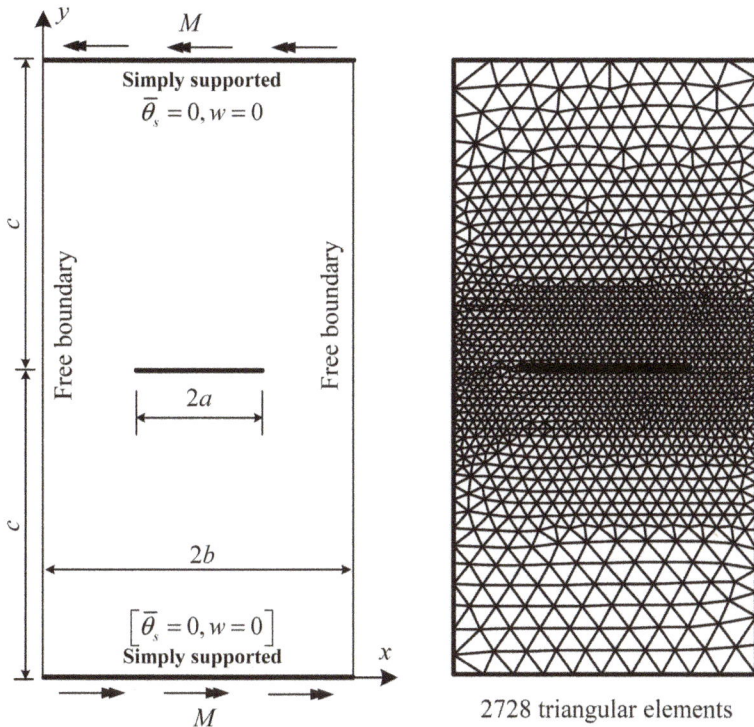

Figure 17. Model of the center cracked plate.

4.5. Symmetric Edge Cracks in a Rectangular Plate

As illustrated in Figure 18, a rectangular plate with symmetric double edge cracks is analyzed in this example. The geometry dimensions, material properties, and boundary conditions are the same as those of the center cracked problem described in Section 4.4. The crack length is a and the plate thickness is h. Division of 2728 triangular elements shown in Figure 17 is also employed for this analysis. The normalized SIFs obtained by the present elements for different values of d/b and b/h are presented in Tables 8 and 9, along with reference solutions [68,69]. The numerical methods and plate theories for solving the problem are the same as the above rectangular plate problem involving a center crack. As expected, the present results are in good agreement with the reference solutions.

Table 8. Normalized SIFs F_1 for the symmetric edge cracks problem ($b/h = 2.0$).

d/b	0.2	0.3	0.4	0.5	0.6
TrSDTPE	1.3429	1.1024	0.9739	0.9016	0.8601
FfSDTPE	1.3353	1.0971	0.9697	0.8990	0.8568
FiSDTPE	1.3502	1.1070	0.9776	0.9028	0.8629
Tanaka [68]	1.3719	1.1201	0.9886	0.9110	0.8706
Boduroglu [69]	1.3689	1.1174	0.9844	0.9086	0.8673

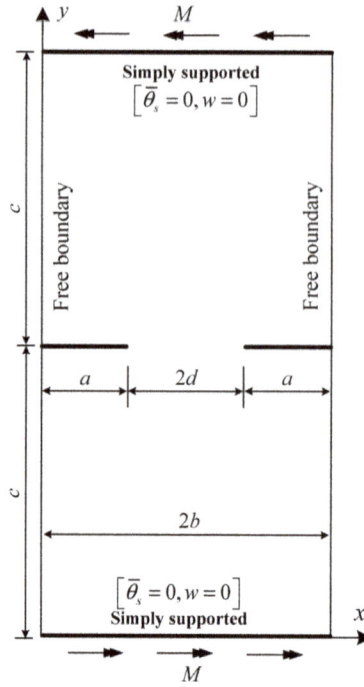

Figure 18. Model of symmetric edge cracks.

Table 9. Normalized SIFs F_1 for the symmetric edge cracks problem ($b/h = 10.0$).

d/b	0.2	0.3	0.4	0.5	0.6
TrSDTPE	1.0966	0.9156	0.8201	0.7687	0.7317
FfSDTPE	1.0941	0.9140	0.8187	0.7690	0.7306
FiSDTPE	1.0995	0.9173	0.8218	0.7656	0.7328
Tanaka [68]	1.1144	0.9225	0.8246	0.7697	0.7377
Boduroglu [69]	1.1140	0.9250	0.8268	0.7692	0.7351

4.6. Simply Supported FG Plate

In this section, a simply supported FG plate subjected to a uniformly distributed load q is analyzed and compared. The FG plate is comprised of aluminum ($E_m = 70\,GPa$, $v_m = 0.3$) and ceramic ($E_c = 151\,GPa$, $v_c = 0.3$). The side length of the plate is $L = 1m$, and the thickness of the plate is h. A quarter of the FG plate is modeled as a result of symmetry, as shown in Figure 19. The displacement boundary conditions for the symmetric and simply supported sides in local coordinates are also illustrated in Figure 19. The 16×16 regular mesh similar to Section 4.1 is employed for computation. Tables 10 and 11 list the normalized defection W_0 of the FG plate for different aspect ratios h/L and different exponents n in Equation (15), where $W_0 = W_c / \frac{qL^4}{E_m h^3}$ and W_c is the central deflection of the plate. In Ferreira et al. [45], the FG plate is solved by the meshless collocation method with multiquadric radial basis functions and a third-order shear deformation theory. The problem is also solved by Talha and Singh [70] using the C^0 isoparametric finite element with 13 degrees of freedom per node, and the power-law similar to Equations (15) and (16) is used to describe the through-the-thickness distribution of FG materials in the HSDT model. The results obtained by the present TrSDTPE, FfSDTPE, and FiSDTPE are in good agreement with the meshless solutions of Ferreira et al. [45], which compute the effective elastic moduli by the rule of mixture.

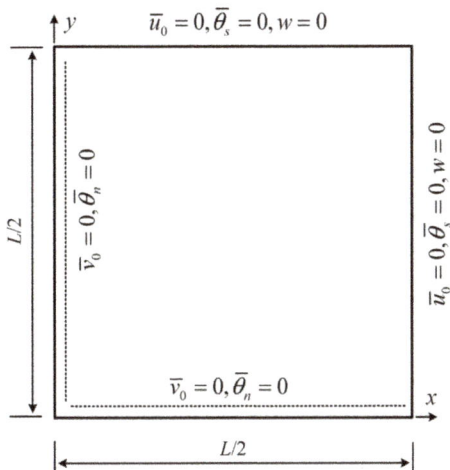

Figure 19. Modal of the FG plate.

Table 10. Normalized deflection for the FG plate ($h/L = 0.2$).

Exponent n	TrSDTPE	FfSDTPE	FiSDTPE	Ferreira et al. [45]	Talha & Singh [70]
0.0 (ceramic)	0.0248	0.0247	0.0248	0.0248	0.0250
0.5	0.0314	0.0313	0.0315	0.0314	0.0319
1.0	0.0352	0.0351	0.0353	0.0352	0.0358
2.0	0.0389	0.0388	0.0387	0.0388	0.0393
Metal	0.0535	0.0534	0.0536	0.0534	0.0541

Table 11. Normalized deflection for the FG plate ($h/L = 0.05$).

Exponent n	TrSDTPE	FfSDTPE	FiSDTPE	Ferreira et al. [45]
0.0 (ceramic)	0.0208	0.0208	0.0208	0.0208
0.5	0.0266	0.0266	0.0266	0.0265
1.0	0.0298	0.0298	0.0298	0.0297
2.0	0.0325	0.0325	0.0325	0.0324
Metal	0.0449	0.0449	0.0449	0.0448

5. Conclusions

In this work, a series of novel HSDTPEs using incompatible polynomial approximation are developed for the analysis of isotropic thick-thin plates and through-thickness functionally graded plates. The HSDTPEs are free from shear-locking, avoid the use of a shear correction factor, and provide stable solutions for thick and thin plates. The present formulation, which defines the element approach with incompatible polynomials and avoids the need to satisfy the requirement of high-order continuity in approximation fields in HSDTs, also provides a new way and methodology to develop simple plate/shell elements based on HSDTs. The accuracy and robustness of the present elements are well demonstrated through various numerical examples.

Only two types of HSDTPEs including TrSDTPE and FfSDTPE, and one special type of first-order shear deformation triangular plate element FiSDTPE, have been studied and discussed in the paper. The present formulation can be further extended to plates and shells with arbitrary shapes of elements, and further applied to more general problems related to the shear deformable effect such as the thermomechanical, vibration, and buckling analysis of functionally graded plates and laminated/sandwich structures.

Author Contributions: Y.G., Y.C., and H.Z. conceived the research idea and the framework of the article; Y.G. and Y.C. performed the numerical experiments and wrote the paper.

Conflicts of Interest: The authors declare no conflict of interest.

References

1. Reissner, E. The effect of transverse shear deformation on the bending of elastic plates. *J. Appl. Mech.* **1945**, *12*, 69–77.
2. Mindlin, R.D. Influence of rotatory inertia and shear on flexural motions of isotropic elastic plates. *J. Appl. Mech.* **1951**, *18*, 31–38.
3. Noor, A.K.; Burton, W.S. Assessment of shear deformation theories for multilayered composite plates. *Appl. Mech. Rev.* **1989**, *42*, 1–13. [CrossRef]
4. Senjanović, I.; Vladimir, N.; Hadžić, N. Modified Mindlin plate theory and shear locking-free finite element formulation. *Mech. Res. Commun.* **2014**, *55*, 95–104. [CrossRef]
5. Ayad, R.; Dhatt, G.; Batoz, J.L. A new hybrid-mixed variational approach for Reissner-Mindlin plates. The MiSP model. *Int. J. Numer. Methods Eng.* **1998**, *42*, 1149–1179. [CrossRef]
6. Belytschko, T.; Stolarski, H.; Carpenter, N. A C^0 triangular plate element with one-point quadrature. *Int. J. Numer. Methods Eng.* **1984**, *20*, 787–802. [CrossRef]
7. Batoz, J.L.; Lardeur, P. A discrete shear triangular nine d.o.f. element for the analysis of thick to very thin plates. *Int. J. Numer. Methods Eng.* **1989**, *29*, 533–560. [CrossRef]
8. Batoz, J.L.; Katili, I. On a simple triangular Reissner/Mindlin plate element based on incompatible modes and discrete constraints. *Int. J. Numer. Methods Eng.* **1992**, *35*, 1603–1632. [CrossRef]
9. Bathe, K.J.; Dvorkin, E.N. A four-node plate bending element based on Mindlin/Reissner plate theory and a mixed interpolation. *Int. J. Numer. Methods Eng.* **1985**, *21*, 367–383. [CrossRef]
10. Onate, E.; Zienkiewicz, O.C.; Suarez, B.; Taylor, R.L. A general methodology for deriving shear constrained Reissner-Mindlin plate elements. *Int. J. Numer. Methods Eng.* **1992**, *33*, 345–367. [CrossRef]
11. Xu, Z.N. A thick-thin triangular plate element. *Int. J. Numer. Methods Eng.* **1992**, *33*, 963–973.
12. Taylor, R.L.; Auriccbio, F. Linked interpolation for Reissner-Mindlin plate elements: Part II—A simple triangle. *Int. J. Numer. Methods Eng.* **1993**, *36*, 3057–3066. [CrossRef]
13. Katili, I. A new discrete Kirchhoff-Mindlin element based on Mindlin-Reissner plate theory and assumed shear strain fields—Part I: An extended DKT element for thick-plate bending analysis. *Int. J. Numer. Methods Eng.* **1993**, *36*, 1859–1883. [CrossRef]
14. Chen, W.J.; Cheung, Y.K. Refined 9-Dof triangular Mindlin plate elements. *Int. J. Numer. Methods Eng.* **2001**, *51*, 1259–1281.
15. Brasile, S. An isostatic assumed stress triangular element for the Reissner-Mindlin plate-bending problem. *Int. J. Numer. Methods Eng.* **2008**, *74*, 971–995. [CrossRef]
16. Thai, H.T.; Kim, S.E. A review of theories for the modeling and analysis of functionally graded plates and shells. *Compos. Struct.* **2015**, *128*, 70–86. [CrossRef]
17. Reddy, J.N. A Simple Higher-Order Theory for Laminated Composite Plates. *J. Appl. Mech.* **1984**, *51*, 745–752. [CrossRef]
18. Reddy, J.N. *Mechanics of Laminated Composite Plates and shells: Theory and Analysis*; CRC Press: Boca Raton, FL, USA, 1997.
19. Nguyen-Xuan, H.; Thai, C.H.; Nguyen-Thoi, T. Isogeometric finite element analysis of composite sandwich plates using a higher order shear deformation theory. *Compos. Part B Eng.* **2013**, *55*, 558–574. [CrossRef]
20. Karama, M.; Afaq, K.S.; Mistou, S. Mechanical behavior of laminated composite beam by new multi-layered laminated composite structures model with transverse shear stress continuity. *Int. J. Solids Struct.* **2003**, *40*, 1525–1546. [CrossRef]
21. Soldatos, K.P. A transverse shear deformation theory for homogeneous monoclinic plates. *Acta Mech.* **1992**, *94*, 195–220. [CrossRef]

22. Mantari, J.; Oktem, A.; Guedes Soares, C. A new higher order shear deformation theory for sandwich and composite laminated plates. *Compos. Part B Eng.* **2012**, *43*, 1489–1499. [CrossRef]

23. Thai, C.H.; Kulasegaram, S.; Tran, L.V.; Nguyen-Xuan, H. Generalized shear deformation theory for functionally graded isotropic and sandwich plates based on isogeometric approach. *Comput. Struct.* **2014**, *141*, 94–112. [CrossRef]

24. Caliri, M.F., Jr.; Ferreira, A.J.M.; Tita, V. A review on plate and shell theories for laminated and sandwich structures highlighting the Finite Element Method. *Compos. Struct.* **2016**, *156*, 63–77. [CrossRef]

25. Reddy, J.N. Analysis of functionally graded plates. *Int. J. Numer. Methods Eng.* **2000**, *47*, 663–684. [CrossRef]

26. Senthilnathan, N.R.; Lim, S.P.; Lee, K.H.; Chow, S.T. Buckling of Shear-Deformable Plates. *AIAA J.* **1987**, *25*, 1268–1271. [CrossRef]

27. Shimpi, R.P. Refined plate theory and its variants. *AIAA J.* **2002**, *40*, 137–146. [CrossRef]

28. Shimpi, R.P.; Shetty, R.A.; Guha, A. A single variable refined theory for free vibrations of a plate using inertia related terms in displacements. *Eur. J. Mech. A/Solids* **2017**, *65*, 136–148. [CrossRef]

29. Thai, H.T.; Nguyen, T.K.; Vo, T.P.; Ngo, T. A new simple shear deformation plate theory. *Compos. Struct.* **2017**, *171*, 277–285. [CrossRef]

30. Tornabene, F.; Fantuzzi, N.; Bacciocchi, M.; Reddy, J.N. An equivalent layer-wise approach for the free vibration analysis of thick and thin laminated and sandwich shells. *Appl. Sci.* **2017**, *7*, 17. [CrossRef]

31. Reddy, J.N. *Mechanics of Laminated Composite Plates and Shells*, 2nd ed.; CRC Press: Boca Raton, FL, USA, 2004.

32. Robbins, D.H.; Reddy, J.N. Modeling of Thick Composites Using a Layer-Wise Laminate Theory. *Int. J. Numer. Methods Eng.* **1993**, *36*, 655–677. [CrossRef]

33. Kumar, A.; Chakrabarti, A.; Bhargava, P. Vibration of laminated composites and sandwich shells based on higher order zigzag theory. *Eng. Struct.* **2013**, *56*, 880–888. [CrossRef]

34. Sahoo, R.; Singh, B.N. A new trigonometric zigzag theory for buckling and free vibration analysis of laminated composite and sandwich plates. *Compos. Struct.* **2014**, *117*, 316–332. [CrossRef]

35. Viola, E.; Rossetti, L.; Fantuzzi, N.; Tornabene, F. Generalized Stress-Strain Recovery Formulation Applied to Functionally Graded Spherical Shells and Panels Under Static Loading. *Compos. Struct.* **2016**, *156*, 145–164. [CrossRef]

36. Malekzadeh, P.; Afsari, A.; Zahedinejad, P.; Bahadori, R. Three-dimensional layerwise-finite element free vibration analysis of thick laminated annular plates on elastic foundation. *Appl. Math. Model.* **2010**, *34*, 776–790. [CrossRef]

37. Boscolo, M.; Banerjee, J.R. Layer-wise dynamic stiffness solution for free vibration analysis of laminated composite plates. *J. Sound Vib.* **2014**, *333*, 200–227. [CrossRef]

38. Li, D.H. Extended layerwise method of laminated composite shells. *Compos. Struct.* **2016**, *136*, 313–344. [CrossRef]

39. Carrera, E. Theories and Finite Elements for Multilayered, Anisotropic, Composite Plates and Shells. *Arch. Comput. Methods Eng.* **2002**, *9*, 87–140. [CrossRef]

40. Tornabene, F.; Viola, E.; Fantuzzi, N. General higher-order equivalent single layer theory for free vibrations of doubly-curved laminated composite shells and panels. *Compos. Struct.* **2013**, *104*, 94–117. [CrossRef]

41. Tornabene, F.; Fantuzzi, N.; Viola, E. Inter-Laminar Stress Recovery Procedure for Doubly-Curved, Singly-Curved, Revolution Shells with Variable Radii of Curvature and Plates Using Generalized Higher-Order Theories and the Local GDQ Method. *Mech. Adv. Mat. Struct.* **2016**, *23*, 1019–1045. [CrossRef]

42. Tornabene, F.; Fantuzzi, N.; Bacciocchi, M. On the Mechanics of Laminated Doubly-Curved Shells Subjected to Point and Line Loads. *Int. J. Eng. Sci.* **2016**, *109*, 115–164. [CrossRef]

43. Tornabene, F. General Higher Order Layer-Wise Theory for Free Vibrations of Doubly-Curved Laminated Composite Shells and Panels. *Mech. Adv. Mat. Struct.* **2016**, *23*, 1046–1067. [CrossRef]

44. Gulshan, T.M.N.A.; Chakrabarti, A.; Sheikh, A.H. Analysis of functionally graded plates using higher order shear deformation theory. *Appl. Math. Model.* **2013**, *37*, 8484–8494. [CrossRef]

45. Ferreira, A.J.M.; Batra, R.C.; Roque, C.M.C.; Qian, L.F.; Martins, P.A.L.S. Static analysis of functionally graded plates using third-order shear deformation theory and a meshless method. *Compos. Struct.* **2005**, *69*, 449–457. [CrossRef]

46. Thai, C.H.; Nguyen, T.N.; Rabczuk, T.; Nguyen-Xuan, H. An improved moving Kriging meshfree method for plate analysis using a refined plate theory. *Comput. Struct.* **2016**, *176*, 34–49. [CrossRef]

47. Thai, C.H.; Do, V.N.V.; Nguyen-Xuan, H. An improved Moving Kriging-based meshfree method for static, dynamic and buckling analyses of functionally graded isotropic and sandwich plates. *Eng. Anal. Bound. Elem.* **2016**, *64*, 122–136. [CrossRef]

48. Nguyen, T.N.; Thai, C.H.; Nguyen-Xuan, H. A novel computational approach for functionally graded isotropic and sandwich plate structures based on a rotation-free meshfree method. *Thin-Walled Struct.* **2016**, *107*, 473–488. [CrossRef]

49. Tran, L.V.; Ferreira, A.J.M.; Nguyen-Xuan, H. Isogeometric analysis of functionally graded plates using higher-order shear deformation theory. *Compos. Part B Eng.* **2013**, *51*, 368–383. [CrossRef]

50. Nguyen-Xuan, H.; Tran, L.V.; Thai, C.H.; Kulasegaram, S.; Bordas, S.P.A. Isogeometric analysis of functionally graded plates using a refined plate theory. *Compos. Part B Eng.* **2014**, *64*, 222–234. [CrossRef]

51. Thai, C.H.; Ferreira, A.J.M.; Bordas, S.P.A.; Rabczuk, T.; Nguyen-Xuan, H. Isogeometric analysis of laminated composite and sandwich plates using a new inverse trigonometric shear deformation theory. *Eur. J. Mech. A/Solids* **2014**, *43*, 89–108. [CrossRef]

52. Farzam-Rad, S.A.; Hassani, B.; Karamodin, A. Isogeometric analysis of functionally graded plates using a new quasi-3D shear deformation theory based on physical neutral surface. *Compos. Part B Eng.* **2017**, *108*, 174–189. [CrossRef]

53. Le-Manh, T.; Huynh-Van, Q.; Phan, T.D.; Phan, H.D.; Nguyen-Xuan, H. Isogeometric nonlinear bending and buckling analysis of variable-thickness composite plate structures. *Compos. Struct.* **2017**, *159*, 818–826. [CrossRef]

54. Nguyen, T.N.; Ngo, T.D.; Nguyen-Xuan, H. A novel three-variable shear deformation plate formulation: Theory and Isogeometric implementation. *Comput. Methods Appl. Mech. Eng.* **2017**, *326*, 376–401. [CrossRef]

55. Cai, Y.C.; Zhu, H.H. A locking-free nine-dof triangular plate element based on a meshless approximation. *Int. J. Numer. Methods Eng.* **2017**, *109*, 915–935. [CrossRef]

56. Nguyen, T.N.; Thai, C.H.; Nguyen-Xuan, H. On the general framework of high order shear deformation theories for laminated composite plate structures: A novel unified approach. *Int. J. Mech. Sci.* **2016**, *110*, 242–255. [CrossRef]

57. Cowper, G.R. Gaussian quadrature formulas for triangles. *Int. J. Numer. Methods Eng.* **1973**, *7*, 405–408. [CrossRef]

58. Endo, M.; Kimura, N. An alternative formulation of the boundary value problem for the Timoshenko beam and Mindlin plate. *J. Sound Vib.* **2007**, *301*, 355–373. [CrossRef]

59. Senjanović, I.; Vladimir, N.; Tomić, M. An advanced theory of moderately thick plate vibrations. *J. Sound Vib.* **2013**, *332*, 1868–1880. [CrossRef]

60. Endo, M. Study on an alternative deformation concept for the Timoshenko beam and Mindlin plate models. *Int. J. Eng. Sci.* **2015**, *87*, 32–46. [CrossRef]

61. Rybicki, E.F.; Kanninen, M.F. A finite element calculation of stress intensity factors by a modified crack closure integral. *Eng. Fract. Mech.* **1977**, *9*, 931–938. [CrossRef]

62. Valvo, P.S. A further step towards a physically consistent virtual crack closure technique. *Int. J. Fract.* **2015**, *192*, 235–244. [CrossRef]

63. Dirgantara, T.; Aliabadi, M.H. Crack growth analysis of plates loaded by bending and tension using dual boundary element method. *Int. J. Fract.* **2000**, *105*, 27–47. [CrossRef]

64. Moran, B.; Shih, C.F. A general treatment of crack tip contour integrals. *Int. J. Fract.* **1987**, *35*, 295–310. [CrossRef]

65. He, K.F.; Yang, Q.; Xiao, D.M.; Li, X.J. Analysis of thermo-elastic fracture problem during aluminium alloy MIG welding using the extended finite element method. *Appl. Sci.* **2017**, *7*, 69. [CrossRef]

66. Giner, E.; Fuenmayor, F.J.; Besa, A.J.; Tur, M. An implementation of the stiffness derivative method as a discrete analytical sensitivity analysis and its application to mixed mode in LEFM. *Eng. Fract. Mech.* **2002**, *69*, 2051–2071. [CrossRef]

67. Long, Y.Q.; Cen, S.; Long, Z.F. *Advanced Finite Element Method in Structural Engineering*; Tsinghua University Press: Beijing, China, 2008.

68. Tanaka, S.; Suzuki, H.; Sadamoto, S.; Imachi, M.; Bui, T.Q. Analysis of cracked shear deformable plates by an effective meshfree plate formulation. *Eng. Fract. Mech.* **2015**, *144*, 142–157. [CrossRef]

69. Boduroglu, H.; Erdogan, F. Internal and edge cracks in a plate of finite width under bending. *J. Appl. Mech. Trans. ASME* **1983**, *50*, 621–627. [CrossRef]

70.   Talha, M.; Singh, B.N. Static response and free vibration analysis of FGM plates using higher order shear deformation theory. *Appl. Math. Model.* **2010**, *34*, 3991–4011. [CrossRef]

applied
sciences

MDPI

Article

Evaluation of the Implicit Gradient-Enhanced Regularization of a Damage-Plasticity Rock Model

Magdalena Schreter *, Matthias Neuner and Günter Hofstetter

Unit for Strength of Materials and Structural Analysis, Institute of Basic Sciences in Engineering Sciences, University of Innsbruck, Technikerstr 13, A-6020 Innsbruck, Austria; Matthias.Neuner@uibk.ac.at (M.N.); guenter.hofstetter@uibk.ac.at (G.H.)
* Correspondence: Magdalena.Schreter@uibk.ac.at; Tel.: +43-512-507-61522

Received: 11 May 2018; Accepted: 29 May 2018; Published: 20 June 2018

Abstract: In the present publication, the performance of an implicit gradient-enhanced damage-plasticity model is evaluated with special focus on the prediction of complex failure modes such as shear failure. Hence, it complements studies on predominant mode I failure frequently found in the literature. To this end, an implicit gradient-enhanced damage-plasticity rock model is presented and validated by means of 2D and 3D finite element simulations of both laboratory tests on intact rock specimens as well as a large-scale structural benchmark related to failure of rock mass. Thereby, a wide range of loading conditions comprising unconfined and/or confined, tensile and/or compressive stress states is considered. The capability of the gradient-enhanced rock model for representing the mechanical response objectively with respect to the finite element discretization and realistically compared to measurement data is assessed. It is shown that complex failure modes and the respective load–displacement curves are predicted in a mesh-insensitive manner.

Keywords: screened-Poisson model; gradient-enhanced model; damage-plasticity model; implicit gradient-enhancement; rock; shear failure

---

1. Introduction

The appropriate representation of complex failure modes of cohesive-frictional materials in numerical simulations is of great interest for many engineering applications, which include, among others, geotechnical applications characterized by material failure playing a dominant role in the overall structural response [1,2]. A prominent example of material failure under highly confined stress states is given in the context of tunnel construction: During the construction of deep tunnels, high geostatic stresses in the surrounding rock mass are redistributed in consequence of the excavation process [3,4]. Depending on the quality of the surrounding rock mass, the installed supporting measures, and the type of excavation process, those stress redistributions can lead to damage in the rock mass, often accompanied by the transition from the rock mass as a continuum to rock blocks moving towards the tunnel center with localized shear bands indicating the sliding interfaces. It follows that an adequate representation of such failure phenomena by means of advanced constitutive models is of great importance and a prerequisite for the risk assessment of potential structural collapse.

Constitutive models based on the combination of the theory of plasticity and continuum damage mechanics, simply denoted as damage-plasticity models, provide a powerful framework for describing inelastic deformations, hardening and softening material behavior, as well as stiffness degradation due to damage. They are well suited for the description of cohesive-frictional materials such as concrete, rock, and soils, since different types of material failure, e.g., cracking in tension, crushing in compression, or failure under mixed stress states, can be described in a realistic manner [5]. On the basis of damage-plasticity models, the material behavior is described in terms of mere continuum relations, i.e., constitutive relations describing the behavior of an infinitesimal material point. However, softening material behavior, described in terms of continuum models, exhibits several theoretical deficiencies, as reported and summarized in [6]:

- the softening process zone is infinitesimally small;
- at structural level, snapback behavior due to the infinitesimally small softening zone is observed;
- the dissipated energy during the failure process is zero due to the infinitesimal zone in which energy is dissipated.

In a mathematical sense, those issues are related to the loss of ellipticity of the initial boundary value problem (IBVP) in static and quasi-static analyses. In finite element analyses, those deficiencies lead to mesh-dependent results, commonly referred to as a pathological mesh sensitivity. Failure patterns like cracks tend to localize into the smallest possible bandwidth, i.e., usually a single layer of finite elements, and upon mesh refinement the localization zone is decreasing to an arbitrarily small domain. Accordingly, the obtained results are not objective with respect to the finite element mesh, i.e., the results are sensitive with respect to the numerical discretization scheme. In the past decades, several techniques to overcome these deficiencies in numerical simulations have been proposed, for instance the crack band approach based on a mesh-adjusted softening modulus [7], nonlocal approaches of the integral type [8], models based on the Cosserat continuum [9], viscoplastic formulations [10], phase field models [11], and explicit and implicit gradient-enhanced formulations [12].

Among these approaches, models based on implicit gradient-enhanced formulations by now form a well-established branch in the literature due to their computational efficiency and numerical stability [12]. Numerous implicit gradient-enhanced damage and plasticity models have been proposed in recent years [13–21], and their performance has been demonstrated based on different examples of material failure. In particular, many gradient-enhanced models have been developed explicitly for concrete [13,16,17,21], and special attention has been paid to the proper representation of cracking under predominantly tensile stress states [17,22]. However, while cracking in tension (cf. Figure 1 left) is an important failure mode for concrete structures, considerably less attention was paid to shear failure (cf. Figure 1 right), i.e., pure mode II failure, or mixed failure under confined stress states. In fact, many of the available models have been validated based on examples of mode I failure, and often it is tacitly assumed that they perform equally well for more complex failure modes.

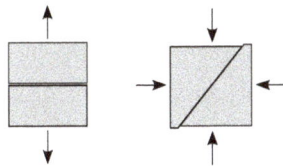

Figure 1. Schematic illustration of two characteristic failure modes (according to [23]): tensile failure (opening mode, **left**) and shear failure (sliding mode, **right**).

The apparent gap in the literature on the assessment of gradient-enhanced models for describing such complex failure modes motivates a systematic investigation of a gradient-enhanced constitutive model applied to different types of material failure. To this end, in the present contribution, a gradient-enhanced damage-plasticity model for rock is proposed and evaluated. This model is considered as a representative for a wide class of gradient-enhanced damage-plasticity models. The model is based on the damage-plasticity model by Unteregger et al. [24], and its damage formulation is extended following the implicit gradient-enhanced approach proposed by Poh and Swaddiwudhipong [17]. Based on numerical examples involving complex failure modes, the capability of the model to capture different types of material failure in a realistic and objective manner in finite element simulations will be demonstrated.

The remainder of the paper is organized as follows: In Section 2, the original damage-plasticity model for intact rock and rock mass, proposed in [24,25], is briefly summarized. In Section 3, the implicit gradient-enhancement of the rock model is presented, and the numerical implementation into a finite element framework is discussed. Section 4 covers numerical simulations of laboratory tests including wedge splitting tests as well as triaxial compression tests and triaxial extension tests with various levels of confining pressure. Additionally, a benchmark example of tunnel excavation will be presented, and the representation of complex failure modes in an objective manner with respect to the employed finite element mesh will be demonstrated. Finally, in Section 5, the paper is closed with a summary and a discussion on recommended future research activities.

2. Damage-Plasticity Model for Intact Rock and Rock Mass

The damage-plasticity model for intact rock and rock mass, denoted as RDP model in the following, was proposed originally for intact rock in [24] and was further extended to rock mass in [25] by incorporating empirical down-scaling factors to account for the influence of discontinuities according to [26–28].

The RDP model is based on the theory of plasticity formulated in the effective stress space combined with continuum damage mechanics. The stress-strain relation is expressed as

$$\sigma = (1 - \omega)\,\mathbb{C} : (\varepsilon - \varepsilon^{\mathrm{p}}) \tag{1}$$

in which σ describes the nominal Cauchy stress tensor (force per total area), ω the scalar isotropic damage parameter ranging from 0 (undamaged material) to 1 (fully damaged material), $\mathbb{C}$ the fourth order elastic stiffness tensor, ε the total strain tensor, and ε^{p} the plastic strain tensor. The effective stress tensor $\bar{\sigma}$ (force per undamaged area) is linked to the nominal stress tensor by

$$\sigma = (1 - \omega)\,\bar{\sigma} \tag{2}$$

The elastic domain is bounded by the smooth Hoek–Brown yield criterion [29,30] formulated in the Haigh–Westergaard coordinates of the effective stress tensor, i.e., the mean stress $\bar{\sigma}_{\mathrm{m}}$, the deviatoric radius $\bar{\rho}$, and the Lode angle in the deviatoric plane θ. In addition, a stress-like hardening variable $q_{\mathrm{h}}(\alpha_{\mathrm{p}})$ is incorporated leading to the definition of the yield function f_{p} as

$$f_{\mathrm{p}}(\bar{\sigma}, q_{\mathrm{h}}(\alpha_{\mathrm{p}})) = \left(\frac{1 - q_{\mathrm{h}}(\alpha_{\mathrm{p}})}{f_{\mathrm{cu}}^2}\right)\left(\bar{\sigma}_{\mathrm{m}} + \frac{\bar{\rho}}{\sqrt{6}}\right)^2 + \sqrt{\frac{3}{2}}\frac{\bar{\rho}}{f_{\mathrm{cu}}}^2 + \frac{q_{\mathrm{h}}^2(\alpha_{\mathrm{p}})}{f_{\mathrm{cu}}}\frac{m_{\mathrm{b}}}{m_0}\,m_0\left(\bar{\sigma}_{\mathrm{m}} + r(\theta, e)\frac{\bar{\rho}}{\sqrt{6}}\right) - s\,q_{\mathrm{h}}^2(\alpha_{\mathrm{p}}) \tag{3}$$

Therein, f_{cu} is the uniaxial compressive strength, m_0 is the friction parameter, $r(\theta, e)$ is the Willam–Warnke function to describe the shape of the yield surface in deviatoric planes, and parameters m_{b}/m_0 and s are empirical down-scaling factors to account for discontinuities in rock mass, the latter depending on the geological strength index GSI and the disturbance factor D according to [26]. A default value for the eccentricity e of 0.51 is proposed in [24]. For representing material behavior of intact rock, m_{b}/m_0 and s are equal to 1.

The flow rule for describing the evolution of plastic strains is defined in the effective stress space as

$$\dot{\varepsilon}^{\mathrm{p}} = \dot{\lambda}\,\frac{\partial g_{\mathrm{p}}(\bar{\sigma}, q_{\mathrm{h}}(\alpha_{\mathrm{p}}))}{\partial \bar{\sigma}} \tag{4}$$

with $\dot{\lambda}$ denoting the consistency parameter and $g_{\mathrm{p}}(\bar{\sigma}, q_{\mathrm{h}}(\alpha_{\mathrm{p}}))$ the non-associated plastic potential function expressed as

$$g_{\mathrm{p}}(\bar{\sigma}, q_{\mathrm{h}}(\alpha_{\mathrm{p}})) = \left(\frac{1 - q_{\mathrm{h}}(\alpha_{\mathrm{p}})}{f_{\mathrm{cu}}^2}\right)\left(\bar{\sigma}_{\mathrm{m}} + \frac{\bar{\rho}}{\sqrt{6}}\right)^2 + \sqrt{\frac{3}{2}}\frac{\bar{\rho}}{f_{\mathrm{cu}}}^2 + \frac{q_{\mathrm{h}}^2(\alpha_{\mathrm{p}})}{f_{\mathrm{cu}}}\left(m_{\mathrm{g1,rm}}\,\bar{\sigma}_{\mathrm{m}} + m_{\mathrm{g2,rm}}\frac{\bar{\rho}}{\sqrt{6}}\right) \tag{5}$$

Therein, volumetric plastic flow is controlled by dilatancy parameters $m_{\mathrm{g1,rm}}$ and $m_{\mathrm{g2,rm}}$. In the expression for $m_{\mathrm{g1,rm}} = (m_{\mathrm{b}}/m_0)\,m_{\mathrm{g1}}$, m_{g1} is calibrated from experimental results (uniaxial tension, uniaxial compression, and triaxial compression tests) of intact rock specimens and $m_{g2,rm}$ is determined from

$$m_{g2,rm} = 2\, m_{g1,rm} - 6 f_{tu}/f_{cu} \tag{6}$$

such that the lateral plastic strain rate in uniaxial tension is zero. Uniaxial tensile strength f_{tu} is calculated from (3) as

$$f_{tu} = -\frac{m_b/m_0\, m_0\, f_{cu}\,(e+1)}{6e} + \sqrt{\left(\frac{m_b/m_0\, m_0\, f_{cu}\,(e+1)}{6e}\right)^2 + s\, f_{cu}^2} \tag{7}$$

Hardening material behavior is described by means of the stress-like internal variable

$$q_h(\alpha_p) = \begin{cases} f_{cy}/f_{cu} + (1 - f_{cy}/f_{cu})\,\alpha_p\,(\alpha_p^2 - 3\,\alpha_p + 3) & \text{if } \alpha_p < 1 \\ 1 & \text{if } \alpha_p \geq 1 \end{cases}, \tag{8}$$

which is conjugate to the strain-like hardening variable α_p and contains the yield stress in uniaxial compression f_{cy}. The evolution law for the strain-like hardening variable α_p is given as

$$\dot{\alpha}_p = \dot{\lambda}\,\frac{E_{rm}}{E_i}\,\frac{1}{x_h(\bar{\sigma}_m)}\left(1 + 3\,\frac{\bar{\rho}^2}{\bar{\rho}^2 + f_{cu}^2 \cdot 10^{-8}}\,\cos^2(3\,\theta/2)\right)\left\|\frac{\partial g_p(\bar{\sigma}, q_h(\alpha_p))}{\partial \bar{\sigma}}\right\| \tag{9}$$

in which $\|\partial g_p(\bar{\sigma}, q_h(\alpha_p))/\partial\bar{\sigma}\|$ is the norm of the gradient of the plastic potential function with respect to effective stress, E_{rm}/E_i denotes the reduction of the Young's modulus of rock mass compared to intact rock [26] ranging from 0 (completely disintegrated) to 1 (intact rock), and $x_h(\bar{\sigma}_m)$ is a measure for describing hardening ductility, defined as

$$x_h(\bar{\sigma}_m) = \begin{cases} (B_h - D_h)\,\exp\left(\frac{R_h\,(A_h - B_h)}{C_h\,(B_h - D_h)}\right) + D_h & \text{if } R_h < 0 \\ A_h - (A_h - B_h)\,\exp\left(-R_h/C_h\right) & \text{if } R_h \geq 0 \end{cases} \tag{10}$$

with $R_h = -\bar{\sigma}_m/f_{cu} - G_h$. In (10), model parameters A_h, B_h, C_h, D_h, and G_h control the hardening behavior. They are calibrated by experimental data from uniaxial tension, uniaxial compression, and triaxial compression tests. In [24,25], default values of $B_h = 10^{-5}$, $D_h = 10^{-6}$ and $G_h = 0$ are suggested in absence of respective experimental data.

Damage is provoked when the hardening variable $q_h(\alpha_p)$ attains its maximum value of 1. At this stage, the scalar isotropic damage parameter ω starts evolving dependent on the strain-like internal softening variable α_d. This relation is described by means of an exponential softening law as

$$\omega(\alpha_d) = 1 - \exp(-\alpha_d/\varepsilon_f) \tag{11}$$

with the softening modulus ε_f controlling the slope of the softening curve. The rate of the strain-like internal softening variable α_d is computed from the volumetric part of the plastic strain rate $\dot{\varepsilon}^{p,vol} = \mathrm{tr}(\dot{\varepsilon}^p)$ as

$$\dot{\alpha}_d = \begin{cases} 0 & \text{if } \alpha_p < 1 \\ \dot{\varepsilon}^{p,vol}/x_s(\dot{\varepsilon}^{p,vol}) & \text{if } \alpha_p \geq 1 \end{cases} \tag{12}$$

Therein, the softening ductility measure

$$x_s(\dot{\varepsilon}^{p,vol}) = 1 + A_s\left(\dot{\varepsilon}_{\ominus}^{p,vol}/\dot{\varepsilon}^{p,vol}\right)^{B_s} \tag{13}$$

accounts for the influence of multi-axial stress states on the softening behavior, with $\dot{\varepsilon}_{\ominus}^{\text{p,vol}} = \text{tr}(\langle -\dot{\varepsilon}^{\text{p}} \rangle)$ describing the compressive part of the volumetric plastic strain rate. Model parameters A_{s} and B_{s} are calibrated from experimental data of uniaxial compression and triaxial compression tests. Again, in absence of respective experimental data for parameters A_{s} and B_{s}, default values of $A_{\text{s}} = 15$ and $B_{\text{s}} = 2$ are proposed in [24].

3. Implicit Gradient-Enhancement of the Damage-Plasticity Model for Intact Rock and Rock Mass

Softening material behavior leads to an ill-posed initial boundary value problem and consequently to pathological mesh-sensitivity in finite element simulations. As a remedy, in [24], the crack band approach was employed for the RDP model. While the crack band approach is a rather simple regularization technique, it is also characterized by a number of shortcomings, which are addressed in [31]. Motivated by those deficiencies, in the following, a more sophisticated regularization technique based on the implicit gradient-enhanced formulation [32] is presented. To this end, the approach by Poh and Swaddiwudhipong [17] proposed for a damage-plasticity model for concrete and based on the gradient of the internal softening variable, is adopted. By incorporating the gradient of an internal variable into the constitutive relations, nonlocality is introduced. Thus, the mechanical response of a material point does not exclusively depend on its local state, but is also influenced by the state in its neighborhood.

Nonlocality is incorporated by replacing the local softening variable α_{d} by a weighted softening variable $\hat{\alpha}_{\text{d}}$ in the exponential damage law (11), which is expressed as

$$\omega(\hat{\alpha}_{\text{d}}) = 1 - \exp(-\hat{\alpha}_{\text{d}}/\varepsilon_{\text{f}}) \tag{14}$$

Therein, the softening modulus ε_{f} is a material parameter. The weighted softening variable $\hat{\alpha}_{\text{d}}$ is calculated from a combination of the local softening variable α_{d} and its nonlocal counterpart $\bar{\alpha}_{\text{d}}$ as

$$\hat{\alpha}_{\text{d}} = m\,\bar{\alpha}_{\text{d}} + (1 - m)\,\alpha_{\text{d}} \tag{15}$$

in which m denotes a weighting parameter. Choosing m larger than 1 yields the over-nonlocal formulation [33] to achieve full regularization of the problem, as proven in [34]. Furthermore, by ensuring that $\hat{\alpha}_{\text{d}}$ can only increase, damage is considered as an irreversible process [35].

According to the implicit approach, the field of the nonlocal softening variable $\bar{\alpha}_{\text{d}}$, henceforth simply denoted as the nonlocal field, is defined implicitly as the solution of a higher-order partial differential equation. Adopting the formulation by Poh and Swaddiwudhipong [17], for the description of the nonlocal field a second order partial differential equation is employed as

$$\bar{\alpha}_{\text{d}} - l^2\,\Delta\bar{\alpha}_{\text{d}} = \alpha_{\text{d}} \qquad \text{in } \Omega \tag{16}$$

in which l denotes a length scale parameter defining the radius of nonlocal interaction, Δ is the Laplace operator and α_{d} represents the local softening variable of (12), and Ω is the spatial domain occupied by the body under consideration. Equation (16) is a screened-Poisson equation, commonly denoted as the Helmholtz equation in the context of implicit gradient-enhanced formulations [12,36]. It is apparent that nonlocality affects only the damage part of the model and the plasticity part of the RDP model remains local.

As suggested in [32,37], homogeneous Neumann boundary conditions are assumed as $\nabla\bar{\alpha}_{\text{d}} \cdot \mathbf{n} = 0$ on the entire boundary Γ of the domain with the normal vector to the boundary $\mathbf{n}$. This boundary condition was interpreted in [38] in the context of phase-field models enforcing cracks to occur perpendicular to the boundary.

The set of governing equations is completed by the equilibrium equation

$$\nabla \cdot \sigma + \bar{\mathbf{f}} = \mathbf{0} \qquad \text{in } \Omega \tag{17}$$

in which $\bar{\mathbf{f}}$ is the vector of body forces. For the boundary conditions, the surface traction vector $\bar{\mathbf{t}} = \sigma \cdot \mathbf{n}$ on Γ_t and prescribed displacements $\mathbf{u} = \bar{\mathbf{u}}$ on Γ_u are assumed.

Partial differential Equations (16) and (17) form a fully coupled system with the unknown displacement vector $\mathbf{u}$ and the nonlocal field $\bar{\alpha}_d$, which is solved by means of the finite element method. To this end, the weak form is formulated, which is subsequently discretized in space and in time. To obtain the weak form, the set of partial differential equations is multiplied by test functions $\delta \mathbf{u}$ for the displacement field and $\delta \bar{\alpha}_d$ for the nonlocal field. Integration over the domain, application of the divergence theorem, incorporation of the infinitesimal strain ε, and consideration of $\delta \mathbf{u} = \mathbf{0}$ on Γ_u yields the weak form of the IBVP expressed in Voigt notation as

$$\int_\Omega \delta \varepsilon^\mathsf{T} \sigma \, d\Omega - \int_{\Gamma_t} \delta \mathbf{u}^\mathsf{T} \bar{\mathbf{t}} \, d\Gamma - \int_\Omega \delta \mathbf{u}^\mathsf{T} \bar{\mathbf{f}} \, d\Omega = 0, \tag{18}$$

$$\int_\Omega \delta \bar{\alpha}_d \, \bar{\alpha}_d \, d\Omega + \int_\Omega l^2 \, (\nabla \delta \bar{\alpha}_d)^\mathsf{T} \, \nabla \bar{\alpha}_d \, d\Omega - \int_\Omega \delta \bar{\alpha}_d \, \alpha_d \, d\Omega = 0. \tag{19}$$

The displacement field $\mathbf{u}$ and the nonlocal field $\bar{\alpha}_d$ are approximated over the domain using a Bubnov-Galerkin approach as

$$\mathbf{u} = \mathbf{N}_u \, \mathbf{q}_u \tag{20}$$

$$\bar{\alpha}_d = \mathbf{N}_{\bar{\alpha}_d} \, \mathbf{q}_{\bar{\alpha}_d} \tag{21}$$

in which $\mathbf{N}_{(\bullet)}$ contains the shape functions and $\mathbf{q}_{(\bullet)}$ are column vectors of the nodal unknown parameters, both expressed in the global form employing the standard assembly procedure, with $(\bullet)$ standing for the displacement field $\mathbf{u}$ and the nonlocal field $\bar{\alpha}_d$, respectively. The infinitesimal strain ε and the gradient of the nonlocal field $\nabla \bar{\alpha}_d$ are discretized as

$$\varepsilon = \mathbf{B}_u \, \mathbf{q}_u \tag{22}$$

$$\nabla \bar{\alpha}_d = \mathbf{B}_{\bar{\alpha}_d} \, \mathbf{q}_{\bar{\alpha}_d} \tag{23}$$

with the strain-displacement matrix $\mathbf{B}_u$ and the row vector of the spatial derivatives of the shape functions for the field of the nonlocal softening variable $\mathbf{B}_{\bar{\alpha}_d}$, again both expressed in the global form employing the standard assembly procedure. It follows that the shape functions for both fields must meet the requirement of C^0-continuity.

Due to the nonlinear and path-dependent character of the IBVP, an incremental solution procedure is employed. For the incremental solution procedure, a discrete (pseudo-)time interval $[t^{(n-1)}, t^{(n)}]$ is considered such that the body under consideration is in equilibrium at time $t^{(n-1)}$. At this time, the nodal values $\mathbf{q}_u^{(n-1)}$ and $\mathbf{q}_{\bar{\alpha}_d}^{(n-1)}$, the stress $\sigma^{(n-1)}$ and the internal variables $\alpha_p^{(n-1)}$ and $\alpha_d^{(n-1)}$ are known. An incremental load is applied such that the traction vector and the vector of body forces at time $t^{(n)}$ are prescribed as $\bar{\mathbf{t}}^{(n)} = \bar{\mathbf{t}}^{(n-1)} + \Delta \bar{\mathbf{t}}^{(n)}$ and $\bar{\mathbf{f}}^{(n)} = \bar{\mathbf{f}}^{(n-1)} + \Delta \bar{\mathbf{f}}^{(n)}$. The updated variables of the constitutive relations at time $t^{(n)}$, i.e., $\sigma^{(n)}$, $\alpha_p^{(n)}$, $\alpha_d^{(n)}$, are evaluated by means of a stress-update algorithm, employing an implicit integration scheme following the return-mapping approach [39] for the plastic regime and a subsequent explicit evaluation of the damage part. Finally, the incremental discretized weak form is obtained as

$$\underbrace{\int_\Omega \left(\delta \mathbf{q}_u^{(n)} \right)^\mathsf{T} \mathbf{B}_u^\mathsf{T} \, \sigma^{(n)} \, d\Omega}_{\left(\delta \mathbf{q}_u^{(n)} \right)^\mathsf{T} \mathbf{f}_{\text{int}}^u \left(\mathbf{q}_u^{(n)}, \mathbf{q}_{\bar{\alpha}_d}^{(n)} \right)} - \underbrace{\int_{\Gamma_t} \left(\delta \mathbf{q}_u^{(n)} \right)^\mathsf{T} \mathbf{N}_u^\mathsf{T} \bar{\mathbf{t}}^{(n)} \, d\Gamma - \int_\Omega \left(\delta \mathbf{q}_u^{(n)} \right)^\mathsf{T} \mathbf{N}_u^\mathsf{T} \bar{\mathbf{f}}^{(n)} \, d\Omega}_{\left(\delta \mathbf{q}_u^{(n)} \right)^\mathsf{T} \mathbf{f}_{\text{ext}}^{u,(n)}} = 0, \tag{24}$$

$$\underbrace{\int_\Omega \left(\delta \mathbf{q}_{\bar{\alpha}_d}^{(n)} \right)^\mathsf{T} \mathbf{N}_{\bar{\alpha}_d}^\mathsf{T} \, \mathbf{N}_{\bar{\alpha}_d} \, \mathbf{q}_{\bar{\alpha}_d}^{(n)} \, d\Omega + \int_\Omega l^2 \left(\delta \mathbf{q}_{\bar{\alpha}_d}^{(n)} \right)^\mathsf{T} \mathbf{B}_{\bar{\alpha}_d}^\mathsf{T} \, \mathbf{B}_{\bar{\alpha}_d} \, \mathbf{q}_{\bar{\alpha}_d}^{(n)} \, d\Omega - \int_\Omega \left(\delta \mathbf{q}_{\bar{\alpha}_d}^{(n)} \right)^\mathsf{T} \mathbf{N}_{\bar{\alpha}_d}^\mathsf{T} \, \alpha_d^{(n)} \, d\Omega}_{\left(\delta \mathbf{q}_{\bar{\alpha}_d}^{(n)} \right)^\mathsf{T} \mathbf{R}_{\bar{\alpha}_d} \left(\mathbf{q}_u^{(n)}, \mathbf{q}_{\bar{\alpha}_d}^{(n)} \right)} = 0. \tag{25}$$

Since Equations (24) and (25) must hold for arbitrary kinematically admissible test functions $\delta\mathbf{q}_u^{(n)}$ and $\delta\mathbf{q}_{\bar{\alpha}_d}^{(n)}$, they can be recast into the residual format as

$$\mathbf{R}_u\left(\mathbf{q}_u^{(n)},\mathbf{q}_{\bar{\alpha}_d}^{(n)}\right) = \mathbf{f}_{\text{int}}^u\left(\mathbf{q}_u^{(n)},\mathbf{q}_{\bar{\alpha}_d}^{(n)}\right) - \mathbf{f}_{\text{ext}}^{u,(n)} = 0 \tag{26}$$

$$\mathbf{R}_{\bar{\alpha}_d}\left(\mathbf{q}_u^{(n)},\mathbf{q}_{\bar{\alpha}_d}^{(n)}\right) = 0 \tag{27}$$

with $\mathbf{f}_{\text{int}}^u$ and $\mathbf{f}_{\text{ext}}^{u,(n)}$ denoting internal and external force vectors related to the displacement field, and $\mathbf{R}_{\bar{\alpha}_d}$ is the residual vector for the nonlocal field. Due to the nonlinear dependence of the system of Equations (26) and (27) on the unknown nodal solution vector

$$\mathbf{q}^{(n)} = \left[\mathbf{q}_u^{(n)} \quad \mathbf{q}_{\bar{\alpha}_d}^{(n)}\right]^{\mathsf{T}}, \tag{28}$$

an iterative Newton–Raphson solution procedure is employed. The nodal unknowns at time $t^{(n)}$ in the i-th iteration step are composed of $\mathbf{q}^{(n,i)} = \mathbf{q}^{(n-1)} + \Delta\mathbf{q}^{(n,i)}$, where $\Delta\mathbf{q}^{(n,i)}$ has to be determined. Linearization of Equations (26) and (27) at the state $\mathbf{q}^{(n,i-1)}$ with the initial guess of the nodal unknowns $\mathbf{q}^{(n,0)} = \mathbf{q}^{(n-1)}$ yields the iterative procedure for the correction of the nodal unknowns $\Delta\Delta\mathbf{q}^{(n,i)}$ for time $t^{(n)}$ after the i-th iteration step

$$\begin{bmatrix} \mathbf{K}_{uu}^{(n,i-1)} & \mathbf{K}_{u\bar{\alpha}_d}^{(n,i-1)} \\ \mathbf{K}_{\bar{\alpha}_d u}^{(n,i-1)} & \mathbf{K}_{\bar{\alpha}_d \bar{\alpha}_d}^{(n,i-1)} \end{bmatrix} \begin{bmatrix} \Delta\Delta\mathbf{q}_u^{(n,i)} \\ \Delta\Delta\mathbf{q}_{\bar{\alpha}_d}^{(n,i)} \end{bmatrix} = \begin{bmatrix} -\mathbf{R}_u\left(\mathbf{q}_u^{(n,i-1)},\mathbf{q}_{\bar{\alpha}_d}^{(n,i-1)}\right) \\ -\mathbf{R}_{\bar{\alpha}_d}\left(\mathbf{q}_u^{(n,i-1)},\mathbf{q}_{\bar{\alpha}_d}^{(n,i-1)}\right) \end{bmatrix}, \tag{29}$$

with the submatrices of the system matrix given as

$$\mathbf{K}_{uu}^{(n,i-1)} = \left.\frac{\partial\mathbf{R}_u}{\partial\mathbf{q}_u^{(n)}}\right|_{\mathbf{q}_u^{(n,i-1)}} = \int_\Omega \mathbf{B}_u^{\mathsf{T}} \left.\frac{\partial\sigma^{(n)}}{\partial\varepsilon^{(n)}}\right|_{\varepsilon^{(n,i-1)}} \mathbf{B}_u \, d\Omega \tag{30}$$

$$\mathbf{K}_{u\bar{\alpha}_d}^{(n,i-1)} = \left.\frac{\partial\mathbf{R}_u}{\partial\mathbf{q}_{\bar{\alpha}_d}^{(n)}}\right|_{\mathbf{q}_{\bar{\alpha}_d}^{(n,i-1)}} = \int_\Omega \mathbf{B}_u^{\mathsf{T}} \left.\frac{\partial\sigma^{(n)}}{\partial\bar{\alpha}_d^{(n)}}\right|_{\bar{\alpha}_d^{(n,i-1)}} \mathbf{N}_{\bar{\alpha}_d} \, d\Omega \tag{31}$$

$$\mathbf{K}_{\bar{\alpha}_d u}^{(n,i-1)} = \left.\frac{\partial\mathbf{R}_{\bar{\alpha}_d}}{\partial\mathbf{q}_u^{(n)}}\right|_{\mathbf{q}_u^{(n,i-1)}} = -\int_\Omega \mathbf{N}_{\bar{\alpha}_d}^{\mathsf{T}} \left.\frac{\partial\alpha_d^{(n)}}{\partial\varepsilon^{(n)}}\right|_{\varepsilon^{(n,i-1)}} \mathbf{B}_u \, d\Omega \tag{32}$$

$$\mathbf{K}_{\bar{\alpha}_d \bar{\alpha}_d}^{(n,i-1)} = \left.\frac{\partial\mathbf{R}_{\bar{\alpha}_d}}{\partial\mathbf{q}_{\bar{\alpha}_d}^{(n)}}\right|_{\mathbf{q}_{\bar{\alpha}_d}^{(n,i-1)}} = \int_\Omega l^2\, \mathbf{B}_{\bar{\alpha}_d}^{\mathsf{T}} \mathbf{B}_{\bar{\alpha}_d} \, d\Omega + \int_\Omega \mathbf{N}_{\bar{\alpha}_d}^{\mathsf{T}} \mathbf{N}_{\bar{\alpha}_d} \, d\Omega \tag{33}$$

in which $\partial\sigma^{(n)}/\partial\varepsilon^{(n)}$, $\partial\sigma^{(n)}/\partial\bar{\alpha}_d^{(n)}$, and $\partial\alpha_d^{(n)}/\partial\varepsilon^{(n)}$ are the consistent tangent stiffness submatrices. They represent the derivatives of the constitutive relations consistent with the numerical algorithm for integrating the path-dependent rate constitutive equations, which is essential for the full Newton–Raphson scheme in order to preserve a quadratic rate of convergence. Due to the non-associated plastic flow rule of the RDP model and the coupling of the displacement field and the nonlocal field, the system matrix is unsymmetric. From the computed correction of the nodal unknowns $\Delta\Delta\mathbf{q}^{(n,i)}$, the updated solutions of the incremental nodal unknowns, $\Delta\mathbf{q}^{(n,i)} = \Delta\mathbf{q}^{(n,i-1)} + \Delta\Delta\mathbf{q}^{(n,i)}$ and the total nodal unknowns $\mathbf{q}^{(n,i)} = \mathbf{q}^{(n-1)} + \Delta\mathbf{q}^{(n,i)}$ are obtained.

4. Numerical Study

The aim of the present numerical study is to evaluate the performance of the implicit gradient-enhanced RDP model for predicting the mechanical response of structures, involving the softening behavior of rock in a realistic and mesh-insensitive manner for a wide range of loading conditions. To this end, a numerical study is presented, related to both laboratory tests and practical applications. It consists of the following parts:

1. 2D modeling of mode I failure in wedge splitting tests on Indiana limestone performed by Brühwiler and Saouma [40],
2. 3D modeling of shear failure in triaxial compression tests performed by Blümel [41] on specimens of Innsbruck quartz phyllite, considering the influence of confined stress states attaining the compressive meridian of the yield surface,
3. 3D numerical simulations of triaxial extension tests performed on the same type of specimens, investigating the influence of confined stress states attaining the tensile meridian of the yield surface, and
4. 2D numerical simulations of the excavation of a deep tunnel in Innsbruck quartz phyllite rock mass leading to the formation of shear bands in the vicinity of the tunnel for demonstrating the capability of the gradient-enhanced RDP model to predict failure of a complex structure.

4.1. Modeling of Mode I Failure, Demonstrated by Analyzing Wedge Splitting Tests on Indiana Limestone

In a first step, the capability of the gradient-enhanced RDP model for predicting mode I failure is assessed. To this end, the experimental study of wedge splitting tests on Indiana limestone performed by Brühwiler and Saouma [40] is considered. The investigated specimen is illustrated in Figure 2.

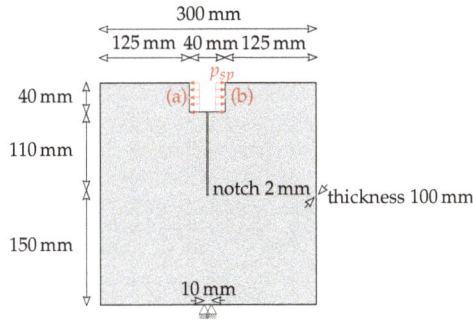

Figure 2. Geometry and boundary conditions of the numerical model of the specimen for the wedge splitting test.

During the experimental test, the splitting force was applied by a vertically driven wedge, exerting a pressure against roller bearings that were mounted on both sides of the groove in the specimen. The vertical (machine) force and the crack mouth opening displacement (CMOD) were recorded, and the latter was controlled during the experimental test. The energy conjugate force to the CMOD, the splitting force F_{sp}, was calculated from the vertical force considering the geometry of the wedge and neglecting any frictional effects. In total, 5 tests were performed, but in [40] detailed load–displacement curves were presented only for Test 3. To investigate the degradation of the stiffness of the specimen during crack propagation, several loading/unloading cycles were performed.

The experimental test is simulated by means of a two-dimensional finite element model assuming plane stress conditions. According to Figure 2, the specimen is supported in the vertical direction at the bottom center over a width of 10 mm and in the horizontal direction at midpoint. The splitting force F_{sp}, transmitted by the wedge, is approximated by the pressure p_{sp} acting on the lateral groove faces. The simulation is performed in a displacement-driven manner by controlling the CMOD (i.e., the relative horizontal displacement between points (a) and (b) in Figure 2) during the loading and unloading cycles. The material parameters for representing the Indiana limestone were determined by a best fit with the recorded experimental results. In the present example of mode I failure, only few parameters have a significant influence on the results: $E = 22000$ MPa, $\nu = 0.15$, $f_{cu} = 20$ MPa, $f_{cy} = 2/3\, f_{cu} = 13.33$ MPa, $m_0 = 6.5$, $m_{g1} = 5$, $A_h = 5 \times 10^{-3}$, $C_h = 20$, $\varepsilon_f = 8 \times 10^{-4}$, $m = 1.05$, and $l = 4$ mm. For the weighting parameter m, any value larger than 1 is sufficient to ensure full regularization of the problem by avoiding spurious localization of plastic strain [33]. A discussion on the influence of the weighting parameter *m* may be found in [42] for an integral-type nonlocal model, where, however, for parameter *m* = 2, an overestimation of the energy dissipation close to a notch was demonstrated. This non-physical effect was also observed by the authors with increasing influence for larger values of *m*. Thus, parameter *m* is chosen just slightly larger than 1, in accordance with proposed values from the literature [17,43]. For the remaining model parameters e, B_h, D_h, G_h, A_s, and B_s, the default values summarized in Section 2 are employed. From Equation (7), the uniaxial tensile strength is calculated as $f_{tu} = 3$ MPa.

To investigate the influence of the finite element discretization on the predicted results, different meshes are employed: Three structured meshes with fully integrated 4-node quadrilateral elements and element sizes of 5 mm, 1 mm, and 0.5 mm in the vicinity of the expected crack path, as well as one unstructured mesh with the same element type and an element size of 1 mm along the crack path. Accordingly, the element size of 5 mm is slightly larger compared to the assumed length scale l of 4 mm for the coarse mesh and considerably smaller for the medium and fine mesh. The purpose of the unstructured mesh is to investigate a potential bias of the crack pattern by following the grid lines of the finite element mesh, since mesh-biased crack paths have been observed for smeared crack models based on the crack band approach [44].

In Figure 3, the resulting load–displacement curves, i.e., splitting force F_{sp} versus CMOD, are shown for the considered experimental test and the numerical simulations. It can be concluded that the qualitative shape of the experimentally obtained curve is approximated quite well in the numerical simulations.

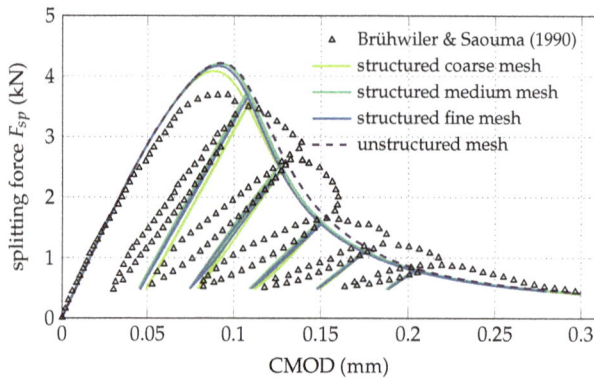

Figure 3. Splitting force vs. crack mouth opening displacement (CMOD) for the wedge splitting test: experimental results (data taken from [40]) and numerical results.

Regarding the influence of the finite element mesh on the computed load–displacement curves, a slight difference between the predicted response based on the structured coarse mesh and the one based on the structured medium mesh is visible. This is explained by the rough approximation of the gradient of the nonlocal field by the coarse mesh. In contrast, almost identical load–displacement curves are obtained for the structured medium and the structured fine mesh, confirming that the gradient of the nonlocal field can be resolved sufficiently by those meshes. The unstructured mesh results in a similar load–displacement curve, demonstrating that the gradient-enhanced approach is not biased by the orientation of the finite element mesh.

The computed deformation of the specimen is depicted in Figure 4 for the three structured meshes. By increasing the splitting force during the numerical simulations, the tensile strength of the material is attained at first in the elements directly below the notch; consequently, damage is initiated. This leads to large, localized deformations in this region. Upon further loading, the increase of these deformations reflects the opening of a crack, propagating towards the bottom of the specimen. This is represented by the gradient-enhanced damage-plasticity approach in a smeared manner. The width of the damage zone is related to the length scale parameter *l* (cf. (16)). Furthermore, it can be seen that an identical symmetrical response with respect to the vertical axis of symmetry is obtained for all three structured meshes.

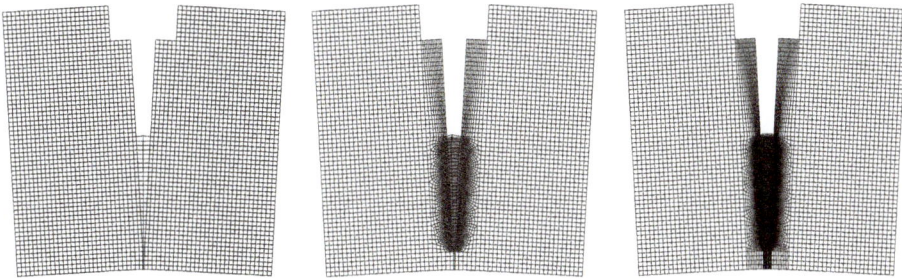

Figure 4. Deformed specimen at CMOD = 0.3 mm with a displacement scale factor of 100 for the three structured finite element meshes: coarse—3580 elements (**left**), medium—6940 elements (**center**), and fine—15430 elements (**right**).

In Figure 5, the distribution of the damage variable ω computed for the three structured meshes is plotted at a CMOD = 0.3 mm. It is visible that damage is also accumulated slightly above the notch tip. This is a consequence of the diffusive character of the gradient-enhanced formulation. Furthermore, the present gradient-enhanced RDP model with the constant length scale *l* predicts a rather broad zone of complete damage (red region in Figure 5). In fact, this behavior does not represent damage localizing into a discrete macrocrack, and has been addressed in [45,46]. However, comparison of the predicted damage distributions for the different meshes demonstrates mesh-insensitivity of the obtained failure patterns. Figures 3 and 6 show an identical load–displacement curve and identical deformation pattern and damage distribution computed by means of the unstructured mesh, which underlines the capability of the gradient approach to produce mesh-insensitive results.

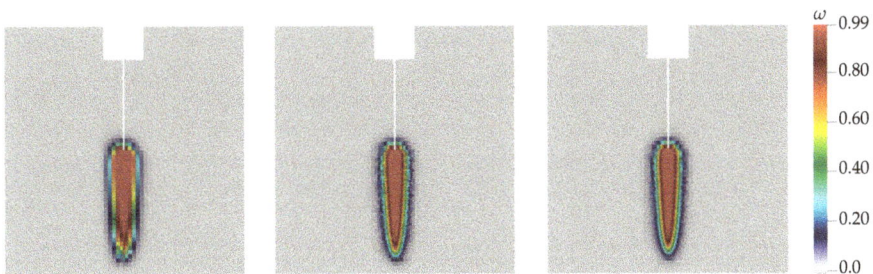

Figure 5. Distribution of the damage variable ω at CMOD = 0.3 mm for the three structured finite element meshes: coarse (**left**), medium (**center**), and fine (**right**).

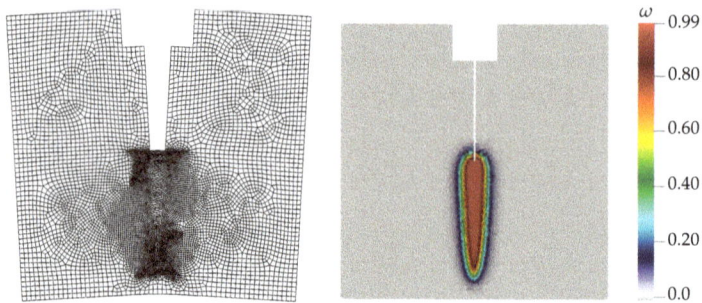

Figure 6. Deformed specimen with a displacement scale factor of 100 (**left**) and distribution of the damage variable ω (**right**) for the unstructured mesh (9504 elements) at a CMOD of 0.3 mm.

4.2. Modeling of Shear Failure, Demonstrated by Analyzing Triaxial Compression Tests on Innsbruck Quartz Phyllite

In a second step, the performance of the gradient-enhanced RDP model for predicting shear failure under confined stress states attaining the compressive meridian of the yield surface is assessed. To this end, numerical simulations of triaxial compression tests on specimens of Innsbruck quartz phyllite performed by Blümel [41] are conducted.

For the experimental program, intact rock specimens with a diameter of 35 mm and a height of 70 mm were taken from drill cores sampled at the construction site of the Brenner Base Tunnel. A series of triaxial compression tests with different levels of confining pressures $p^{(0)} = 0$ MPa (uniaxial compression), 12.5 MPa, 25.0 MPa, and 37.5 MPa was conducted. The experiments were performed in a sequential manner: Firstly, the confining pressure was applied, resulting in an initial hydrostatic stress state in the specimen, and subsequently, an axial pressure was applied displacement-driven up to failure. The mechanical response in the post-peak regime was also recorded.

Finite element simulations of triaxial compression tests are often performed in an approximate manner. Commonly, such tests are simplified as single element tests in which the non-homogeneous deformation of the specimen observed in the experiments cannot be captured, or by two-dimensional models in which three-dimensional effects are neglected, e.g., [24,47–49]. By contrast, in the present study, a three-dimensional finite element model is employed in order to capture the three-dimensional deformations in a realistic manner. The numerical model with the prescribed boundary conditions and loads is illustrated in Figure 7. Since the failure mode is expected to be symmetric with respect to a vertical plane through the center axis of the specimen, symmetry is exploited by considering only one half of the specimen. By analogy to the experimental tests, the numerical simulations are performed in two sequential steps: firstly, the specimen is supported vertically at the bottom face and the confining pressure is applied; secondly, the axial loading is applied by imposing a uniform vertical displacement at the top of the specimen.

The specimens are discretized with 20-node hexahedral elements employing reduced numerical integration. For both the displacement field and the nonlocal field, quadratic shape functions are used. To analyze potential mesh-sensitivity of the gradient-enhanced RDP model, three different structured finite element meshes are examined. A coarse mesh with 828 elements (an element size of 4 mm), a medium mesh with 5950 elements (an element size of 2 mm), and a fine mesh with 13,160 elements (element size of 1.5 mm) are employed. To trigger localized failure, at the center of the specimen, a small zone of slightly weakened elements is introduced, as indicated in Figure 7. In the numerical simulations, snapback behavior may occur, i.e., a simultaneous decrease of the load and the displacement after attaining the peak load. At this stage, displacement-controlled experiments become unstable. To overcome potential snapback behavior in the numerical simulations, the indirect displacement control technique [50,51] is employed. For the present example, this technique can be applied by enforcing a monotonic decrease of the vertical distance between two nodes, with each node located at one boundary of the expected shear band.

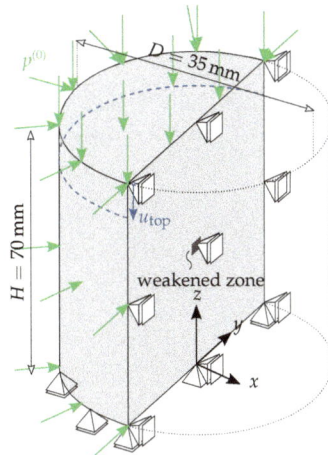

Figure 7. Geometry and boundary conditions of the specimen for the triaxial compression test.

The material parameters required for the elastic-plastic part of the RDP model were identified from single element simulations, as discussed in [52]. The additional parameters for the softening regime A_s, ε_f, l, and m are calibrated from the present numerical simulations for a best fit with the experimental results for the confining pressure of $p^{(0)} = 37.5$ MPa. The employed parameters for Innsbruck quartz phyllite are $E = 56670$ MPa, $v = 0.2$, $f_{cu} = 42$ MPa, $f_{cy} = 29.5$ MPa, $m_0 = 12.0$, $m_{gl} = 9.9$, $A_h = 0.0045$, $C_h = 8.8$, $A_s = 4$, $\varepsilon_f = 4 \times 10^{-4}$, $m = 1.05$, $l = 2$ mm. For the remaining model parameters e, B_h, D_h, G_h, and B_s, the default values summarized in Section 2 are employed. The experimental results for confining pressures of $p^{(0)} = 0$ MPa, 12.5 MPa, and 25 MPa, which have not been used for calibration, serve for validation of the numerical model.

Figure 8 shows the load–displacement curves obtained from the experiments and the numerical simulations for the different confining pressures. Note that the non-zero axial force at the beginning is the consequence of the initial hydrostatic stress state due to the applied confining pressure. Expectedly, for $p^{(0)} = 37.5$ MPa and for the uniaxial compression test ($p^{(0)} = 0$ MPa), the experimentally obtained peak load is represented very well since the test results were used for calibration. For $p^{(0)} = 12.5$ MPa and 25 MPa, the peak loads are predicted satisfactorily, slightly underestimating the experimental results. The qualitative shape of the softening branch is also represented quite well. For the uniaxial compression test, no meaningful experimental results after the peak stress were recorded during the experiments. This unstable behavior is also manifested in the numerical simulations, for which strong snapback behavior is observed. The numerical results computed on the basis of the different finite element meshes reveal the capability of the gradient-enhanced RDP model to regularize the underlying IBVP: Once the mesh is sufficiently fine, mesh-insensitive load–displacement curves are obtained.

Figure 9a shows the deformed specimen with the confining pressure $p^{(0)} = 37.5$ MPa in the final stage of the triaxial compression tests, computed by means of the three different meshes. While the displacements in the lower and the upper part of the specimen are almost uniform, the displacements localize into a single inclined shear band in the center part of the specimen. In the experiments, localization into a shear band was found to be the dominant failure mode, as shown in Figure 9b for a specimen tested with $p^{(0)} = 37.5$ MPa. The formation of the shear band is also reflected by the distribution of the damage variable ω shown in Figure 10. Comparing the predicted damage distributions for the medium and the fine mesh confirms this distribution as insensitive with respect to the discretization.

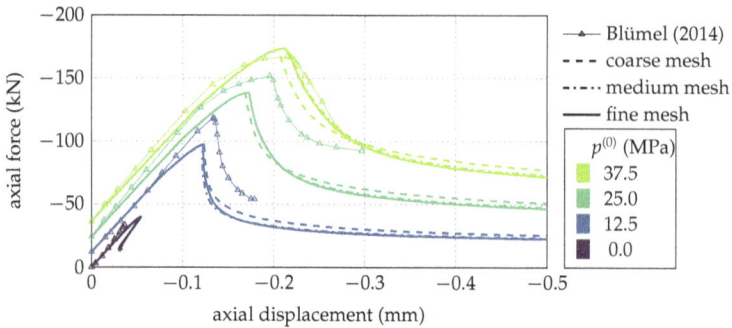

Figure 8. Load–displacement curves (axial force versus axial displacement) for triaxial compression tests: experimental and numerical results for different levels of confining pressures $p^{(0)}$.

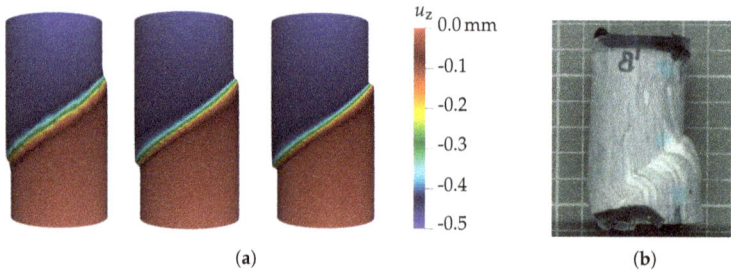

Figure 9. (**a**) Distribution of the vertical displacement u_z (scale factor 5) in the final stage of the triaxial compression test with $p^{(0)} = 37.5\,\text{MPa}$ for the three finite element meshes: coarse (**left**), medium (**center**), and fine (**right**). (**b**) Corresponding deformed rock specimen after a triaxial compression test with $p^{(0)} = 37.5\,\text{MPa}$, reproduced with permission from M. Bluemel taken from the report [41].

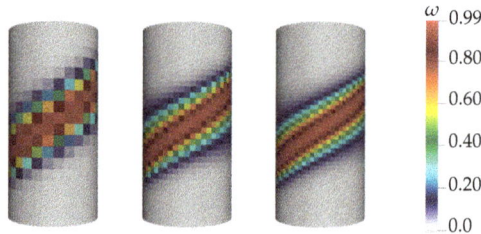

Figure 10. Distribution of the damage variable ω in the final stage of the triaxial compression test with $p^{(0)} = 37.5\,\text{MPa}$ for the three finite element meshes: coarse (**left**), medium (**center**), and fine (**right**).

The influence of the level of confining pressure on the inclination of the shearing zone is demonstrated in Figure 11 based on the predicted distribution of the vertical displacement for $p^{(0)} = 0\,\text{MPa}, 12.5\,\text{MPa}, 25\,\text{MPa}$, and $37.5\,\text{MPa}$, respectively. For the uniaxial compression test ($p^{(0)} = 0\,\text{MPa}$), the zone of localized displacements is strongly inclined. With increasing confining pressure, the inclination angle of the shear band is decreasing, which is best visible by comparing the results for $p^{(0)} = 0\,\text{MPa}$ and $p^{(0)} = 12.5\,\text{MPa}$. Upon further increase of the confining pressure, the inclination of the shear band becomes slightly smaller. The represented dependence of the inclination angle on the level of confining pressure is explained by the curvature of the compressive meridian of the yield surface and of the employed plastic potential function of the RDP model.

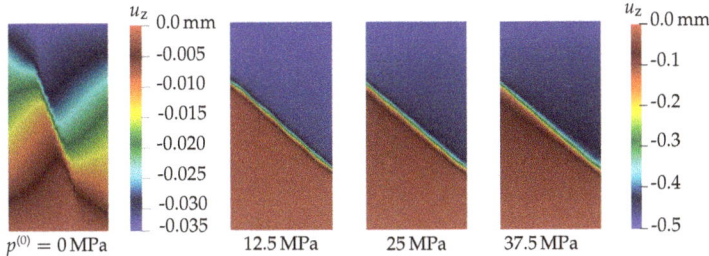

Figure 11. Triaxial compression tests: distribution of the vertical displacement u_z in the symmetry plane for the four different levels of confining pressure.

4.3. Modeling of Shear Failure, Demonstrated by Analyzing Triaxial Extension Tests on Innsbruck Quartz Phyllite

In a third step, the ability of the model for predicting shear failure under confined stress states attaining the tensile meridian of the yield surface is demonstrated. To this end, numerical simulations of triaxial extension tests on specimens with geometric and material properties identical to those of the previously presented triaxial compression tests are performed. Similar to the triaxial compression tests, confining pressures $p^{(0)}$ of $0\,\text{MPa}$, $12.5\,\text{MPa}$, $25\,\text{MPa}$, and $37.5\,\text{MPa}$ are investigated. The same finite element meshes are employed for investigating the influence of the discretization. In contrast to the triaxial compression tests, subsequent to the application of the initial hydrostatic stress state, generated by the confining pressure, a displacement in positive vertical direction is applied at the top surface of the specimens. Since for triaxial extension tests experimental results are not available, the present study focuses on the assessment of the influence of the finite element mesh on the obtained results and serves as further verification of the gradient-enhanced RDP model.

The predicted load–displacement curves are shown in Figure 12. While nearly identical results are obtained for the medium mesh and the fine mesh, the load–displacement curves for the coarse mesh are somewhat more ductile. This discrepancy indicates the coarse mesh as insufficiently fine for the accurate resolution of the gradient of the nonlocal field. Compared to the results from the triaxial compression tests, a more brittle structural response is predicted, resulting in strong snapback behavior in the post-peak regime for all three confining pressures and, in particular, for the uniaxial tension test. In contrast to the triaxial compression tests, the structural response becomes increasingly brittle as confining pressure increases, and the peak load decreases gradually. This phenomenon was also observed in experiments on Berea sandstone in [53], and it is characteristic of brittle and quasi-brittle materials.

Figure 13 shows the computed deformation of the specimen for an applied axial displacement of $0.1\,\text{mm}$ and a confining pressure of $37.5\,\text{MPa}$. Compared to the triaxial compression tests, a smaller inclination angle of the shear band is predicted for the medium mesh and the fine mesh, whereas for the coarse mesh a considerably steeper inclination angle of the shear band is obtained. Again, this discrepancy indicates the insufficient representation of the gradient of the nonlocal field by the coarse mesh.

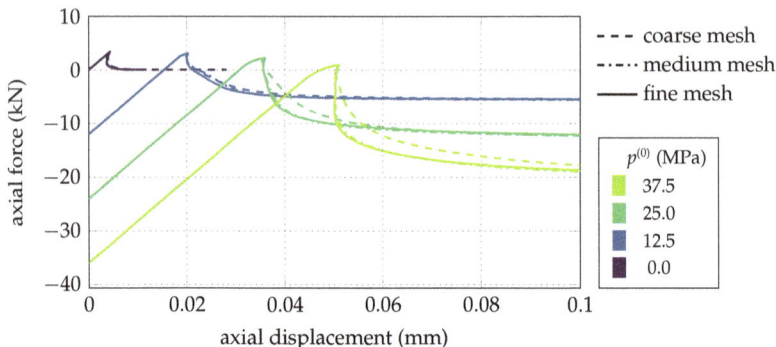

Figure 12. Computed load–displacement curves for triaxial extension tests for different levels of confining pressure $p^{(0)}$.

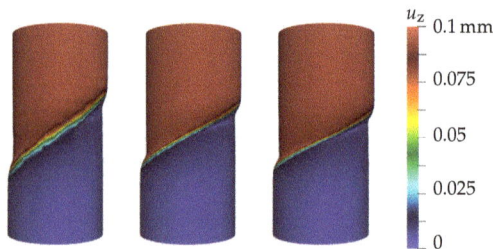

Figure 13. Distribution of the vertical displacement u_z (deformation scale factor 10) at an applied top displacement of 0.1 mm in the triaxial extension test with $p^{(0)} = 37.5$ MPa for the three finite element meshes: coarse (**left**), medium (**center**), and fine (**right**).

4.4. Numerical Simulation of Localizing Deformations in Deep Tunnel Excavation

Finally, the performance of the gradient-enhanced RDP model for predicting the formation of multiple shear bands during the excavation of a deep tunnel is assessed. This benchmark is derived from a stretch of the Brenner Base Tunnel constructed by the drill, blast, and secure procedure, which has already been the subject of investigations in previous publications [52,54]. An analogy to the present problem can be found in the context of petrol engineering, where the formation of shear bands has been observed and reported for borehole breakout [47,55,56].

Since in this contribution the major focus is on the assessment of the gradient-enhanced rock model, the tunnel excavation is approximated by means of a simplified two-dimensional model. Supporting measures like a shotcrete shell or rock anchors are neglected. Potential time-dependent effects of the mechanical behavior of rock mass due to the excavation procedure are not considered. For the excavation of the tunnel profile by means of drill and blast, either a full-face or a sequential excavation procedure can be employed. The chosen excavation sequence may have a considerable impact on the stability of the tunnel. In this numerical model, the worst case scenario is considered by assuming full-face excavation of the circular tunnel profile without any supporting measures, which results in the maximum loading of the rock mass in the vicinity of the tunnel.

The IBVP of tunnel excavation with its geometry, initial conditions, and boundary conditions is illustrated in Figure 14. Within the discretized domain of rock mass, a hydrostatic geostatic stress state characterized by a pressure of $p_i^{(0)} = 25.7$ MPa is assumed, corresponding to the overburden at the tunnel axis of 950 m. In the numerical simulations, initial equilibrium is established by applying the geostatic stress state together with the internal pressure $p_i^{(0)}$ acting on the excavation boundary.

The excavation procedure is simulated by gradually decreasing the internal pressure $p_i^{(0)}$ to zero. Since the analyzed tunnel section is located in Innsbruck quartz phyllite rock mass, most of the material parameters of Section 4.2 are adopted. The material behavior of rock mass in contrast to intact rock is considered by empirical down-scaling factors based on the geological strength index GSI and the disturbance factor D proposed by Hoek and Brown [27] accounting for the influence of distributed discontinuities. They are taken from the geological survey, reported in [52]. The additional material parameters for Innsbruck quartz phyllite rock mass are $GSI = 40$, $D = 0$, $A_s = 15$, $\varepsilon_f = 7 \times 10^{-4}$, and $l = 50$ mm. To trigger the formation of shear bands in spite of the axisymmetric problem, non-uniformly distributed rock mass properties are employed by introducing zones in which the strength of the rock mass is slightly weakened (indicated in Figure 14). It was verified that these weakened zones do not affect the predicted mechanical response before the onset of strain softening. For investigating the influence of the finite element mesh on the predicted results, three structured meshes with fully integrated and reduced integrated 8-node quadrilateral elements with element sizes of 300 mm (8078 elements), 140 mm (21,414 elements), and 70 mm (55828 elements) in the close vicinity of the tunnel are employed. Both the displacement field and the nonlocal field are approximated by quadratic shape functions.

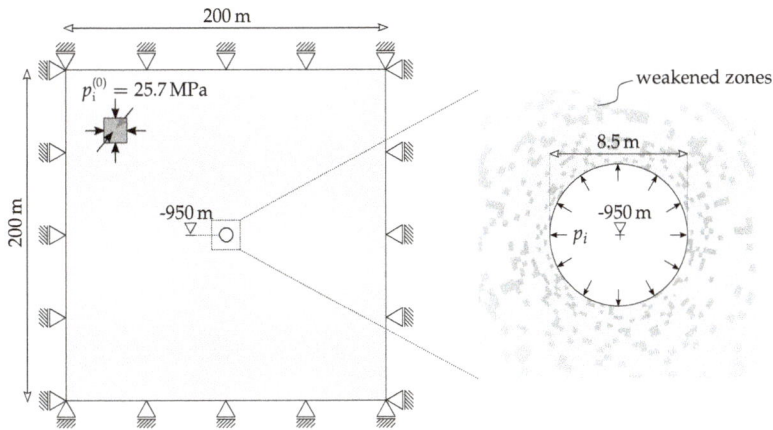

Figure 14. 2D initial boundary value problem of deep tunnel excavation: full model (**left**) and the detail center view (**right**).

Upon decreasing the internal pressure, initially the rock mass behavior remains in the linear elastic regime, followed by the formation of plastic zones emerging from the excavation boundary. Once the strength of the rock mass is attained, strain softening is initiated. From the onset of strain softening, the strains are localizing into narrow zones of the rock mass, and, eventually, large displacements accumulate. Finally, at a certain release level of the internal pressure, equilibrium is lost.

Figure 15 shows the deformed rock mass at the level of 12% of the internal pressure for the three meshes with fully integrated elements. For the medium and the fine mesh, the localization of displacements into narrow zones has already reached a very progressed stage close to failure, and the formation of shear bands is clearly visible. The non-axisymmetric displacement field indicates the transition from an initially quasi-continuous rock mass to quasi-discontinuous rock mass. The latter is characterized by blocks of rock mass moving towards the tunnel center, so potential failure of the tunnel is imminent. Concerning the influence of the finite element discretization, for the medium and the fine mesh, an almost identical displacement field is obtained. Slight differences between those two solutions can be observed only for the upper right quadrant.

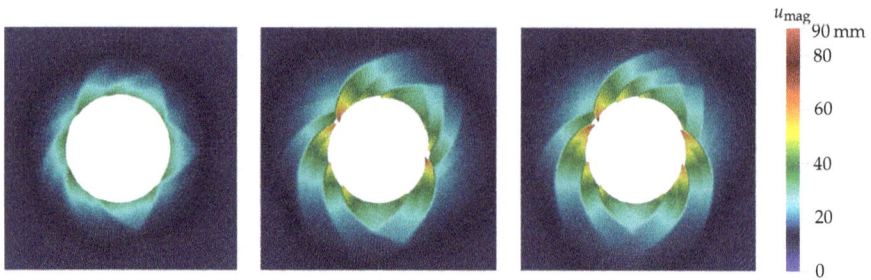

Figure 15. Distribution of the magnitude of the displacement vector in the vicinity of the tunnel surface at the level of 12 % of the initial internal pressure with a displacement scale factor of 10 for the three finite element meshes with full numerical integration: coarse (**left**), medium (**center**), and fine (**right**).

Figure 16 shows the load–displacement curves, i.e., the normalized internal pressure versus the mean displacement magnitude along the excavation boundary computed for each mesh. The load–displacement curves predicted by the three meshes are very close to each other, and, in particular, the results for the medium and the fine mesh are almost identical. The results for the coarse mesh are slightly different due to the already discussed required mesh size for a sufficient resolution of the gradient of the nonlocal field.

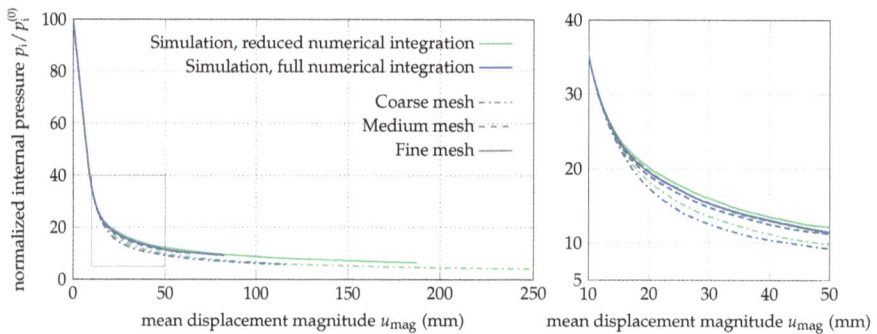

Figure 16. Load–displacement curves: normalized internal pressure versus the mean displacement magnitude at the tunnel surface for the three finite element meshes with full and reduced numerical integration: total view (**left**), detailed view (**right**).

Figure 17 depicts the damaged zones in the rock mass at the level of 12% of the initial internal pressure for the three meshes employing full numerical integration. The highly damaged zones correspond to large shear deformations, which were shown previously in Figure 15. A similar shape of failure zones was obtained by Addis et al. [57] in laboratory tests on a bore hole in weak sandstone. Hence, the potential of the gradient-enhanced approach to predict the onset of failure of a structure was demonstrated in spite of the rather complex failure mode.

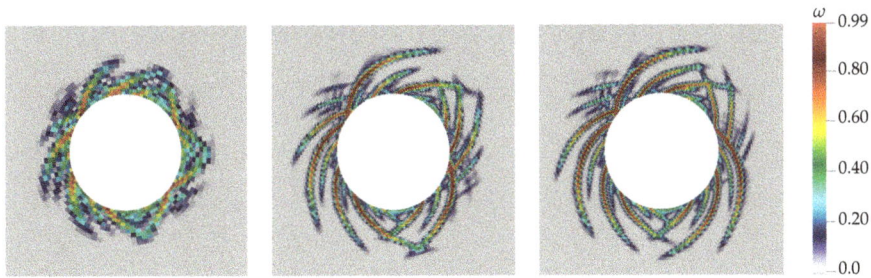

Figure 17. Distribution of the damage variable ω in the rock mass in the vicinity of the tunnel surface at the level of 12 % of the internal pressure for the three finite element meshes employing full numerical integration: coarse (**left**), medium (**center**), and fine (**right**).

5. Conclusions

The present contribution addressed the prediction of complex failure modes in numerical simulations by means of a new implicit gradient-enhanced damage-plasticity model for intact rock and rock mass. It was derived from the damage-plasticity rock model presented by Unteregger et al. [24,25] and extended by adopting the implicit gradient-enhancement proposed in [17]. For the assessment of the model, a comprehensive numerical study was presented. In particular, numerical simulations of wedge splitting tests for evaluating the representation of mode I failure, simulations of triaxial compression and extension tests for evaluating the representation of shear failure under confined stress states, and finally simulations of deep tunneling for examining the prediction of failure mechanisms of a complex structure were performed. From the obtained results of the numerical study, the following conclusions can be drawn:

- In numerical simulations of wedge splitting tests, the formation of a crack propagating from the notch is modeled by the implicit gradient-enhancement in a smeared manner over a width related to the assumed length scale parameter. The capability of the gradient-enhanced damage-plasticity rock model of representing the experimentally observed material behavior was realistically demonstrated.
- Finite element analyses with both structured and unstructured meshes confirmed the regularizing effect of the implicit gradient-enhancement in mode I failure and thus revealed mesh-insensitive results. In particular, it was demonstrated that the crack direction is not biased by the orientation of the finite element mesh.
- Regarding the simulations of triaxial compression tests, a proper representation of shear failure under confined stress states, attaining the compressive meridian of the yield surface, was shown. After attaining peak strength upon initiation of damage, localization into a distinct shear band was observed. Furthermore, reasonable agreement with experimental results, and in particular the experimentally observed increasingly ductile material behavior with increasing confining pressure was obtained.
- The simulations of triaxial extension tests demonstrated that failure under confined stress states attaining the tensile meridian of the yield surface is represented reasonably well, i.e., the localization of damage into a distinct shear band is predicted. In contrast to the triaxial compression tests, a more brittle structural response was observed.
- For both the triaxial compression and extension tests, a mesh study confirmed mesh-insensitive results for shear failure under confined stress states.
- In the numerical simulations of deep tunnel excavation, the rock mass was subjected to softening material behavior due to the excavation procedure, which leads to localization of strains into multiple distinct shear bands. In spite of the complex structural failure mechanism involving the

45

formation of multiple shear bands, mesh insensitive load–displacement curves were obtained. By analyzing the convergence of the damage patterns upon mesh refinement, only slight differences were recognized.

- For the adopted formulation of the implicit gradient-enhanced rock model, the length scale l is assumed as a constant parameter. Thus, a rather broad zone of completely damaged material is predicted by the model. Possible remedies were proposed in the literature, for instance, based on variable length scale parameters [58,59] or the so-called concept of decreasing interactions [45], which is motivated by a physically based micromechanical homogenization theory [60], cf. [46] for a discussion of these approaches. For a more realistic representation of sharp cracks within the present gradient-enhanced rock model, further investigations of these concepts are recommended.

- Regarding the identification of the parameters controlling the gradient-enhanced softening part of the model, i.e., softening modulus ε_f, length scale parameter l, and weighting parameter m, an ad hoc approach was employed: While the length scale parameter l was treated as a model parameter conforming to the size of the employed finite element mesh to ensure a sufficient resolution of the nonlocal gradient, ε_f was identified by a best fit with experimental results for a chosen l and m. A more systematic approach for identifying these parameters based on experimental results is an open issue. In particular, a scheme to compute ε_f based on a prescribed value of l and a typical material parameter, such as the specific mode I fracture energy, is desirable.

Summarizing, it can be concluded that the presented damage-plasticity model is capable of representing a wide range of failure mechanisms in numerical simulations in a realistic and objective manner.

Author Contributions: This paper was jointly conceived by M.S., M.N., and G.H.; M.S. and M.N. implemented the numerical framework for the implicit gradient-enhanced model, performed the numerical simulations, and evaluated the results; M.S., M.N., and G.H. wrote the paper.

Funding: This research received no external funding.

Conflicts of Interest: The authors declare no conflict of interest.

References

1. Swoboda, G.; Shen, X.P.; Rosas, L. Damage model for jointed rock mass and its application to tunnelling. *Comput. Geotech.* **1998**, *22*, 183–203. [CrossRef]
2. Martin, C.; Kaiser, P.; McCreath, D. Hoek–Brown parameters for predicting the depth of brittle failure around tunnels. *Can. Geotech. J.* **1999**, *36*, 136–151. [CrossRef]
3. Egger, P. Design and construction aspects of deep tunnels (with particular emphasis on strain softening rocks). *Tunn. Undergr. Space Technol.* **2000**, *15*, 403–408. [CrossRef]
4. Jing, L. A review of techniques, advances and outstanding issues in numerical modelling for rock mechanics and rock engineering. *Int. J. Rock Mech. Min. Sci.* **2003**, *40*, 283–353. [CrossRef]
5. Grassl, P.; Jirásek, M. Damage-plastic model for concrete failure. *Int. J. Solids Struct.* **2006**, *43*, 7166–7196. [CrossRef]
6. Bažant, Z.P. Instability, ductility, and size effect in strain-softening concrete. *ASCE J. Eng. Mech. Div.* **1976**, *102*, 331–344.
7. Bažant, Z.P.; Oh, B.H. Crack band theory for fracture of concrete. *Mater. Struct.* **1983**, *16*, 155–177. [CrossRef]
8. Bažant, Z.P.; Jirásek, M. Nonlocal integral formulations of plasticity and damage: Survey of progress. *J. Eng. Mech.* **2002**, *128*, 1119–1149. [CrossRef]
9. De Borst, R. Simulation of strain localization: A reappraisal of the Cosserat continuum. *Eng. Comput.* **1991**, *8*, 317–332. [CrossRef]
10. Needleman, A. Material rate dependence and mesh sensitivity in localization problems. *Comput. Methods Appl. Mech. Eng.* **1988**, *67*, 69–85. [CrossRef]
11. Miehe, C.; Welschinger, F.; Hofacker, M. Thermodynamically consistent phase-field models of fracture: Variational principles and multi-field FE implementations. *Int. J. Numer. Methods Eng.* **2010**, *83*, 1273–1311. [CrossRef]

12. Peerlings, R.H.J.; Geers, M.G.D.; de Borst, R.; Brekelmans, W.A.M. A critical comparison of nonlocal and gradient-enhanced softening continua. *Int. J. Solids Struct.* **2001**, *38*, 7723–7746. [CrossRef]

13. De Borst, R.; Pamin, J. Gradient plasticity in numerical simulation of concrete cracking. *Eur. J. Mech. A/Solids* **1996**, *15*, 295–320.

14. Geers, M.; de Borst, R.; Brekelmans, W.; Peerlings, R. Validation and internal length scale determination for a gradient damage model: Application to short glass-fibre-reinforced polypropylene. *Int. J. Solids Struct.* **1999**, *36*, 2557–2583. [CrossRef]

15. Peerlings, R.H.J.; Massart, T.J.; Geers, M.G.D. A thermodynamically motivated implicit gradient damage framework and its application to brick masonry cracking. *Comput. Methods Appl. Mech. Eng.* **2004**, *193*, 3403–3417. [CrossRef]

16. Pearce, C.J.; Nielsen, C.V.; Bićanić, N. Gradient enhanced thermo-mechanical damage model for concrete at high temperatures including transient thermal creep. *Int. J. Numer. Anal. Methods Geomech.* **2004**, *28*, 715–735. [CrossRef]

17. Poh, L.H.; Swaddiwudhipong, S. Over-nonlocal gradient enhanced plastic-damage model for concrete. *Int. J. Solids Struct.* **2009**, *46*, 4369–4378.

18. Verhoosel, C.V.; Scott, M.A.; Hughes, T.J.R.; de Borst, R. An isogeometric analysis approach to gradient damage models. *Int. J. Numer. Methods Eng.* **2011**, *86*, 115–134. [CrossRef]

19. Zreid, I.; Kaliske, M. Regularization of microplane damage models using an implicit gradient enhancement. *Int. J. Solids Struct.* **2014**, *51*, 3480–3489. [CrossRef]

20. Hosseini, H.; Horák, M.; Zysset, P.; Jirásek, M. An over-nonlocal implicit gradient-enhanced damage-plastic model for trabecular bone under large compressive strains. *Int. J. Numer. Methods Biomed. Eng.* **2015**, *31*. [CrossRef] [PubMed]

21. Zreid, I.; Kaliske, M. A gradient enhanced plasticity–damage microplane model for concrete. *Comput. Mech.* **2018**, 1–19. [CrossRef]

22. De Borst, R.; Pamin, J.; Geers, M.G. On coupled gradient-dependent plasticity and damage theories with a view to localization analysis. *Eur. J. Mech.-A/Solids* **1999**, *18*, 939–962. [CrossRef]

23. Hoek, E. Brittle fracture of rock. In *Rock Mechanics in Engineering Practice*; Wiley: London, UK, 1968; Volume 130.

24. Unteregger, D.; Fuchs, B.; Hofstetter, G. A damage plasticity model for different types of intact rock. *Int. J. Rock Mech. Min. Sci.* **2015**, *80*, 402–411. [CrossRef]

25. Unteregger, D. Advanced Constitutive Modeling of Intact Rock and Rock Mass. Ph.D. Thesis, Innsbruck University: Innsbruck, Austria, 2015.

26. Hoek, E.; Diederichs, M.S. Empirical estimation of rock mass modulus. *Int. J. Rock Mech. Min. Sci.* **2006**, *43*, 203–215. [CrossRef]

27. Hoek, E.; Brown, E.T. Practical estimates of rock mass strength. *Int. J. Rock Mech. Min. Sci.* **1997**, *34*, 1165–1186. [CrossRef]

28. Hoek, E.; Carranza-Torres, C.; Corkum, B. Hoek–Brown failure criterion—2002 edition. In Proceedings of the 5th North American Rock Mechanics Symposium, 17th Tunnelling Association of Canada, Toronto, ON, Canada, 7–10 July 2002; Hammah, R., Ed.; University of Toronto Press: Toronto, ON, Canada, 2002; pp. 267–273.

29. Hoek, E.; Brown, E.T. Empirical strength criterion for rock masses. *J. Geotech. Geoenviron. Eng.* **1980**, *106*, 1013–1035.

30. Menétrey, P.; Willam, K.J. Triaxial failure criterion for concrete and its generalization. *ACI Struct. J.* **1995**, *92*, 311–318.

31. Jirásek, M.; Bauer, M. Numerical aspects of the crack band approach. *Comput. Struct.* **2012**, *110*, 60–78. [CrossRef]

32. Peerlings, R.H.J.; de Borst, R.; Brekelmans, W.A.M.; De Vree, J.H.P. Gradient enhanced damage for quasi-brittle materials. *Int. J. Numer. Methods Eng.* **1996**, *39*, 68. [CrossRef]

33. Vermeer, P.A.; Brinkgreve, R.B.J. A new effective non-local strain measure for softening plasticity. In *Localisation and Bifurcation Theory for Soils and Rocks*; Chambon, R., Desrues, J., Vardoulakis, I., Eds.; CRC Press: Boca Raton, FL, USA, 1994; pp. 89–100.

34. Di Luzio, G.; Bažant, Z.P. Spectral analysis of localization in nonlocal and over-nonlocal materials with softening plasticity or damage. *Int. J. Solids Struct.* **2005**, *42*, 6071–6100. [CrossRef]

35. Charlebois, M.; Jirásek, M.; Zysset, P.K. A nonlocal constitutive model for trabecular bone softening in compression. *Biomech. Model. Mechanobiol.* **2010**, *9*, 597–611. [CrossRef] [PubMed]
36. Areias, P.; Rabczuk, T.D.; De Sá, J.C. A novel two-stage discrete crack method based on the screened Poisson equation and local mesh refinement. *Comput. Mech.* **2016**, *58*, 1003–1018. [CrossRef]
37. Peerlings, R.H.J.; de Borst, R.; Brekelmans, W.A.M.; Geers, M.G.D. Gradient-enhanced damage modelling of concrete fracture. *Mech. Cohesive-Frict. Mater.* **1998**, *3*, 323–342. [CrossRef]
38. De Borst, R.; Verhoosel, C.V. Gradient damage vs phase-field approaches for fracture: Similarities and differences. *Comput. Methods Appl. Mech. Eng.* **2016**, *312*, 78–94. [CrossRef]
39. Simo, J.C.; Hughes, T.J.R. *Computational Inelasticity, Interdisciplinary Applied Mathematics 7*; Springer: New York, NY, USA, 1998.
40. Brühwiler, E.; Saouma, V.E. Fracture testing of rock by the wedge splitting test. In *Rock Mechanics Contributions and Challenges, Proceedings of the 31st US Symposium on Rock Mechanics, CO, USA, 18–20 June 1990*; Hustrulid, W., Johnson, G., Eds.; CRC Press: Boca Raton, FL, USA, 1990; pp. 287–294.
41. Blümel, M. *Prüfprotokolle Laborversuche Druckversuche Anfahrtstutzen Ahrntal*; Technical Report; Institut für Felsmechanik und Tunnelbau TU Graz: Graz, Austria, 2014.
42. Grassl, P.; Xenos, D.; Jirásek, M.; Horák, M. Evaluation of nonlocal approaches for modelling fracture near nonconvex boundaries. *Int. J. Solids Struct.* **2014**, *51*, 3239–3251. [CrossRef]
43. Pourhosseini, O.; Shabanimashcool, M. Development of an elasto-plastic constitutive model for intact rocks. *Int. J. Rock Mech. Min. Sci.* **2014**, *66*, 1–12. [CrossRef]
44. Jirásek, M.; Grassl, P. Evaluation of directional mesh bias in concrete fracture simulations using continuum damage models. *Eng. Fract. Mech.* **2008**, *75*, 1921–1943. [CrossRef]
45. Poh, L.H.; Sun, G. Localizing gradient damage model with decreasing interactions. *Int. J. Numer. Methods Eng.* **2017**, *110*, 503–522. [CrossRef]
46. Jirásek, M. Regularized continuum damage formulations acting as localization limiters. In *Computational Modelling of Concrete Structures, Proceedings of the Conference on Computational Modelling of Concrete and Concrete Structures (EURO-C 2018), Bad Hofgastein, Austria, 26 February–1 March 2018*; Meschke, G., Pichler, B., Rots, J.G., Eds.; Taylor & Francis Group: London, UK, 2018; pp. 25–42.
47. Zervos, A.; Papanastasiou, P.; Cook, J. Elastoplastic finite element analysis of inclined wellbores. In Proceedings of the SPE/ISRM Rock Mechanics in Petroleum Engineering, Trondheim, Norway, 8–10 July 1998; Society of Petroleum Engineers: Richardson, TX, USA, 1998; pp. 1–10.
48. Fang, Z.; Harrison, J.P. Application of a local degradation model to the analysis of brittle fracture of laboratory scale rock specimens under triaxial conditions. *Int. J. Rock Mech. Min. Sci.* **2002**, *39*, 459–476. [CrossRef]
49. Golshani, A.; Okui, Y.; Oda, M.; Takemura, T. A micromechanical model for brittle failure of rock and its relation to crack growth observed in triaxial compression tests of granite. *Mech. Mater.* **2006**, *38*, 287–303. [CrossRef]
50. De Borst, R. Computation of post-bifurcation and post-failure behavior of strain-softening solids. *Comput. Struct.* **1987**, *25*, 211–224. [CrossRef]
51. Jirásek, M.; Bažant, Z.P. *Inelastic Analysis of Structures*; John Wiley & Sons, Ltd.: Chichester, UK, 2002.
52. Schreter, M.; Neuner, M.; Unteregger, D.; Hofstetter, G.; Reinhold, C.; Cordes, T.; Bergmeister, K. Application of a damage plasticity model for rock mass to the numerical simulation of tunneling. In Proceedings of the IV International Conference on Computational Methods in Tunneling and Subsurface Engineering (EURO:TUN 2017), Innsbruck, Austria, 18–20 April 2017, Hofstetter, G., Bergmeister, K., Eberhardsteiner, J., Meschke, G., Schweiger, H.F., Eds.; Innsbruck University: Innsbruck, Austria, 2017; pp. 549–556.
53. Bobich, J. Experimental Analysis of the Extension to Shear Fracture Transition in Berea Sandstone. Master's Thesis, Texas A&M University, College Station, TX, USA, 2005.
54. Neuner, M.; Schreter, M.; Unteregger, D.; Hofstetter, G. Influence of the Constitutive Model for Shotcrete on the Predicted Structural Behavior of the Shotcrete Shell of a Deep Tunnel. *Materials* **2017**, *10*, 577. [CrossRef] [PubMed]
55. Vardoulakis, I.; Sulem, J.; Guenot, A. Borehole instabilities as bifurcation phenomena. *Int. J. Rock Mech. Min. Sci. Geomech. Abstr.* **1988**, *25*, 159–170. [CrossRef]
56. Crook, T.; Willson, S.; Yu, J.G.; Owen, R. Computational modelling of the localized deformation associated with borehole breakout in quasi-brittle materials. *J. Pet. Sci. Eng.* **2003**, *38*, 177–186. [CrossRef]

57. Addis, M.A.; Barton, N.R.; Bandis, S.C.; Henry, J.P. Laboratory studies on the stability of vertical and deviated boreholes. In Proceedings of the SPE Annual Technical Conference and Exhibition, New Orleans, LA, USA, 23–26 September 1990.
58. Geers, M.G.D. Experimental Analysis and Computational Modeling of Damage and Fracture. Ph.D. Thesis, Einhoven University of Technology, Eindhoven, The Netherlands, 1997.
59. Geers, M.G.D.; de Borst, R.; Brekelmans, W.A.M.; Peerlings, R.H.J. Strain-based transient-gradient damage model for failure analyses. *Comput. Methods Appl. Mech. Eng.* **1998**, *160*, 133–153. [CrossRef]
60. Sun, G.; Poh, L. Homogenization of intergranular fracture towards a transient gradient damage model. *J. Mech. Phys. Solids* **2016**, *95*, 374–392. [CrossRef]

applied
sciences

MDPI

Article

A Nonlinear Crack Model for Concrete Structure Based on an Extended Scaled Boundary Finite Element Method

Jian-bo Li [1,*], Xin Gao [1], Xing-an Fu [1], Chenglin Wu [2] and Gao Lin [1]

[1] State Key Laboratory of Coastal and Offshore Engineering, Institute of Earthquake Engineering, Dalian University of Technology, Dalian 116024, China; gxgaoxin@mail.dlut.edu.cn (X.G.); xinganfu@mail.dlut.edu.cn (X.-a.F.); gaolin@dlut.edu.cn (G.L.)
[2] Department of Civil, Architectural and Environmental Engineering, Missouri University of Science and Technology, Rolla, MO 65409, USA; wuch@mst.edu
* Correspondence: jianboli@dlut.edu.cn

Received: 27 April 2018; Accepted: 26 June 2018; Published: 29 June 2018

Abstract: Fracture mechanics is one of the most important approaches to structural safety analysis. Modeling the fracture process zone (FPZ) is critical to understand the nonlinear cracking behavior of heterogeneous quasi-brittle materials such as concrete. In this work, a nonlinear extended scaled boundary finite element method (X-SBFEM) was developed incorporating the cohesive fracture behavior of concrete. This newly developed model consists of an iterative procedure to accurately model the traction distribution within the FPZ accounting for the cohesive interactions between crack surfaces. Numerical validations were conducted on both of the concrete beam and dam structures with various loading conditions. The results show that the proposed nonlinear X-SBFEM is capable of modeling the nonlinear fracture propagation process considering the effect of cohesive interactions, thereby yielding higher precisions than the linear X-SBFEM approach.

Keywords: elastoplastic behavior; extended scaled boundary finite element method (X-SBFEM); stress intensity factors; fracture process zone (FPZ)

1. Introduction

With the development of numerical analysis technology, structural fracture mechanics is an important approach to structural safety evaluation. Fracture process zone (FPZ) is defined as the intermediate space between cracked and uncracked portions of concrete [1]. Different from real cracks, the FPZ can still transmit stress, and the stress, σ, that FPZ transmits decreases with increasing crack open displacement, w. When the crack open displacement reaches a certain critical value, w_c, the surface force of the crack surface becomes zero, as illustrated in Figure 1. The FPZ consists of microcracks, which are minute individual cracks; this gives rises to the cohesive tractions ahead of the crack tip, which comes from the aggregate interlocking and surface friction. Therefore, a nonlinear fracture-mechanics-based method needs to be applied to account for the effect of cohesive tractions during the fracture propagations.

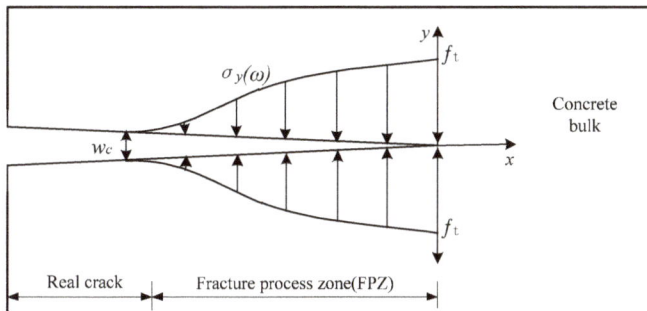

Figure 1. Fracture process zone (FPZ) in concrete.

Two main approaches are often taken to model the FPZ, which are: the smeared crack models and the discrete crack models. The smeared crack models proposed by Rashid [2] are based on the continuum approach, where the computational mesh of the FPZ remains constant while the fracture propagation is modeled by the growth of a number of parallel cracks smeared over the elements within the FPZ. Using this approach, Bhattacharjee et al. [3] and Calayir et al. [4] successfully carried out dynamic cracking analyses of the Koyna gravity dam under the influence of nonorthogonal cracks. Cai et al. [5] also follow a similar approach incorporating the linear or bilinear softening dispersion crack models to predict the crack response of concrete gravity dams. The damage-based fracture mechanics was also developed sharing the similar concept by Bazant [6]. These methods have shown tremendous success in modeling concrete fractures under complex loading and boundary conditions. However, the strong mesh dependency often causes complications in determining the characteristic length scale, fracture strength, and fracture toughness to accurately describe the fracturing propagation process. The smeared crack models mainly combine with damage mechanics for crack propagation studies.

On the other hand, the discrete crack model proposed by Dugdale [7] and Barenblatt [8] utilizes a predefined fracture path as a part of the computational domain boundary, which reduces the mesh dependency comparing to the smeared crack model. Hillerborg et al. [9] firstly implemented the cohesive zone model (CZM) to describe the FPZ based on the discrete crack approach. Following this approach, numerous scholars have improved and implemented the cohesive zone method to model the fracture propagation process (Skrikerud et al. [10], Ayari et al. [11], Xie et al. and Yang et al. [12–14]). Among these improvements, the scaled boundary element method (SBFEM) was developed to model the complex fracture growth in terms of fracture branching and coalescing under complex loading conditions. However, the SBFEM fails to capture the nonlinearity brought by the cohesive interactions within the FPZ. Therefore, the mesoscale and atomic-scale-inserted CZMs were developed to model the FPZ [15–18]. However, the predetermined fracture paths often cause computational complexities and inaccuracies.

Recently, the extended scaled boundary finite element method (X-SBFEM) based on the level set method was developed on the basis of both the SBFEM [19,20] and the extended finite elements (XFEM) [21,22]. Capitalizing on the advantages of both methods, X-SBFEM [23,24] can make full use of XFEM to describe the discontinuous displacement field and SBFEM to solve the problem of the stress singularity with higher precision. Simulating the crack body section using XFEM and the crack tip using SBFEM, the method finally establishes the total equilibrium equation of the crack body and solves the equation, thereby overcoming the disadvantages of XFEM, such as obtaining the analytical form of the displacement and the stress asymptotic fields of the crack tip in advance and constructing the complex enhanced functions, which can express nonsmooth behaviors near the crack tip. In some special circumstances, the enrichment functions are discontinuous or have nonpolynomial forms to specially address the issue when the stiffness matrix is constructed using numerical integration.

Currently, the application of X-SBFEM in fractures [23] based on linear elastic fracture mechanics (LEFM) mainly focuses on the linear fracture. Due to the complexity of the three-dimensional analysis model, the theoretical researches of fracture mechanics still focus on the state of the two-dimensional analysis model. Therefore, based on the X-SBFEM algorithm, this paper introduces a nonlinear crack model which adopts the linear superposition of iterative methods to incorporate the cohesive interactions within the FPZ. The proposed approach was then implemented to model mixed-mode fracture of concrete beam and gravity dam structures. The results show improvements comparing to other methods. Close agreements were found between the numerical and experimental results.

The contents of this paper are arranged as follows. In Section 2, we explain the principle of X-SBFEM. In Section 3, a nonlinear crack model with iterative method for cohesive interactions in the FPZ is introduced and a flowchart for solving the cohesion is given. In Section 4, four numerical simulations (a three-point bending beam, a four-point shear beam, an experimental concrete gravity dam with single crack expansion, and a static cohesive crack propagation simulation of Koyna Dam) are modeled to validate the nonlinear model. In Section 5, we conclude that this paper has developed the X-SBFEM with the nonlinear model to improve the modeling of crack propagation.

2. Extended Scaled Boundary Finite Element Method

The core content of X-SBFEM based on the level set method focuses on the simulation of the nonsmooth behavior near the crack tip using semianalytical SBFEM in the form of super-elements and the simulation of the crack body using XFEM. The key lies in the way the algorithm addresses the boundary conditions at the joint. Figure 2 below shows the topological relationship in the model domain including a crack based on X-SBFEM.

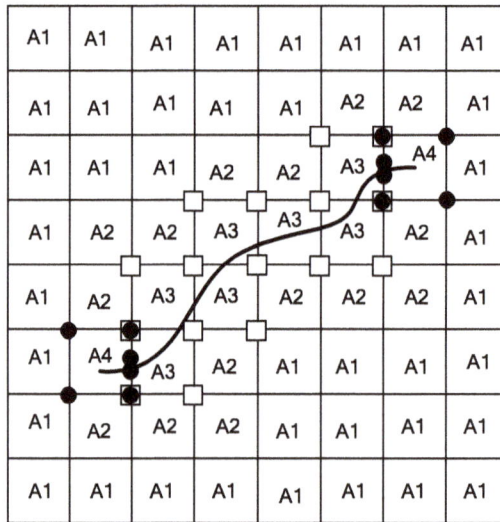

Figure 2. Topological relation in the domain including a crack based on X-SBFEM.

2.1. Extended Finite Element Method

Based on the partition of unity methods simulating the crack body by XFEM, the general formula of a displacement field [25,26] is

$$u^h(x) = \sum_{I \in N^{fem}} N_I(x)q_I + \sum_{J \in N^e} N_J(x)\vartheta(x)a_J \tag{1}$$

where N^{fem} and N^e respectively represent a node in a general element and an enriched node in an internal split crack. Correspondingly, q_I is a normal degree-of-freedom, a_J is a generalized degree-of-freedom related to ϑ (as shown in the square node in Figure 2), and $\vartheta(x)$ is the Heaviside step function.

The equilibrium equation is

$$\begin{bmatrix} K_{aa}^e & K_{ab}^e \\ K_{ba}^e & K_{bb}^e \end{bmatrix} \begin{Bmatrix} q_I \\ a_J \end{Bmatrix} = \begin{Bmatrix} P_a \\ P_b \end{Bmatrix} \quad P_b = \iint\limits_{\Omega/\Gamma d} (H\cdot N)^T \cdot p_v d\Omega + \int_{\Gamma t} (H\cdot N)^T \cdot \bar{t} d\Gamma + \int_{\Gamma d} (H\cdot N)^T \cdot p d\Gamma \quad (2)$$

where K_{aa}^e and K_{bb}^e represent the stress matrix respectively related to a normal degree-of-freedom and a generalized degree-of-freedom. Moreover, K_{ab}^e and K_{ba}^e are the coupling matrices, and P_a and P_b are the equivalent nodal forces in accordance to general degrees-of-freedom and generalized degrees-of-freedom, respectively.

$$K_{aa}^e = \iint\limits_{\Omega/\Gamma d} B^T DB d\Omega \quad (3)$$

$$K_{ab}^e = (K_{ba}^e)^T = \iint\limits_{\Omega/\Gamma d} B^T D(HB) d\Omega \quad (4)$$

$$K_{bb}^e = \iint\limits_{\Omega/\Gamma d} (HB)^T D(HB) d\Omega \quad (5)$$

$$P_a = \iint\limits_{\Omega/\Gamma d} N^T \cdot p_v d\Omega + \int_{\Gamma t} N^T \cdot \bar{t} d\Gamma \quad (6)$$

$$P_b = \iint\limits_{\Omega/\Gamma d} (H\cdot N)^T \cdot p_v d\Omega + \int_{\Gamma t} (H\cdot N)^T \cdot \bar{t} d\Gamma + \int_{\Gamma d} (H\cdot N)^T \cdot p d\Gamma \quad (7)$$

where Γd and Γt respectively represent the crack face and the force interface. Moreover, p_v and $\bar{t}$ respectively represent the body force in computational domain and the surface force on the force interface.

According to the advantage of the description of discontinuous displacement field description, XFEM is used to simulate the main body of the crack.

2.2. Scaled Boundary Finite Element Method

X-SBFEM is used to simulate the crack tip for the high efficiency and high precision of stress singular field simulations. As shown in Figure 3, there are side-face forces at the face of the super-element at the crack tip in the case of SBFEM. Without taking the body force into account, the displacement field and the stress field given by the SBFEM are [21]

$$\{u(\xi, \eta)\} = [N(\eta)]\{u(\xi)\} = \sum_{i=1}^{n} c_i \xi^{\lambda_i} [N(\lambda)]\{\varphi_i\} \quad (8)$$

$$\{\sigma(\xi, \eta)\} = [D][L][N(\eta)]\{u(\xi)\} = [D]\left[B^1(\eta)\right]\{u(\xi)\}_{,\xi} + \frac{1}{\xi}[D]\left[B^2(\eta)\right]\{u(\xi)\} \quad (9)$$

where $N(\eta)$ is the interpolation shape function for one-dimensional line elements and $\{\varphi_i\}$ and λ_i are the displacement modes and the eigenvalues, respectively. Furthermore, $[B^1(\eta)]$ and $[B^2(\eta)]$ are determined by the geometric shape of the boundary of the element. The formulas calculated by the virtual work principle are

$$\{P\} = [K]\{u_h\} = \left[E^0\right][\Phi][\lambda][\Phi]^{-1} + \left[E^1\right]^T \{u_h\} \quad (10)$$

$$\left[E^0\right]\xi^2\{u(\xi)\}_{,\xi\xi}+\left(\left[E^0\right]+\left[E^1\right]^T-\left[E^1\right]\right)\{u(\xi)\}_{,\xi}-\left[E^2\right]\{u(\xi)\}=0 \tag{11}$$

where $\{P\}$ is the equivalent boundary nodal force and $\left[E^0\right]$, $\left[E^1\right]$, and $\left[E^2\right]$ are the coefficient matrices of the SBFEM governing equations.

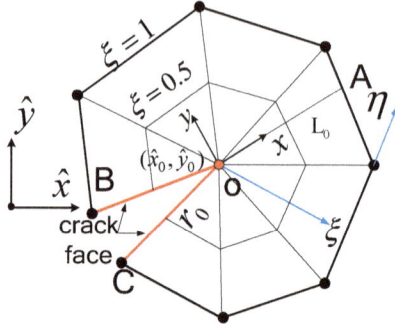

Figure 3. Discretization on boundary with element and scaled transformation of coordinates of bounded media.

Equation (11) is a second-order homogeneous ordinary differential equation and its solution is

$$\{u(\xi)\}=[\varphi]\xi^\lambda\{c\}=\sum_{i=1}^{n}c_i\xi^{\lambda_i}\varphi_i \tag{12}$$

where c_i is the weight of this mode.

Substitute Equation (12) into Equation (10) and get a quadratic eigenvalue problem.

$$[\lambda_i^2\left[E^0\right]+\lambda_i\left(\left[E^1\right]^T-\left[E^1\right]\right)-\left[E^2\right]]\varphi_i=0 \tag{13}$$

Substitute Equation (12) into Equation (10) and get the equivalent node force q_i on the boundary corresponding to the displacement mode.

$$q_i=\left(\left[E^1\right]^T+\lambda_i\left[E^0\right]\right)\varnothing_i \tag{14}$$

Combining Equations (13) and (14) and introducing auxiliary variables can transform quadratic eigenvalue problems into standard linear eigenvalue problems.

$$\begin{bmatrix} -\left[E^0\right]^{-1}\left[E^1\right]^T & \left[E^0\right]^{-1} \\ -\left[E^1\right]\left[E^0\right]^{-1}\left[E^1\right]^T+\left[E^2\right] & \left[E^1\right]\left[E^0\right]^{-1} \end{bmatrix}\begin{bmatrix} \Phi \\ Q \end{bmatrix}=\lambda\begin{bmatrix} \Phi \\ Q \end{bmatrix} \tag{15}$$

where $[\Phi]$ and $[Q]$ are modal matrices of the displacement modal matrix and the force modal matrix.

For the square root singular problem for homogeneous materials, the stress intensity factors can be defined as

$$\begin{Bmatrix} K_I \\ K_{II} \end{Bmatrix}=\sqrt{2\pi L_0}=\sum_{i=1,2}\left(c_i\begin{Bmatrix} \psi_{yy}(\eta=\eta_A) \\ \psi_{xy}(\eta=\eta_A) \end{Bmatrix}_i\right) \tag{16}$$

where L_0 is the distance between the scaling center and the point of the crack surface (the segment OA in Figure 3). From Equation (8), it can be deduced that if $\lambda_i \geq 1$, the distribution of the stress mode tends to 0 when $\xi \to 0$ [27].

2.3. Force Balance and Displacement Coordination

Figure 4 shows a typical method simulating the crack with the coupling of the XFEM and the SBFEM [28] domain. In the process of XFEM and X-SBFEM coupling, there is a problem of the balance of virtual and real freedom, continuous displacement, and force. To solve the problem, the SBFEM simulates the crack surface using the boundary of the element, while the XFEM introduces an additional degree-of-freedom based on the use of the step function to describe the discontinuous displacement field. To coordinate the displacement of the two types of elements along the boundary, a transition matrix T is used to match the nodal displacements in SBFEM with those in XFEM. A transformation matrix, which can be derived from the previous equations that translates the unknown SBFEM nodal displacements (u_E, u_F, u_A, and u_B) to the unknown XFEM nodal displacements (q_2, q_3, a_2, and a_3) can be described as

$$
\left\{ \begin{array}{c} u_B \\ u_E \\ u_A \\ u_F \end{array} \right\} = \left[\begin{array}{cccc} I & 0 & 0 & 0 \\ 0 & I & 0 & 0 \\ N_2(x_A) & N_3(x_A) & 0 & -2N_3(x_A) \\ N_2(x_F) & N_3(x_F) & 2N_2(x_F) & 0 \end{array} \right] \left\{ \begin{array}{c} q_2 \\ q_3 \\ a_2 \\ a_3 \end{array} \right\}
\tag{17}
$$

where I is a unit matrix. In order to ensure the compatibility of the displacement and to integrate the element stiffness matrix, it is necessary to rearrange the column displacement vectors of the SBFEM and the stiffness matrix according to whether the nodes are on the common boundary,

$$
\left[\begin{array}{cc} K_{aa} & K_{ab} \\ T^T K_{ba} & T^T K_{bb} T \end{array} \right] \left\{ \begin{array}{c} u_{qS} \\ u_{xF} \end{array} \right\} = \left\{ \begin{array}{c} F_a \\ T^T F_b \end{array} \right\}
\tag{18}
$$

where the subscript a represents the nodes on the common boundary of SBFEM and XFEM and subscript b represents the nodal degree-of-freedom of the noncommon boundary. The transformation matrix T is only related to the interpolation shape function at the crack opening at the common boundary of the SBFEM and the XFEM.

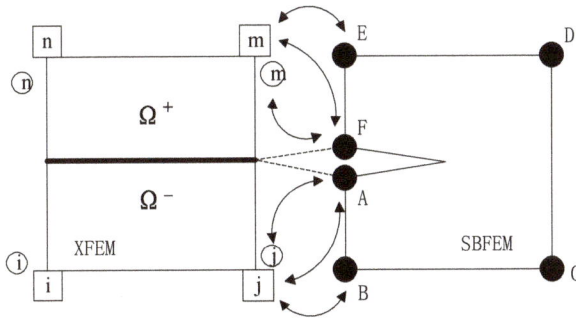

Figure 4. Coupling the XFEM and the SBFEM domains.

3. A Nonlinear Crack Model with Iterative Method for Cohesive Interactions in the FPZ

The relative displacement of the crack surface, including the opening displacement (COD) and the sliding displacement (CSD) of the crack surface, does not exceed the limits shown in Figure 5. When the COD and CSD of the crack surface are not beyond the limits shown in Figure 5, the total loads generated by the structure include the external loads and the cohesive forces at the virtual crack surfaces. However, if the load exceeds the limit, the cohesive force is 0. In the case of cohesive forces, considering the stress intensity factor *I* as an example, the stress intensity factor consists of two parts [29],

$$K_I = K_I^P + K_I^C \tag{19}$$

where K_I is the total stress intensity factor and K_I^P and K_I^C are the components related to the external and cohesive forces, respectively. All three stress intensity factors can be calculated by the standard SBFEM solution stress intensity factor formula. Thus, $K_I^P > 0$ when the crack opens as a result of the external force of the model, while $K_I^C < 0$ when the crack tends to close owing to the cohesive force. Equivalently, $K_I = 0$ when force balance is achieved as a result of the roles of the external and cohesive forces. Therefore, $K_I \geq 0$ can be used as the criterion for judging whether the crack will continue to propagate or not [15].

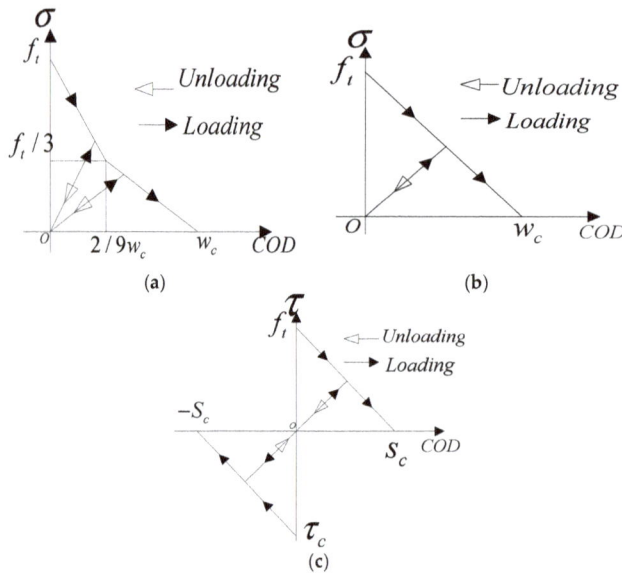

Figure 5. Relations between the relative displacements of crack and cohesive tractions: (**a**) $\sigma = COD$ bilinear curve; (**b**) $\sigma = COD$ linear curve; and (**c**) $\tau = CSD$ curve.

The positive component of the cohesive tractions in the fracture process zone is determined by the bilinear softening curve [30] of Figure 5a or the linear softening curve of Figure 5b. The tangential component is determined by the curve of Figure 5c. The areas below the curve of Figure 5a,b are the mode I fracture energy, G_{fI}. The area between the curves in Figure 5c is twice that of the mode II fracture energy, G_{fII} [30]. The key concept of this method is based on the relative displacement of the crack surface and its application to the linear superposition of an iterative scheme to solve and estimate the cohesive tractions on the crack surface. Specifically,

A. Assume that the structure is only affected by the external force so that the relative displacement Δu_1 of the super-element crack surface Δu_1 can be obtained based on the linear elastic assumptions of X-SBFEM, and that the corresponding cohesive traction t_1 can be obtained according to Figure 5.

B. As shown in Figure 6, the external force and the cohesive force obtained in the previous step are applied to the structure, wherein the cohesive traction t_1 is applied in the form of a side-face force, formulated in accordance to the following equation:

$$\int_S \{\delta u(\xi, \eta)\}^T \{p_n(\xi, \eta)\} dS = \int_0^1 \{\delta u(\xi)\}^T \{p_n(\xi)\} \sqrt{x'^{(\xi)^2} + y'^{(\xi)^2}} d\xi$$

$$= \int_0^1 \{\delta u(\xi)\}^T F_t(\xi) d\xi \qquad (20)$$

where

$$F_t(\xi) = \left\{ \begin{array}{c} \sqrt{x_1^2 + y_1^2} \{p_n(\xi)\}\Big|_{\eta=-1} \\ \sqrt{x_1^2 + y_1^2} \{p_n(\xi)\}\Big|_{\eta=1} \end{array} \right\} \qquad (21)$$

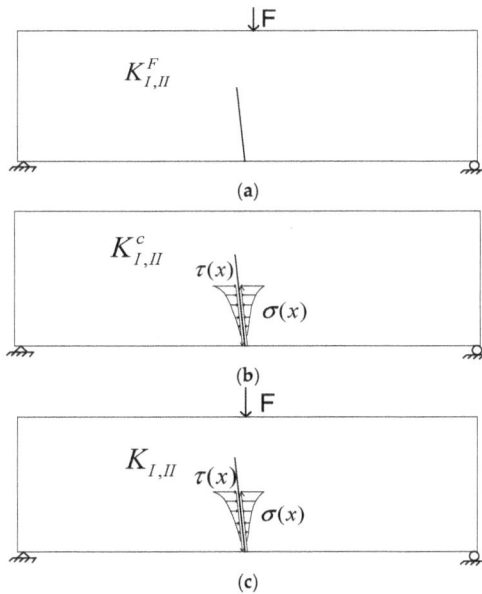

Figure 6. Superposition method for calculating $K_{I,II}$: (**a**) external traction only; (**b**) cohesive tractions only; (**c**) superposition of external traction and cohesive tractions.

The SBFEM nonhomogeneous control equation can be easily obtained in accordance to

$$\left[E^0\right]\xi^2\{u(\xi)\}_{,\xi\xi} + \left(\left[E^0\right] + \left[E^1\right]^T - \left[E^1\right]\right)\{u(\xi)\}_{,\xi} - \left[E^2\right]\{u(\xi)\} + \xi\{F_t(\xi)\} = 0 \qquad (22)$$

Assume that the load can be expressed by the power series,

$$\{F_t(\xi)\} = \sum_{i=1}^n \xi^{t_i}\{F_{t_i}\} \qquad (23)$$

The corresponding displacement mode is

$$\{u_t(\xi)\} = \sum_{i=1}^{n} \xi^{t_i}\{\Phi_{t_i}\} \tag{24}$$

Substituting Equation (24) into Equations (23) and (11) results in

$$\{\Phi_{t_i}\} = \left[(t_i+1)^2\left[E^0\right] + (t_i+1)\left[\left[E^1\right]^T - \left[E^1\right]\right] - \left[E^2\right]\right]^{-1}\{F_{t_i}\} \tag{25}$$

$$\{q_{t_i}\} = \left[(t_i+1)\left[E^0\right] + \left[E^1\right]^T\right]\{\Phi_{t_i}\} \tag{26}$$

Thus, the complete displacement of the boundary nodes and the equivalent nodal force are respectively,

$$\{u_h\} = \sum_{i=1}^{n}\{\Phi_{t_i}\} + [\Phi]\{c\} \tag{27}$$

$$\{P\} = \sum_{i=1}^{n}\{q_{t_i}\} + [Q]\{c\} \tag{28}$$

$[\Phi]$ and $[Q]$ are modal matrices of the displacement modal matrix and the force modal matrix solved by Equation (15), respectively. The following equation can be obtained using Equations (27) and (28):

$$[K]\{u_h\} = \{P\} - \sum_{i=1}^{n}\{q_{t_i}\} + [K]\sum_{i=1}^{n}\{\Phi_{t_i}\} \tag{29}$$

The equivalent nodal force of the super-element boundary node generated by the distributed load of the crack surface is

$$R_F = -\sum_{i=1}^{n}\{q_{t_i}\} + [K]\sum_{i=1}^{n}\{\Phi_{t_i}\} \tag{30}$$

When the crack tip is solved characteristically, $\{u^h_{sb}\}$ and $\{u_{xf}\}$ can be obtained after solving the linear equations. Substituting $\{u_{xf}\}$ into Equation (17), the displacement of all nodes in the crack tip of the super-element $\{u^h_{sb}\}$ can be obtained. Substituting $\{u^h_{sb}\}$ into Equation (27) yields

$$\{c\} = [\Phi]^{-1}\left(\{u^h_{sb}\} - \{\Phi_t\}\right) \tag{31}$$

The displacement field of the crack tip element is

$$\{u(\xi,\eta)\} = N(\eta)\left(\sum_{i=1}^{n} c_i\xi^{\lambda_i}\{\varphi_i\} - \xi^{t+1}\{\Phi_t\}\right) \tag{32}$$

where $\{\psi_i\}$ is the stress mode solved based on SBFEM. The relative displacement Δu_{i+1} is solved using Equation (32).

C. Repeat the steps until the relationship between t_i and Δu_{i+1} becomes consistent with the pattern of variation plotted in Figure 5.

A simplified flowchart of the solution of the cohesive tractions is shown in Figure 7.

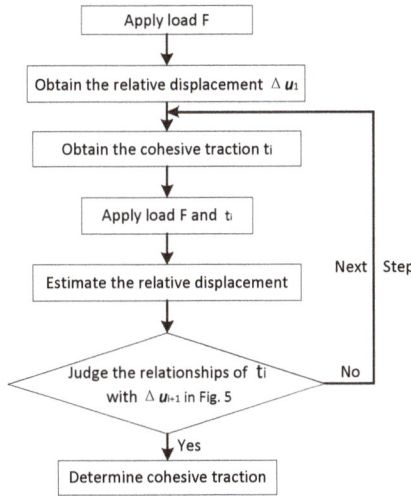

Figure 7. Key steps of the solution of cohesion.

4. Numerical Examples

4.1. A Three-Point Bending Beam

This model was first studied by Petersson (1981) as an experimental study of the mode I fracture propagation problem [31]. The geometry, boundary conditions, and material parameters of the beam are shown in Figure 8. The tensile strength f_t is 3.33 MPa and the mode I fracture energy G_{fI} is 137 N/m. The present example predicts the crack propagation path based on the LEFM maximum circumferential tensile stress criterion. Analyzing the single linear softening curve (Figure 5b) and based on the mode I fracture energy G_{fI}, the limit value of the linear softening curve w_c is 0.0823 mm. The results of three crack propagation steps a = 10 mm, 20 mm, and 30 mm, for a 20 × 200 grid density, are calculated and compared with the results based on the linear elasticity method [24].

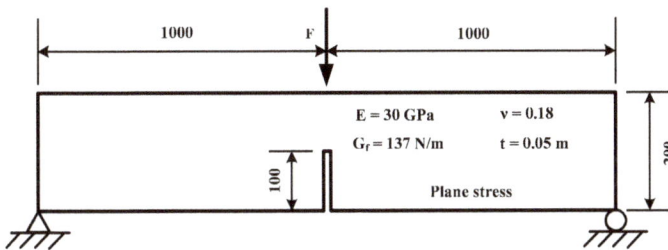

Figure 8. Three-point bending beam for crack propagation (unit: mm).

Figure 9a depicts the relationship of the load for three different crack propagation steps and load point displacements (Load–LPD) based on LEFM work. The results are compared with Yang's work based on the linear elasticity method [12] and experimental results published by Petersson [27], as shown in Figure 9. We can see that the elicited results based on LEFM are very different from the experimental data. This is particularly evident for the peak load, which is much higher than the experimental peak. This is because the LEFM-based method is incapable of simulating the energy dissipation of the fracture zone. Based on X-SBFEM, this study has used different methodologies to solve the cohesive tractions based on the iterative method of linear superposition by simulating the energy dissipation of FPZ. Figure 9b shows the Load–LPD curves for three different crack propagation steps considering FPZ nonlinearities. It can be seen from Figure 9 that the calculated results are in good agreement with the experimental results of Petersson, which shows that the method used in this study can simulate the energy dissipation of FPZ. Moreover, it can be seen from Figure 9 that the results of the three crack propagation steps are in good agreement with the experimental curve, which shows that different steps have minor effects on the calculated results.

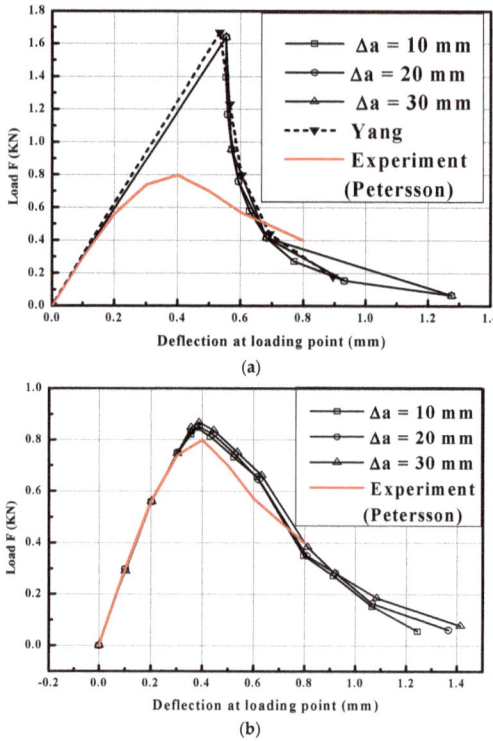

Figure 9. Load–LPD curves for different crack increment lengths: (**a**) LEFM-based and (**b**) NFM-based.

4.2. A Four-Point Shear Beam

Arrea and Ingraffea first tested and analyzed the four-point unilateral shear beam [28]. The geometry and boundary conditions of the beam are shown in Figure 10. Assuming that the structure is in the plane stress state, the Young's modulus E is 24.8 GPa, Poisson's ratio v is 0.18, tensile strength f_t is 3.0 MPa, mode I fracture energy G_{fI} is 100 N/m, and mode II fracture energy G_{fII} is 10 N/m. The crack path is also predicted by using the LEFM-based maximum circumferential stress criterion. Analyzing the linear softening curve of Figure 5b, the limit value of CODs w_c obtained by the mode I fracture energy G_{fI} is 0.067 mm and the limit value of CSDs s_c obtained by the mode II fracture energy G_{fII} is 0.02 mm. The results of the three crack propagation steps a = 10 mm, 20 mm, and 30 mm, using a 20 × 200 grid density, are calculated and compared with the results based on the linear elasticity method [24].

Figures 11a and 12a show the relationship between the load calculated by LEFM, the crack mouth sliding displacement (Load–CMSD), and the relationship between the load and the loading point displacement (Load–LPD). It can be seen from the figure that the calculated results are close to the numerical solutions of Yang et al. [12] and the effect of different steps on the calculation results is not considerable, which proves the applicability of the X-SBFEM algorithm to complex crack propagation problems. However, the FPZ energy dissipation of the crack tip makes the LEFM-based method slightly different compared to previous calculations and experimental data [15]. In this study, the iterative method used to simulate the cohesive tractions is used to simulate the energy dissipation of the FPZ. The linear softening curve in Figure 5b and the curve of Figure 5c are used to solve the cohesive tractions of the vertical crack surface and the parallel crack surface. As shown in Figure 11b, it can be seen that the Load–CMSD curve calculated herein is in good agreement with Yang's experimental data (NFM-based). Figure 12b shows that the method considered herein can describe the snap-back phenomenon of the Load–LPD curve. It can be concluded that the iterative, linear superposition method based on X-SBFEM can yield high-precision results without coupling the interface unit (CIEs) near the crack tip.

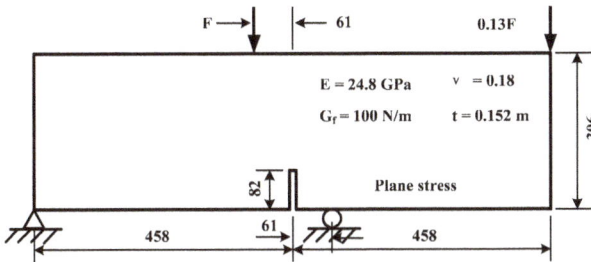

Figure 10. Four-point notched shear beam for crack propagation (unit: mm).

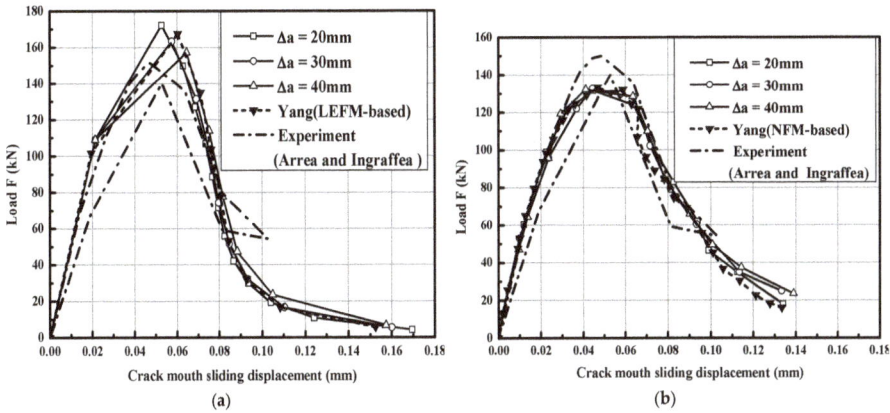

Figure 11. Load-CMSD curves as a function of changes of the crack increment length: (**a**) LEFM-based and (**b**) NFM-based.

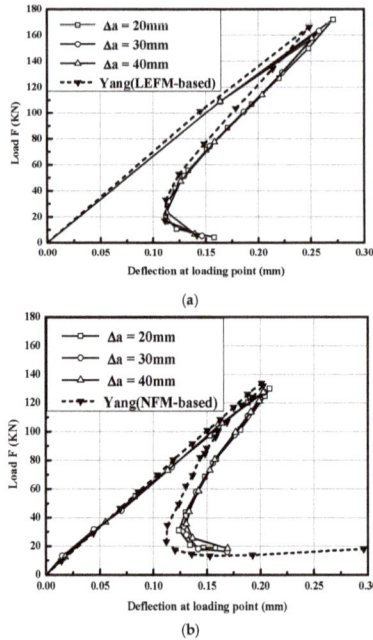

Figure 12. LEFM-based Load–LPD curves with respect to the crack increment length changes: (a) LEFM-based and (b) NFM-based.

4.3. An Experimental Concrete Gravity Dam with Single-Crack Expansion

Carpinteri [32], Barpi [33], and Shi [34] tested and analyzed the single-crack expansion gravity dam. A 1:40 scale concrete gravity dam model is used herein. The geometry, boundary conditions, and material parameters of the gravity dam are shown in Figure 13. Consisting of concrete, the gravity dam is analyzed based on a plane strain assumption and on the bilinear softening curve, where $w_c = 0.256$ mm, $G_f = 184$ N/m, and $f_t = 3.6$ MPa. Suppose that the indentation length is $1/10$ W (0.15 m), where W is the width of the dam at the elevation of the indentation. Hydrostatic pressure exists in the upstream face of the dam. This hydrostatic pressure acts on the upstream face and can be equivalently replaced by four concentrated loads, as shown in Figure 13. The hydrostatic pressure gradually increases until the dam breaks.

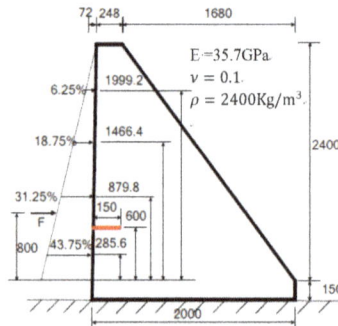

Figure 13. Dam model with single-crack propagation (unit: mm).

In Figure 14, the results of the Load–COD curve in this study are compared with reference solutions, the experimental data of Carpinteri et al. [32], and the numerical simulation data of Barpi et al. [33] and Shi et al. [34]. It can be seen that the elicited results before the peak, load peak, and experimental data obtained by this method are in good agreement with all the other numerically elicited data. It can also be seen that the post-peak curve of the experiment was significantly higher than the numerical results, which means that the crack opening displacement (COD) of the experiment was greater under the same loading conditions. This phenomenon may be owing to the unanticipated rigid rotation in the experiment of Carpinteri et al. [32] which results in premature failure of the prefabricated crack. Compared with other numerical results shown in Figure 14a, the initial stiffness obtained by the X-SBFEM method in this study is the closest to that obtained from the results of Carpinteri et al. [32]. The results obtained by Barpi et al. [34] were smoothened before the peak, while the large stiffness elicited in the pre-peak response obtained by Shi et al. [35] resulted in a crack opening displacement (COD) that was smaller in value compared to the experimental results.

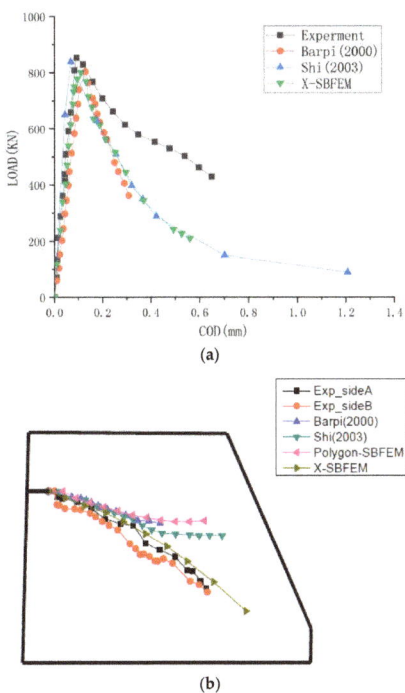

(a)

(b)

Figure 14. Comparison with reference solutions: (**a**) Load–COD curves and (**b**) crack paths.

The crack trajectories obtained by the above experiment and numerical method are shown in Figure 14b. The crack trajectory obtained by the X-SBFEM method in this study is better matched with the experimental and with all the other numerical results. Among them, the crack trajectory obtained from the experiment directly developed towards the dam site before the dam ruptured. At the same time, the trajectories obtained by the numerical simulations of Barpi et al. [33] and Shi et al. [34] horizontally penetrated the dam body, which is quite different compared to the experimental results. The results generated in this study obviously match more closely the experimental results.

Figure 15 shows the variation of the crack opening displacement and the distribution of cohesive traction with the crack path in this example model, based on the X-SBFEM method.

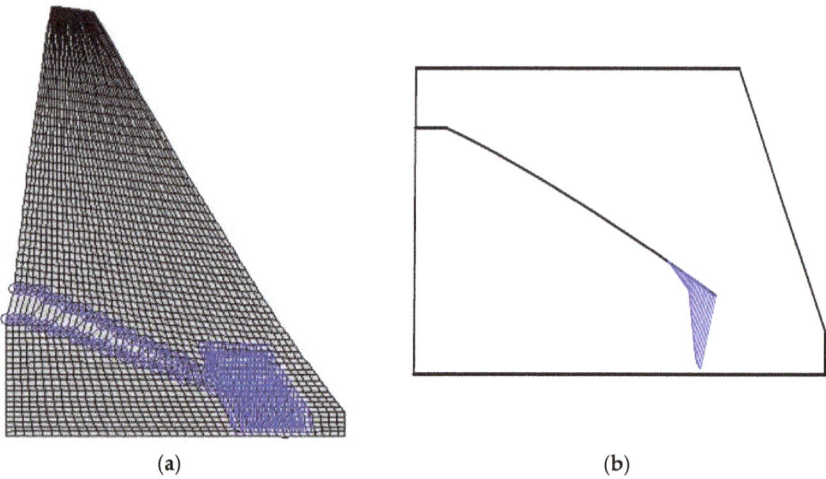

(a)   (b)

Figure 15. Variation of crack opening displacement and distribution of cohesive traction: (**a**) crack opening displacement and (**b**) the distribution of cohesive traction.

4.4. Static Cohesive Crack Propagation Simulation of Koyna Dam

After a severe earthquake in 1967, the neck of the Koyna Dam suffered serious damages. Gioia et al. [35] (1992) simulated the generated crack based on linear elastic fracture mechanics and Bhattacharjee et al. [36] applied the smeared model to analyze the crack propagation. The geometry and material parameters of the Koyna Dam are shown in Figure 16. The dam concrete was assumed to be homogeneous and the Koyna Dam was analyzed based on the planar strain assumption and the bilinear softening curve, where $\omega_c = 0.256$ mm, $\omega_0 = 0.04$ mm, $\omega_1 = 0.075$ mm, $f_t = 1$ MPa, $G_f = 100$ N/m, and $f_t = 0.25$ MPa. A crack was set at a horizontal orientation at an elevation of 66.5 m on the upstream face of the dam in advance. The initial length of the crack was 1.93 m, which equaled 1/10 of the width of the dam at an elevation of 66.5 m. The loads considered included the body load of the dam, hydrostatic pressure of the full reservoir, and the overloading head load applied at the ultimate fracture of the dam. The crack expansion step assumed that $\Delta a = 2$ m.

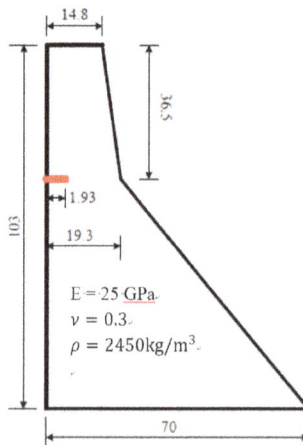

Figure 16. The geometry and material parameters of the Koyna Dam.

During the numerical simulation, the crack initially expanded horizontally. When the overflow increased gradually, the crack trajectory gradually expanded downward owing to the increase of the compressive stress in the downstream area of the dam. Figure 17 is the schematic diagram of the ultimate crack propagation path and corresponding cohesive tractions when the overflow reached 10.35 m.

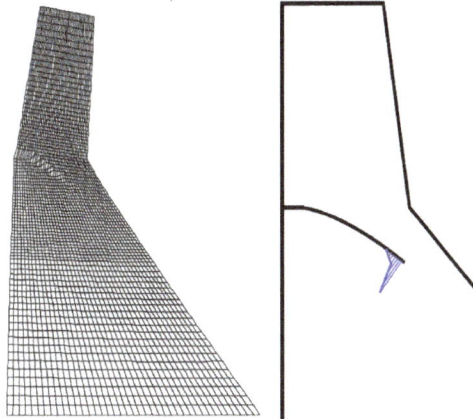

Figure 17. Ultimate crack propagation path and corresponding cohesive tractions for the Koyna Dam (step 11, overflow = 10.35 m).

Figure 18 compares the crack paths corresponding to an overflow of 10.35 m using the present method with the results of Zhong [34], Bhattacharjee and Léger [36], and Gioia and Bazant [35]. In the results reported in these published studies, the overflow was approximately 10.2 m, 10 m, and 14 m, respectively. The crack path predicted by Bhattacharjee and Léger [36] was initially nearly horizontal. It then turned downwards when the crack became equal to half of the width of the dam neck. The crack path based on X-SBFEM in this study is consistent with those predicted by Zhong [34] and Gioia and Bazant [35]. These studies reported the formation of very short horizontal extensions that initially curved downwards and towards the dam's heel.

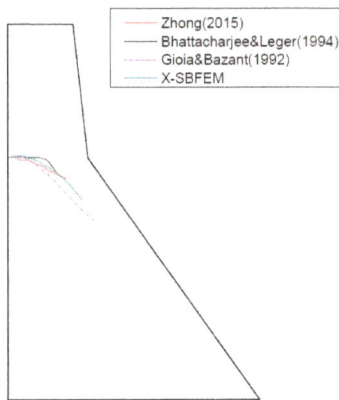

Figure 18. Comparison of crack paths with existing results.

Figure 19 shows the relationships between the overflow and the crest displacement obtained herein and four previously published numerical simulations [34–36]. It can be seen from Figure 19 that the results of Gioia et al. [35] are consistent with those of Zhong et al. [34], Bhattacharjee and Leger et al. [36], and Li et al. [23] and the method presented herein during the initial loading, before the obvious nonlinearity took place. Subsequently, the results obtained by the method presented in this study are comparatively different from those reported by Gioia et al. [34] and Li et al. [23] based on the linear elastic fracture mechanics method. Nevertheless, they match closely to the results of Zhong et al. [35] and Bhattacharjee and Leger et al. [36]. In Figure 19, the resistance of the dam increases as a function of the crack length and there is no postpeak region owing to the stabilization effect of the self-weight of the dam. The results from these examples also reveal the minor difference encountered between the numerical simulation of the scaled-down dam model and the actual dam.

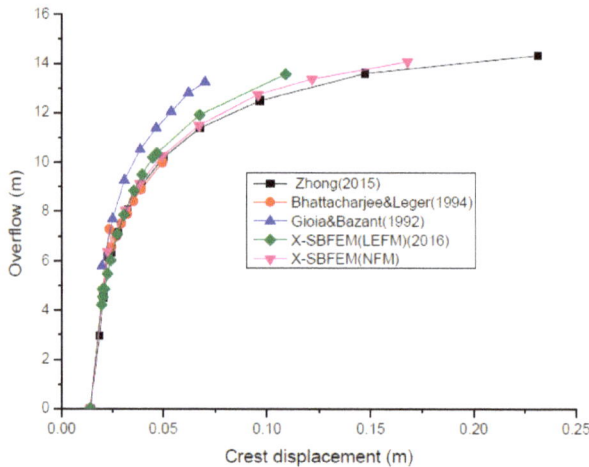

Figure 19. Plots of overflow as a function of crest displacement.

Meanwhile, it can be noted that the extended scaled boundary finite (NFEM-based) method is numerically stable and robust in modeling the nonlinear postpeak response of the dam up to a state where severe fractures and significant deformation start to occur.

5. Conclusions

The following conclusions are drawn based on the presented results:

(1)   A nonlinear X-SBFEM model using the linear superposition of the iterative method was developed and validated to include the cohesive tractions and the fracture energy from FPZ.
(2)   The proposed model can be applied to complex structures without inserting CIEs.
(3)   The accuracy of the proposed model was in close agreement with the experiments showing improvement over the linear SBFEM method.
(4)   The numerical procedure is easily implemented within the finite element method software and can be compatible with various nonlinear constitutive relations.

Author Contributions: J.-b.L. designed, analyzed and guided the full text. X.G. conducted the numerical simulations, analyzed the data and wrote the paper. X.-a.F. conducted the numerical simulations. C.W. and G.L. reviewed and edited the paper. All authors have contributed to and given approval of the manuscript.

Acknowledgments: This research was supported by Grant 51779222 from the National Natural Science Foundation of China, Grant 2016YFB0201000 from National Major Scientific Research Program of China, and DUT17LK16 from the Fundamental Research Funds for the Central Universities.

Conflicts of Interest: The authors declare no conflict of interest.

References

1. Bazant, Z.P.; Cedolin, L. Blunt crack band propagation in finite element analysis. *J. Eng. Mech. Div.* **1979**, *105*, 297–315.
2. Rashid, Y.R. Analysis of prestressed concrete pressure vessels. *Nucl. Eng. Des.* **1968**, *7*, 334–344. [CrossRef]
3. Bhattacharjee, S.S.; Leger, P. Seismic cracking and energy dissipation in concrete gravity dams. *Earthq. Eng. Struct. Mech.* **1993**, *22*, 991–1007. [CrossRef]
4. Calayir, Y.; Karaton, M. Seismic fracture analysis of concrete gravity dams including dam-reservoir interaction. *Comput. Struct.* **2005**, *83*, 1595–1606. [CrossRef]
5. Cai, Q.; Robberts, J.M.; van Rensburg, B.W.J. Finite element fracture modeling of concrete gravity dams. *J. S. Afr. Inst. Civ. Eng.* **2008**, *50*, 13–24.
6. Bazant, Z.P.; Planas, J. *Fracture and Size Effect in Concrete and Other Quasibrittle Materials*; CRC Press: Boca Raton, FL, USA, 1998.
7. Dugdale, D.S. Yielding of steel sheets containing slits. *J. Mech. Phys. Solids* **1960**, *8*, 100–104. [CrossRef]
8. Barenblatt, G.I. The mathematical theory of equilibrium crack in the brittle fracture. *Adv. Appl. Mech.* **1962**, *7*, 100–104.
9. Hillerborg, A.; Modéer, M.; Petersson, P.E. Analysis of crack formation and crack growth in concrete by means of fracture mechanics and finite elements. *Cem. Concr. Res.* **1976**, *6*, 773–782. [CrossRef]
10. Skrikerud, P.E.; Bachmann, H. Discrete crack modeling for dynamically loaded, unreinforced concrete structures. *Earthq. Eng. Struct. Dyn.* **1986**, *14*, 297–315. [CrossRef]
11. Ayari, L.M.; Saouma, V.E. A fracture mechanics based seismic analysis of concrete gravity dams using discrete cracks. *Eng. Fract. Mech.* **1990**, *35*, 587–598. [CrossRef]
12. Yang, Z.J.; Chen, J.F. Finite element modelling of multiple discrete cohesive crack propagation in reinforced concrete beams. *Eng. Fract. Mech.* **2005**, *72*, 280–297. [CrossRef]
13. Xie, M.; Gerstle, W.H. Energy-based cohesive crack propagation modelling. *ASCE J. Eng. Mech.* **1995**, *121*, 1349–1458. [CrossRef]
14. Yang, Z.J.; Chen, J.F. Fully automatic modelling of cohesive discrete crack propagation in concrete beams using local arc-length methods. *Int. J. Solids Struct.* **2004**, *41*, 801–826. [CrossRef]
15. Benkemoun, N.; Poullain, P.; Al Khazraji, H.; Choinska, M.; Khelidj, A. Meso-scale investigation of failure in the tensile splitting test: Size effect and fracture energy analysis. *Eng. Fract. Mech.* **2016**, *168*, 242–259. [CrossRef]
16. Grassla, P.; Jirásek, M. Meso-scale approach to modelling the fracture process zone of concrete subjected to uniaxial tension. *Int. J. Solids Struct.* **2010**, *47*, 957–968. [CrossRef]
17. Mai, N.T.; Phi, P.Q.; Nguyen, V.P.; Choi, S.T. Atomic-scale mode separation for mixed-mode intergranular fracture in polycrystalline metals. *Theor. Appl. Fract. Mech.* **2018**, *96*, 45–55. [CrossRef]
18. Mai, N.T.; Choi, S.T. Atomic-scale mutual integrals for mixed-mode fracture: Abnormal fracture toughness of grain boundaries in grapheme. *Int. J. Solids Struct.* **2018**, *138*, 205–216. [CrossRef]
19. Rabczuk, T.; Bordas, S.; Zi, G. On three-dimensional modelling of crack growth using partition of unity methods. *Comput. Struct.* **2009**, *88*, 1391–1411. [CrossRef]
20. Belytschko, T.; Gracie, R.; Ventura, G. A review of extended/generalized finite element methods for material model. *Model. Simul. Mater. Sci. Eng.* **2009**, *17*, 1–24. [CrossRef]
21. Song, C.M.; Wolf, J.P. Semi-analytical representation of stress singularities as occurring in cracks in anisotropic multi-materials with the scaled boundary finite element method. *Comput. Struct.* **2002**, *80*, 183–197. [CrossRef]
22. Song, C.M.; Francis, T.; Wei, G. A definition and evaluation procedure of generalized stress intensity factors at cracks and multi-material wedges. *Eng. Fract. Mech.* **2010**, *72*, 2316–2336. [CrossRef]
23. Li, J.; Fu, X.; Chen, B.; Wu, C.; Lin, G. Modeling crack propagation with the extended scaled boundary finite element method based on the level set method. *Comput. Struct.* **2016**, *167*, 50–68. [CrossRef]
24. Ooi, E.T.; Song, C.M.; Tin-Loi, F.; Yang, Z.J. Automatic modelling of cohesive crack propagation in concrete using polygon scaled boundary finite elements. *Eng. Fract. Mech.* **2012**, *93*, 13–33. [CrossRef]

25. Huang, R.; Sukumar, N.; Prevost, J.H. Modelling quasi-static crack growth with the extended finite element method part I: Computer implementation. *Int. J. Solids Struct.* **2003**, *40*, 7513–7539. [CrossRef]
26. Huang, R.; Sukumar, N.; Prevost, J.H. Modelling quasi-static crack growth with the extended finite element method part II: Numerical applications. *Int. J. Solids Struct.* **2003**, *40*, 7539–7552. [CrossRef]
27. Natarajan, S.; Song, C. Representation of singular fields without asymptotic enrichment in the extended finite element method. *Int. J. Numer. Meth. Eng.* **2013**, *96*, 813–841. [CrossRef]
28. Yang, Z.J. Fully automatic modeling of mixed-mode crack propagation using scaled boundary finite element method. *Eng. Fract. Mech.* **2006**, *73*, 1711–1731. [CrossRef]
29. Xu, S.; Reinhardt, H.W. Determination of double-K criterion for crack propagation in quasi-brittle fracture. Part II: Analytical evaluating and practical measuring methods for three-point bending notched beams. *Int. J. Fract.* **1999**, *98*, 151–177. [CrossRef]
30. Petersson, P.E. *Crack Growth and Development of Fracture Zone in Plain Concrete and Similar Materials*; Report TVBM-1006; Lund Institute of Technology: Lund, Sweden, 1981.
31. Arrea, M.; Ingraffea, A.R. *Mixed-Mode Crack Propagation in Mortar and Concrete*; Department of Structural Engineering, Cornell University: Ithaca, NY, USA, 1982.
32. Carpinteri, A. Size Effects on Strength, Toughness, and Ductility. *J. Eng. Mech.* **1989**, *115*, 1375–1392. [CrossRef]
33. Barpi, F.; Valente, S. Numerical Simulation of Prenotched Gravity Dam Models. *J. Eng. Mech.* **2000**, *126*, 611–619. [CrossRef]
34. Shi, M.; Zhong, H.; Ooi, E.T.; Zhang, C.; Song, C. Modelling of crack propagation of gravity dams by scaled boundary polygons and cohesive crack model. *Int. J. Fract.* **2013**, *183*, 29–48. [CrossRef]
35. Gioia, G.; Bazant, Z.; Pohl, B.P. Is no-tension dam design always safe?—A numerical study. *Dam Eng.* **1992**, *3*, 23–34.
36. Bhattacharjee, S.S.; Léger, P. Application of NFEM models to predict cracking in concrete gravity dams. *J. Struct. Eng.* **1994**, *120*, 1255–1271. [CrossRef]

*applied
sciences*

MDPI

Article

Numerical Analysis of a Moderate Fire inside a Segment of a Subway Station

Rodrigo Díaz [1], Hui Wang [1,2], Herbert Mang [1,2], Yong Yuan [2] and Bernhard Pichler [1,*]

[1] Institute for Mechanics of Materials and Structures, TU Wien—Vienna University of Technology, Karlsplatz 13/202, 1040 Vienna, Austria; rodrigo_diaz92@hotmail.com (R.D.); hui.wang@tuwien.ac.at (H.W.); herbert.mang@tuwien.ac.at (H.M.)

[2] College of Civil Engineering, Tongji University, Shanghai 200092, China; yuany@tongji.edu.cn

* Correspondence: bernhard.pichler@tuwien.ac.at

Received: 25 September 2018; Accepted: 26 October 2018; Published: 1 November 2018

Abstract: A 1:4 scaled fire test of a segment of a subway station is analyzed by means of three-dimensional Finite Element simulations. The first 30 min of the test are considered to be representative of a moderate fire. Numerical sensitivity analyses are performed. As regards the thermal boundary conditions, a spatially uniform surface temperature history and three different piecewise uniform surface temperature histories are used. As regards the material behavior of concrete, a temperature-independent linear-elastic model and a temperature-dependent elasto-plastic model are used. Heat transfer within the reinforced concrete structure is simulated first. The computed temperature evolutions serve as input for thermomechanical simulations of the fire test. Numerical results are compared with experimental measurements. It is concluded that three sources of uncertainties render the numerical simulation of fire tests challenging: possible damage of the structure prior to testing, the actual distribution of the surface temperature during the test and the time-dependent high-temperature behavior of concrete. In addition, the simulations underline that even a moderate fire represents a severe load case, threatening the integrity of the reinforced concrete structure. Tensile cracking is likely to happen at the inaccessible outer surface of the underground structure. Thus, careful inspection is recommended even after non-catastrophic fires.

Keywords: thermomechanical analysis; moderate fire; finite element simulations

1. Introduction

Structural engineers are interested in the load-carrying behavior of reinforced concrete (RC) structures exposed to fires. Research on the high-temperature performance of RC structures involves the entire field of engineering sciences, dealing with the thermal degradation of concrete [1–3], the underlying hygro-thermo-chemo-mechanical couplings [4,5], transport of heat and moisture [6–8], the transient thermal strain of concrete [9–11], also referred to as load-induced thermal strains [12], spalling of concrete [13–15], the interaction between concrete and steel rebars [16–18] and the interaction between different RC elements that are connected to form RC structures.

These interactions were studied by means of experiments and/or numerical analyses. As for structural elements, related studies have been devoted to normal-strength concrete columns with circular cross-sections [19], high-strength concrete columns with quadratic cross-sections [20–22], RC beams, either unprotected [23] or protected by fiber-reinforced polymer laminates [24], high-performance self-compacting concrete slabs with superabsorbent polymers and polypropylene fibers [25], T-shaped beams, made of high-strength reinforced concrete [26] and prefabricated RC segments for linings of shield tunnels [27]. As for entire RC structures, testing and/or simulations were carried out for RC slabs resting on steel frames [28], slab-column connections, made of reinforced concrete [29], composite slabs with, as well as without a supporting secondary beam [30],

RC frame structures [31–33], the Channel Tunnel between France and the United Kingdom [2,13], other monolithic tunnel linings with cross-sections in the form of a segment of a circle [34,35] , of an ellipse [36] and in the form of a double box [37], a segmental tunnel ring of a metro tunnel [38] and the twin-tube cross-section of the immersed tunnel of the Hong Kong-Zhuhai-Macao-Bridge [39].

The present study refers to another interesting structure: a segment of a subway station. It represents a statically indeterminate RC structure, consisting of a top slab, a bottom slab, two lateral walls and two columns connecting the top and bottom slabs. A 1:4-scale model of this structure was tested by Lu et al. [40]. Before the fire test, the structure was subjected to mechanical loads, simulating a combination of in situ ground pressure, water pressure, as well as dead and live loads that occur in the tunnel under service conditions. These loads were kept constant throughout the subsequent fire test. During this test, the temperature of the air in the interior of the tested structure was increased according to a prespecified temperature history. The performance of the structure was monitored during the test by means of temperature sensors ("thermocouples") and strain gauges.

The aforementioned studies refer to catastrophic fire scenarios with maximum temperatures typically larger than 1000 °C. Such disasters are fortunately rare events, whereas moderate fires happen much more frequently [41]. Thereby, the expression "moderate fire" refers to a scenario that develops initially like a catastrophic fire disaster, but is stopped by fire fighters early enough so that the structure is not damaged severely. Consequently, moderate fires are, at least from the structural viewpoint, non-catastrophic events. Because of their frequency, however, they deserve more attention from structural engineers. This is setting the scene for the present study.

Here, the first 30 min of the fire test by Lu et al. [40] are analyzed, based on three-dimensional Finite Element simulations using the commercial software Abaqus FEA 2016 [42]. This period of time is chosen since it is representative of a moderate fire. Heat transfer in the analyzed reinforced concrete structure is simulated first. The obtained temperature field histories are subsequently used as the basis for thermomechanical analyses of the load-carrying behavior of the tested structure during the fire test. The specific challenges of these two types of simulations refer to:

- the boundary conditions required for the analysis of the heat transfer problem and
- the material behavior of concrete, subjected to both mechanical loading and elevated temperatures.

Both items involve significant uncertainties. This provides the motivation for corresponding sensitivity analyses. They are described in the following.

Regarding the thermal boundary conditions, the first approach is based on prescribing one specific temperature history at the entire interior surface of the structure. The prescribed temperature evolution is set equal to the average of the readings of two temperature sensors, which were positioned at a distance of 2 mm from the heated surface. One sensor was located inside the top slab and the other one inside the right wall. In the second approach, the heated inner surface of the structure is subdivided into three sub-regions. Each of them is subjected to a uniform distribution of a specific temperature history. The three required temperature histories are selected, in the context of model updating, such that satisfactory agreement between simulation results and temperature measurements is obtained.

The material behavior of concrete is simulated as either temperature-independent and linear-elastic or temperature-dependent and elasto-plastic. For the latter simulation, the Concrete Damaged Plasticity (CDP) model of Abaqus FEA is used [42]. Thereby, the nonlinear constitutive behavior of concrete and steel, including their temperature-induced degradation, agree with the recommendations by the Eurocode 2 [43] and the fib Model Code [44].

Results obtained from the described structural sensitivity analyses will allow for an assessment of the relative importance of thermal boundary conditions and the material behavior of concrete. This is important for structural engineers whose research efforts are devoted to dealing with large uncertainties that still exist in this research area. In addition, the simulation results will allow for providing recommendations regarding the inspection of RC structures after non-catastrophic fire events.

The present paper is organized as follows. In Section 2, experimental data from the scaled fire test of an underground reinforced concrete structure are presented. In Section 3, transient simulations of the non-stationary heat transfer in the reinforced concrete structure are described. This includes sensitivity analyses with two different types of thermal boundary conditions. In Section 4, the load-carrying behavior of the structure during the fire test is analyzed, based on thermomechanical numerical simulations. This includes sensitivity analyses with two different types of material models for concrete. In addition, the two temperature field histories of Section 3 serve as input. Thus, altogether, four structural simulations are described. The comparison of simulation results and experimental measurements focuses on two time instants: (i) the time instant right before the fire test (at that time, the structure was already subjected to mechanical loading) and (ii) 30 min after the start of the fire test. Finally, Section 5 contains a summary and conclusions.

2. Experimental Results from a Scaled Fire Test

The tested structure was inspired by a three-span two-floor reinforced concrete frame, as is frequently used in China for subway lines; see Figure 1. The height of the floors amounts to 5950 mm and 6190 mm, respectively, the total internal span to 20,700 mm and the cross-sectional area of the columns, which subdivide the frame into three cells, to 1200 mm × 800 mm. In the real structure, the distance of neighboring columns in the axial direction of the tunnel amounts to 7500 mm.

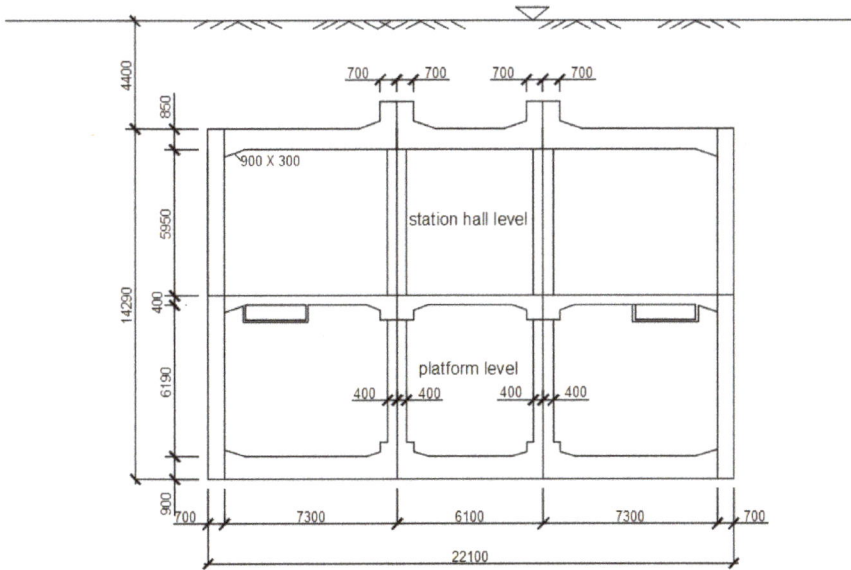

Figure 1. Cross-sectional view (vertical cut through the structure, normal to the axis of the tunnel) of a three-span two-floor reinforced concrete frame, providing the inspiration for fire testing in [40].

2.1. Production of the Tested Structure

Inspired by the structure illustrated in Figure 1, a model of the upper floor was tested at a scale of 1:4; see Figure 2 and [40]. The width of the tested structure was 5260 mm; its height was 1880 mm; and its axial length was 1200 mm; see Figure 3.

Figure 2. Setup of the large-scale fire test, taken from [40]. The specimen was placed sidelong on top of the furnace and closed with a fire-resistant cover.

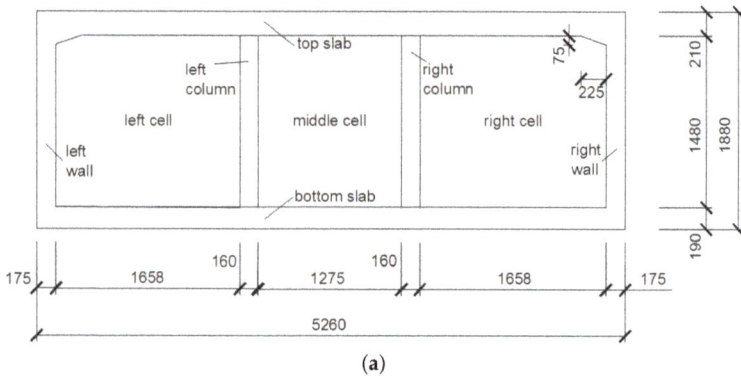

(a)

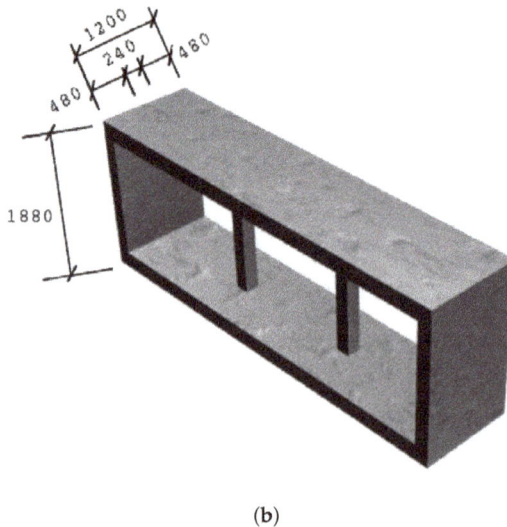

(b)

Figure 3. Geometric dimensions of the tested structure (mm): (**a**) cross-section and (**b**) perspective representation; adapted from [40].

The dimensions of the top slab and the lateral walls amounted to one fourth of the real dimensions. The total length of the inner span of the structure amounted to 20.7 m/4 = 5.175 m. However, it had to be slightly adjusted to fit the dimensions of the furnace. Thus, the inner span of the model is 4.91 m, which is equal to that of the furnace. The design of the columns was the result of the following considerations: The scaled distance of neighboring columns in the axial direction of the tunnel amounted to 7.5 m/4 = 1.875 m. This is by a factor of 1.5652 larger than the axial length of the tested structure. Thus, the 1:4 scaled cross-sectional area of the columns, amounting to 300 mm × 200 mm, had to be divided by a factor of 1.5652; see [40] for details. In order to obtain a geometrically similar cross-section, both scaled dimensions were divided by $\sqrt{1.5652} = 1.25$. Therefore, the cross-sectional area of the columns of the tested structure amounted to 240 mm × 160 mm. Finally, the thickness of the bottom slab was set equal to 190 mm, in order to account for the influence of the lower floor on the stiffness of the modeled upper floor; see [40] for details.

The reinforcement ratio of the top slab, the bottom slab, the columns and the walls amounted, by analogy to the real structure, to 1.22%, 1.19%, 2.95% and 1.76%, respectively. The reinforcement bars had diameters of 10 mm, 12 mm and 14 mm, respectively; see Figure 4.

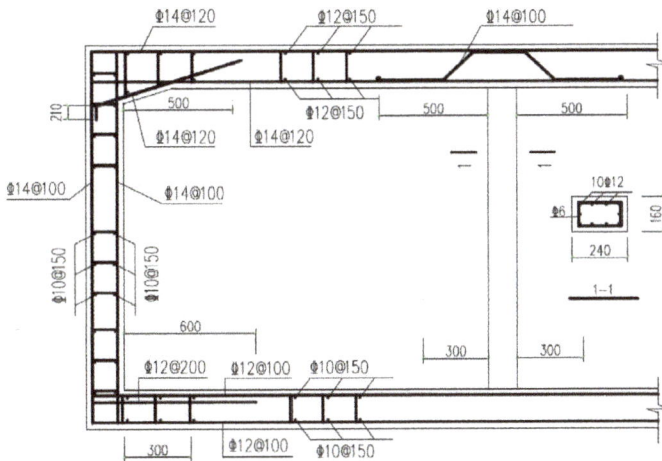

Figure 4. Reinforcement drawing of the tested structure (mm), taken from [40].

Temperature and strain sensors were put in place already before casting of the concrete. In this way, the sensors were finally embedded inside the tested structure. As to be expected in fire testing [45], some of the sensors failed during the experiment. Therefore, the following description is limited to measurement equipment, the readings of which are considered in the present work. Thermocouples were placed at six positions within the tested structure (Figure 5): three in the top slab (one at the midspan of each one of the three cells), one in the right column, one in the right wall and one in the middle cell of the bottom slab. At the selected positions, several sensors were placed at different distances from the heated inner surface, in order to monitor the ingress of heat into the structure; see Figure 5. The minimum cover depth of the thermocouples amounted to 2 mm.

Strain gauges were mounted to steel bars of the inner and the outer reinforcement layer. Measurement positions were located at the top slab, at the midspan of the left cell and at the center of the right wall; see Figure 5.

Normal concrete "C40", with a mass density of 2373 kg/m³, was used for the production of the tested structure; see Table 1 for the composition of the material. The stiffness and the strength of both concrete and steel were quantified before the test, following protocols from the Chinese Standard for

Test Method of Mechanical Properties on Ordinary Concrete [46]; see Table 2. Notably, the concrete was tested at an age of 28 days.

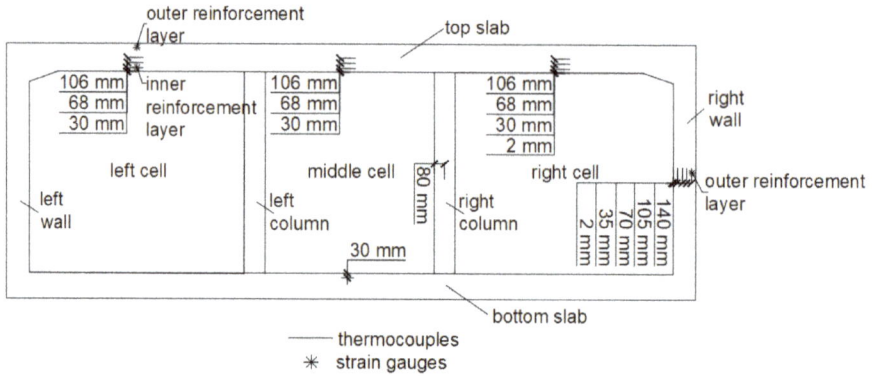

Figure 5. Elements of the tested structure and positions of thermocouples and strain gauges.

Table 1. Composition of the concrete used for the tested structure.

Raw Material	Content (kg/m^3)
Cement (42.5 PO)	249
Tap water	176
Sand 1 (middle size)	306
Sand 2 (middle size)	458
Gravel (5–25 mm)	1013
Fly ash (Level II)	70
Admixture (ZK 904-3)	6
Blast furnace slag S95	95

Table 2. Stiffness and strength properties of concrete (age = 28 days) and steel at room temperature.

Material	Compressive Strength (MPa)	Yield Stress (MPa)
Concrete	36.5	–
Steel (diameter = 12 mm)	–	531.9
Steel (diameter = 14 mm)	–	530.2

As for the fire experiment, the tested structure was rotated by 90° and placed in the furnace, such that the axial direction of the modeled tunnel segment was equal to the vertical direction. Additional supports were used to avoid rigid body motions in horizontal planes. Four supports prevented the displacements of the bottom ends of the walls and of the columns; see Figure 6. As for the displacements in the horizontal direction, orthogonal to the axes of the columns, two supports were positioned at the left wall; see Figure 6.

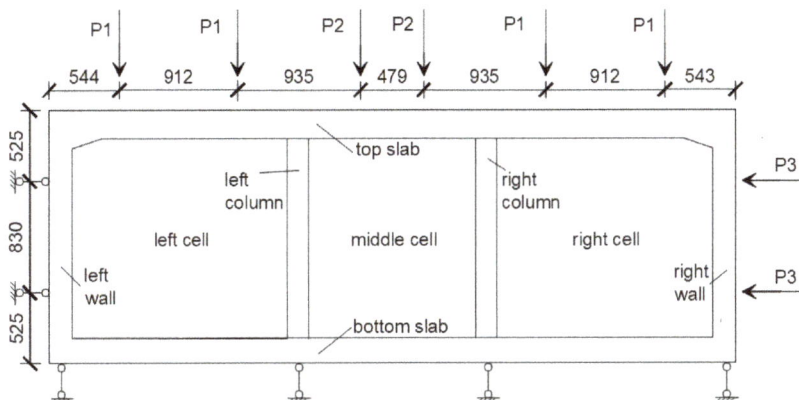

Figure 6. Support and loading conditions of the tested structure, adapted from [40].

2.2. Mechanical and Thermal Loading

In order to simulate service conditions, the structure was mechanically loaded in both horizontal directions according to the recommendations by the Chinese Standard for Metro Design [47] and Eurocode 2 [48]. The imposed loads accounted for the effective traffic load at the surface above the real structure (20 kN/m), the pedestrian load on both stories (4 kN/m), the earth pressure resulting from a 3.5 m-thick layer of covering soil (specific gravity $= 19$ kN/m^3), and the water pressure resulting from a 0.5 m-thick layer of groundwater (specific gravity $= 9.8$ kN/m^3). These loads were combined, using safety factors for dead load and live loads, amounting to 1.35 and 1.40, respectively. The resulting loading scenario was simulated by three sets of concentrated loads, referred to as P1, P2 and P3; see Figure 6 and Table 3. They were applied in nine steps. This took 70 min. Subsequently, the fire test was started.

Table 3. Applied mechanical loads.

	P1	P2	P3
Load (kN)	192.0	151.2	120.0

During the fire test, the temperature of the air inside the cross-section was increased according to a time-dependent temperature history (Figure 7). The latter was the result of a statistical analysis of documented fire accidents. Within 25 min, the temperature was increased to a target value of approximately 525 °C. It was kept constant thereafter. The prescribed fire load accounted for the ventilation of the real structure, with a speed of 2.5 m/s, automatic sprinkler devices and a heat release rate of 5 MW [40], in accordance with the Chinese Standard for Metro Design [47]. Two controlled heat sources of the furnace were used to produce the thermal loading. The heat was transferred to the tested structure by means of natural (unforced) convection.

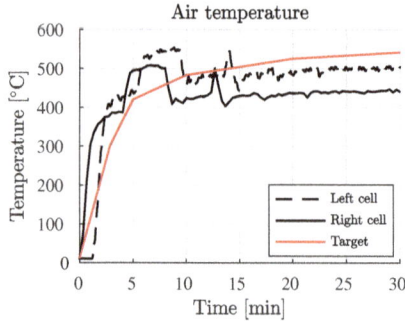

Figure 7. Temperature history of the air inside the tested structure.

2.3. Results from Structural Monitoring

The thermocouples and the strain gauges undertook readings every 20 s. Thus, the first 30 min of the fire test were documented by 90 readings of each sensor.

The temperature close to the inner surface of the top slab, at the midspan of the right cell, rose by approximately 90 °C during the first 30 min of the fire test; see Figure 8a. The temperature decreased with increasing distance from the heated inner surface (see the four graphs in Figure 8a), which shows the measured evolutions of temperature changes at distances of 2 mm, 30 mm, 68 mm and 106 mm from the heated inner surface. At a depth equal to or greater than 106 mm, the temperature remained practically constant throughout the analyzed part of the fire test. A qualitatively similar, but quantitatively different behavior was measured at the other measurement positions of the top slab, the right wall and the bottom slab; see Figure 8d,e. At the center of the right wall, e.g., the temperature close to the inner surface rose by about 80 °C, while no temperature increase was measured at depths equal to or greater than 105 mm. The temperature of the right column was measured at its core, at a distance of 80 mm to the nearest heated surface. Notably, the core temperature of the column rose significantly more than that at the same depth in the slabs and the right wall. The measured temperature evolution is, in fact, comparable with that at a depth of some 30 mm in the top slab.

The strain gauges measured deformations resulting from the mechanical loading (see the values labeled as 0-min readings in Figure 9) and the progressive thermomechanical loading (see the evolution of the total strains after 0 min in Figure 9). For instance, at the outer reinforcement layer of the top slab, the strains resulting from the mechanical loading amounted to approximately -50×10^{-6}; see the ordinate of the dash-dotted curve at $t = 0$ min. During the subsequent 30 min of the fire test, the reading increased by approximately 150×10^{-6} to approximately 100×10^{-6}; see the dashed-dotted graph in Figure 9. This underlines that strains resulting from the thermal loading became the dominant part of the total strains soon after the start of the fire test. Notably, strain changes of the inner rebar resulted from both the eigenstrains, caused by the temperature increase, and the structural load redistributions, induced by the thermal loading. The strain changes of the outer rebar, where the temperature remained constant throughout the analyzed part of the fire test, resulted exclusively from redistribution of the structural load.

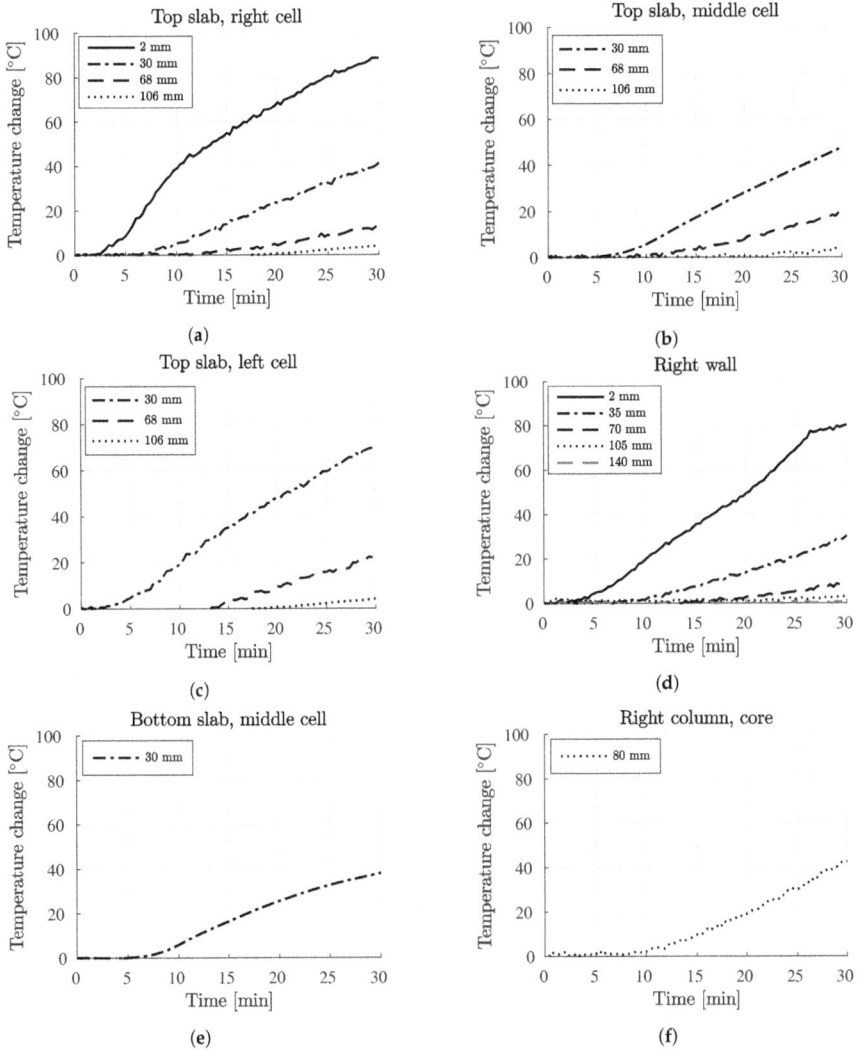

Figure 8. Measured evolutions of temperature changes: (**a**) top slab, at the midspan of the right cell, (**b**) top slab, at the midspan of the middle cell, (**c**) top slab, at the midspan of the left cell, (**d**) center of the right wall, (**e**) bottom slab, at the midspan of the middle cell, and (**f**) core of the right column.

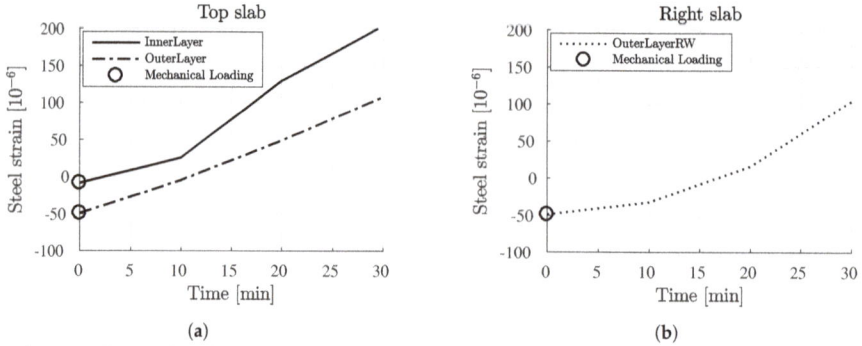

Figure 9. Measured evolutions of total strains of steel rebar: (**a**) top slab, at the midspan of the left cell, and (**b**) center of the right wall.

3. Transient Simulation of Non-Stationary Heat Conduction

Heat conduction within the tested structure was simulated by means of three-dimensional non-stationary Finite Element simulations using the commercial software Abaqus FEA 2016. The Finite Element mesh (Figure 10) was the result of a convergence study. It was a satisfactory trade-off between simulation accuracy and computational effort [49]. The mesh consisted of 139,040 linear hexahedral brick finite elements, with eight nodes and one temperature degree of freedom per node. These elements are referred to as "DC3D8" by Abaqus [42]. The characteristic size of the finite elements amounted to 3 cm. Notably, there are studies [23,50] showing that the steel bars have an insignificant influence on the heat conduction problem. For this reason, the thermal properties of concrete were assigned to all finite elements for the simulation of the non-stationary heat conduction problem. This analysis was only made for the analysis of heat conduction. For the subsequent thermomechanical analysis, the specific properties of concrete and steel were assigned to the corresponding elements.

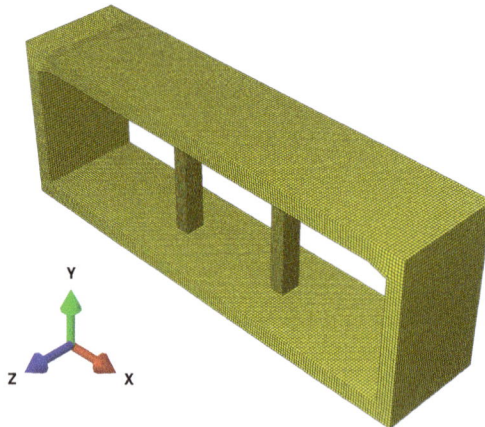

Figure 10. Three-dimensional Finite Element mesh of the analyzed structure and the Cartesian coordinate system used.

The thermal properties of the concrete of the tested structure were unknown. As a remedy, the values of the specific heat capacity and the thermal conductivity were estimated in accordance with building codes [43,44,51,52] and scientific studies [53–55]; see Table 4. These values refer to room temperature. Notably, the experimental measurements suggest that the temperature of the structure remained below 100 °C during the first 30 min of the fire test (see Figure 8) and, thus, that the evaporable water of the concrete was not released [53,56]. This served as a motivation to assume, in the spirit of a reductionist approach, that the thermal properties of concrete at room temperature are a reasonable approximation throughout the entire analysis, although the thermal conductivity of concrete decreases to some 90% of its room-temperature value, provided that the material is heated up to 100 °C; see the building codes [43,44,52] and the scientific studies [1,53,55,57,58].

Table 4. Thermal properties of concrete at room temperature.

Property	Value
Specific heat capacity $(J/(kgK))$	900
Thermal conductivity $(W/(mK))$	1.6

As regards the boundary conditions, the simulations were based on histories of temperature fields that were prescribed at the inner and outer surfaces of the simulated structure. This approach was appealing as it rendered a computational fluid dynamics simulation of heat transfer from the hot air to the simulated structure dispensable. Such simulations are rather challenging, because of thermal instabilities occurring inside the highly turbulent air flow [34,59,60]. Herein, the temperature at the outer surface of the simulated structure was set equal to the initial temperature, $T_{ini} = 10$ °C, throughout the entire simulation. This agrees with the experimental measurements. As for the time-dependent temperature field prescribed at the heated inner surface, two different strategies were implemented within the framework of a sensitivity analysis: (i) spatially-uniform heating of the entire inner surface of the structure and (ii) piecewise spatially-uniform heating of three sub-regions of the inner surface.

3.1. Uniform Prescription of One Temperature History

At the entire inner surface of the simulated structure, the same temperature history was prescribed; see the blue dotted graphs, labeled "BC", in Figure 11. The prescribed evolution of the surface temperature was obtained by averaging the readings of the two thermocouples that were positioned at a distance of 2 mm from the heated inner surface; see the dotted black lines in Figure 11a,d. The other graphs shown in Figure 11a refer to the simulated (label "Sim") or measured (label "Meas") evolutions of temperature changes at positions 30 mm, 68 mm and 105 mm away from the heated inner surface; see also Figure 5.

The results of the numerical simulations (Figures 12 and 13) overestimated the experimentally measured temperatures at the right wall, while underestimations were obtained at all other positions that were equipped with thermocouples; see Figure 11. The largest percent underestimation, obtained 30 min after the start of the fire test, amounted to some 70% and referred to the temperature evolution of the column; see Figure 11f. This was the motivation to refine the numerical simulations by means of updating of the model.

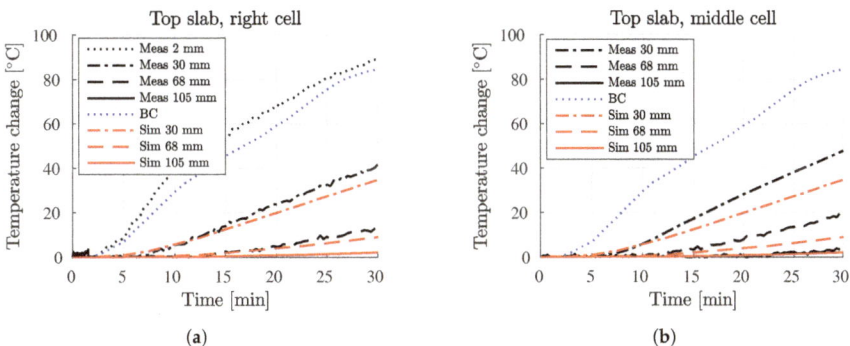

(a) (b)

Figure 11. *Cont.*

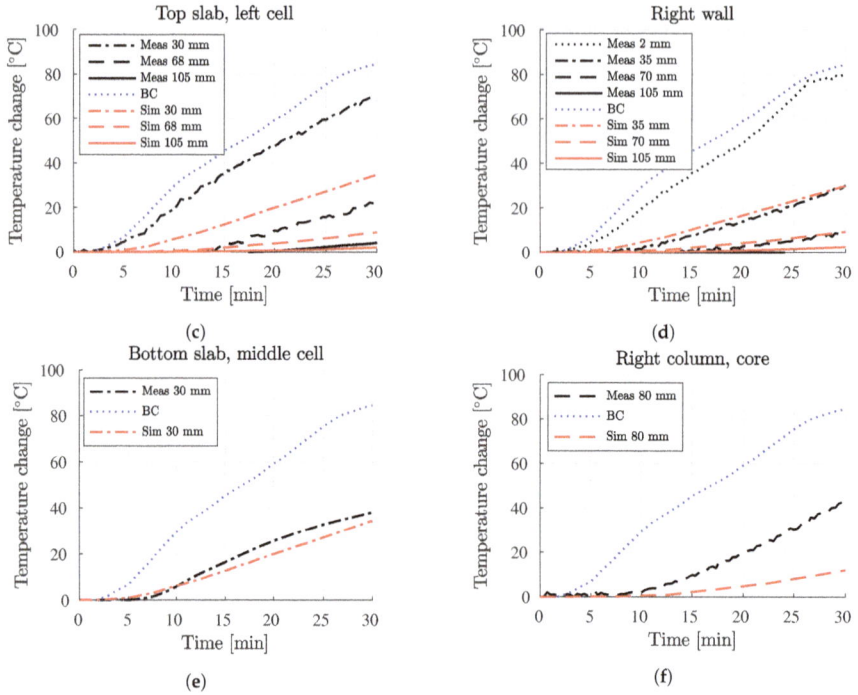

Figure 11. Comparison of simulated evolutions of the temperature inside the analyzed structure, obtained by prescribing a spatially-uniform temperature history along the entire inner surface, with experimental data: (**a**) top slab, at the midspan of the right cell, (**b**) top slab, at the midspan of the middle cell, (**c**) top slab, at the midspan of the left cell, (**d**) center of the right wall (**e**) bottom slab, at the midspan of the middle cell, and (**f**) core of the right column. Sim, simulated; Meas, measured.

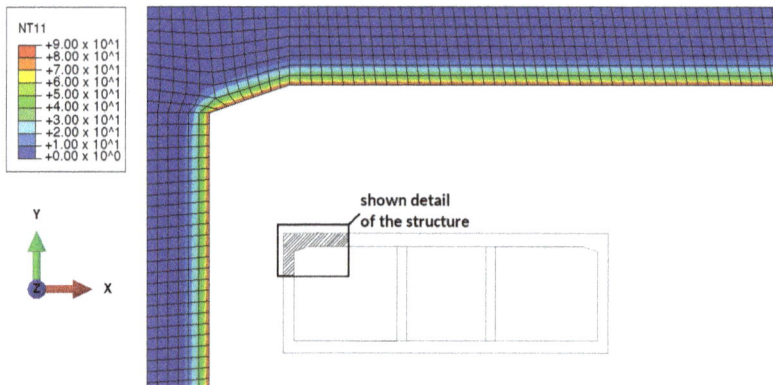

Figure 12. Temperature distribution obtained by prescribing a uniform surface temperature history, 30 min after the start of the thermal loading: detail showing the connection of the left wall to the top slab.

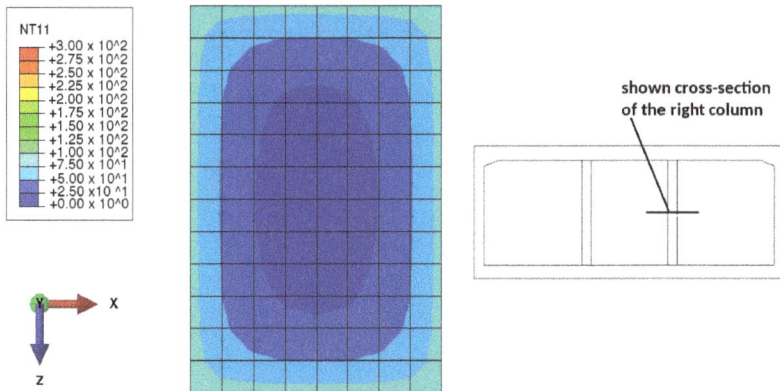

Figure 13. Temperature distribution obtained by prescribing a uniform surface temperature history, 30 min after the start of the thermal loading: detail showing the central cross-section of the right column.

3.2. Piecewise Uniform Prescription of Three Specific Temperature Histories

The results of the preceding section suggested that the right column was exposed to higher surface temperatures compared to the near-surface-measurements at the top slab and the right wall. In order to improve the agreement between simulation results and experimental measurements, the inner surface of the analyzed structure was subdivided into three sub-regions. In each of them, a specific temperature history was prescribed at the inner surface in a spatially-uniform fashion. This was done such that the symmetry of the simulated heat conduction problem was preserved.

1. The top and bottom slabs were subjected to the temperature history, measured at the midspan of the left cell of the top slab, at a depth of 2 mm from the heated surface.
2. Both walls were subjected to the temperature history, measured at the center of the right wall, at a depth of 2 mm from the heated surface.
3. Both columns were subjected to the temperature history imposed on the top and bottom slabs, amplified by a fitting-factor. This was done, noting that the temperature was not measured near the surface of the columns and that, because of their position and their larger exposed-surface-to-volume ratio, the columns were expected to heat up faster than the slabs and the walls. This was confirmed by the results from the previous section. Furthermore, since the same source that heated the slabs also heated the columns, similar qualitative temperature evolutions were assumed to occur at both positions. Setting the amplifying factor equal to 3.2 provided satisfactory agreement between the simulated and the measured temperature evolutions.

The three different prescribed temperature histories are shown as blue dotted graphs, labeled "BC", in Figure 14.

The simulated and measured temperatures agree well at the positions where near-surface-measurements are available, provided that the temperature history measured at a depth of 2 mm is prescribed as the boundary condition at the surface; see Figure 14a,d. At these positions, the maximum difference between the simulated and the measured temperatures amounts to 3.6 °C and 1.4 °C, respectively. This underlines the fact that the heat transfer was predominantly one-dimensional at these positions and took place in the direction orthogonal to the heated surface.

The simulated and measured temperatures also agree well at the middle cell of the bottom slab. Here, the maximum difference between the simulated and the measured temperatures is 2.5 °C. This suggests that the two surfaces at the midspan positions of the right cell of the top slab and of the middle cell of the bottom slab were exposed to very similar temperature histories; see Figure 14a,e.

The updated simulation (Figure 15) suggests that the temperature increase at the surface of the right column was significantly larger than the available near-surface-measurements at the top slab and the right wall. In the updated simulation, the maximum surface temperature of the columns amounts to 284 °C, and the difference between the simulated and the measured temperatures amounts to 3.1 °C. In this context, it is emphasized that the updated simulation was simply based on constant thermal properties of concrete (Table 4).

As for the top slab, at the midspan positions both in the middle and the left cell, also the updated simulation does not deliver satisfactory results; see Figure 14b,c. This underlines the fact that the thermal loading of the top slab was characterized by significant gradients of the surface temperature across the three cells.

Good agreement between simulated and measured temperatures would likely be achieved by amplifying, separately for the middle and the left cell, the history of the surface temperature that was so far prescribed. The corresponding subdivision of the heated surface of the simulated structure into five sub-regions, with a specific temperature history for each of them, could be the target of another refinement step. Still, temperature measurements remain unavailable for the right and the left cell of the bottom slab, the left wall and the left column. Thus, the re-analysis of the fire test inevitably requires assumptions concerning the specific histories of the surface temperature in these regions. Alternatively, one could fit them such that the subsequent thermomechanical simulations deliver strains that agree well with the available strain measurements. This would result in the best-possible reproduction of the available experimental data. However, the involved fitting process would render the assessment of the sensitivity of simulation results with respect to the simulated material behavior of concrete very difficult. Since this sensitivity is a central focus of the present contribution, it was decided to stay with the two described strategies of prescribing thermal boundary conditions along the heated inner surface of the analyzed structure and to combine them with two strategies of accounting for the material behavior of concrete. A discussion of the simulation results in the regions of the structure that were not equipped with thermocouples is given in Appendix A.

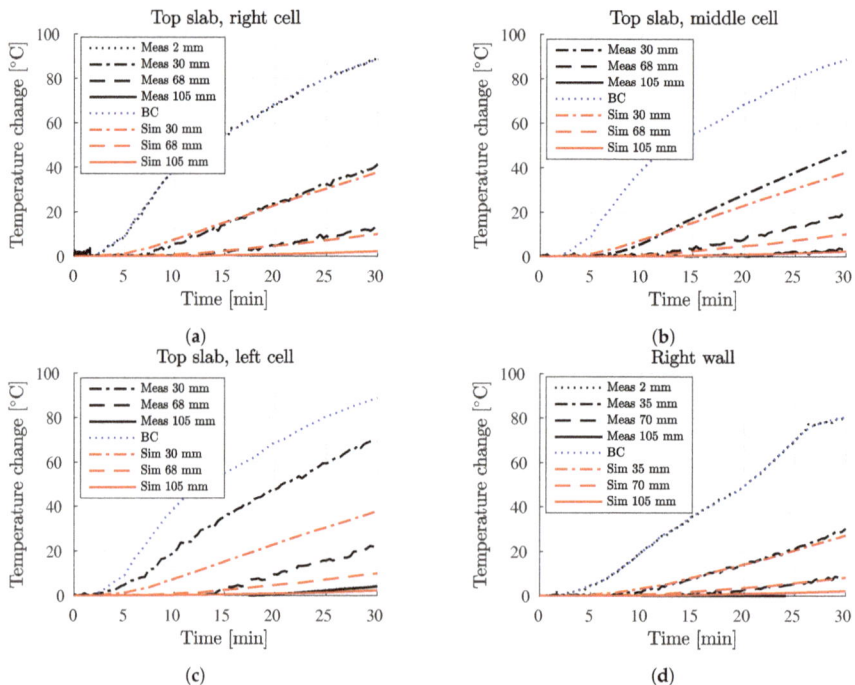

Figure 14. *Cont.*

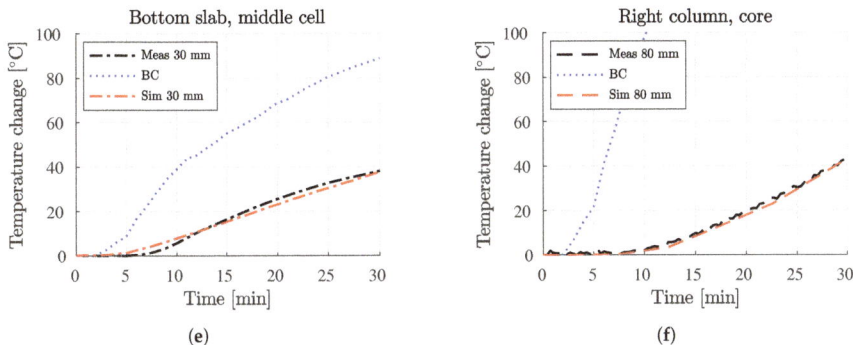

(e)   (f)

Figure 14. Comparison of the simulated evolutions of the temperature inside the analyzed structure, obtained by prescribing three piecewise spatially-uniform temperature histories along the inner surface, with experimental data: (**a**) top slab, at the midspan of the middle cell, (**b**) top slab, at the midspan of the right cell, (**c**) top slab, at the midspan of the left cell, (**d**) center of the right wall (**e**) bottom slab, at the midspan of the middle cell, and (**f**) core of the right column.

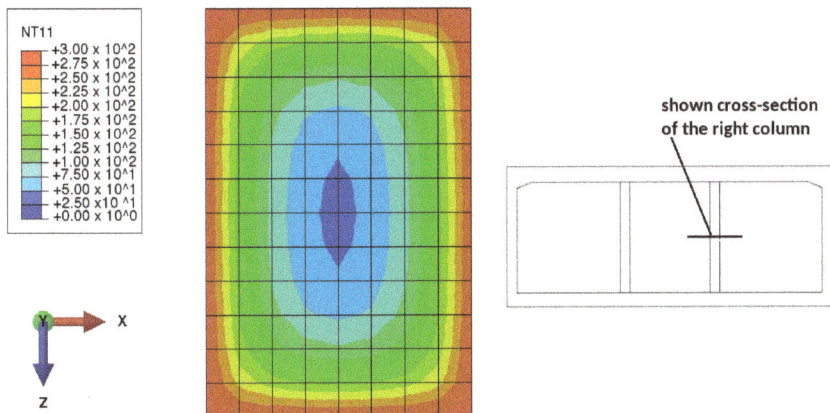

Figure 15. Temperature distribution obtained by prescribing three piecewise uniform surface temperature histories, 30 min after the start of the thermal loading: detail showing the central cross-section of the right column.

4. Thermomechanical Simulation at Selected Time Instants

The load-carrying behavior of the tested structure was analyzed by means of three-dimensional Finite Element simulations, using Abaqus FEA 2016. The chosen Finite Element mesh was similar to the one used for the thermal simulations; see Figure 10. The concrete structure was modeled by 139,040 hexahedral elements of type "C3D8R", i.e., eight-node linear elements with reduced integration and hourglass control, with three translational degrees of freedom per node. The steel rebar was considered by 18,035 line elements of type "T3D2", i.e., two-node linear three-dimensional truss elements, with an axial displacement degree of freedom per node. Perfect bond between concrete and steel rebar was assumed by attaching the nodes of the steel elements to the nodes of the concrete elements.

The supports of the structure were accounted for by means of prescribing displacement boundary conditions. Four sets of such boundary conditions were prescribed in order to prevent displacements of the bottom ends of the walls and of the columns, in the direction of the axes of the columns; see Figure 6. Two additional sets of displacement boundary conditions were prescribed in order to prevent horizontal displacements of the left wall, in the direction orthogonal to the axes of the columns; see Figure 6.

The external mechanical loads, imposed on the structure, were accounted for by means of prescribing traction boundary conditions. The six concentrated forces, acting on the top slab, and two loads, acting on the right wall (Figure 6), were prescribed by means of equivalent pressures amounting to 13.3 MPa.

The thermomechanical simulation was organized by analogy to the sequence of actions during the analysis of the experiment. At first, the external forces were applied. While they were kept constant thereafter, the temperature fields, computed in Section 3, were prescribed as a time-dependent input, resulting in a transient thermomechanical loading of the structure.

Combined sensitivity analyses were carried out, in order to study the sensitivity of simulation results with respect to uncertainties regarding the thermal loading and the material behavior of concrete. As for the first type of sensitivity analysis, the two different temperature field histories, computed in Section 3, were imposed on the structure, by way of two different simulations. As for the second type of sensitivity analysis, two different types of material models were used for the simulation of concrete. It was either modeled as a linear-elastic material with temperature-independent material properties or as an elasto-plastic material with temperature-dependent material properties, as described in the following.

4.1. Material Behavior of Concrete and Steel

4.1.1. Thermoelastic Properties at Room Temperature

The expected value of Young's modulus of concrete, E_c, was estimated based on the mean value of the experimentally-determined values of the compressive strength, $f_{c,p} = 36.5$ MPa; see Table 2. Notably, these strength values were obtained by crushing prismatic specimens with the following dimensions: 150 mm × 150 mm × 300 mm. The corresponding mean value of the cube compressive strength (referring to specimens with dimensions 150 mm × 150 mm × 150 mm), $f_{c,cu}$, was estimated, based on the following regulation of the Chinese Practice Manual for Design of Concrete Structures [61],

$$f_{c,cu} = \frac{f_{c,p}}{\alpha_1 \, \alpha_2} = 48.03 \text{ MPa} , \tag{1}$$

where $\alpha_1 = 0.76$ and $\alpha_2 = 1$ for the investigated concrete.

The corresponding mean value of the cylinder strength, $f_{c,cy}$, was estimated, following Eurocode 2 [48]:

$$f_{c,cy} = \frac{f_{c,cu}}{1.2} = 40.03 \text{ MPa} . \tag{2}$$

The sought value of E_c follows from [48] as:

$$E_c = 22 \text{ GPa} \times \left(\frac{f_{c,cy}}{10 \text{ MPa}} \right)^{0.3} = 33.35 \text{ GPa} . \tag{3}$$

Poisson's ratio of the concrete was set equal to 0.2, and the coefficient of thermal expansion of concrete was chosen as $9.03 \times 10^{-6}/°C$; see also Table 5. This value was obtained as follows: The Eurocode [43] provides formulae for the thermal strain as a function of the temperature for different types of concrete. The considered thermal expansion coefficient refers to concrete made of siliceous aggregates, and it was computed as the slope of the described function, evaluated at room temperature $T_{room} = 20\,°C$.

The values of the thermomechanical properties of steel were taken from [40]. Young's modulus, Poisson's ratio and the coefficient of thermal expansion amount to 195 GPa, 0.3 and $12.2 \times 10^{-6}/°C$, respectively; see Table 5.

Table 5. Mechanical properties of concrete and steel at room temperature.

Property	Concrete	Steel
Young's modulus (GPa)	33.4	195
Poisson's ratio (-)	0.2	0.3
Thermal expansion coefficient $(°C^{-1})$	9.03×10^{-6}	12.2×10^{-6}

4.1.2. Evolution of Thermoelastic Properties resulting from Thermal Loading

The evolution of the elastic stiffness, the strength and the coefficient of thermal expansion of both concrete and steel, resulting from thermal loading up to 300 °C, are discussed, based on regulations of the Eurocode [43], recommendations of the fib Model Code 2010 [44] and the results from scientific studies [53]. This was motivated by results from the thermal simulation with piecewise spatially-uniform histories of surface temperatures, which suggests that the maximum temperature of the columns rose up to 280 °C; see Figure 14f. As regards the thermal degradation of Young's modulus of concrete and steel, scientific results from Bažant et al. [53] and regulations of the Eurocode [43] were used. The reductions of Young's modulus of concrete and steel are expressed relative to their reference values at room temperature; see the first two columns of Tables 6 and 7, respectively. Linear interpolation was used to quantify these moduli between the listed values. Poisson's ratio of concrete and steel were assumed to be temperature-independent. Thus, they are set equal to the values listed in Table 5.

Table 6. Temperature-dependent thermoelastic properties of concrete.

Temperature T (°C)	Young's Modulus $E_c(T)/E_c(T_{room})$ (−)	Compressive Strength $f_c(T)/f_c(T_{room})$ (−)	Coefficient of Thermal Expansion α_T (°C^{-1})
20	1.00	1.00	9.03×10^{-6}
100	0.85	1.00	9.70×10^{-6}
200	0.72	0.95	11.7×10^{-6}
300	0.60	0.85	15.2×10^{-6}

Table 7. Temperature-dependent thermoelastic properties of steel.

Temperature T (°C)	Young's Modulus $E_s(T)/E_s(T_{room})$ (−)	Yield Stress $f_y(T)/f_y(T_{room})$ (−)	Coefficient of Thermal Expansion α_T (°C^{-1})
20	1.00	1.00	12.2×10^{-6}
100	1.00	1.00	12.8×10^{-6}
200	0.90	1.00	13.6×10^{-6}
300	0.80	1.00	14.4×10^{-6}

As for the thermal degradation of the uniaxial compressive strength of concrete and the yield stress of hot rolled steel, regulations of the Eurocode [43] were used. The reduction of the compressive strength of concrete relative to its reference values at room temperature is listed in the first and the third columns of Table 6. Linear interpolation was used to quantify the compressive strength between the listed values.

The temperature-dependent coefficients of thermal expansion of both concrete and steel were chosen according to the regulations of the Eurocode 2 [43]; see Tables 6 and 7 for specific values at characteristic temperatures. The continuous representations of the underlying nonlinear evolutions are illustrated in Figure 16.

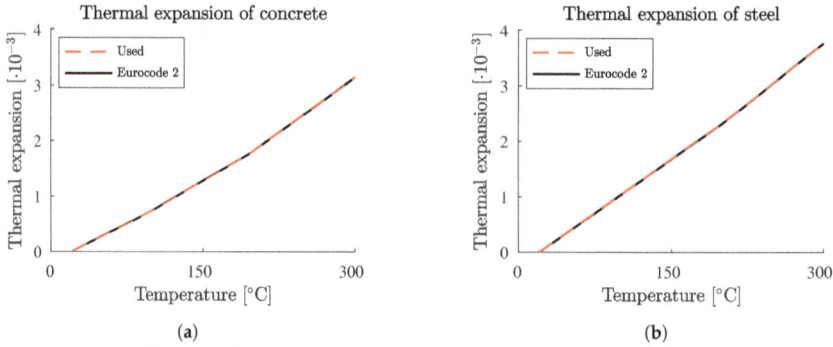

Figure 16. Coefficients of thermal expansion as a function of the temperature, according to Eurocode 2 [43], for: (**a**) concrete and (**b**) steel.

4.1.3. Constitutive Behavior: Elasto-Plastic Material Models for Concrete and Steel

The multiaxial elasto-plastic constitutive behavior of concrete was accounted for by means of the "Concrete Damaged Plasticity" model of Abaqus [42]; see Appendix B for more information regarding the model and the specific input parameters. As for uniaxial compression, stress-strain relations based on the regulations of the Eurocode were used; see Figure 17a. As for uniaxial tension, the stress-strain relations are based on trilinear behavior including a linear-elastic loading branch, a linear softening branch and a residual stress plateau, required for the stability of the numerical solution; see Figure 17b.

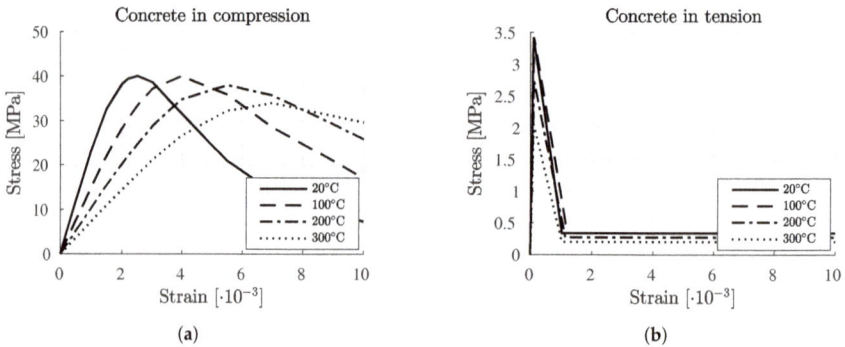

Figure 17. Temperature-dependent stress-strain relations recommended by Eurocode 2 for concrete subjected to uniaxial (**a**) compression and (**b**) tension.

As for the steel rebar, the employed Finite Element program requires specification of the material behavior under uniaxial tension. Up to a temperature of 100 °C, the material is linear-elastic and ideal-plastic; see Figure 18. At higher temperatures, the elastic limit of steel is decreasing, resulting in a trilinear stress-strain curve [43], representing linear-elastic, linear-hardening and ideal-plastic behavior; see Figure 18. The behavior of steel in compression is obtained by multiplying the numbers on the ordinate in Figure 18 by −1.

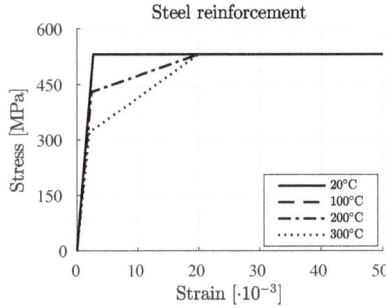

Figure 18. Temperature-dependent behavior of steel rebar subjected to uniaxial tension, as recommended by Eurocode 2.

4.2. Simulation Results and Comparison with Experimental Data

The comparison of experimental data and simulation results concerns the total strains at the inner and the outer reinforcement at the midspan of the left cell of the top slab and at the outer reinforcement at the center of the right wall; see Figure 19.

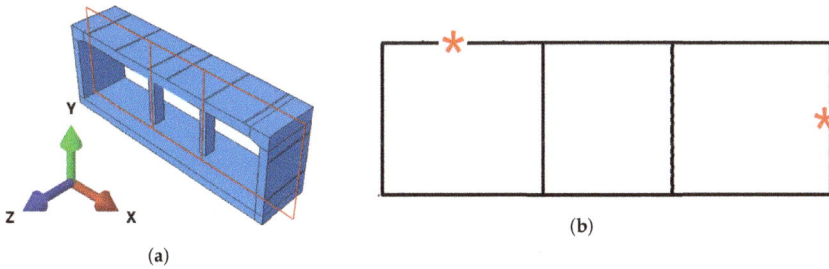

Figure 19. Positions at which measurements of strains were carried out during testing: (**a**) perspective representation and (**b**) cross-section, coinciding with the plane of symmetry of the structure that contains the columns.

4.3. Structural Response under Mechanical Loading

Experimental data and simulation results referring to the state after the application of the external loading, but before the application of the thermal loading are compared in the following. As regards the numerical results, there are strains from two simulations: one is based on the linear-elastic (label: LE) and the other one on the elasto-plastic (label: EP) material behavior of concrete.

The two simulations produce similar results; see Table 8. The EP simulation delivers strains that are by less than 3.5% larger than the strains from the LE simulation. This indicates that linear-elastic behavior of concrete governed the structural performance during and right after the application of the external loading, while inelastic effects played a significantly less important role.

Both types of simulations delivered strains that agree with the measurements in terms of the mathematical sign, referring to tensile and compressive strains, respectively; see Table 8. Satisfactory agreement between simulation results and experimental data was obtained for the right wall. There, the difference between simulated and measured strains amounted to 3.0×10^{-6} (LE simulation) and 1.4×10^{-6} (EP simulation). This is equivalent to prediction errors of 6.1% and 2.8%, respectively. The absolute and relative differences at the other two positions, however, were significantly larger; see the last two columns of Table 8. These differences could be the result either of inaccurate measurements or of damage of the structure prior to testing. The latter could have resulted from restrained shrinkage of concrete or from the transport and maneuvering of the structure.

Table 8. Measured and simulated strains of steel rebar, positioned at the inner and the outer reinforcement at midspan of the left cell of the top slab and at the outer reinforcement at the center of the right wall: additional strains, resulting from application of the mechanical loading, considering linear-elastic (LE) or elasto-plastic (EP) material behavior of concrete.

Analysis Type	RW Outer Layer (10^{-6})	TS Outer Layer (10^{-6})	TS Inner Layer (10^{-6})
Measured	−48.9	−49.4	3.72
LE	−45.9	−68.9	29.2
EP	−47.6	−69.4	30.1

4.4. Structural Response under Thermomechanical Loading

The comparison of experimental data and simulation results refers to the strain increments caused by the application of the thermal loading. As regards the numerical results, there are strain increments from four different simulations. They refer either to the linear-elastic (label: LE) or to the elasto-plastic (label: EP) material behavior of concrete and either to a spatially-uniform prescription of one surface temperature history (label: U) or to a piecewise spatially-uniform prescription of three different surface temperature histories (label: P).

As regards the outer reinforcement both at the midspan of the left cell of the top slab and at the center of the right wall, the two different types of thermal boundary conditions have little influence on the simulation results; see Figure 20a,c. This follows from the fact that the outer reinforcement layers were not experiencing a temperature change during the experiment. The material model used for concrete, in turn, has a much larger influence. The linear-elastic model results in larger strain increments as compared to the elasto-plastic model. The differences increase with increasing duration of the thermal loading, and they are larger at the top slab as compared to the right wall; compare Figure 20a,c. In addition, both types of simulations underestimate the measured strain increments.

In this context, the uncertainties regarding the thermal boundary conditions and the material behavior of concrete must be mentioned. On the one hand, there are strong indications that the inner surface of the tested structure was exposed to significant temperature gradients. Given that there were only two thermocouples positioned very close to the heated surface, the thermal boundary conditions are affected by considerable uncertainties. On the other hand, the material behavior of concrete was accounted for as suggested by current standards and pertinent guidelines. Nevertheless, some aspects deserve special attention. It is likely that concrete was damaged prior to testing because of restrained shrinkage. Moreover, time-dependent viscous material behavior in the form of creep and load induced-thermal strains (LITS) could also have influenced the structural performance considerably.

As regards the inner reinforcement at the midspan of the left cell of the top slab, the differences of the strain increments obtained with different thermal boundary conditions are significantly larger than described above; compare Figure 20b with Figure 20a,c. This follows from the fact that the inner reinforcement layers were experiencing a significant temperature change during the experiment. It is also interesting to note that the elasto-plastic simulations deliver larger strains than the linear-elastic simulations. This indicates that inelastic material behavior of concrete, in particular tensile cracking, had a significant influence on the structural performance during the fire test.

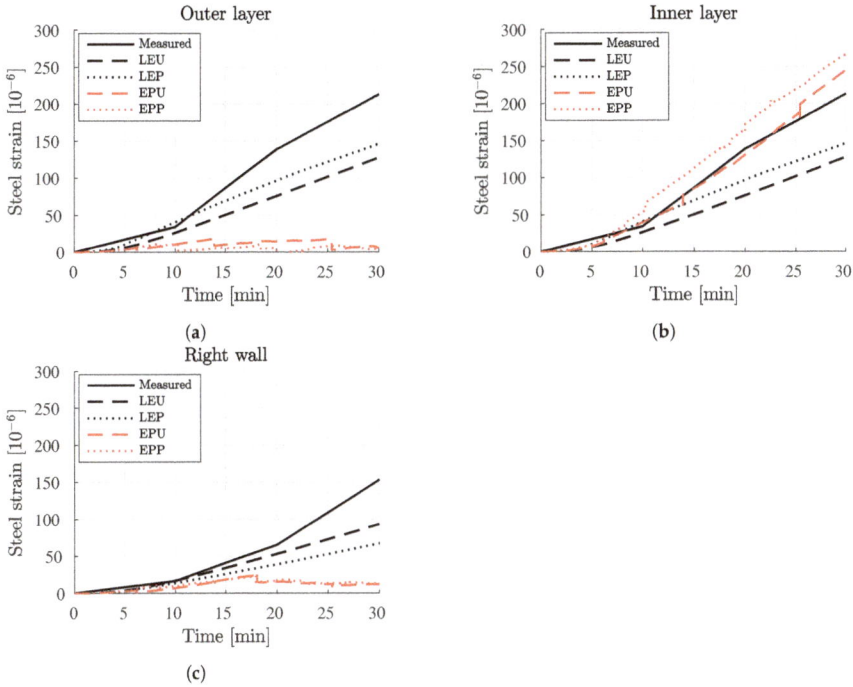

Figure 20. Measured and simulated strains of steel rebar: (**a**) outer and (**b**) inner reinforcement at the midspan of the left cell of the top slab, and (**c**) outer reinforcement at the center of the right wall: strain increments, resulting from application of the thermal loading, considering linear-elastic (LE) or elasto-plastic (EP) material behavior of concrete and a prescription of one uniform surface temperature history (U) or of three different piecewise uniform surface temperature histories (P).

The elasto-plastic material model for concrete improved the quality of the simulations, albeit with a significant increase in computational effort. The brittle nature of the structure caused abrupt local failures in specific regions where the tensile stresses were particularly large; see Figures 21–26. The figures show states of the structure separated by less than 0.2 s. This sudden loss of strength caused numerical problems. In order to overcome these problems by means of numerical simulations, fictitious stabilizing viscous forces were introduced, following the recommendations of Abaqus FEA, in order to redistribute the stresses near the affected regions smoothly. The quality of the results was nevertheless verified by comparing the magnitude of these fictitious viscous forces with the total magnitude of the acting forces. Additionally, the ratio of the energy introduced by these forces (also referred to as "static dissipation energy" [42]) to the total internal energy was verified as remaining below 5% at all times.

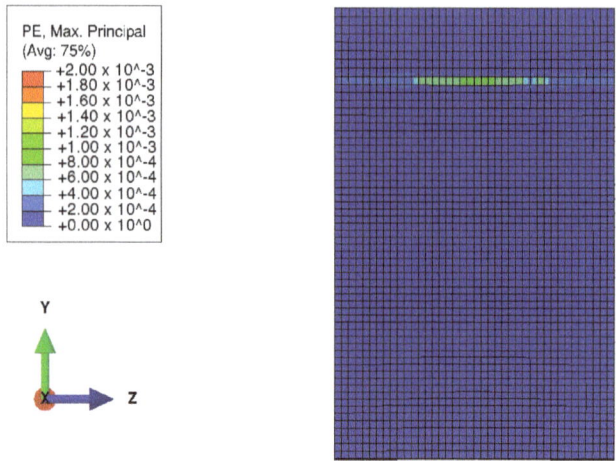

Figure 21. Plastic strains, developed on the outer surface of the top slab at its connection to the lateral wall right before the onset of cracking, as obtained from a thermomechanical simulation based on temperature-dependent and elasto-plastic material behavior, as well as on the prescription of one uniform surface temperature history; the results refer to the time instant six minutes and 3.4 s after the beginning of the fire test.

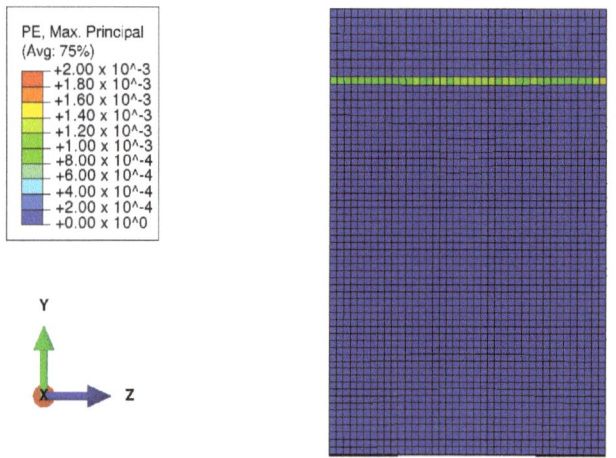

Figure 22. Plastic strains, developed on the outer surface of the top slab at its connection to the lateral wall right after the onset of cracking, as obtained from a thermomechanical simulation based on temperature-dependent and elasto-plastic material behavior, as well as on the prescription of one uniform surface temperature history; the results refer to the time instant 6 min and 3.6 s after the beginning of the fire test.

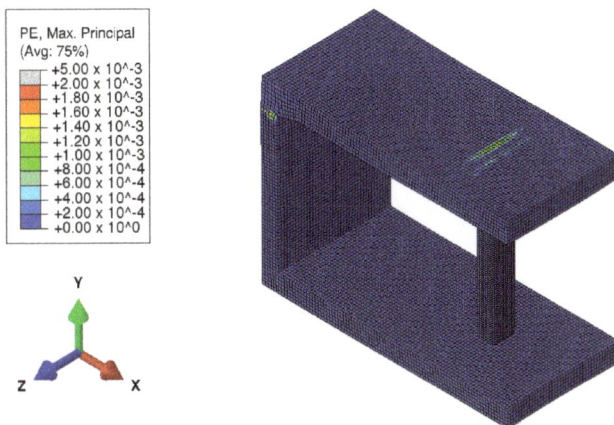

Figure 23. Plastic strains, developed on the outer surface of the top slab at its connection to the column right before the onset of cracking, as obtained from a thermomechanical simulation based on temperature-dependent and elasto-plastic material behavior, as well as on the prescription of one uniform surface temperature history; the results refer to the time instant 13 min and 50.5 s after the beginning of the fire test.

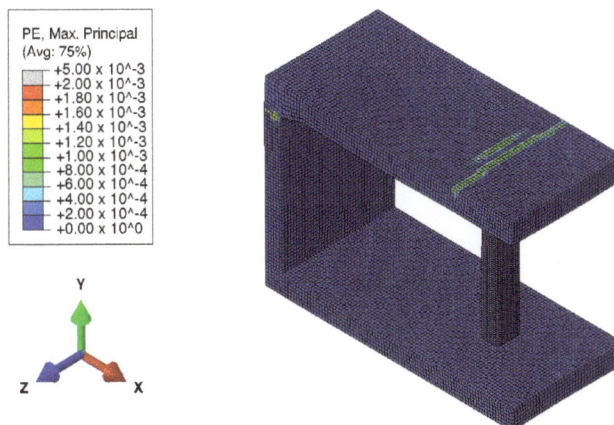

Figure 24. Plastic strains, developed on the outer surface of the top slab at its connection to the column right after the onset of cracking, as obtained from a thermomechanical simulation based on temperature-dependent and elasto-plastic material behavior, as well as on the prescription of one uniform surface temperature history; the results refer to the time instant 13 min and 50.7 s after the beginning of the fire test.

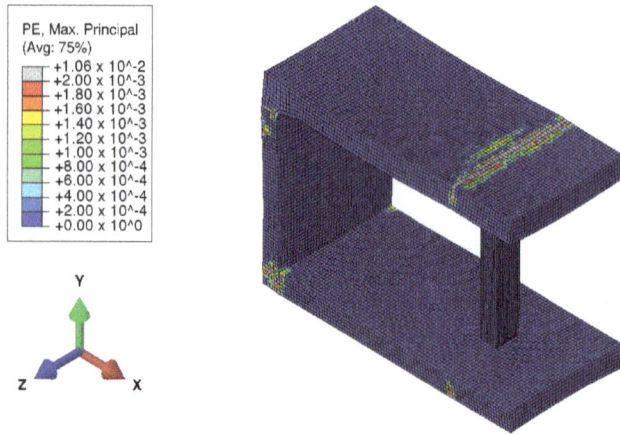

Figure 25. Plastic strains, developed on the outer surface of the top slab at its connection to the column right before the onset of cracking, as obtained from a thermomechanical simulation based on the temperature-dependent and elasto-plastic material behavior, as well as on the prescription of one uniform surface temperature history; the results refer to the time instant 25 min and 1.0 s after the beginning of the fire test.

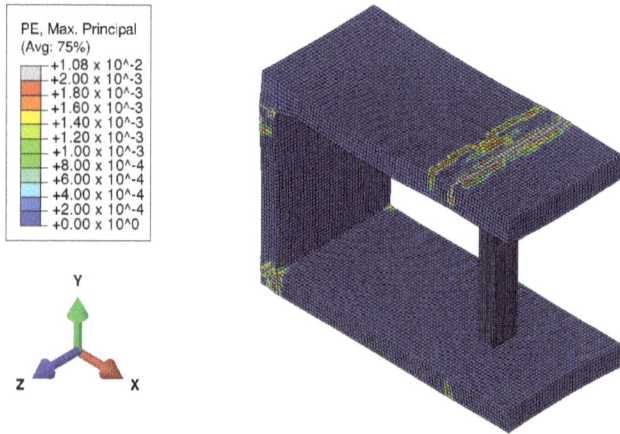

Figure 26. Plastic strains, developed on the outer surface of the top slab at its connection to the column right after the onset of cracking, as obtained from a thermomechanical simulation based on temperature-dependent and elasto-plastic material behavior, as well as on the prescription of one uniform surface temperature history; the results refer to the time instant 25 min and 1.2 s after the beginning of the fire test.

5. Conclusions

Because moderate fires happen much more frequently than fire disasters [41], the first 30 min of a fire test were analyzed. Sensitivity analyses by means of three-dimensional Finite Element simulations were used to address two specific challenges related to the prediction of the structural performance of a reinforced concrete segment of a subway station. These challenges are:

1. the estimation of the spatial and temporal development of the temperature within the reinforced concrete structure and

2. the choice of a suitable material model for concrete subjected to mechanical loads and elevated temperatures.

Experimental data from a scaled fire test of an underground substructure by Lu et al. [40] were used as the reference for comparison with the numerical results. From these results, the following conclusions and remarks may be extracted:

- Even under controlled laboratory conditions, different temperatures were measured at equal depths, but at different positions of the tested structure. This underlines significant uncertainties related to the interaction of the hot gas and the surface of the exposed structure. As emphasized by Achenbach et al. [62], these uncertainties can hardly be reduced.
- The temperature within the concrete and the steel elements at all positions at which thermocouples were placed, both in the slabs and the walls, remained below 100 °C during the first 30 min of the fire test. Thus, the material properties of concrete and steel remained approximately constant at these positions. This was the motivation to consider the temperature-independent and linear-elastic material behavior of concrete.
- The numerical simulations clarified that the monitored column was exposed to significantly higher temperatures than the slabs and the walls. This was probably a result of the fact that the columns were positioned above the two heat sources of the furnace. Because of the elevated temperatures, a considerable thermal degradation of the material properties of concrete and steel took place in the columns.
- The strain measurements clarified that strains resulting from the thermal loading dominate the total strains soon after the beginning of the fire. This underlines that even moderate fires do represent a considerable threat to the integrity and durability of RC structures.

The challenge of the three-dimensional simulation was met in two parts. At first, the non-stationary heat transfer inside the reinforced concrete structure was simulated. The obtained temperature field histories were used as input for subsequent simulations of the load-carrying behavior of the reinforced concrete structure. Two types of thermal boundary conditions were combined with two types of material models for concrete.

The following conclusions are drawn from the two different types of heat transfer simulations:

- The prescription of one uniform temperature history at all exposed surfaces does not produce convincing results. Temperature measurements are particularly underestimated inside the right column and at the midspan of the left cell of the top slab.
- The prescription of three different piecewise uniform temperature histories showed that, provided the temperature history at a specific surface is known and used as the local thermal boundary condition, the ingress of heat at that position of the structure is predicted accurately.
- The surface temperature histories were only measured at two positions of the tested structure. Thus, assumptions concerning the thermal boundary conditions were indispensable for the numerical simulations. They are anyway questionable, because the unsolved problem of the dynamics of flames remains an important source of uncertainty; see, e.g., the study by Blanchard et al. [34].
- The simulation with the improved thermal boundary conditions underlined that the temperature increase at the surface of the columns is expected to be larger than 280 °C. Thus, thermal degradation of the mechanical properties of concrete and steel occurs. This was the motivation to consider the temperature-dependent and elasto-plastic material behavior of concrete.

The following conclusions are drawn from the four different types of thermomechanical simulations of the load-carrying behavior of the analyzed structure:

- The load-carrying behavior of the structure, when subjected to mechanical loads only, was governed by linear-elastic material behavior. Although tensile cracking took place at several positions of the structure, inelastic material behavior did not play an important role.

- The numerical simulations referring to the mechanical loading reproduced strain measurements that were only reliable at the right wall, whereas the absolute values of the strains at midspan of the left cell of the top slab were overestimated significantly. This has revealed two additional sources of uncertainty, namely the reliability of measurements and the initial state of the structure. It is likely that the highly statically indeterminate structure was damaged already before the test, by restrained shrinkage of concrete and by the transport of the structure from the production site to the test furnace.
- The numerical simulations referring to the mechanical and the thermal loading reproduced strain measurements that were only reliable at the inner reinforcement at the midspan of the left cell of the top slab, whereas the absolute values of the strains at the other two positions at which strain measurements were available were underestimated significantly. The best results were obtained from the most realistic simulation, based on the prescription of three different piecewise uniform surface temperature histories, as well as on temperature-dependent elasto-plastic behavior of concrete.

The Finite Element simulations provide insight into the structural behavior of the tested segment of a subway station. From the numerical results obtained with the nonlinear material model for concrete, the following conclusions are drawn:

- Already moderate fires threaten the integrity of RC structures. As for the analyzed structure, the Finite Element simulations indicate localized tensile cracking at the outer surface of the structure, in regions where the top and the bottom slabs are connected to the walls and the columns; see Figures 21–26. In more detail, the Finite Element simulations indicate that the corners, i.e., the connections between the walls and the slabs, were damaged first (Figures 21 and 22), followed by a localized damage of the slabs, in the immediate vicinity of the connections to the columns (Figures 23–26).
- Tensile cracking of concrete and the associated redistribution of stresses within the RC structure take place in a quasi-instantaneous fashion. This renders numerical stabilization approaches indispensable in order to achieve convergence in nonlinear Finite Element simulations.
- Careful inspection of RC structures is strongly recommended also after moderate fire events (see, e.g., the methods proposed by Felicetti [63]), even if the accessible interior surface of an underground structure appears to be undamaged.
- Connections between structural elements with strong differences regarding the ratio between their heat-exposed surface and their volume are prone to suffer from localized damage. Notably, columns tend to exhibit larger surface-to-volume ratios than slabs and walls. Thus, connections between columns and slabs should be thoroughly inspected.
- Tensile cracking of the inaccessible exterior surface of an underground structure is a serious threat for the long-term durability of the structure after the fire, because the cracks represent pathways for substances that promote the corrosion risk/rate of the steel rebar.

As for future fire tests, the following recommendations can be made based on the results of the present study:

- It is recommended to carry out test repetitions, also in the context of structural experiments, even though such tests are time-consuming and expensive. Test repetitions render the desirable assessment of the experimental scatter associated with the chosen testing method possible; see, e.g., the experimental approach by Schlappal et al. [64].
- It is recommended to carry out redundant measurements of key quantities. This implies that key quantities shall be measured by two independent test methods. The availability of redundant measurements allows for the assessment of the reliability of the used measurement equipment; see, e.g., the experimental approach by Wyrzykowski et al. [65].
- When it comes to the design of a large fire test, it is recommended to position many thermocouples close to the heated surface in order to gain access to the spatial distribution

of the surface temperature histories. This way, the uncertainties of the actual fire load can be reduced significantly.

- It is recommended to produce test structures at the place and in the position of subsequent fire testing, such that possible damage associated with the transport and the maneuvering of the structure can be excluded.
- It is recommended to equip the tested reinforced concrete structure with embedded sensors that allow for quantification of possible damage of the structure resulting from restrained shrinkage. As for the design of such embedded sensors, multi-physics simulations of the structure, providing insight into the performance of the structure from its production, throughout all early-age stages, all the way up to the time of testing, are needed. This requires a strong investment of the global scientific community into basic research, aimed at a better understanding of structures made from modern concretes.

Author Contributions: Conceptualization, Bernhard Pichler. Formal analysis, R.D. Investigation, R.D. and H.W. Supervision, B.P. Writing, original draft, R.D. Writing, review and editing, H.W., H.M., Y.Y. and B.P.

Funding: Financial support was provided by the Austrian Science Fund (FWF), within the project P 281 31-N32 "Bridging the Gap by Means of Multiscale Structural Analyses". The second author also gratefully acknowledges financial support by the China Scholarship Council (CSC).

Conflicts of Interest: The authors declare no conflict of interest. The founding sponsors had no role in the design of the study; in the collection, analyses or interpretation of data; in the writing of the manuscript; nor in the decision to publish the results

Appendix A. Temperature Distribution Inside the Concrete Structure

Simulation results in regions of the structure which were not equipped with thermocouples are presented and discussed in the following. As regards the heat transfer in the analyzed structure, it can be noted that:

- During the first 30 min of the fire test, the temperature of the reinforced concrete structure increased up to a distance of some 10 cm from the heated surface, see Figures A1–A3.
- One-dimensional heat conduction occurs in the top slab, the bottom slab, and the lateral walls, except in the immediate vicinity of the columns and the walls, see Figures A1 and A2. Notably, the one-dimensional heat transfer problems could also be solved in a semi-analytical fashion, following e.g., Wang et al. [66].
- Two-dimensional heat conduction occurs in the columns (Figures 13 and 15) and in the vicinity of the connections of the slabs with the columns and the walls. As for studying the heat transfer in these regions, Finite Element simulations are indispensable for reliable results.

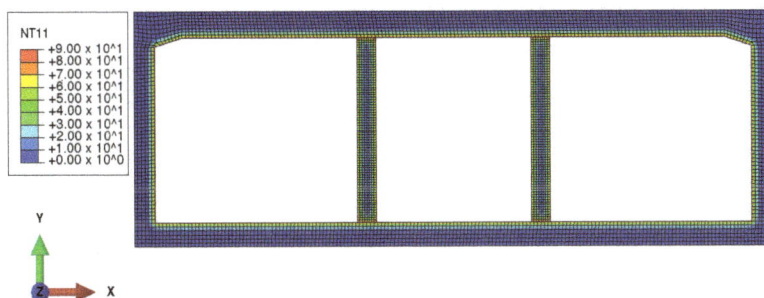

Figure A1. Temperature distribution obtained by prescribing a uniform surface temperature history, 30 min after the start of the thermal loading: overview over the whole structure.

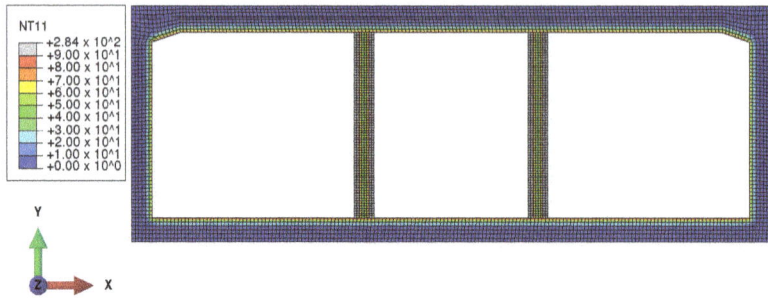

Figure A2. Temperature distribution obtained by prescribing three piecewise uniform surface temperature histories, 30 min after the start of the thermal loading: overview over the whole structure.

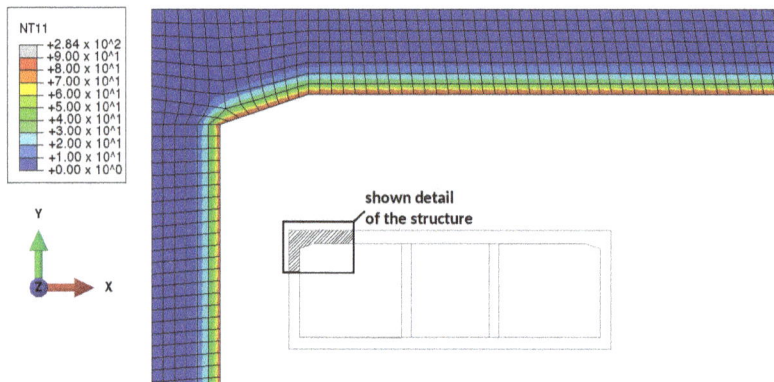

Figure A3. Temperature distribution obtained by prescribing three piecewise uniform surface temperature histories, 30 min after the start of the thermal loading: detail showing the connection of the left wall to the top slab.

Appendix B. Elasto-Plastic Material Model for Concrete

The "Concrete Damaged Plasticity" model of Abaqus [42] was used. A theoretical description of the model is presented in this appendix. The terms used by Abaqus are adopted in order to facilitate the application of this computer program. The constitutive model is essentially based on work by Lubliner al al. [67]. The plasticity part of the model controls the evolution of the yield stress of concrete. Thus, it refers to the strength of concrete. The damage part refers to the stiffness of concrete. It is designed to model the degradation of Young's modulus, resulting from mechanical loading. This is particularly important if concrete is unloaded in the process of softening. In this context, it is noteworthy that the challenge of the present simulation refers to an experiment, in which an increase of mechanical loading is followed by an increase of thermal loading. Thus, the damage part of the model remains inactive without reduction of the informative content of the simulation. Additionally, because the damage part is not used, there is no need to distinguish between Cauchy stresses and "effective stresses", between the "tensile cracking strain" and the plastic strain, as well as the "compressive inelastic strain" and the plastic strain, mentioned in the manual of Abaqus FEA.

The employed material model accounts for two failure mechanisms: tensile cracking and compressive crushing. As input, it requires the definition of stress-strain relationships for uniaxial compression and uniaxial tension, respectively.

As for uniaxial compression, the input required by Abaqus FEA consists of look-up tables that list pairs of values of the "yield stress" and the "inelastic strain", and these listings are specified for different temperatures. The "yield stress" is equal to (i) the imposed stress, in the region of initial elastic loading, (ii) the uniaxial elastic-limit stress, in the region of pre-peak hardening, (iii) the uniaxial compressive strength, at the peak, and (iv) the residual uniaxial compressive strength, in the region of post-peak softening. In the present context, the "inelastic strain" is simply equal to the plastic strain, given that the damage part remains unused. The used look-up tables are based on stress-strain regulations of the Eurocode, see Figure 17a. Thereby, the term "strain" refers to the total strain, ε, which is the sum of the elastic strain ε^{el}, and the plastic strain, ε^{pl}:

$$\varepsilon = \varepsilon^{el} + \varepsilon^{pl} . \tag{A1}$$

The elastic strain follows from Hooke's law as:

$$\varepsilon^{el} = \frac{\sigma}{E_c(T)} , \tag{A2}$$

where $E_c(T)$ denotes the temperature-dependent Young's modulus of concrete. The plastic strains, required for the look-up tables, result from inserting Equation (A2) into Equation (A1), and solving the resulting expression for ε_c^{pl} as:

$$\varepsilon^{pl} = \varepsilon - \frac{\sigma}{E_c(T)} , \tag{A3}$$

see Table A1.

As for uniaxial tension, the used look-up tables are based on trilinear stress-strain behavior including a linear-elastic loading branch, a linear softening branch, and a residual stress plateau, required for the stability of the numerical simulation, see Figure 17b. Thereby, the term "strain" refers to the total strain. By analogy to the previously described procedure for uniaxial compression, the plastic tensile strains required for the look-up tables results from Equation (A3).

Table A1. Input values for compressive behavior at (a) room temperature, (b) 100 °C, (c) 200 °C, (d) 300 °C.

Yield Stress (MPa)	Inelastic Strain (10^{-3})
a	
23.6	0.0
32.5	0.5
38.2	0.9
39.4	1.0
40.0	1.3
38.6	1.8
31.5	3.1
20.9	4.9
14.3	6.6
7.3	9.8
b	
21.9	0.7
28.2	1.0
30.5	1.1
33.4	1.3
37.1	1.7
40.0	2.6
35.9	4.2
28.5	6.0
17.0	9.4

Table A1. *Cont.*

Yield Stress (MPa)	Inelastic Strain (10^{-3})
c	
10.3	0.0
15.4	0.9
20.2	1.2
22.1	1.3
24.7	1.5
28.8	1.8
34.8	2.6
38.0	3.9
35.7	5.5
25.9	8.9
d	
7.3	0.0
10.9	1.0
14.4	1.3
15.8	1.4
17.8	1.6
21.0	1.9
26.7	2.7
32.2	3.9
34.0	5.3
29.6	8.5

Table A2. Input values for tensile behavior at a) room temperature, (b) 100 °C, (c) 200 °C, (d) 300 °C.

Yield Stress (MPa)	Cracking Strain (10^{-3})
a	
3.41	0.0
0.34	1.02
0.34	10
b	
3.41	0.0
0.34	1.20
0.34	10
c	
2.73	0.0
0.27	1.14
0.27	10
d	
2.05	0.0
0.21	1.02
0.21	10

The elasto-plastic classification of multiaxial stress states, defined in terms of Cauchy stress tensors $\underline{\underline{\sigma}}$, is based on the failure surface F, reading as [42]:

$$F(\underline{\underline{\sigma}}) = \frac{1}{1-\alpha}\left[q - 3\,\alpha\,p + \beta(\varepsilon_c^{pl},\varepsilon_t^{pl})\cdot\langle\sigma_{max}\rangle - \gamma\,\langle\sigma_{max}\rangle\right] - \sigma_c(\varepsilon_c^{pl}) \le 0. \tag{A4}$$

In Equation (A4) p denotes the hydrostatic pressure:

$$p = -\frac{\text{tr}\underline{\sigma}}{3},$$

(A5)

where the symbol "tr" stands for the "trace"-function; q denotes the von Mises stress, given as:

$$q = \sqrt{\frac{3}{2}(\underline{s}:\underline{s})},$$

(A6)

where $\underline{s}$ denotes the stress deviator:

$$\underline{s} = \underline{\sigma} + p\underline{1},$$

(A7)

with $\underline{1}$ as the second-order identity tensor. σ_{max} is the maximum principal normal stress. $\sigma_c(\varepsilon_c^{pl})$ is the current value of the "compressive cohesion stress" [42], i.e., the current value of the uniaxial compressive strength. The brackets $\langle \rangle$ stand for the Macaulay brackets, defined as:

$$\langle x \rangle = \frac{1}{2}\left(|x| + x\right).$$

(A8)

The dimensionless constant α is related to the ratio of the elastic limit stresses under symmetric biaxial and uniaxial compressive loading:

$$\alpha = \frac{(\sigma_{b0}/\sigma_{c0}) - 1}{2(\sigma_{b0}/\sigma_{c0}) - 1},$$

(A9)

The default value of σ_{b0}/σ_{c0} amounts to 1.16, such that $\alpha = 0.12$. The dimensionless parameter $\beta(\varepsilon_c^{pl}, \varepsilon_t^{pl})$ is defined as:

$$\beta(\varepsilon_c^{pl}, \varepsilon_t^{pl}) = (1 - \alpha)\frac{\sigma_c(\varepsilon_c^{pl})}{\sigma_t(\varepsilon_t^{pl})} - (1 + \alpha),$$

(A10)

where $\sigma_t(\varepsilon_t^{pl})$ is the current value of the "tensile cohesion stress" [42], i.e., the current value of the uniaxial tensile strength. γ is a constant, controlling the anisotropy of the failure surface in the deviatoric planes:

$$\gamma = \frac{3(1 - K_c)}{2K_c - 1},$$

(A11)

where K_c denotes a dimensionless parameter. Its default value for concrete amounts to 0.67, such that $\gamma = 3$.

The flow rule represents the evolution law for the plastic strains:

$$\underline{\dot{\varepsilon}}^{pl} = \dot{\lambda}\frac{\partial G}{\partial \underline{\sigma}},$$

(A12)

where $\underline{\dot{\varepsilon}}^{pl}$ is the rate of the plastic strain tensor, G is the plastic potential function, and λ is the consistency factor. The plastic potential G of the non-associated plasticity approach reads as [42]:

$$G(\underline{\sigma}) = \sqrt{(\epsilon\,\sigma_{t0}\,\tan\psi)^2 + q^2} - p\,\tan\psi,$$

(A13)

where ψ is the dilatation angle, σ_{t0} is the initial value of the uniaxial tensile strength, and ϵ is an "eccentricity" parameter. The default values of ψ and ϵ amount to 30° and 0.10, respectively. The present simulations are based on the described default input values, recommended by Lubliner et al. [67], Abaqus FEA [42], and others [68,69], see Table A3.

Table A3. Values used for the Concrete Damaged Plasticity model from Abaqus FEA [42].

ψ	ϵ	σ_{b0}/σ_{c0}	K_c	α	γ
30°	0.10	1.16	0.67	0.12	3

References

1. Schneider, U. Concrete at high temperatures—A general review. *Fire Saf. J.* **1988**, *13*, 55–68. [CrossRef]
2. Ulm, F.J.; Coussy, O.; Bažant, Z. The "Chunnel" fire. I: Chemoplastic softening in rapidly heated concrete. *J. Eng. Mech.* **1999**, *125*, 272–282. [CrossRef]
3. Gawin, D.; Pesavento, F.; Schrefler, B. Modelling of hygro-thermal behaviour of concrete at high temperature with thermo-chemical and mechanical material degradation. *Comput. Methods Appl. Mech. Eng.* **2003**, *192*, 1731–1771. [CrossRef]
4. Gawin, D.; Pesavento, F.; Schrefler, B. What physical phenomena can be neglected when modelling concrete at high temperature? A comparative study. Part 2: Comparison between models. *Int. J. Solids Struct.* **2011**, *48*, 1945–1961. [CrossRef]
5. Bažant, Z.; Jirásek, M. Temperature effect on water diffusion, hydration rate, creep and shrinkage. In *Creep and Hygrothermal Effects in Concrete Structures*; Springer: Berlin, Germany, 2018; pp. 607–686.
6. Lamont, S.; Usmani, A.; Drysdale, D. Heat transfer analysis of the composite slab in the Cardington frame fire tests. *Fire Saf. J.* **2001**, *36*, 815–839. [CrossRef]
7. Tenchev, R.; Li, L.; Purkiss, J. Finite element analysis of coupled heat and moisture transfer in concrete subjected to fire. *Numer. Heat Transf. A Appl.* **2001**, *39*, 685–710. [CrossRef]
8. Chung, J.; Consolazio, G. Numerical modeling of transport phenomena in reinforced concrete exposed to elevated temperatures. *Cem. Concr. Res.* **2005**, *35*, 597–608. [CrossRef]
9. Khoury, G.; Grainger, B.; Sullivan, P. Transient thermal strain of concrete: Literature review, conditions within specimen and behaviour of individual constituents. *Mag. Concr. Res.* **1985**, *37*, 131–144. [CrossRef]
10. Terro, M. Numerical modeling of the behavior of concrete structures in fire. *ACI Struct. J.* **1998**, *95*, 183–193.
11. Mindeguia, J.C.; Hager, I.; Pimienta, P.; Carré, H.; La Borderie, C. Parametrical study of transient thermal strain of ordinary and high performance concrete. *Cem. Concr. Res.* **2013**, *48*, 40–52. [CrossRef]
12. Torelli, G.; Mandal, P.; Gillie, M.; Tran, V.X. Concrete strains under transient thermal conditions: A state-of-the-art review. *Eng. Struct.* **2016**, *127*, 172–188. [CrossRef]
13. Ulm, F.J.; Acker, P.; Lévy, M. The "Chunnel" fire. II: Analysis of concrete damage. *J. Eng. Mech.* **1999**, *125*, 283–289. [CrossRef]
14. Kalifa, P.; Chéné, G.; Gallé, C. High-temperature behaviour of HPC with polypropylene fibres—From spalling to microstructure. *Cem. Concr. Res.* **2001**, *31*, 1487–1499. [CrossRef]
15. Hertz, K. Limits of spalling of fire-exposed concrete. *Fire Saf. J.* **2003**, *38*, 103–116. [CrossRef]
16. Diederichs, U.; Schneider, U. Bond strength at high temperatures. *Mag. Concr. Res.* **1981**, *33*, 75–84. [CrossRef]
17. Haddad, R.; Al-Saleh, R.; Al-Akhras, N. Effect of elevated temperature on bond between steel reinforcement and fiber reinforced concrete. *Fire Saf. J.* **2008**, *43*, 334–343. [CrossRef]
18. Gao, W.; Dai, J.G.; Teng, J.; Chen, G. Finite element modeling of reinforced concrete beams exposed to fire. *Eng. Struct.* **2013**, *52*, 488–501. [CrossRef]
19. Franssen, J.M.; Dotreppe, J.C. Fire tests and calculation methods for circular concrete columns. *Fire Technol.* **2003**, *39*, 89–97. [CrossRef]
20. Kodur, V.; Cheng, F.P.; Wang, T.C.; Sultan, M. Effect of strength and fiber reinforcement on fire resistance of high-strength concrete columns. *J. Struct. Eng.* **2003**, *129*, 253–259. [CrossRef]
21. Kodur, V.; Mcgrath, R. Fire endurance of high strength concrete columns. *Fire Technol.* **2003**, *39*, 73–87. [CrossRef]
22. Kodur, V.; Wang, T.; Cheng, F. Predicting the fire resistance behaviour of high strength concrete columns. *Cem. Concr. Compos.* **2004**, *26*, 141–153. [CrossRef]
23. Kodur, V.; Dwaikat, M. A numerical model for predicting the fire resistance of reinforced concrete beams. *Cem. Concr. Compos.* **2008**, *30*, 431–443. [CrossRef]

24. Kodur, V.; Yu, B.; Solhmirzaei, R. A simplified approach for predicting temperatures in insulated RC members exposed to standard fire. *Fire Saf. J.* **2017**, *92*, 80–90. [CrossRef]

25. Lura, P.; Terrasi, G. Reduction of fire spalling in high-performance concrete by means of superabsorbent polymers and polypropylene fibers: Small scale fire tests of carbon fiber reinforced plastic-prestressed self-compacting concrete. *Cem. Concr. Compos.* **2014**, *49*, 36–42. [CrossRef]

26. Xu, Q.; Han, C.; Wang, Y.C.; Li, X.; Chen, L.; Liu, Q. Experimental and numerical investigations of fire resistance of continuous high strength steel reinforced concrete T-beams. *Fire Saf. J.* **2015**, *78*, 142–154. [CrossRef]

27. Yan, Z.G.; Shen, Y.; Zhu, H.H.; Li, X.J.; Lu, Y. Experimental investigation of reinforced concrete and hybrid fibre reinforced concrete shield tunnel segments subjected to elevated temperature. *Fire Saf. J.* **2015**, *71*, 86–99. [CrossRef]

28. Nguyen, T.T.; Tan, K.H.; Burgess, I. Behaviour of composite slab-beam systems at elevated temperatures: Experimental and numerical investigation. *Eng. Struct.* **2015**, *82*, 199–213. [CrossRef]

29. Annerel, E.; Lu, L.; Taerwe, L. Punching shear tests on flat concrete slabs exposed to fire. *Fire Saf. J.* **2013**, *57*, 83–95. [CrossRef]

30. Li, G.Q.; Zhang, N.; Jiang, J. Experimental investigation on thermal and mechanical behaviour of composite floors exposed to standard fire. *Fire Saf. J.* **2017**, *89*, 63–76. [CrossRef]

31. Ring, T.; Zeiml, M.; Lackner, R. Underground concrete frame structures subjected to fire loading: Part I—Large-scale fire tests. *Eng. Struct.* **2014**, *58*, 175–187. [CrossRef]

32. Ring, T.; Zeiml, M.; Lackner, R. Underground concrete frame structures subjected to fire loading: Part II—Re-analysis of large-scale fire tests. *Eng. Struct.* **2014**, *58*, 188–196. [CrossRef]

33. El-Tayeb, E.; El-Metwally, S.; Askar, H.; Yousef, A. Thermal analysis of reinforced concrete beams and frames. *HBRC J.* **2017**, *13*, 8–24. [CrossRef]

34. Blanchard, E.; Boulet, P.; Desanghere, S.; Cesmat, E.; Meyrand, R.; Garo, J.; Vantelon, J. Experimental and numerical study of fire in a midscale test tunnel. *Fire Saf. J.* **2012**, *47*, 18–31. [CrossRef]

35. Maraveas, C.; Vrakas, A. Design of concrete tunnel linings for fire safety. *Struct. Eng. Int.* **2014**, *24*, 319–329. [CrossRef]

36. Pichler, C.; Lackner, R.; Mang, H. Safety assessment of concrete tunnel linings under fire load. *J. Struct. Eng.* **2006**, *132*, 961–969. [CrossRef]

37. Feist, C.; Aschaber, M.; Hofstetter, G. Numerical simulation of the load-carrying behavior of RC tunnel structures exposed to fire. *Finite Elem. Anal. Des.* **2009**, *45*, 958–965. [CrossRef]

38. Yan, Z.G.; Zhu, H.H.; Ju, J.; Ding, W.Q. Full-scale fire tests of RC metro shield TBM tunnel linings. *Constr. Build. Mater.* **2012**, *36*, 484–494. [CrossRef]

39. Guo, J.; Jiang, S.; Zhang, Z. Fire thermal stress and its damage to subsea immersed tunnel. *Procedia Eng.* **2016**, *166*, 296–306. [CrossRef]

40. Lu, L.; Qiu, J.; Yuan, Y.; Tao, J.; Yu, H.; Wang, H.; Mang, H. Large-scale test as the basis of investigating the fire-resistance of underground RC substructures. *Eng. Struct.* **2019**, *178*, 12–23. [CrossRef]

41. Rattei, G.; Lentz, A.; Kohl, B. How frequent are fire in tunnels-Analysis from Austrian tunnel incident statistics. In Proceedings from the Seventh International Conference on Tunnel Safety and Ventilation, Graz, Austria, 12–13 May 2014; pp. 5–11.

42. ABAQUS (2016). *ABAQUS Documentation*; Dassault Systèmes: Vélizy-Villacoublay, France, 2016.

43. CEN. Eurocode 2: Design of Concrete Structures—Part 1–2: General rules—Structural Fire Design (EN 1992-1-2). 2005. Available online: https://www.phd.eng.br/wp-content/uploads/2015/12/en.1992.1.2.2004.pdf (accessed on 30 October 2018).

44. International Federation for Structural Concrete (FIB). *FIB Model Code for Concrete Structures 2010*; International Federation for Structural Concrete: Lausanne, Switzerland, 2013.

45. Bailey, C. Holistic behaviour of concrete buildings in fire. *Proc. Inst. Civ. Eng.-Struct. Build.* **2002**, *152*, 199–212. [CrossRef]

46. Ministry of Construction of China. *Standard for Test Method of Mechanical Properties on Ordinary Concrete (GB/T 50081-2002)*; Ministry of Construction of China: Beijing, China, 2002.

47. Ministry of Construction of China. Standard for Metro Design (GB50157-2003); Ministry of Construction of China: Beijing, China, 2003.

48. CEN. Eurocode 2: Design of Concrete Structures—Part 1-1: General Rules and Rules for Buildings (EN 1992-1-1). 2005. Available online: https://www.phd.eng.br/wp-content/uploads/2015/12/en.1992.1.1. 2004.pdf (accessed on 30 October 2018).
49. Díaz, R. Thermomechanical Analysis of a Fire Test of a Segment of a Subway Station by Means of Three-Dimensional Finite Element Simulations. Master's Thesis, TU Wien, Vienna, Austria, 2018.
50. Lie, T.; Erwin, R. Method to calculate the fire resistance of reinforced concrete columns with rectangular cross section. *ACI Struct. J.* **1993**, *90*, 52–60.
51. American Society for Testing and Materials (ASTM). *Standard Test Methods for Fire Tests of Building Construction And Materials (ASTM E119)*; ASTM: West Conshohocken, PA, USA, 1988.
52. American Society of Civil Engineers (ASCE). *Structural Fire Protection*; ASCE: Reston, VA, USA, 1982.
53. Bažant, Z.; Kaplan, M. *Concrete at High Temperatures: Material Properties and Mathematical Models*; Longman Group Limited: Harlow, UK, 1996.
54. Fletcher, I.; Borg, A.; Hitchen, N.; Welch, S. Performance of concrete in fire: A review of the state of the art, with a case study of the Windsor Tower fire. In Proceedings of the 4th International Workshop in Structures in Fire, Aveiro, Portugal, 10–12 May 2006.
55. Kodur, V. Properties of concrete at elevated temperatures. *ISRN Civ. Eng.* **2014**, *2014*, 1–15. [CrossRef]
56. Anderberg, Y.; Forsen, N. *Fire Resistance of Concrete Structures*; Division of Building Fire Safety and Technology, Lund Institute of Technology: Lund, Sweden, 1982.
57. Naus, D. *The Effect of Elevated Temperature on Concrete Materials And Structures—A literature Review*; U.S. Nuclear Regulatory Commission Office of Nuclear Regulatory Research Under Interagency Agreement No. 1886-N674-1Y; U.S. Nuclear Regulatory Commission: Rockville, MD, USA, 2015.
58. Ren, H. Theoretical Analysis and Fire-Resistance Design of HPC Shear Walls. Master's Thesis, Tongji University, Shanghai, China, 2006.
59. Fletcher, I.; Welch, S.; Torero, J.; Carvel, R.; Usmani, A. Behaviour of concrete structures in fire. *Therm. Sci.* **2007**, *11*, 37–52. [CrossRef]
60. Law, A. The Assessment and Response of Concrete Structures Subject to Fire. Ph.D. Thesis, University of Edinburgh, Edinburgh, Scotland, 2010.
61. Ministry of Construction of China. *Practice Manual for Design of Concrete Structures (GB 50010-2002)*; Ministry of Construction of China: Beijing, China, 2002.
62. Achenbach, M.; Lahmer, T.; Morgenthal, G. Identification of the thermal properties of concrete for the temperature calculation of concrete slabs and columns subjected to a standard fire—Methodology and proposal for simplified formulations. *Fire Saf. J.* **2017**, *87*, 80–86. [CrossRef]
63. Felicetti, R. Assessment methods of fire damages in concrete tunnel linings. *Fire Technol.* **2013**, *49*, 509–529. [CrossRef]
64. Schlappal, T.; Schweigler, M.; Gmainer, S.; Peyerl, M.; Pichler, B. Creep and cracking of concrete hinges: Insight from centric and eccentric compression experiments. *Mater. Struct.* **2017**, *50*, 244. [CrossRef] [PubMed]
65. Wyrzykowski, M.; Lura, P. Moisture dependence of thermal expansion in cement-based materials at early ages. *Cem. Concr. Res.* **2013**, *53*, 25–35. [CrossRef]
66. Wang, H.; Binder, E.; Mang, H.; Yuan, Y.; Pichler, B. Multiscale structural analysis inspired by exceptional load cases concerning the immersed tunnel of the Hong Kong-Zhuhai-Macao Bridge. *Undergr. Space* **2018**. [CrossRef]
67. Lubliner, J.; Oliver, J.; Oller, S.; Oñate, E. A plastic-damage model for concrete. *Int. J. Solids Struct.* **1989**, *25*, 299–326. [CrossRef]
68. Birtel, V.; Mark, P. Parameterised Finite Element modelling of RC beam shear failure. In Proceedings of the 2006 ABAQUS Users' Conference, Boston, MA, USA, 23–25 May 2006; pp. 95–108.
69. Jankowiak, T.; Lodygowski, T. Identification of parameters of Concrete Damage Plasticity constitutive model. *Found. Civ. Environ. Eng.* **2005**, *6*, 53–69.

Article

2D Micromechanical Modeling and Simulation of Ta-Particles Reinforced Bulk Metallic Glass Matrix Composite

Pei-Liang Bian, Tian-Liang Liu, Hai Qing * and Cun-Fa Gao

State Key Laboratory of Mechanics and Control of Mechanical Structures, Nanjing University of Aeronautics and Astronautics, Nanjing 210026, China; bplcn@nuaa.edu.cn (P.-L.B.); cfgao@nuaa.edu.cn (C.-F.G.)
* Correspondence: qinghai@nuaa.edu.cn; Tel.: +86-025-8489-6410

Received: 29 September 2018; Accepted: 5 November 2018; Published: 8 November 2018

Abstract: The influence of particle shape, orientation, and volume fractions, as well as loading conditions, on the mechanical behavior of Ta particles reinforced with bulk metallic glass matrix composite is investigated in this work. A Matlab program is developed to output the MSC.Patran Command Language (PCL) in order to generate automatically two-dimensional (2D) micromechanical finite element (FE) models, in which particle shapes, locations, orientations, and dimensions are determined through a few random number generators. With the help of the user-defined material subroutine (UMAT) in ABAQUS, an implicit numerical method based on the free volume model has been implemented to describe the mechanical response of bulk metallic glass. A series of computational experiments are performed to study the influence of particle shapes, orientations, volume fractions, and loading conditions of the representative volume cell (RVC) on its composite mechanical properties.

Keywords: metallic glass matrix composite; finite element analysis; shear band; microstructure; ductility

---

1. Introduction

Bulk metallic glass (BMG) as an amorphous alloy has attracted much attention due to its extreme high strengths, superior elastic limits, etc. [1]. Although there has been much progress in the development of metallic glass with the development of manufacturing technologies, unlike common alloy, BMGs show nearly no ductility under loading at room temperature, which severely limits their structural application [2,3]. The micromechanical research shows that the plastic deformation localization is the main reason for brittle failure [4,5].

In order to avoid the brittle failure and improve the ultimate plastic extensibility of BMGs, metallic glass matrix composites (MGMCs) were firstly synthesized at the California Institute of Technology (Caltech) in the United States (USA) in 1998 [6,7]. A "soft" second phase, which has lower yield stress than BMGs, has been introduced for particle-reinforced composite [8]. Unlike pure BMGs, the second reinforced phase can block the propagation of the main shear band and generate a few of minor shear deformation zone, which can make the plastic deformation be distributed widely instead of the localization that is seen in pure BMGs [9,10]. The secondary phases themselves can absorb the plasticity, and profuse shear bands induced by the secondary phases can accommodate more plasticity [11].

Generally, the macroscopic response of MGMCs is some kind of statistical average of the microstructural factors, such as particle volume fraction, shape, orientation, and spatial spacing, as well as interface strength and residual stress, etc. It is extremely expensive and time-consuming, if it is not impossible, to establish the relationship between the macroscale mechanical properties of MGMCs and their microscale structures through either the experimental method or theoretical analysis. A number of computational models have been developed to predict the MGMCs' mechanical behaviors. Different numerical methods, including the molecular dynamics method [12–14] and phase-field method [15,16], as well as the finite element method (FEM) [17–23] are utilized to reveal the deformation mechanisms at the microscale.

Shi and Falk [12] identified the location of the plastic deformation, the deformation mechanism of the crystallites, and the interaction between the shear band and the crystalline inclusions through the molecular dynamics method. Albe et al. [13] applied the molecular dynamics method to study the shear band formation in homogenous bulk metallic glasses, nanocomposites, and nanoglasses. Zhou et al. performed a comprehensive study of the structural evolution of MGMCs through large-scale atomistic simulations [14]. The simulation results showed that slender crystalline second phases are better at suppressing shear band propagation than those with spherical shapes, and that increasing the volume fraction of the crystalline second phase will enhance the global plasticity. Abdeljawad et al. applied a two-dimensional (2D) phase-field model to examine the effects of BMG composite microstructures, e.g., the area fraction and the characteristic length scale of the ductile dendritic particles, on the mechanical properties of MGMCs [15]. Zhang and Zheng applied the phase-field simulation approach to investigate the formation mechanisms of shear bands in MGMCs containing dendrite particles [16].

Compared with molecular dynamics and phased field approaches, finite element analysis is more widely employed to investigate the mechanical behavior of MGMCs. Ott et al. [17] examined the microscale deformation mechanisms of MGMCs under uniaxial compression, combining high-energy X-ray scattering and finite element modeling. Lee et al. [18] investigated the effect of the crystalline phase on shear band initiation, interaction, and propagation in MGMCs with a unit finite element model. Biner [19] studied the influences of the mechanical properties, volume fraction, and morphology of ductile reinforcements on the ductility of MGMCs. Wu et al. [20] studied the influence of sample dimension on the toughness of MGMCs through scaling a microscale finite element model. Zhu et al. revealed that the plastic deformation of reinforced particles creates a shear stress concentration on the interface, and shear stress distribution leads to the formation of multiple shear bands through a combination of in situ SEM observations and finite element simulation [21]. Qiao et al. [22] quantitatively described the macroscopic MGMC deformation mechanics through a two-phase finite element model. The simulation of Hardin and Homer [23] showed that increasing the volume fraction alone is insufficient to promote strain delocalization in the case of a crystalline phase with a high relative yield stress, which is different from the results of Zhou et al. [14]. Jiang et al. [24,25] applied finite element methods to analyze the shear banding evolution and elucidate the relationship between the microstructure and ductility of MGMCs subjected to uniaxial tension. Unlike the results of Zhou et al. [14], they found that the particle shape has almost no effect in improving the tensile ductility of MGMCs. In addition, network second-phase was reported to be more efficient in improving the extensibility of composites [26].

Although there are many studies on the microscale deformation mechanism of MGMCs, it is still far away from a complete and thorough understanding of fundamental synergic mechanisms. For example, there are two contradictions in the literature about the influence of particle volume fraction and shape on the toughening mechanism of MGMCs [14,22,25]. Therefore, more detailed study is necessary in order to clarify the effect of the microstructure on the mechanical response of MGMCs. In this work, we study the influence of the microstructures and loading conditions on the mechanical properties of MGMCs through the periodic boundary conditions, which are realized with a multi-point constraint subroutine MPC in ABAQUS. The different kinds of shapes have been simplified to ellipses with different respect ratios. A program based on Matlab has been developed to generate ellipses with a random distribution of locations and orientations. The nodes and elements are generated with MSC.Patran and a free volume constitutive model for bulk metallic glasses is implied with the user-defined material (UMAT) in ABAQUS. A series of numerical studies are performed to analyze the influence of particle volume, shape, and orientation, as well as the loading conditions on the macroscopic stress–strain relationships and damage evolution in MGMCs.

2. Micromechanical Finite Element Model

2.1. Automatic Generation of the Representative Elementary Cell (RVC)

It is commonly recognized that the microstructure of the composite plays an important role in the macroscale response. Therefore, in order to study the effect of the microstructure on its deformation and failure process, the microstructure of the composite material under consideration should be able to vary according to the requirement. Here, the microstructure contains the randomly distributed elliptic particles with different shapes, but the same area is controlled through a few random number generators.

It's assumed that the RVC of the composite microstructure is a square with length of L_0, and the particle number and area fraction are n_p and v_p, respectively. Therefore, the area of each particle A_e can be expresses as:

$$A_e = L_0^2 v_p / n_p \qquad (1)$$

Notice that the area of ellipse A_e with the lengths of the semi-major and semi-minor axes (a_i and b_i, $i = 1, 2, \ldots, n_p$) can be expressed as:

$$A_e = \pi a_i b_i \qquad (2)$$

A random number stream Rand_1 controlled by the generator seed $s1$ is introduced to define the shape ratio between the lengths of two semi-axes:

$$r_i = a_i / b_i = 1 + r_0 \text{Rand}_{1i} \qquad (3)$$

Meanwhile, another random number stream Rand_2 controlled by random number generator seed s_2 is applied to define the alignment of each particle:

$$\theta_i = \pi \, \text{Rand}_{2i} \qquad (4)$$

The particle centers are determined independently and sequentially with two random number streams (Rand_3 and Rand_4) that are controlled by two random number generator seeds (s_3 and s_4). If an introduced particle cuts any of the RVC boundaries, the particle is copied to the opposite side of the square unit, as shown in Figure 1. Furthermore, the introduced particle should not be too close to the square boundaries or the existing particles. An iteration algorithm [27] is applied to calculate the distance between two elliptic particles, and is schematically shown in Figure 2. If any condition above is not met, the location of the new particle is determined by the next random number.

A Matlab program is developed to realize the above algorithm, and after running the Matlab program, a command file for the commercial software MSC/Patran can be obtained. A 2D microstructural finite element model with predefined parameters, such as the volume fraction and number of particles, mesh dimension, and so on, can be obtained through playing the command file with MSC/Patran. A multi-point constraint subroutine MPC in ABAQUS is developed to apply periodic boundary conditions (PBC) [28,29].

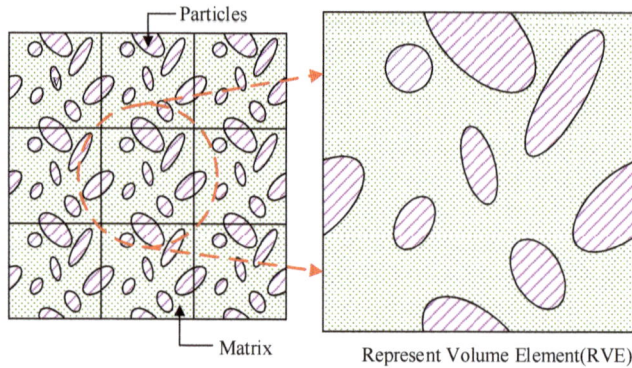

Figure 1. A representative cell (RVC) for two-dimensional (2D) particle-reinforced composites.

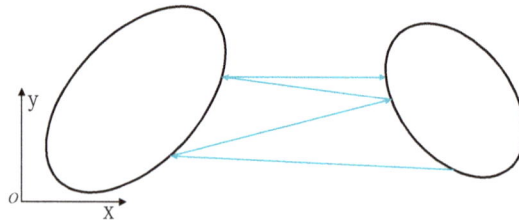

Figure 2. Process of the iteration to calculate the distance between two ellipses on a plane.

2.2. Constitutive Response of Ta Particles and BMGs

So far, several constitutive models are developed to describe the deformation of the BMGs. Argon proposed the shear transformation model using Eshelby's insightful theory, in which he assumes that the plastic deformation only occurs in the region called the shear transformation zone [30]. Furthermore, Jiang et al. developed a tension transformation zone model to describe the quasi-brittle dilatation deformation in metallic glass [31]. Viewing the basic "flow event" as an individual atom jump driven by the shear stress, Spaepem [32] developed the free volume model to analyze the plastic deformation in the BMGs. Compared other constitutive models, the free volume model has a clear physical meaning, and has been widely used to evaluate the mechanical properties of the BMGs.

According to Spaepem, the general free volume model flow equation is written as:

$$\frac{\partial \gamma^p}{\partial t} = 2f \exp\left(-\frac{\alpha v^*}{\overline{v}_f}\right) \exp\left(-\frac{\Delta G^m}{k_B T}\right) \sinh\left(\frac{\tau \Omega}{2 k_B T}\right) \tag{5}$$

where f is the frequency of atomic vibration, α is a geometrical factor, v^* is the critical volume, $\overline{v}_f$ is the average free volume per atom, ΔG^m is the activation energy, Ω is the atomic volume, τ is the shear stress, k_B is the Boltzmann constant, and T is the absolute temperature. Unlike the equivalent plastic strain in the usual metal plastic model, the free volume is used as a parameter to describe plastic deformation. The net rate of the free-volume increase is:

$$\frac{\partial \overline{v}_f}{\partial t} = v^* f \exp\left(-\frac{\alpha v^*}{\overline{v}_f}\right) \exp\left(-\frac{\Delta G^m}{k_B T}\right) \left\{ \frac{2\alpha k_B T}{\overline{v}_f C_{eff}} \left(\cosh\left(\frac{\tau \Omega}{2 k_B T} - 1\right)\right) - \frac{1}{n_D} \right\} \tag{6}$$

Here n_D is a constant varying from three to 10. $C_{eff} = \frac{E}{3(1-v)}$ is the effective elastic module for isotropic materials with Young's modulus E and Poisson's ratio v.

According to the flow rule, the strain can be decomposed into elastic and plastic parts:

$$\dot{\varepsilon}_{ij} = \dot{\varepsilon}_{ij}^e + \dot{\varepsilon}_{ij}^p \tag{7}$$

Moreover, the elastic and plastic strain rate can be expressed as follows:

$$
\begin{aligned}
\dot{\varepsilon}_{ij}^e &= \frac{1+v}{E}\left(\dot{\sigma}_{ij} - \frac{v}{1+v}\dot{\sigma}_{kk}\delta_{ij}\right) \\
\dot{\varepsilon}_{ij}^p &= \exp\left(-\frac{1}{v_f}\right)\sinh\left(\frac{\sigma_e}{\sigma_0}\right)\frac{S_{ij}}{\sigma_e}
\end{aligned}
\tag{8}
$$

where S_{ij} is the deviatoric stress tensor, and σ_e is the equivalent stress. According to the J_2 flow rule, the evolution of plastic strain is a function depending on the deviatoric stress. So, the flow equation can be written as follows:

$$\dot{v}_f = \frac{1}{\alpha}\exp\left(-\frac{1}{v_f}\right)\left\{\frac{3(1-v)}{E}\left(\frac{\sigma_0}{\beta v_f}\right)\left[\cosh\left(\frac{\sigma_e}{\sigma_0}\right)-1\right]-\frac{1}{n_d}\right\} \tag{9}$$

in which, $\sigma_0 = \frac{2k_BT}{\Omega}$ is the reference stress, and $\beta = \frac{v^*}{\Omega}$ and $v_f = \frac{\bar{v}_f}{\alpha v^*}$ are the normalized free volume. A user-defined material subroutine (UMAT) in ABAQUS code is developed to implement the free-volume model [33].

For reinforcing Ta particles, an exponentiation isotropic hardening relationship according to Zhang et al. [16] has been used to describe the plastic deformation in the crystal metal here:

$$g(\varepsilon^p) = \sigma_y\left(1 + \frac{\varepsilon^p}{\varepsilon_0^p}\right)^{\frac{1}{n}} \tag{10}$$

where ε_0^p is the reference plastic strain, n is the hardening exponent, and σ_y is the yield stress under uniaxial loading. ε^p is the equivalent strain, and is defined by:

$$\varepsilon^p = \int_0^t d\varepsilon^p = \int_0^t \sqrt{\frac{2}{3}\varepsilon_{ij}^p\varepsilon_{ij}^p}dt \tag{11}$$

3. Simulation Results and Discussion

3.1. Verification of Numerical Model

To verify the basic property of mechanics, a RVC reinforced by a 10% volume fraction of identical circle particles is generated through the developed Matlab program. The uniaxial compressive load along the *y*-axis is applied, and the strain rate of applied load in this paper is fixed to be 0.0005/s. Figure 3A shows the nominal stress–strain response of the MGMC and pure BMG model, and Figure 3B show the distribution of equivalent plastic strain in BMG corresponding to point (I–IV) in Figure 3A. It can be seen from Figure 3 that the stiffness of the MGMC is a bit higher than that of the BMG, and obviously, the MGMC has better ductile property than pure BMG. Furthermore, there are four distinguishable stages in the process of the composite deformation, which is different from that of the BMGs. In first stage from the loading start to point (I), there is only elastic deformation in both the BGM matrix and Ta particles. In the second stage from point (I) to point (II), Ta particles enter the plastic deformation, while BMG is still in an elastic state. In the third stage from point (II) to point (III), both the matrix and particles deform with plastic strain. However, the response compressive stress is kept nearly as a constant, which is different from the steep descent of a pure BMG specimen. The main shear band in BMG is blocked by soft particles because plastic deformation is "absorbed" by them. The steep stress fall occurs in the fourth stage from point (III) to point (IV). The main shear band penetrates whole RVC, and both the particles and the matrix slide along it. As a result, the MGMC loses the load capacity.

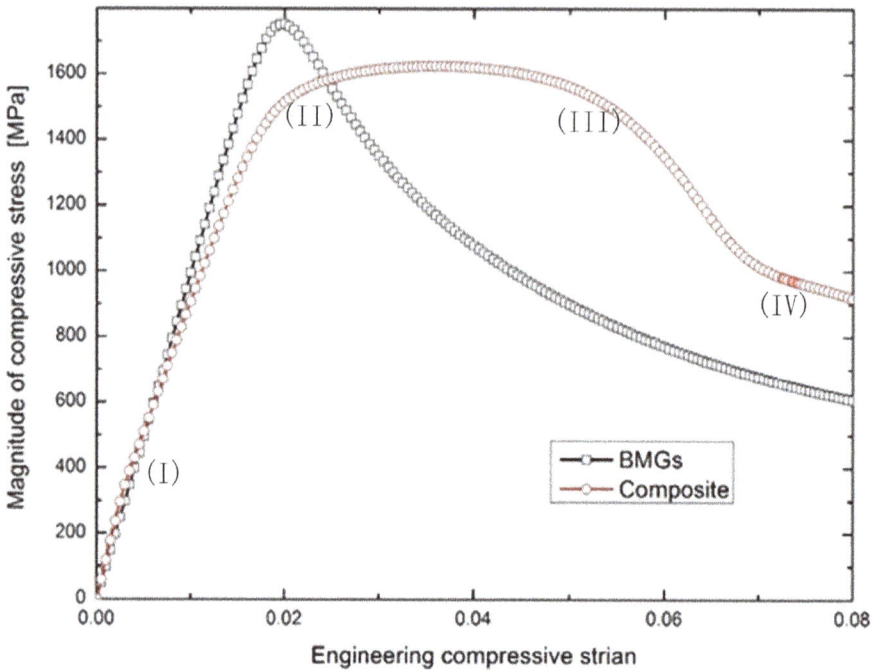

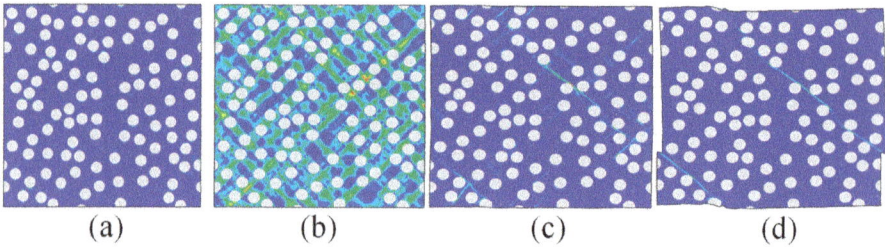

Figure 3. Simulation results of model with 100 identical circular particles under uniaxial compress loading. (**A**) Nominal stress–strain relations of bulk metallic glass composite (BMGC) and bulk metallic glass (BMG). (**B**) (**a–d**) equivalent plastic strain distributions in BMG corresponding to the positions (I–IV).

3.2. Effects of Particle Orientation

In this subsection, the influence of particle orientation varying between 0 degrees, 22.5 degrees, and 45 degrees related to the x-axis on the mechanical properties of MGMCs with a 25% particle volume fraction is studied under uniaxial compressive loading along the y-axis. Five different microstructural finite element models with a constant particle shape ratio of two are generated through the developed program for different particle orientations, respectively. It can be seen from Figure 4 that the particle orientation plays a small role on the stiffness of the MGMC, while the strength of the MGMCs increases with the increase of particle orientation. Meanwhile, when MGMCs enter the yield stages, the hardening stages increase with the increase of the particle orientation angle. Zero-degree orientation will cause biggest stress concentration among the three different orientations; thus, there is much more plastic deformation in MGMCs with zero-degree particle arrangement than those with 45 degrees, as shown in Figure 5. From Figure 5, one can also see that the particles in MGMCs block the propagation of shear bands in BMG.

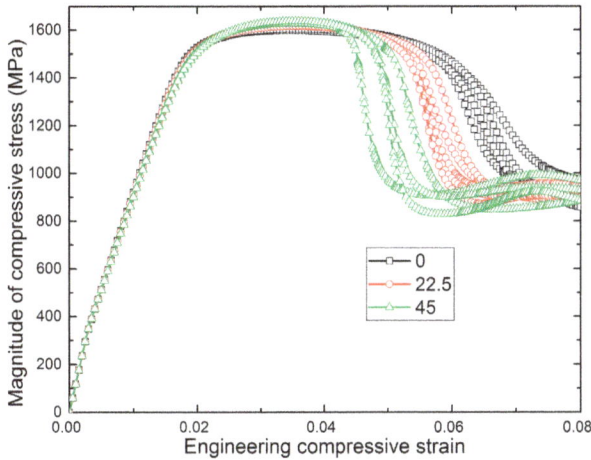

Figure 4. The influence of particle orientation on the stress-strain curves of MGMCs (metallic glass matrix composites).

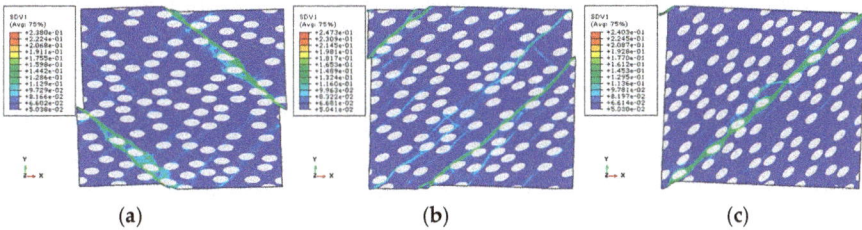

Figure 5. The influence of particle orientations on normalized free volume in BMG. (**a**) $\alpha = 0°$; (**b**) $\alpha = 22.5°$; (**c**) $\alpha = 45°$.

3.3. Effects of Particle Shape

In this subsection, the influence of particle shape varying between one (circular particle), two, and three on the mechanical properties of MGMCs with a 25% particle volume fraction is studied under uniaxial compressive loading along the y-axis. Five different microstructural finite element models containing ellipse particles with random orientation are generated through the developed program for different particle shapes, respectively. It can be seen from Figure 6 that particle shape plays a small role on the stiffness and strength of MGMCs. However, circle particles can bring more ductility than ellipse particles. The extensibility brought by particles decreases with the increasing aspect ratio of the particle. In Figure 7, it can be seen that the shear band is always generated from the tips of ellipses. Particles that have a large respect ratio may have more stress concentration and stress misfit, which will promote the generation of shear bands.

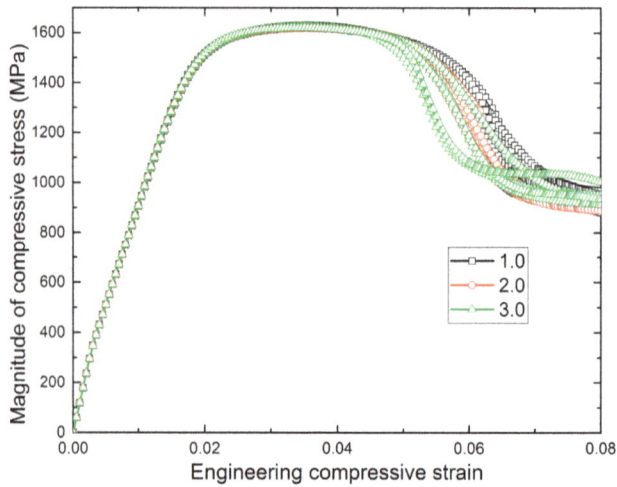

Figure 6. The influence of particle shape on the stress–strain curves of metallic glass matrix composites (MGMCs).

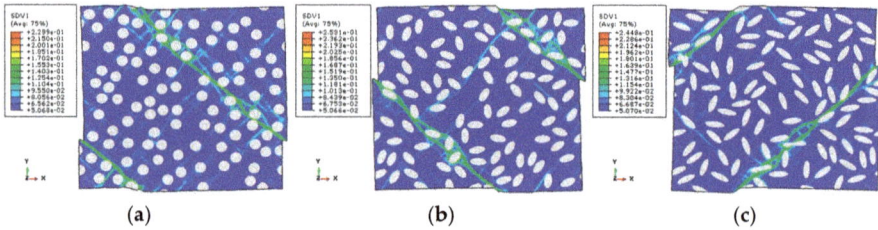

Figure 7. The influence of particle shapes on normalized free volume in BMG: (a) $r_i = 1$; (b) $r_i = 2$ and (c) $r_i = 3$.

3.4. Effects of Volume Fraction

In this subsection, the influence of particle volume fraction ranging between 15%, 25%, and 35% on the mechanical properties of MGMCs is studied under uniaxial compressive loading along the y-axis. Five different microstructural finite element models containing ellipse particles with random orientations and shapes are generated through the developed program for different particle volume fractions, respectively. It can be seen from Figure 8 that the volume fraction of the particle plays a small role on the stiffness of MGMCs, because the Ta particle and BMG have almost the same stiffness. When an MGMC containing 15% particles reaches the yield stress, the nominal stress decreases slowly for about 4% strain, and then loses its carrying capacity rapidly. In other words, the yield stress of an MGMC with 15% particles is also its strength. When an MGMC containing 35% particles reaches the yield stress, a hardening stage can be clearly seen with the increase of the loading. Shear bands intersect with each other and generate major and minor bands. The composites with 15 vol.% particles have the same shear band direction with pure BMGs due to the scale limitation of the particles in them. Nevertheless, the existing minor shear bands share plastic deformation with the major band. As a result, there will be no obvious difference between the different shear bands, which will improve the macroscopic ductility of the composites. Furthermore, with the increase of the particle volume fraction, shear bands cannot propagate in original direction because of the particle block. Ultimate shear bands are wavy, as shown in Figure 9.

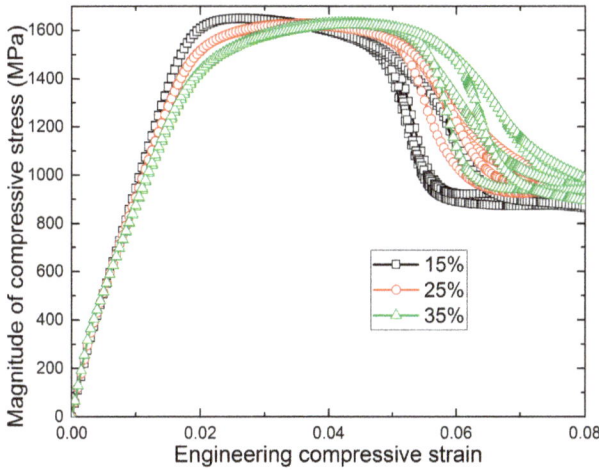

Figure 8. The influence of particle volume fractions on the stress–strain curves of MGMCs.

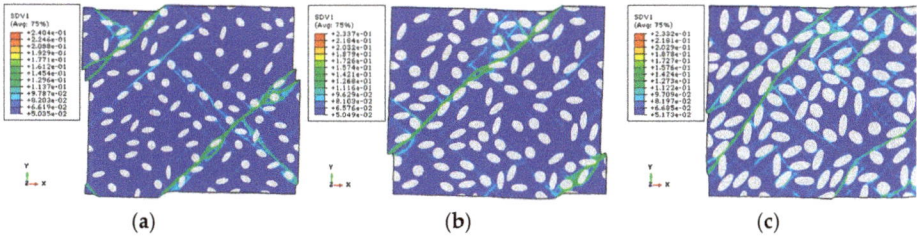

(a)   (b)   (c)

Figure 9. The influence of particle volume fractions on normalized free volume in BMG. (a) V_f = 15%; (b) V_f = 25%; (c) V_f = 35%.

3.5. Effects of Load Condition

The real shear band directions of propagation are determined by local stress. Therefore, unlike uniaxial compress, different load cases coupling compression and shear loading have been tested. The ratio between two kinds of loads is defined as $k = |\varepsilon_{12}/\varepsilon_{22}|$. Here, the MGMC models with 25 vol.% particles that were generated in the previous subsection are adopted to study the influence of loading condition on the mechanical properties of MGMCs. The equivalent stress and strain are respectively defined as:

$$\overline{\sigma} = \sqrt{\frac{(\sigma_{11}-\sigma_{22})^2+\sigma_{11}^2+\sigma_{22}^2+6\tau_{12}^2}{2}}$$

$$\overline{\varepsilon} = \frac{2}{3}\sqrt{\varepsilon_{11}^2 + \varepsilon_{22}^2 - \varepsilon_{11}\varepsilon_{22} + 3\varepsilon_{12}^2} \tag{12}$$

Figure 10 illustrates the relationship between equivalent stress and strain under different load conditions. Similarly, the load condition almost plays no role on the stress–strain relationship in the elastic stage. Nevertheless, the yield strengths decease with the increase of the shear part in the applied load. In addition, the ductility of MGMCs decreases with the increase of the shear part in the applied load. Figure 11 shows the plastic distribution of BMG in MGMCs under different load conditions. It can be seen from Figure 11 that with the decrease of the compressive part in the load, the direction of the main shear band approaches the y-axis.

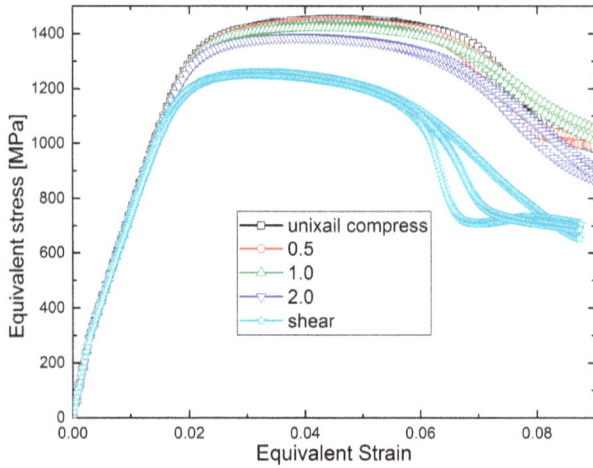

Figure 10. The influence of particle shape on the stress–strain curves of MGMCs.

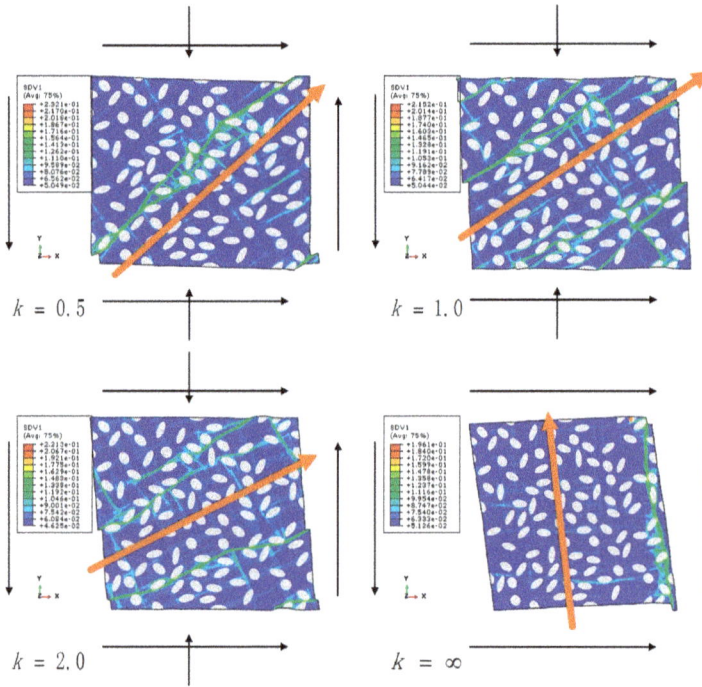

Figure 11. The influence of load conditions on the normalized free volume in BMG.

4. Conclusions

A new method and a software code are developed for the automatic generation of two-dimensional (2D) micromechanical FE models of Ta particle-reinforced metallic glass matrix composite with a random distribution of elliptic shape and orientation, as well as location arrangement. A series of computational experiments are performed to study the influence of the microstructure of MGMC on its stiffness and strength properties. Four deformation stages can be distinguished during the external load: both particles and BMG are in elastic states, particles are in a plastic-hardening state while BMG is still in an elastic state, both particles and BMG are in plastic deformation states, and the diffuse shear band in BMG emerges into a main shear band, and MGMC loses its carrying capability. The following conclusions can be obtained from the computational experiments:

- A larger angle between the load axis and the particle orientation leads to better ductile properties for MGMCs.
- The extensibility of MGMCs decreases with the increase of the respect ratio of the particle. Meanwhile, particle shapes play a small role in the ultimate strength of MGMCs.
- Particles with higher volume fraction can bring a greater improvement of ductility, but less ultimate strength.
- The yield strengths decease with the increase of the shear part in the combined tensile and shear loading.

Author Contributions: Conceptualization, H.Q. and C.-F.G.; Methodology, P.-L.B. and H.Q.; Software, P.-L.B., T.-L.L. and H.Q.; Validation, P.-L.B., T.-L.L. and H.Q.; Formal Analysis, H.Q. and C.-F.G.; Investigation, P.-L.B. and T.-L.L.; Resources, P.-L.B. and T.-L.L.; Data Curation, P.-L.B. and H.Q.; Writing-Original Draft Preparation, P.-L.B.; Writing-Review & Editing, H.Q.; Visualization, P.-L.B., T.-L.L. and C.-F.G.; Supervision, H.Q. and C.-F.G.; Project Administration, H.Q.; Funding Acquisition, H.Q.

Funding: This research was funded by the National Natural Science Foundation of China grant number 11672131, the Research Fund of State Key Laboratory of Mechanics and Control of Mechanical Structures (Nanjing University of Aeronautics and Astronautics) grant number MCMS-0217G02, and the Priority Academic Program Development of Jiangsu Higher Education Institutions and the Scientific Research Foundation for the Returned Overseas Chinese Scholars, State Education Ministry.

Conflicts of Interest: The authors declare no conflicts of interest.

References

1. Klement, W.; Willens, R.H.; Duwez, P. Non-crystalline Structure in Solidified Gold–Silicon Alloys. *Nature* **1960**, *187*, 869–870. [CrossRef]
2. Qiao, J.; Jia, H.; Liaw, P.K. Metallic glass matrix composites. *Mater. Sci. Eng. R-Rep.* **2016**, *100*, 1–69. [CrossRef]
3. Trexler, M.M.; Thadhani, N.N. Mechanical properties of bulk metallic glasses. *Prog. Mater. Sci.* **2010**, *55*, 759–839. [CrossRef]
4. Chen, M. Mechanical Behavior of Metallic Glasses: Microscopic Understanding of Strength and Ductility. *Annu. Rev. Mater. Res.* **2008**, *38*, 445–469. [CrossRef]
5. Zhang, Y.; Greer, A.L. Thickness of shear bands in metallic glasses. *Appl. Phys. Lett.* **2006**, *89*, 71907. [CrossRef]
6. Hofmann, D.C.; Johnson, W.C. Bulk metallic glass matrix composites. *Appl. Phys. Lett.* **1997**, *71*, 3808–3810.
7. Conner, R.D.; Dandliker, R.B.; Johnson, W.L. Mechanical properties of tungsten and steel fiber reinforced $Zr_{41.25}Ti_{13.75}Cu_{12.5}Ni_{10}Be_{22.5}$ metallic glass matrix composites. *Acta Mater.* **1998**, *46*, 6089–6102. [CrossRef]
8. Park, E.S.; Kim, D.H. Formation of Ca–Mg–Zn bulk glassy alloy by casting into cone-shaped copper mold. *J. Mater. Res.* **2004**, *19*, 685–688. [CrossRef]
9. Chen, M.; Inoue, A.; Fan, C.; Sakai, A.; Sakurai, T. Fracture behavior of a nanocrystallized $Zr_{65}Cu_{15}Al_{10}Pd_{10}$ metallic glass. *Appl. Phys. Lett.* **1999**, *74*, 2131–2133. [CrossRef]
10. Qiao, J.; Sun, A.C.; Huang, E.; Zhang, Y.; Liaw, P.K.; Chuang, C. Tensile deformation micromechanisms for bulk metallic glass matrix composites: From work-hardening to softening. *Acta Mater.* **2011**, *59*, 4126–4137. [CrossRef]
11. Ma, X.Z.; Ma, D.Q.; Xu, H.; Zhang, H.Y.; Ma, M.Z.; Zhang, X.Y.; Liu, R.P. Enhancing the compressive and tensile properties of Ti-based glassy matrix composites with Nb addition. *J. Non. Cryst. Solids* **2017**, *463*, 56–63. [CrossRef]
12. Shi, Y.; Falk, M.L. Strain localization and percolation of stable structure in amorphous solids. *Phys. Rev. Lett.* **2005**, *95*, 95502. [CrossRef] [PubMed]

Appl. Sci. **2018**, *8*, 2192

13. Albe, K.; Ritter, Y.; Şopu, D. Enhancing the plasticity of metallic glasses: Shear band formation, nanocomposites and nanoglasses investigated by molecular dynamics simulations. *Mech. Mater.* **2013**, *67*, 94–103. [CrossRef]

14. Zhou, H.; Qu, S.; Yang, W. An atomistic investigation of structural evolution in metallic glass matrix composites. *Int. J. Plast.* **2013**, *44*, 147–160. [CrossRef]

15. Abdeljawad, F.; Fontus, M.; Haataja, M. Ductility of bulk metallic glass composites: Microstructural effects. *Appl. Phys. Lett.* **2011**, *98*, 31909. [CrossRef]

16. Zhang, H.; Zheng, G. Simulation of shear banding in bulk metallic glass composites containing dendrite phases. *J. Alloys Compd.* **2014**, *586*, S262–S266. [CrossRef]

17. Ott, R.T.; Sansoz, F.; Molinari, J.; Almer, J.; Ramesh, K.T.; Hufnagel, T.C. Micromechanics of deformation of metallic-glass-matrix composites from in situ synchrotron strain measurements and finite element modeling. *Acta Mater.* **2005**, *53*, 1883–1893. [CrossRef]

18. Lee, J.C.; Kim, Y.; Ahn, J.P.; Kim, H.S. Enhanced plasticity in a bulk amorphous matrix composite: Macroscopic and microscopic viewpoint studies. *Acta Mater.* **2005**, *53*, 129–139. [CrossRef]

19. Biner, S.B. Ductility of bulk metallic glasses and their composites with ductile reinforcements: A numerical study. *Acta Mater.* **2006**, *54*, 139–150. [CrossRef]

20. Wu, F.F.; Zhang, Z.F.; Mao, S.X.; Eckert, J. Effect of sample size on ductility of metallic glass. *Philos. Mag. Lett.* **2009**, *89*, 178–184. [CrossRef]

21. Zhu, Z.Z.; Zhang, H.F.; Hu, Z.Q.; Zhang, W.; Inoue, A. Ta-particulate reinforced Zr-based bulk metallic glass matrix composite with tensile plasticity. *Scr. Mater.* **2010**, *62*, 278–281. [CrossRef]

22. Qiao, J.W.; Zhang, T.; Yang, F.; Liaw, P.K.; Pauly, S.; Xu, B.S. A Tensile Deformation Model for In-situ Dendrite/Metallic Glass Matrix Composites. *Sci. Rep.* **2013**, *3*, 2816. [CrossRef] [PubMed]

23. Hardin, T.J.; Homer, E.R. Microstructural factors of strain delocalization in model metallic glass matrix composites. *Acta Mater.* **2015**, *83*, 203–215. [CrossRef]

24. Jiang, Y.; Qiu, K. Computational micromechanics analysis of toughening mechanisms of particle-reinforced bulk metallic glass composites. *Mater. Des.* **2015**, *65*, 410–416. [CrossRef]

25. Jiang, Y.; Sun, L.; Wu, Q.; Qiu, K. Enhanced tensile ductility of metallic glass matrix composites with novel microstructure. *J. Non. Cryst. Solids* **2017**, *459*, 26–31. [CrossRef]

26. Jiang, Y.; Shi, X.; Qiu, K. Numerical study of shear banding evolution in bulk metallic glass composites. *Mater. Des.* **2015**, *77*, 32–40. [CrossRef]

27. Kim, I. An algorithm for finding the distance between two ellipses. *Commun. Korean Math. Soc.* **2006**, *21*, 559–567. [CrossRef]

28. Xia, Z.; Zhang, Y.; Ellyin, F. A unified periodical boundary conditions for representative volume elements of composites and applications. *Int. J. Solids Struct.* **2003**, *40*, 1907–1921. [CrossRef]

29. Qing, H. Automatic generation of 2D micromechanical finite element model of silicon–carbide/aluminum metal matrix composites: Effects of the boundary conditions. *Mater. Des.* **2013**, *44*, 446–453. [CrossRef]

30. Argon, A.S. Plastic deformation in metallic glasses. *Acta Metall.* **1979**, *27*, 47–58. [CrossRef]

31. Jiang, M.; Ling, Z.; Meng, J.; Dai, L.H. Energy dissipation in fracture of bulk metallic glasses via inherent competition between local softening and quasi-cleavage. *Philos. Mag.* **2008**, *88*, 407–426. [CrossRef]

32. Spaepen, F. A microscopic mechanism for steady state inhomogeneous flow in metallic glasses. *Acta Metall.* **1977**, *25*, 407–415. [CrossRef]

33. Gao, Y. An implicit finite element method for simulating inhomogeneous deformation and shear bands of amorphous alloys based on the free-volume model. *Model. Simul. Mater. Sci. Eng.* **2006**, *14*, 1329–1345. [CrossRef]

![applied sciences logo] *applied sciences*

MDPI

Article

Peridynamic Analysis of Rail Squats

Andris Freimanis [1] and Sakdirat Kaewunruen [2],*

[1] Institute of Transportation Engineering, Riga Technical University, Kipsalas iela 6A, Riga LV-1048, Latvia; andris.freimanis_1@rtu.lv

[2] Birmingham Centre for Railway Research and Education, University of Birmingham, Birmingham B15 2TT, UK

* Correspondence: s.kaewunruen@bham.ac.uk

Received: 18 October 2018; Accepted: 14 November 2018; Published: 19 November 2018

Featured Application: This study highlights a novel application of peridynamics to rail surface defects by improving the computational method for fracture mechanics. Rail squats is a rail surface defect that can consume over 70% of track maintenance budget. This novel application of peridynamics will enable a better preventative and predictive track maintenance strategy, enhancing public safety while saving hundreds of million euros annually.

Abstract: Rail surface defects are a serious concern for railway infrastructure managers all around the world. They lead to poor ride quality due to excess vibration and noise; in rare cases, they can result in a broken rail and a train derailment. Defects are typically classified as 'rail studs' when they initiate from the white etching layer, and 'rail squats' when they initiate from rolling contact fatigue. This paper presents a novel investigation into rail squat initiation and growth simulations using peridynamic theory. To the best of the authors' knowledge, no other comprehensive study of rail squats has been carried out using this approach. Peridynamics are well-suited for fracture problems, because, contrary to continuum mechanics, they do not use partial-differential equations. Instead, peridynamics use integral equations that are defined even when discontinuities (cracks, etc.) are present in the displacement field. In this study, a novel application of peridynamics to rail squats is verified against a finite element solution, and the obtained simulation results are compared with in situ rail squat measurements. Some new insights can be drawn from the results. The outcome exhibits that the simulated cracks initiate and grow unsymmetrically, as expected and reported in the field. Based on this new insight, it is apparent that peridynamic modelling is well-applicable to fatigue crack modeling in rails. Surprisingly, limitations to the peridynamic analysis code have also been discovered. Future work requires finding an adequate solution to the matter-interpenetration problem.

Keywords: peridynamics; fatigue; rolling contact; damage; rail squats; cracks

---

1. Introduction

Rail surface defects are a critical safety concern for railway infrastructure owners and operators all over the world. They undermine the safety and operational reliability of both moderate- and high-speed trains in passenger suburban, metro, urban, mixed-traffic, and freight rail systems. Furthermore, the cost of rail replacements due to such defects has become a significant portion of the whole track maintenance costs, especially in European countries, e.g., Austria, Germany, and France [1].

Traditionally, two different defects are classified: rail studs and rail squats [2]. Rail studs initiate from the white etching layer (WEL) due to wheel slides or excessive traction and grow horizontally 3–6 mm below the rail surface. Rail squats propagate from surface cracks initiated by rolling contact fatigue (RCF), and grow at similar depth of 3–6 mm below the rail surface. Both defects are shown in

Figure 1. As a result, the rail surface becomes depressed and passing wheels create excess vibration, noise, and impact loads. This leads to uncomfortable rides for passengers [3], and in cases where impact forces exceed acceptable limits the safety of track components can be compromised [3–7]. Rail squats and studs have been observed in all arrays of track geometries and gradients, in all types of track structures, and in all operational rail traffics. Squats are often found in tangent tracks, in high rails of moderate-radius curves, and in turnouts with vertical, unground rails. Due to the high potential damage caused by rail squats and studs, several research and development projects have been initiated around the world to investigate the causes of, and feasible solutions to, these defects.

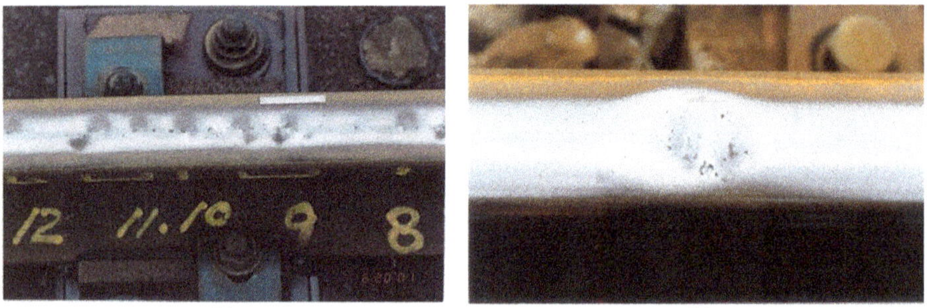

(a)   (b)

Figure 1. Rail surface defects: (a) white etching layer (WEL)-related rail studs (multiple studs); (b) a rolling contact fatigue (RCF)-related rail squat (single squat).

Computational rail squat and stud modeling has been the topic of several studies. A finite-element (FE) analysis with a two-dimensional (2D) elastic-plastic model under the assumption of plane strain was used to investigate crack growth from the WEL in [8]. The researchers found that the crack growth direction in the interface between the base material and the WEL is determined by the discontinuity of a material rather than the stress state, and that cracks tend to grow along the interface between the WEL and the rail material, because it is comparatively hard for a crack to propagate into the rail material. Field observations and a numerical analysis in [9] showed that squats initiate as a result of differential wear and differential plastic deformation. Numerical simulations in [10] have also shown that the growth of squats is related to some eigenmodes of the wheel–track interaction system and the high-frequency vibration at wheel–rail contact plays an important role. The probability of rail squat initiation from surface defects based on a transient stress analysis was studied in [11] using an FE model of the vehicle–track interaction. The results showed that when a defect is smaller than 6 mm, its chance to grow into a squat is very small, and when it is larger than 8 mm and in the middle of the running band, the chance is large. RCF occurring on Chinese high-speed rails and wheels was investigated in [12]. Based on field observations and a numerical simulation, it was concluded that indentations seem to be the main cause of RCF. If relatively small but deep indentations exist, then peak von Mises stress can occur both on the surface and at the bottom of the crack, but stress at the bottom is likelier to create RCF cracks [13,14].

The development of rail squats is most commonly studied using the finite-element method, which is based on the classical continuum mechanics theory. It uses spatial derivatives, which do not exist when the displacement field is discontinuous, i.e., when cracks are present. So, as a remedy, techniques of fracture mechanics must be used; however, their major drawback is that the crack path must be known a priori. Due to such limitations, independent crack branching is difficult to implement.

Peridynamic (PD) theory [15,16] was created as an alternative to continuum mechanics for problems with cracks, voids, and other discontinuities. PD uses integral, not partial-differential, equations and deformation instead of strain to compute internal forces. Since integral equations are defined even when the displacement field is discontinuous, this theory is well-suited for fracture studies. Contrary to FE analysis, in PD cracks initiate automatically and grow according to some prescribed damage law. The crack path does not have to be set at the beginning of a simulation, resulting in a more natural crack growth with branching. This theory has a large potential in fracture problems, and has been used to study damage in fiber-reinforced laminated composites [17–19], glass [20,21], wood [22], concrete [23–25], and steel [26].

In this study, rail squats are simulated using ordinary state-based peridynamic theory (PD). This technique is the recent fundamental development from the original bond-based PD theory. The state-based PD has made a significant advancement in capturing sufficient behaviors of real materials. To the best of the authors' knowledge, and based on a critical review of the open literature, this paper is the first to present a comprehensive study of rail squat initiation and growth using peridynamic theory. We have presented some initial findings in [13], but this paper describes the full development of the model, the calibration of its parameters, and an application of coordinate-variable loads. Simulation results are compared to field measurements from [14]. The insight from this study is novel and can help further improve the technique for applications of the new theory of peridynamics to real-world problems, and help to enhance better prognostics of rail squats. Overall, the insight enhances an alternative computational method for fracture mechanics.

2. Methods

2.1. State-Based Peridynamic Theory

A brief overview of state-based peridynamic theory is presented in the following paragraphs. An extended overview can be found in [27–29]. A peridynamic body consists of some number of nodes each uniquely described by its volume V_i, density ρ_i, and position vector in the reference configuration x_i. An example of a 2D body is shown in Figure 2. Node x_i interacts with other nodes x_j through bonds (relative position vectors) $\xi_{ij} = x_j - x_i$. These interactions are limited to a range called the horizon δ. Nodes x_j that are connected to x_i are called the family of x_i, H_{x_i}. When a body deforms, node x_i experiences displacement u_i and moves to its deformed position $y_i = x_i + u_i$. The bond in the deformed configuration is $y_j - y_i$. This deformation creates a bond force density vector t_{ij} that depends on the collective deformation of all nodes in H_{x_i} and an opposite bond force density vector t_{ji} that depends on the collective deformation of H_{x_j}. Bond forces are force densities (force per volume), not stresses (force per area), because each node describes some volume. The bond deformation vectors are stored in an array called the deformation state

$$Y_{x_i} = \left\{ \begin{array}{c} y_1 - y_i \\ \vdots \\ y_n - y_i \end{array} \right\}, \tag{1}$$

similarly, the force density vectors are stored in an array called the force state

$$T_{x_i} = \left\{ \begin{array}{c} t_{i1} \\ \vdots \\ t_{in} \end{array} \right\}. \tag{2}$$

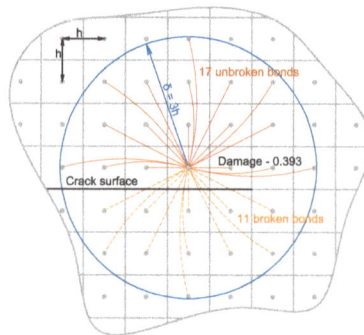

Figure 2. The most conservative two-dimensional (2D) case when it could be thought that a crack has appeared. Some bonds are drawn curved to avoid overlapping.

The bond force density vectors are computed using bond deformations:

$$T(x_i) = T(Y(x_i)),\tag{3}$$

where the function $T(x_i)$ is a material model. It is common to state $T(x_i)\langle x_j - x_i\rangle$ or $T(x_i)\langle \xi_{ij}\rangle$ and when referring to the force density vector t_{ij} in a bond $\xi_{ij} = x_j - x_i$, and similarly for deformation state and deformed bond vectors. The peridynamic equation of motion in the integral form is

$$\rho(x_i)\ddot{u}(x_i,\ t) = \int_{H_{x_i}} \left(T(x_i)\langle x_j - x_i\rangle - T(x_j)\langle x_i - x_j\rangle\right)dV_{x_j} + b(x_i)\tag{4}$$

where $\rho(x_i)$ is the density, $\ddot{u}(x_i,\ t)$ is the acceleration, and $b(x_i)$ is the external force density.

The contribution of a bond to the force density at a node can be weighed using an influence function $\omega(x_i)$. They have been introduced in [16], and their role is further explored in [30]. The value of an influence function can depend on the length, direction, or other bond properties. It can also be used to introduce damage; remove the interaction between two nodes by setting the influence function to 0, i.e., break the bond, when some damage criterion is reached. The simplest damage criterion could be the critical stretch, in which a bond breaks when it is stretched past some critical value s_c:

$$\omega(x_i) = \begin{cases} 1, & \text{if } s_{ij} < s_c \\ 0, & \text{if } s_{ij} \geq s_c \end{cases}, s_{ij} = \frac{\left|y_j - y_i\right| - \left|x_j - x_i\right|}{\left|x_j - x_i\right|} = \frac{\left|Y\langle\xi_{ij}\rangle\right| - \left|\xi_{ij}\right|}{\left|\xi_{ij}\right|},\tag{5}$$

where s_{ij} is the bond stretch. Then, the damage at a node can be defined as a ratio between the broken and the initial number of bonds [31]:

$$\phi(x_i) = 1 - \frac{\int_{H_{x_i}} \omega(x_i)dV_{x_j}}{\int_{H_{x_i}} dV_{x_j}}.\tag{6}$$

The PD fatigue damage model used in this study was introduced in [32] and used in [33–35]. Other researchers have also developed fatigue damage models [36,37]; however, these models use bond-based peridynamic theory and simulate only the crack growth phase. A small overview of the model is given here for completeness; Equations (7) through (11) were first presented in [32].

A body undergoes some cyclic deformation between two extremes + and −, then bond strains at each extreme are s_{ij}^+, s_{ij}^- and the cyclic bond strain ε_{ij} is:

$$s_{ij}^+ = \frac{\left|\mathbf{Y}^+\langle\boldsymbol{\xi}_{ij}\rangle\right| - \left|\boldsymbol{\xi}_{ij}\right|}{\left|\boldsymbol{\xi}_{ij}\right|}, s_{ij}^- = \frac{\left|\mathbf{Y}^-\langle\boldsymbol{\xi}_{ij}\rangle\right| - \left|\boldsymbol{\xi}_{ij}\right|}{\left|\boldsymbol{\xi}_{ij}\right|}, \varepsilon_{ij} = \left|s_{ij}^+ - s_{ij}^-\right|. \tag{7}$$

For each bond, a variable called the "remaining life" $\lambda_{ij}(x_i, \boldsymbol{\xi}_{ij}, N)$ is defined. It degrades at each loading cycle N, and a bond breaks when the remaining life is reduced to zero:

$$\lambda_{ij}(N) \leq 0. \tag{8}$$

At the beginning, when $N = 0$:

$$\lambda_{ij}(0) = 1, \tag{9}$$

at each cycle in the crack nucleation phase (phase I), the change of λ is given by

$$\frac{d\lambda_{ij}}{dN}(N) = \begin{cases} -A_I(\varepsilon_{ij} - \varepsilon_\infty)^{m_I} & , \; if \; \varepsilon_{ij} > \varepsilon_\infty \\ 0 & , \; if \; \varepsilon_{ij} \leq \varepsilon_\infty \end{cases}, \tag{10}$$

where ε_∞ is the fatigue limit under which no fatigue damage occurs, and A_I, m_I are parameters for phase I. In phase II, the remaining life changes according to:

$$\frac{d\lambda_{ij}}{dN}(N) = -A_{II}\varepsilon_{ij}^{m_{II}}, \tag{11}$$

where A_{II}, m_{II} are parameters for phase II.

The transition from phase I to phase II is handled by applying the phase I model to bonds connected to x_i until there is a node x_j in H_{x_i} with damage

$$\phi(x_j) \geq \phi_c, \tag{12}$$

where ϕ_c is the damage at which phase II begins. Then, reset the remaining life of bonds connected to x_i to 1 and switch to the phase II model.

2.2. Computational Model

The fatigue damage model was implemented in the open-source PD program Peridigm [38,39]. If the quasi-static analysis acceleration term in (4) is zero, then the peridynamic equation of motion in the discreet form is approximated as:

$$\sum_{H_{x_i}} (T[x_i, t]\langle x_j - x_i\rangle - T[x_j, t]\langle x_i - x_j\rangle)\Delta V_{x_j} + b(x_i, t) = 0. \tag{13}$$

Two techniques introduced in [32], including implicit strain simulation and time mapping, have been used to speed up the simulations. They are illustrated in Equations (14) through (17). In the case of high-cycle fatigue, the bond strains are below the elastic limit, so an elastic material model can be used. In such cases, the strain in a bond would change linearly between + and − loading conditions, so it is possible to simulate only the + loading condition and compute

$$s^- = Rs^+, R = \frac{P^-}{P^+}, \tag{14}$$

where R is the loading ratio, and P is the applied load at each extreme. Then, the cyclic strain is given by:

$$\varepsilon = \left|s^+ - s^-\right| = \left|(1 - R)s^+\right|. \tag{15}$$

The loading ratio $R = 0$ was used for all simulations. Using linear time mapping ([32] also introduces exponential time mapping, but it was not used in this study), the simulation time t relates to the current cycle through

$$N = \frac{t}{\tau}, \tag{16}$$

where τ is a constant. Then, the remaining life at step n in a bond ξ_{ij} is given by

$$\begin{array}{l} \frac{\lambda^n - \lambda^{n-1}}{\Delta t} = -A \left(\varepsilon_{ij}^n \right)^m \\ \frac{\Delta \lambda}{\Delta t} = \frac{\Delta \lambda}{\Delta N} \frac{\Delta N}{\Delta t} \end{array} \rightarrow \lambda^n = \lambda^{n-1} - \frac{t^n - t^{n-1}}{\tau} A \left(\varepsilon_{ij}^n \right)^m. \tag{17}$$

In PD, unlike in a fracture mechanics model with a pre-crack, it is almost impossible to develop a sharp crack surface; instead, a damaged zone is developed. For example, if damage at a node is 0.5, it could mean that 25% of the bonds on the opposite sides of a node are broken, and it could also mean that 15% are broken on one side and 35% on the other.

Crack growth in phase II is faster than that in phase I, so switching to phase II sooner would lead to faster crack growth and more conservative results. In this study, the emphasis is placed on the consideration of the most conservative case, i.e., switching to phase II at the lowest damage when it can be thought that a crack has appeared. Such a situation would happen when all of the bonds on one side of a node are broken, but other bonds remain intact. The 2D case, if the horizon is 3 times the distance between the nodes, is shown in Figure 2. An equivalent case in three dimensions (3D) would have 47 broken bonds and 75 unbroken bonds, i.e., damage of 0.385. Therefore, the fatigue model transitions from phase I to phase II when the damage at a node reaches $\phi_c = 0.385$.

2.3. Model of a Rail

Initially, a model of a whole UIC60 rail head had been developed. However, due to the required fine discretization, computational resources, and time constraints, only a part of the rail head has been modeled. Mesh was firstly created in the Ansys FE program using 3D eight-node solid elements. Afterwards, element centroid coordinates and their volumes were exported and converted to Peridigm's mesh file, and both models are shown in Figure 3. The dimensions of the model were $0.03 \times 0.024 \times 0.03$ m with a node size of $h = 0.0005$ m and the horizon of $\delta = 0.0015$ m. The load area, see Section 2.5. *Boundary conditions*, was about 0.013×0.013 m, which means that the distance between the edge of the model and the load area was about 8δ. A cartesian left-handed coordinate system was used, and the center of the model's bottom face was located at the origin. The top face was made of $R = 300$ mm and $R = 80$ mm arcs, as in the specifications of the UIC60 rail.

Since the top surface was curved and mapped meshing was used, the nodes were not perfectly cubic. However, the difference between the average node volume and the volume of a cubic node is only 1.17% (see Table 1). This is relevant when converting the applied loads from stress to force density, but since the difference is small, the impact is negligible.

A Linear Peridynamic Solid (LPS) [16] material model was used. The material properties were: density 7850 kg/m^3, Poisson's ratio 0.3, and Young's modulus 189.9 GPa, obtained from [40]. The LPS model is the peridynamic equivalent to the elastic material model in continuum mechanics. It has been selected because the applied loads do not cause the material to exceed its yield strength.

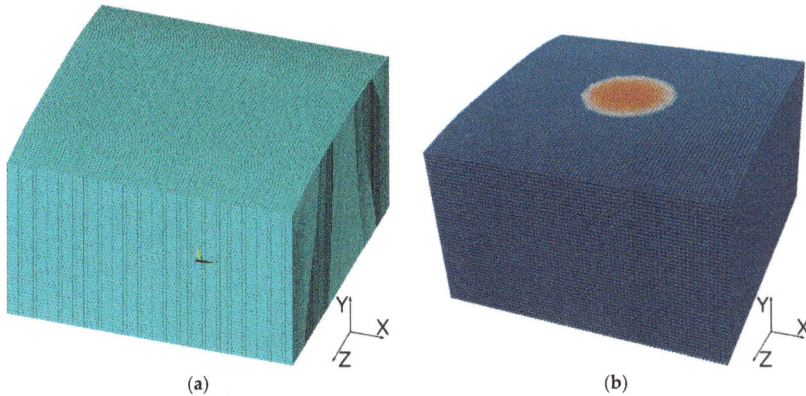

Figure 3. A model of a rail: (**a**) an Ansys model with solid elements; (**b**) peridynamic (PD) mesh-free discretization with the load area highlighted.

Table 1. The cubic versus the smallest, largest, and average node volume.

Parameter	Volume, m^3	% Difference
Cubic	1.25000×10^{-10}	0.00%
Min	1.18960×10^{-10}	−4.83%
Max	1.29750×10^{-10}	3.80%
Average	1.26464×10^{-10}	1.17%

To verify the peridynamic model, the displacement in the X and Y directions in an undamaged state has been compared against an FE model. The same model was used for verification of the Peridigm's mesh creation. Movements are restricted in all directions for all nodes within 1δ from the bottom. Two loadings—vertical pressure and surface shear traction—are applied to the load area at the top of the rail head. Both loadings are applied as functions in terms of node x and z coordinates; for the exact functions, please see Section 2.5 *Boundary conditions*. In the FE model, the pressure has been applied as distributed pressure on the top face of solid elements in the load area, and traction has been applied as horizontal forces acting on the top three layers of nodes within the load area.

Figure 4 shows the X and Y displacement in the cross-section along the centerline of the rail, and Table 2 presents the maximum and minimum displacement values. It is clear that a very good agreement between models can be found. In fact, if the difference in extreme displacement values would be within ±10%, the displacements would be similar in the cross-sections. Figure 4 definitely shows a similar displacement distribution between the FE and PD models. The maximum displacement values between the FE and PD models are −6.95% and 7.92% for the X and Y directions, respectively. The difference in the minimum Y displacement is 7.48%. In the X displacement, however, it is 20.61%, which is more than what should be considered a good agreement. Though the relative difference in the X displacement is large, it should not negatively influence the simulation results, because the region with low X displacement is far from the load area, and, therefore, far from the area of interest with the growing cracks. Additionally, the absolute difference is very small, i.e., only 1.71×10^{-7} m. It should have no effect on the PD model's accuracy.

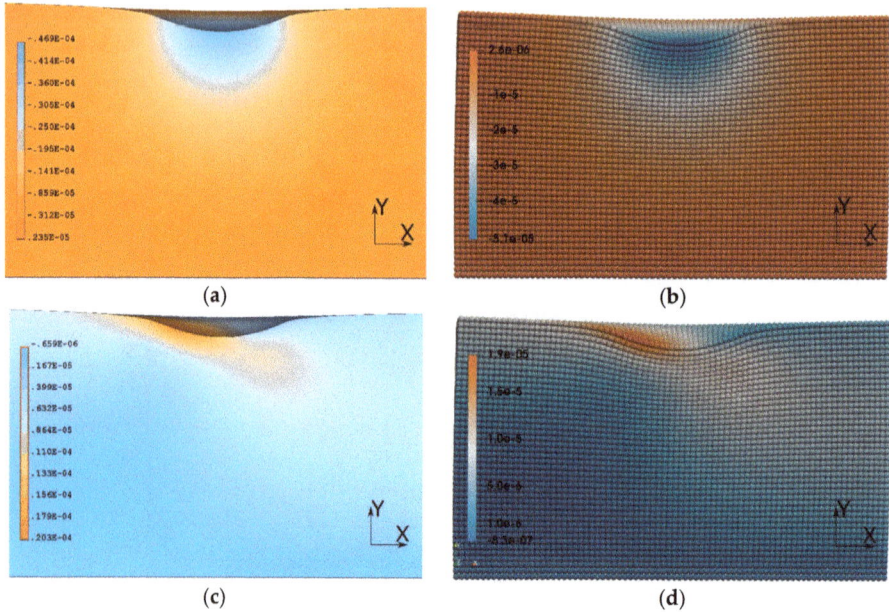

Figure 4. The displacement in the cross-section of an undamaged model: (a) the Y displacement finite element (FE) model; (b) the Y displacement PD model; (c) the X displacement FE model; (d) the X displacement PD model. Deformations are increased 50 times.

Table 2. Maximum and minimum displacement values in the *X* and *Y* directions from the finite-element (FE) and Peridynamic (PD) simulations.

Value	X			Y		
	FE, m	PD, m	Difference	FE, m	PD, m	Difference
Max	2.03×10^{-5}	1.90×10^{-5}	-6.95%	2.35×10^{-6}	2.55×10^{-6}	7.92%
Min	-6.59×10^{-7}	-8.30×10^{-7}	20.61%	-4.69×10^{-5}	-5.07×10^{-5}	7.48%

2.4. Fatigue Damage Model Parameters

This study adopts the rail steel data from Figures 4 and 5 in [40], and follows the procedure to obtain damage model parameters in [32]. Although [40] presents quite old data, it contains ε-N (strain-life), K-da/dN (Paris law), and material properties data. This is beneficial, because it assures that the data are for the same material. Other fatigue data sources have been considered [41–46], but either have only the S-N curves available, contain less data points, or do not have both ε-N and K-da/dN plots. The fatigue damage model parameters are shown in Table 3.

Table 3. Fatigue damage model parameters.

	Phase I	Phase II
A	426.00	25,237.48
m	2.77	4.00
ε_∞	0.00186	–

The parameters for phase I (A_I, m_I, ε_∞) are found from functions fitted to the ε-N plot (see Figure 5). The fitted power law function takes the same form, $y = ax^b$, as the phase I damage model in (10). So, parameter m_I is the inverse of slope b:

$$b = -\frac{1}{m_I} \tag{18}$$

and parameter A_I is calculated from the value of intercept:

$$a = \frac{-\log A_I}{m_I} \Rightarrow A_I = \frac{1}{a^{m_I}}. \tag{19}$$

The fatigue limit of rail steel was determined from the function:

$$\left(\frac{\Delta\varepsilon}{2} - \varepsilon_\infty\right)^{\xi} N = C, \tag{20}$$

where $\frac{\Delta\varepsilon}{2}$ is the strain amplitude, ε_∞ is the fatigue limit, N is the number of cycles, and ξ, C are constants.

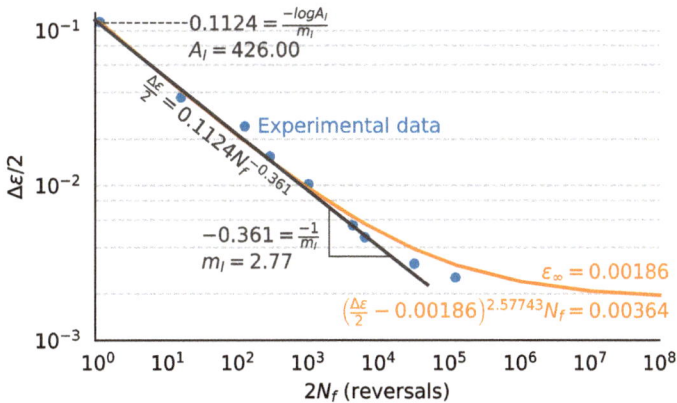

Figure 5. The ε-N data, fitted functions, and damage model parameters for phase I.

A Paris law plot is required to find the parameters for phase II. In this study, $R = 0.05$, and moist air data from Figure 5 in [40] were used. The plot is replicated in Figure 6. The fatigue damage model in (11) has the same form as the Paris law for crack growth:

$$\frac{da}{dN} = c\Delta K^M, \tag{21}$$

where $\frac{da}{dN}$ is the crack growth speed, c, M are constants, and ΔK is the cyclic stress intensity factor. ΔK is proportional to the bond strain at the crack tip (in [32] called ε_{core}); therefore, $m_{II} = M$, so this parameter can be obtained directly from a Paris law plot. The remaining parameter A_{II}, however, cannot. Instead, a simulation with some trial value A'_{II} has to be run to obtain the trial crack growth speed $\left(\frac{da}{dN}\right)'$. Then, the real A_{II} value can be found from [32]:

$$A_{II} = A'_{II} \frac{\frac{da}{dN}}{\left(\frac{da}{dN}\right)'}. \tag{22}$$

To find A_{II}, a single edge notch (SEN) specimen with a pre-crack in uniaxial tension is simulated. The stress intensity at a crack tip is given by:

$$K_I = \sigma\sqrt{\pi a}F\left(\frac{a}{b}\right),$$

(23)

$$F\left(\frac{a}{b}\right) = 1.122 - 0.231\left(\frac{a}{b}\right) + 10.550\left(\frac{a}{b}\right)^2 - 21.710\left(\frac{a}{b}\right)^3 + 30.382\left(\frac{a}{b}\right)^4,$$

(24)

where σ is the applied stress, a is the crack length, and b is the specimen width. The crack tip's location was defined as the maximum x coordinate at which all nodes through the depth of the model have damage of at least 0.385.

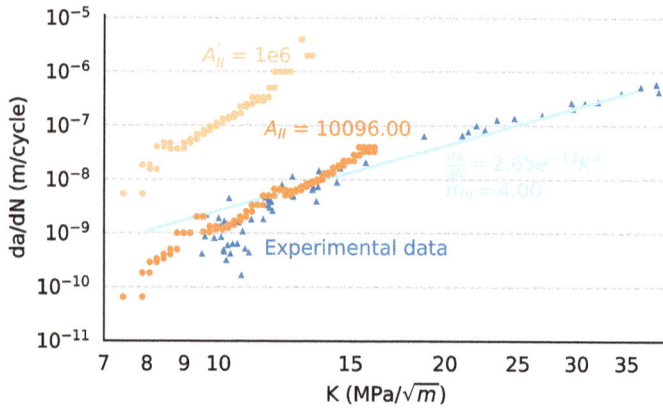

Figure 6. The experimental and simulated crack speed growth data and phase II parameters.

The model's size is 0.05 × 0.008 × 0.003 m. It has been discretized with 150,000 nodes with a spacing of 0.0002 m using the mesh-free method described in [31]. The horizon is set to a little over 3 times the nodal spacing: 0.0006001. The model has a 0.005-m-long pre-crack on the left side to ensure that Equation (23) is applicable. A force density of 6.25 × 10^10 N/m³ (equivalent to 50 MPa) has been applied to nodes within one δ of both the top and bottom, and damage is disabled for nodes within 3δ from the top and bottom, to avoid unphysical behavior near the boundary conditions. Crack growth speed data only from phase II are required, so switching to phase II at low damage reduces the simulation time. The damage required for transition from phase I to phase II has been, therefore, set to 0.017. For the trial simulation, A'_{II} = 1e6 and m_{II} = 4.00. An LPS material model with the same parameters as for the rail head simulation is used. The first simulation (with A'_{II}) ran for 163,100 cycles, after which the crack turned upward, so Equation (23) is no longer accurate; the second simulation runs for 13,275,999 cycles until the crack splits in two. Figure 7 shows the simulation with A_{II} at cycle 309,999 (top) and step 13,275,999 (bottom). The number of cycles is large because a low applied stress causes fatigue damage to increase slowly.

Since a crack grows in discrete jumps between nodes, the crack growth speed between two cycles m and n with such jumps has been calculated as the difference in crack length divided by the difference between the current cycle and the cycle at which the previous jump occurred:

$$\left(\frac{da}{dN}\right)' = \frac{a_n - a_m}{N_n - N_m},$$

(25)

where a is the crack length, and N is the number of cycles. Then, the $\left(\frac{da}{dN}\right)'$ values are interpolated to match the ΔK values from the experimental data and A_{II} is calculated using Equation (22). In total, 22 A_{II} values have been calculated. These values vary greatly, and the coefficient of variation is 0.7134;

therefore, an average value has been used. A repeated simulation with A_{II} and not A'_{II} (see Figure 6) exhibits a very good agreement with the first part of the experimental data. It was not possible to determine agreement with the latter part of the data, because the simulated crack splits into two and Equation (18) could no longer be used. A better approach (with less variance between the calculated A_{II} values) might be to use the real crack growth rate $\frac{da}{dN}$ not from experimental data, but from the fitted Paris Law function. This approach will be explored in future research.

The model of a rail head uses coarser discretization than the model of an SEN specimen. Since the horizon has been kept at three node spacings for both, the actual value of the horizon is different in both simulations: 0.0015001 and 0.0006001, respectively. A change in horizon does not change the A_I, m_I, m_{II} parameters (see chapter 4.3 in [32] for details), but A_{II} has to be scaled with the horizon. Equation (29) in [32] provides the means to do that:

$$A_{II}(\delta) = \hat{A}_{II}\delta^{\frac{m_{II}-2}{2}} \rightarrow A_{II}(\delta) = 16,823,863 \times 0.0015001^{\frac{4-2}{2}} = 25,237.48, \tag{26}$$

where $\hat{A}_{II}$ is independent of δ.

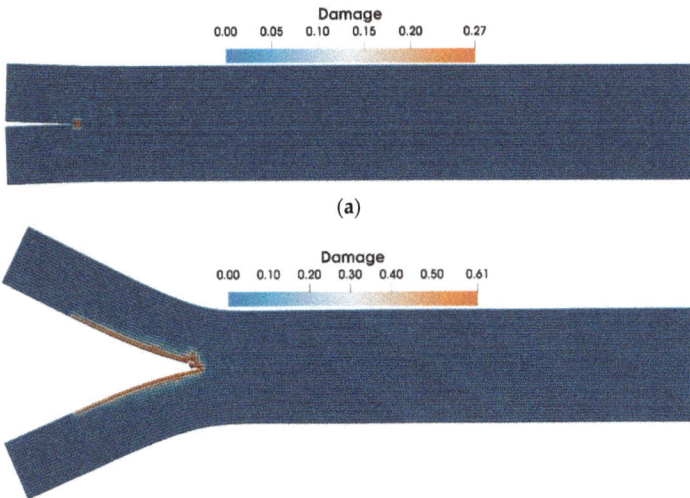

(a)

Figure 7. A single edge notch (SEN) specimen at: (**a**) 3,099,999 cycles; (**b**) 13,275,999 cycles. Displacements are increased 10 times.

2.5. Boundary Conditions

Since nodes describe some volume, boundary conditions (BCs) must also be applied to some volumes. A BC layer thickness equal to the horizon was recommended in [47]. In [13], the researchers similarly applied loads only to a single layer of nodes on top of the rail head, and such an approach lead to poor results. BC nodes separated from the rest just after 26 thousand cycles, due to the low number of bonds over which the applied loads were distributed.

Displacement in all directions was fixed for nodes within one δ from the bottom. Additionally, damage was disabled for nodes within 3δ from the bottom to avoid the concentration of unphysical damage near the BC layer.

Train wheel load data from [48] have been used in this study. The wheel–rail contact area (Figure 4f in [48]) is centered at the coordinate origin, see Figure 3b, and approximated with an ellipse with a half-axis $a = 0.0066$ m, $c = 0.006386$ m. Two different train wheel loadings are used: vertical pressure and surface shear traction. They are applied to a 1δ thick layer on top of the rail head.

The vertical force density, in N/m³, from the elastic pressure (data from Figure 5f in [48]) can be computed from a modified ellipsoid's formula:

$$p = \frac{1.116 \times 10^9}{h} \sqrt{1 - \frac{x^2}{0.0066^2} - \frac{z^2}{0.006386^2}},$$

(27)

where p is the force density (N/m³), x, z are node coordinates (m), and h is the node size (m). Since loads are applied to a 1δ (three node spacings) thick layer, the computed value at a position (x, z) has been divided by 3 and applied to each of three nodes under this position.

Shear traction forces are taken from Figures 5 and 6 in [48], where they are given as a stress distribution over an area. Mesh-free discretization requires that loads are applied to discrete nodes, so the shear traction data over the whole load area had to be described by some function from which an exact value at a node could be calculated. Only half of the load area is considered, because the traction data were symmetric. Half of the load area is divided into four parts along the z axis, see Figure 8, and the shear traction values in each part are described by a tri-linear function, see Figures 8 and 9. Stress values from [48] can be plotted with symbols in Figure 8 and fitted with tri-linear functions from which the exact shear traction force value at each node could be calculated. Since loads are applied to a three-node-thick layer, the calculated stress values must be converted to force density and divided by 3 before being applied to nodes.

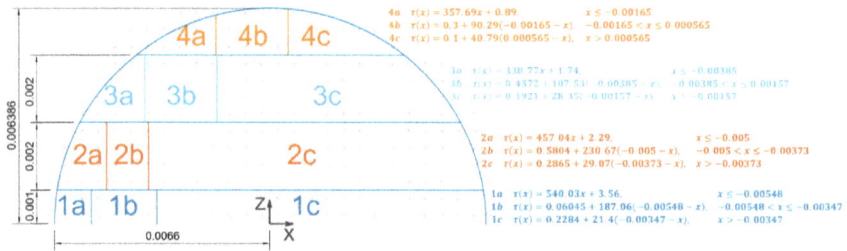

Figure 8. Half of the load area divided into four parts with a tri-linear function for each part. Functions describe the shear traction stress values in the load area. The other half of the load area is a mirror image. The axis directions and node size are the same as in the rail head's model.

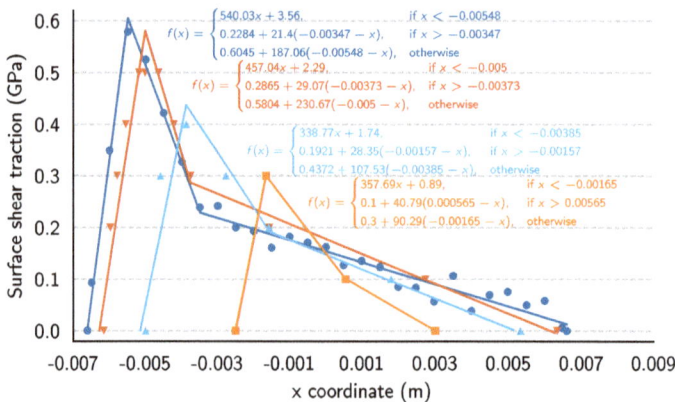

Figure 9. Surface shear traction data from [48] (shown with symbols) and the tri-linear functions used to describe the shear traction values in the load area.

3. Results

This problem has been simulated on a computing cluster at Riga Technical University using 4 × 36 cores. Each simulation was run for 42,884 cycles, after which the solver failed to converge. The results are shown in Figure 10 (cross-section in the longitudinal direction) and Figure 11 (cross-section in the transversal direction).

The simulation results are compared with the rail squat field measurements in [14]. Cracks have been measured at specified grid points on a tangent rail over a span of 3 years using a handheld ultrasonic testing device with the accuracy range of ±0.1 mm. The field measurement results can be seen in Figures 12 and 13.

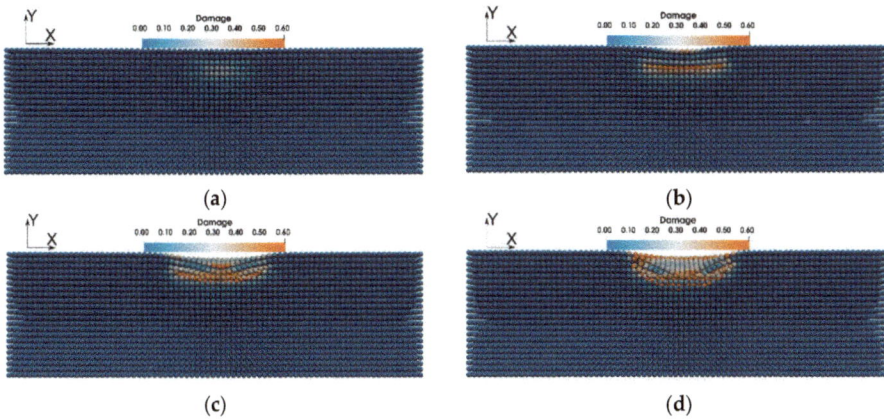

Figure 10. The cross-section (x-y plane) along the middle of the rail head in the longitudinal direction. Damage is shown in the top part of the model after: (**a**) 37,000; (**b**) 42,500; (**c**) 42,850; and (**d**) 42,884 cycles.

Two loadings—pressure and shear traction—have been applied to the model. The magnitude of the shear traction is not symmetric around the coordinate origin, even although the load area is. Traction is then applied in the positive x direction, and the values change as shown in Figures 8 and 9. This can cause damage to develop slightly asymmetrically against the y-z plane, which is best seen in Figure 10a. Damage develops faster on the positive side of the x axis (the right side in Figure 10). Against the x-z plane, in Figure 11, damage developed symmetrically, because both the pressure and the shear traction are symmetric. The same asymmetric crack growth has been observed in the field measurements (see Figure 13). In reality, such asymmetry happens because the shear traction from a wheel rolling forward is applied in the rolling direction.

Figure 11a clearly shows that damage first develops close to the location of maximum pressure (the middle of the rail in transversal direction). In addition, the maximum damage remains under the same area (see Figure 11b–d). This is consistent with the field measurements shown in Figure 12. Cracks are deeper closer to the center of the rail head and shallower closer to the sides. This shows that they first initiated and have been growing for longer under the rolling surface.

The simulation ended unexpectedly quickly, because the fatigue resistance of a rail without any defects should definitely be above 42,884 cycles. Loads have thus been applied to a three node (one horizon) thick layer on top of the rail head. As bonds extending to nodes below this layer are broken, the applied loads are no longer transferred downwards and the loaded nodes simply moved through the layers below them; this can be seen in the c and d parts of Figures 10 and 11. This is the so-called "matter-interpenetration" problem, and usually it is solved through different contact models. The simplest model—short-range force—was introduced in [31], and a better description is available in [39,49]. Other contact models and properties are presented in [50–53].

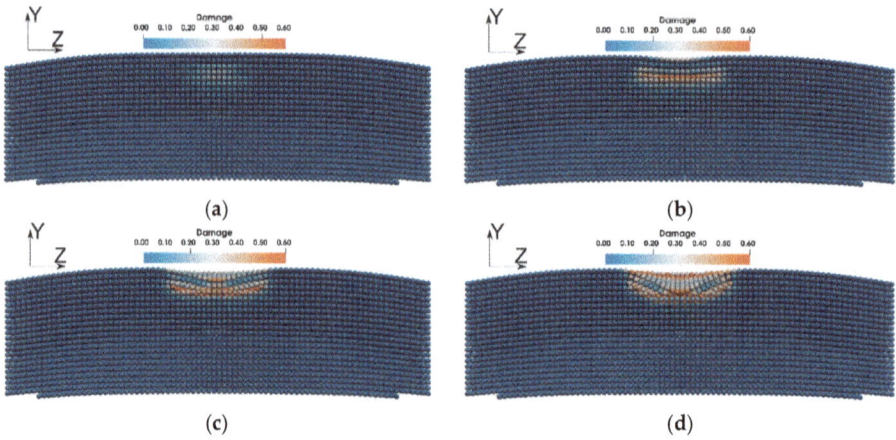

Figure 11. The cross-section (x-z plane) along the middle of the rail head in the transversal direction. Damage is shown in the top half of the model after: (**a**) 37,000; (**b**) 42,500; (**c**) 42,850; and (**d**) 42,884 cycles.

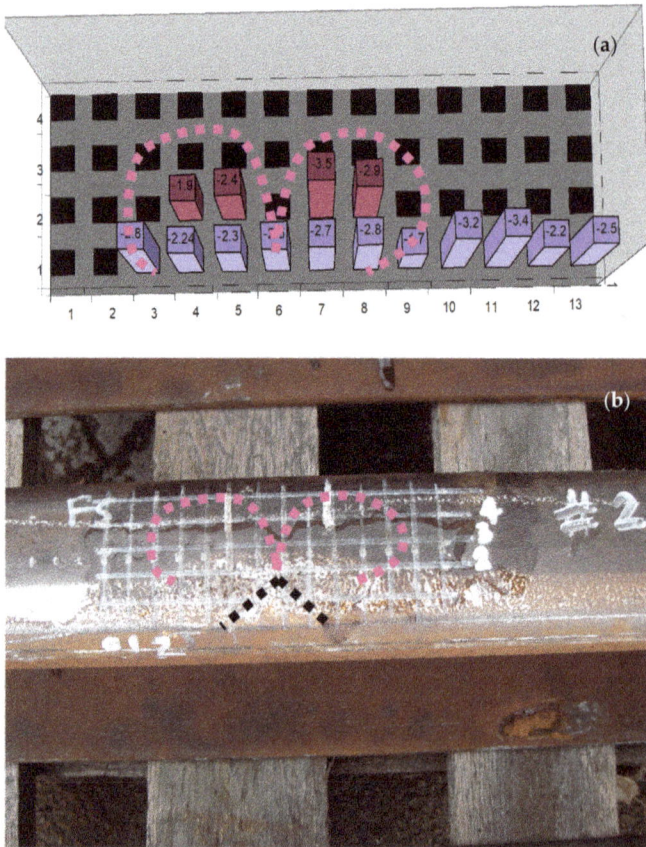

Figure 12. An ultrasonic rail squat measurement: (**a**) crack depths at each grid point; (**b**) top view of the rail surface.

While a shot-range force contact mode is available in Peridigm, it has been implemented only for explicit and not quasi-static simulations and it does not consider contact between nodes that are bonded initially. As damage develops, it is important that the contact model reconsiders the contact between two nodes that were bonded initially but are not anymore. This limitation has been explored, and it is possible to resolve it. Future work will concentrate on how to efficiently pass data between Peridigm's damage and contact models.

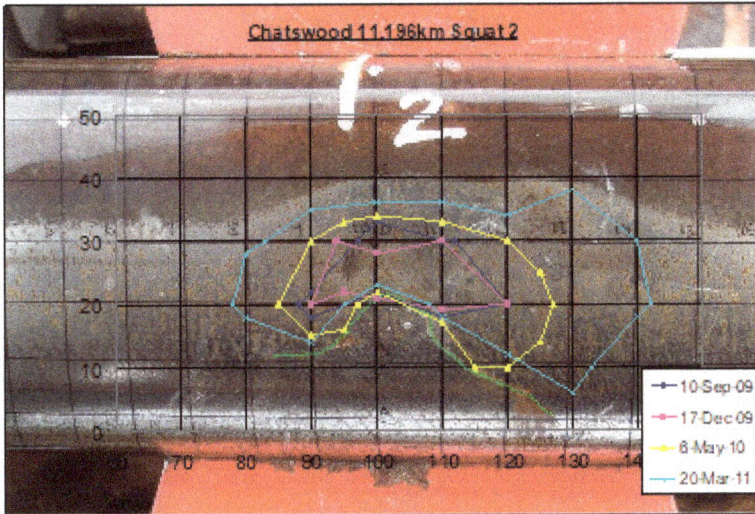

Figure 13. Rail squat growth.

4. Discussion and Concluding Remarks

This study used a new approach to rail squat simulation: the peridynamic theory. It describes the derivation of model's parameters, and illustrates how to apply a variable loading that is dependent on a node's location. The simulation successfully captures the initiation of, and initial, rail squat growth. Due to limitations of the simulation, a larger crack at this stage could not be simulated. However, the simulation results are in excellent agreement with field measurements for the crack initiation phase.

Damage initiates and grows faster close to the location of maximum pressure; similar crack growth has been measured in the field. Additionally, the computational model reveals that the squat damage first grows in the direction of the applied shear traction, and the same has been shown in field measurements.

The computational model experiences a matter-interpenetration problem, where damaged nodes, no longer connected with bonds, move freely through each other, without considering possible contacts. This problem can be solved by applying a contact model; however, contact models in Peridigm do not consider the contact between nodes that were bonded initially. To solve this problem, bond damage data needs to be passed between the damage model and the contact model. Future work will focus on re-developing parts of Peridigm's code, so that data can be passed between its damage and contact models.

Author Contributions: Conceptualization, A.F. and S.K.; Methodology, A.F. and S.K.; Software, A.F.; Validation, A.F.; Formal analysis, A.F.; Investigation, A.F.; Resources, A.F.; Data Curation, A.F.; Writing—Original Draft Preparation, A.F.; Writing—Review and Editing, A.F. and S.K.; Visualization, A.F.; Supervision, S.K.; Project Administration, S.K.; Funding Acquisition, A.F. and S.K.

Funding: This study was partially funded by the Riga Technical University under the project 34-24000-DOK.BIF/17. The authors are also sincerely grateful to the European Commission for the financial sponsorship of the H2020-RISE Project No. 691135.

Acknowledgments: The authors wish to express gratitude to the high-performance computing center's team at Riga Technical University for the many helpful discussions and advice. The corresponding author wishes to thank the Australian Academy of Science and the Japan Society for the Promotion of Sciences for his Invitation Research Fellowship (Long-term), Grant No. JSPS-L15701 at the Railway Technical Research Institute and The University of Tokyo, Japan. The authors are also sincerely grateful to the European Commission for the financial sponsorship of the H2020-RISE Project No. 691135 "RISEN: Rail Infrastructure Systems Engineering Network", which enables a global research network that tackles the grand challenge of railway infrastructure resilience and advanced sensing in extreme environments (www.risen2rail.eu) [54].

Conflicts of Interest: The authors declare no conflict of interest.

References

1. Grassie, S.L.; Fletcher, D.I.; Gallardo Hernandez, E.A.; Summers, P. Studs: A squat-type defect in rails. *Proc. Inst. Mech. Eng. Part F J. Rail Rapid Transit* **2012**, *226*, 243–256. [CrossRef]
2. Grassie, S.L. Squats and squat-type defects in rails: The understanding to date. *Proc. Inst. Mech. Eng. Part F J. Rail Rapid Transit* **2012**, *226*, 235–242. [CrossRef]
3. Remennikov, A.M.; Kaewunruen, S. A review of loading conditions for railway track structures due to train and track vertical interaction. *Struct. Control Health Monit.* **2008**, *15*, 207–234. [CrossRef]
4. Kaewunruen, S.; Remennikov, A.M. Progressive failure of prestressed concrete sleepers under multiple high-intensity impact loads. *Eng. Struct.* **2009**, *31*, 2460–2473. [CrossRef]
5. Kaewunruen, S.; Remennikov, A.M. Dynamic properties of railway track and its components: Recent findings and future research direction. *Insight Non-Destr. Test. Cond. Monit.* **2010**, *52*, 20–22. [CrossRef]
6. Carden, E.P. Vibration Based Condition Monitoring: A Review. *Struct. Health Monit.* **2004**, *3*, 355–377. [CrossRef]
7. Kaewunruen, S.; Remennikov, A.M. On the residual energy toughness of prestressed concrete sleepers in railway track structures subjected to repeated impact loads. *Electron. J. Struct. Eng.* **2013**, *13*, 41–61.
8. Seo, J.; Kwon, S.; Jun, H.; Lee, D. Numerical stress analysis and rolling contact fatigue of White Etching Layer on rail steel. *Int. J. Fatigue* **2011**, *33*, 203–211. [CrossRef]
9. Li, Z.; Zhao, X.; Dollevoet, R.; Molodova, M. Differential wear and plastic deformation as causes of squat at track local stiffness change combined with other track short defects. *Veh. Syst. Dyn.* **2008**, *46*, 237–246. [CrossRef]
10. Li, Z.; Zhao, X.; Esveld, C.; Dollevoet, R.; Molodova, M. An investigation into the causes of squats-Correlation analysis and numerical modeling. *Wear* **2008**, *265*, 1349–1355. [CrossRef]
11. Li, Z.; Zhao, X.; Dollevoet, R. An approach to determine a critical size for rolling contact fatigue initiating from rail surface defects. *Int. J. Rail Transp.* **2017**, *5*, 16–37. [CrossRef]
12. Zhao, X.; An, B.; Zhao, X.; Wen, Z.; Jin, X. Local rolling contact fatigue and indentations on high-speed railway wheels: Observations and numerical simulations. *Int. J. Fatigue* **2017**, *103*, 5–16. [CrossRef]
13. Freimanis, A.; Kaewunruen, S.; Ishida, M. Peridynamics Modelling of Rail Surface Defects in Urban Railway and Metro Systems. *Proceedings* **2018**, *2*, 1147. [CrossRef]
14. Kaewunruen, S.; Ishida, M. In Situ Monitoring of Rail Squats in Three Dimensions Using Ultrasonic Technique. *Exp. Tech.* **2016**, *40*, 1179–1185. [CrossRef]
15. Silling, S.A. Reformulation of elasticity theory for discontinuities and long-range forces. *J. Mech. Phys. Solids* **2000**, *48*, 175–209. [CrossRef]
16. Silling, S.A.; Epton, M.; Weckner, O.; Xu, J.; Askari, E. Peridynamic states and constitutive modeling. *J. Elast.* **2007**, *88*, 151–184. [CrossRef]
17. Hu, Y.L.L.; De Carvalho, N.V.V.; Madenci, E. Peridynamic modeling of delamination growth in composite laminates. *Compos. Struct.* **2015**, *132*, 610–620. [CrossRef]
18. Hu, Y.; Madenci, E.; Phan, N. Peridynamics for predicting damage and its growth in composites. *Fatigue Fract. Eng. Mater. Struct.* **2017**, *40*, 1214–1226. [CrossRef]
19. Kilic, B.; Agwai, A.; Madenci, E. Peridynamic theory for progressive damage prediction in center-cracked composite laminates. *Compos. Struct.* **2009**, *90*, 141–151. [CrossRef]
20. Bobaru, F.; Ha, Y.D.; Hu, W. Damage progression from impact in layered glass modeled with peridynamics. *Cent. Eur. J. Eng.* **2012**, *2*, 551–561. [CrossRef]

21. Bobaru, F.; Zhang, G. Why do cracks branch? A peridynamic investigation of dynamic brittle fracture. *Int. J. Fract.* **2016**, *196*, 1–40. [CrossRef]
22. Perré, P.; Almeida, G.; Ayouz, M.; Frank, X. New modelling approaches to predict wood properties from its cellular structure: Image-based representation and meshless methods. *Ann. For. Sci.* **2015**, *73*. [CrossRef]
23. Gerstle, W.; Sau, N.; Sakhavand, N. *On Peridynamic Computational Simulation of Concrete Structures*; American Concrete Institute, ACI Special Publication: Detroit, MI, USA, 2009; pp. 245–264. ISBN 9781615678280.
24. Shen, F.; Zhang, Q.; Huang, D. Damage and Failure Process of Concrete Structure under Uniaxial Compression Based on Peridynamics Modeling. *Math. Probl. Eng.* **2013**, *2013*, 631074. [CrossRef]
25. Yaghoobi, A.; Chorzepa, M.G. Meshless modeling framework for fiber reinforced concrete structures. *Comput. Struct.* **2015**, *161*, 43–54. [CrossRef]
26. De Meo, D.; Diyaroglu, C.; Zhu, N.; Oterkus, E.; Amir Siddiq, M. Modelling of stress-corrosion cracking by using peridynamics. *Int. J. Hydrogen Energy* **2016**, *41*, 6593–6609. [CrossRef]
27. Bobaru, F.; Foster, J.T.; Geubelle, P.H.; Silling, S.A. *Handbook of Peridynamic Modeling*; CRC Press: Boca Raton, FL, USA, 2016; ISBN 9781482230437.
28. Madenci, E.; Oterkus, E. *Peridynamic Theory and Its Applications*; Springer: New York, NY, USA, 2014; ISBN 978-1-4614-8464-6.
29. Silling, S.A.; Lehoucq, R.B. Peridynamic theory of solid mechanics. Advances in Applied Mechanics. *Adv. Appl. Mech.* **2010**, *44*, 73–168.
30. Seleson, P.; Parks, M. On the Role of the Influence Function in the Peridynamic Theory. *Int. J. Multiscale Comput. Eng.* **2011**, *9*, 689–706. [CrossRef]
31. Silling, S.A.; Askari, E. A meshfree method based on the peridynamic model of solid mechanics. *Comput. Struct.* **2005**, *83*, 1526–1535. [CrossRef]
32. Silling, S.; Askari, A. *Peridynamic Model for Fatigue Cracks SANDIA REPORT SAND2014-18590*; Sandia National Laboratories: Albuquerque, NM, USA, 2014.
33. Zhang, G.; Le, Q.; Loghin, A.; Subramaniyan, A.; Bobaru, F. Validation of a peridynamic model for fatigue cracking. *Eng. Fract. Mech.* **2016**, *162*, 76–94. [CrossRef]
34. Jung, J.; Seok, J. Mixed-mode fatigue crack growth analysis using peridynamic approach. *Int. J. Fatigue* **2017**, *103*, 591–603. [CrossRef]
35. Jung, J.; Seok, J. Fatigue crack growth analysis in layered heterogeneous material systems using peridynamic approach. *Compos. Struct.* **2016**, *152*, 403–407. [CrossRef]
36. Oterkus, E.; Guven, I.; Madenci, E. Fatigue failure model with peridynamic theory. In Proceedings of the ITherm 2010: 12th IEEE Intersociety Conference on Thermal and Thermomechanical Phenomena in Electronic Systems, Las Vegas, NV, USA, 2–5 June 2010.
37. Baber, F.; Guven, I. Solder joint fatigue life prediction using peridynamic approach. *Microelectron. Reliab.* **2017**, *79*, 20–31. [CrossRef]
38. Parks, M.L.; Littlewood, D.J.; Mitchell, J.A.; Silling, S.A. *Peridigm Users' Guide*; Digital Library: Piscataway, NJ, USA, 2012.
39. Littlewood, D.J. *Roadmap for Peridynamic Software Implementation*; Sandia National Laboratories: Albuquerque, NM, USA, 2015.
40. Scutti, J.J.; Pelloux, R.M.; Fuquen-Moleno, R. Fatigue behavior of a rail steel. *Fatigue Fract. Eng. Mater. Struct.* **1984**, *7*, 121–135. [CrossRef]
41. Ahlström, J.; Karlsson, B. Fatigue behaviour of rail steel—A comparison between strain and stress controlled loading. *Wear* **2005**, *258*, 1187–1193. [CrossRef]
42. Schone, D.; Bork, C.-P. Fatigue Investigations of Damaged Railway Rails of UIC 60 PROFILE. In Proceedings of the 18th European Conference on Fracture: Fracture of Materials and Structures from Micro to Macro Scale, Dresden, Germany, 30 August–3 September 2010.
43. Eisenmann, J.; Leykauf, G. The Effect of Head Checking on the Bending Fatigue Strength of Railway Rails. In *Rail Quality and Maintenance for Modern Railway Operation*; Springer: Dordrecht, The Netherlands, 1993; pp. 425–433.
44. Christodoulou, P.I.; Kermanidis, A.T.; Haidemenopoulos, G.N. Fatigue and fracture behavior of pearlitic Grade 900A steel used in railway applications. *Theor. Appl. Fract. Mech.* **2016**, *83*, 51–59. [CrossRef]
45. Deshimaru, T.; Kataoka, H.; Abe, N. Estimation of Service Life of Aged Continuous Welded Rail. *Q. Rep. RTRI* **2006**, *47*, 211–215. [CrossRef]

46. Cannon, D.F.; Pradier, H. Rail rolling contact fatigue Research by the European Rail Research Institute. *Wear* **1996**, *191*, 1–13. [CrossRef]

47. Bobaru, F.; Yang, M.; Alves, L.F.; Silling, S.A.; Askari, E.; Xu, J. Convergence, adaptive refinement, and scaling in 1D peridynamics. *Int. J. Numer. Methods Eng.* **2009**, *77*, 852–877. [CrossRef]

48. Wei, Z.; Li, Z.; Qian, Z.; Chen, R.; Dollevoet, R. 3D FE modelling and validation of frictional contact with partial slip in compression–shift–rolling evolution. *Int. J. Rail Transp.* **2016**, *4*, 20–36. [CrossRef]

49. Freimanis, A.; Kaewunruen, S.; Ishida, M. Peridynamic Modeling of Rail Squats. In *Sustainable Solutions for Railways and Transportation Engineering*; El-Badawy, S., Valentin, J., Eds.; GeoMEast 2018. Sustainable Civil Infrastructures; Springer: Cham, Switzerland, 2018. [CrossRef]

50. Ye, L.Y.; Wang, C.; Chang, X.; Zhang, H.Y. Propeller-ice contact modeling with peridynamics. *Ocean Eng.* **2017**, *139*, 54–64. [CrossRef]

51. Tupek, M.R.; Rimoli, J.J.; Radovitzky, R. An approach for incorporating classical continuum damage models in state-based peridynamics. *Comput. Methods Appl. Mech. Eng.* **2013**, *263*, 20–26. [CrossRef]

52. Ferdous, W.; Manalo, A.; Van Erp, G.; Aravinthan, T.; Kaewunruen, S.; Remennikov, A.M. Composite railway sleepers–Recent developments, challenges and future prospects. *Comp. Struct.* **2015**, *134*, 158–168. [CrossRef]

53. Kaewunruen, S.; Chiengson, C. Railway track inspection and maintenance priorities due to dynamiccoupling effects of dipped rails and differential track settlements. *Eng. Fail. Anal.* **2018**, *93*, 157–171. [CrossRef]

54. Kaewunruen, S.; Sussman, J.M.; Matsumoto, A. Grand challenges in transportation and transit systems. *Front. Built Environ.* **2016**, *2*, 4. [CrossRef]

applied
sciences

MDPI

Article

Progressive Collapse Analysis of SRC Frame-RC Core Tube Hybrid Structure

Xingxing Chen [1,*], Wei Xie [1], Yunfeng Xiao [2], Yiguang Chen [3] and Xianjie Li [1]

[1] School of Urban Construction, Yangtze University, Jingzhou 434023, China;
    201672334@yangtzeu.edu.cn (W.X.); 201671363@yangtzeu.edu.cn (X.L.)
[2] School of Civil Engineering and Mechanics, Huazhong University of Science and Technology,
    Wuhan 430074, China; yunfengxiao@hust.edu.cn
[3] Wuhan Construction Engineering Group CO., LTD, Wuhan 430056, China; 201572315@yangtzeu.edu.cn
* Correspondence: 500135@yangtzeu.edu.cn

Received: 25 October 2018; Accepted: 15 November 2018; Published: 20 November 2018

Abstract: Steel reinforced concrete (SRC) frame-reinforced concrete (RC) core tube hybrid structures are widely used in high-rise buildings. Focusing on the progressive collapse behavior of this structural system, this paper presents an experiment and analysis on a 1/5 scaled, 10-story SRC frame-RC core tube structural model. The finite element (FE) model developed for the purpose of progressive collapse analysis was validated by comparing the test results and simulation results. The alternate load path method (APM) was applied in conducting nonlinear static and dynamic analyses, in which key components including columns and shear walls were removed. The stress state of the beams adjacent to the removed component, the structural behavior including inter-story drift ratio and shear distribution between frame and tube were investigated. The demand capacity ratio (DCR) was applied to evaluate the progressive collapse resistance under loss of key components scenarios. The results indicate that the frame and the tube cooperate in a certain way to resist progressive collapse. The core tube plays a role as the first line of defense against progressive collapse, and the frame plays a role as the second line of defense against progressive collapse. It is also found that the shear distribution is related to the location of the component removed, especially the corner column and shear walls.

Keywords: steel reinforced concrete frame; reinforced concrete core tube; progressive collapse analysis; loss of key components

1. Introduction

Compared with traditional structural systems composed of steel or reinforced concrete members, steel reinforced concrete (SRC) frame-reinforced concrete (RC) core tube hybrid structure has a better combination of small sectional dimensions, higher strength, higher rigidity and resistance to corrosion, abrasion and fire. The SRC columns and RC core tube are rigidly connected by steel beams and composite floors. Benefiting from this connection and high stiffness of the core tube, most of the shear force caused by the horizontal load is assumed to be resisted by the core tube, and the lateral deformation can be restricted to an acceptable level. The vertical load of the building and partial overturning moment aroused by horizontal load is undertaken by the frame. Moreover, with the stiffness and resistance degeneration of the core tube under a strong earthquake, the frame will play the role of the second line of defence to resist shear force and avoid collapse. Hence, as a suitable and economical structure form of high-rise buildings, it is widely applied in the US, China, and Japan, especially in earthquake prone regions.

Due to the advantages of concrete-encased composite structure [1], there are extensive studies [2–6] focusing on its seismic performance. The research on progressive collapse has become of increasing interest in recent years, especially since the World Trade Center towers collapse following the terrorist attacks of 11 September 2001. The progressive collapse of a structure can be caused by the failure of structural components under unexpected loads including, car accidents, earthquakes, or explosions and so on. [7]. To study the response of structures under extreme conditions such as explosions and fire, some research has been carried out. A new analysis method for the progressive collapse analysis of a structure with consideration of both the non-zero initial condition and existing damage in structural members [8], and a new finite element model was proposed as a feasible tool to evaluate the fire response of composite floor systems [9]. Recently, the alternate load path method (APM) recommended by the current codes and manuals of practice [10,11] for anti-collapse design and analysis has been popular. The alternate load path method is easy to implement. In this method, the robustness of a structure is evaluated through removal of the key vertical components to determine whether the local damage may be absorbed by the remaining structural members and whether the structural system can bridge over the removed components.

In the literature, there have been extensive experimental studies focusing on progressive collapse behavior of reinforced concrete frame and steel frame structures subjected to the loss of key components [12–17]. The compressive arch action and catenary action were clearly observed in the experiments. Due to exorbitant cost and safety issues, numerical simulations are preferred for studying the progressive collapse resistance of structures [18–23], in which nonlinear static and dynamic analyses had been conducted [24–29].

As the two main components resisting the horizontal force aroused by earthquake, the frame and the core tube operate cooperatively with different stiffness. An unequal distribution of shear in the two components is produced and will affect the behavior of building structures against progressive collapse. To effectively prevent earthquake-induced structural collapse it is necessary to study a widely used structural system in high-rise buildings, SRC frame-RC core tube hybrid structures, focusing on progressive collapse behavior. Nevertheless, to date, limited related experimental and analysis work has been carried out on this particular structural system.

Based on the experiment recommend by reference [6], using the general purpose finite element package OpenSees [30,31], a numerical model is first developed in this paper which enables the non-linear progressive collapse analysis of high rise building. The proposed numerical model was validated by a pseudo-static test. The alternate load path method was applied in conducting nonlinear static and dynamic analyses, and robustness was studied under column and shear wall removal scenarios. The model accurately displayed the overall behavior, including inter-story drift ratio and shear distribution under sudden loss of key components and seismic waves input, which provided important information for additional design guidance on progressive collapse for the SRC frame-RC core tube hybrid structures.

2. Experiment Program

2.1. Details of Specimen

Based on the currently design codes and specifications in China [32–34], a 1/5 scaled, 10-story prototype building was designed and built [6]. The plane layout is shown in Figure 1, the dimension and reinforcement of core tube, columns and beams is shown in Figures 2 and 3. The properties of concrete and steel are listed in Tables 1 and 2. The similarity ratios of properties (elastic modulus, stress and strain) of concrete and steel are 1, which means the specimen and the prototype are in an equal strain state. So, the failure mode of specimen can truly reflect that of the prototype.

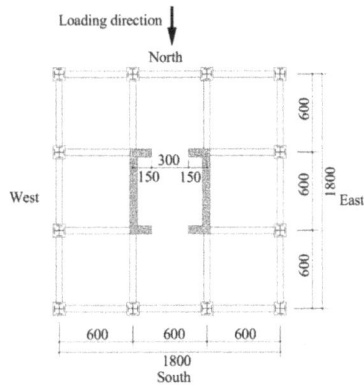

Figure 1. Plan arrangement (units: mm).

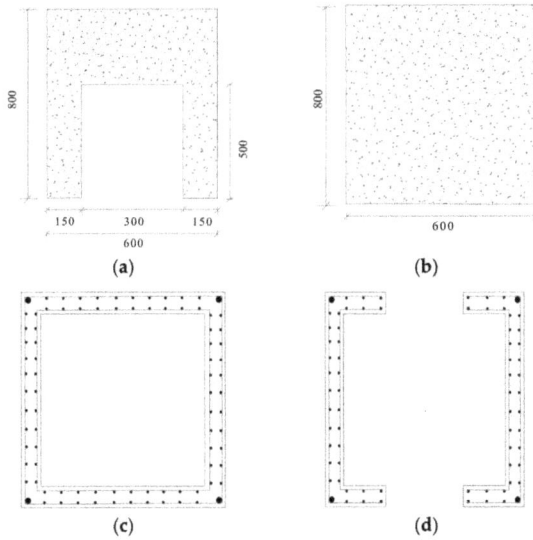

Figure 2. Details of core tube: (**a**) the south and north shear wall, (**b**) the west and east shear wall, (**c**) the reinforcement in the shear wall and (**d**) the reinforcement in the opening. (units: mm).

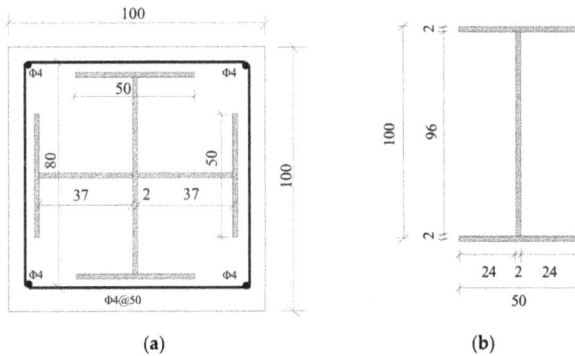

Figure 3. Details of components: (**a**) steel reinforced concrete (SRC) column and (**b**) steel beam. (units: mm).

135

Table 1. Properties of Concrete.

Concrete Strength Grade	Concrete Compressive Strength/MPa	Elastic Modulus/MPa
C40	41.5	3.03×10^4

Table 2. Properties of steel.

Material	Yield Strength/MPa	Ultimate Strength/MPa	Elastic Modulus/MPa
Φ4 bar	305	424	2.1×10^5
Steel plate	327	463	2.0×10^5

2.2. Test Setup and Procedure

The test setup is shown in Figure 4. By controlling displacement, a pseudo static horizontal cyclic loading scheme was implemented in this test. The lateral load was applied by two hydraulic servo actuators at 4th and 9th floor respectively. Mode-superposition response spectrum method was adopted to control the amplitude of displacement and finally $\Delta_9/\Delta_4 = 1.5:1$. The vertical load was applied by sandbags on the floors. Based on the Load Code for the Design of Building Structures (GB50009-2012) [35], it was necessary to take the weight of infill walls into account, and the live load and dead load were taken as 2.0 kN/m^2 and 1.6 kN/m^2 respectively. The history of the loading program is shown in Figure 5.

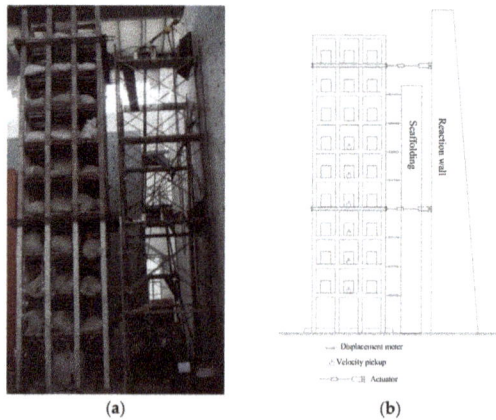

(a)   (b)

Figure 4. Test setup: (**a**) Test specimen and (**b**) Arrangement of instruments.

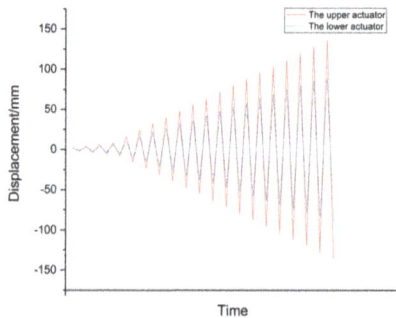

Figure 5. History of loading program.

2.3. Test Results

Figure 6a shows the hysteretic curve of the test results, which is the relationship of the building base shear and the displacement at the top loading point. The curve is full and in a spindle shape, the yielding, limiting and failure stages are obvious. Figure 6b shows the test value of lateral displacement at the 1st, 2nd, 3rd, 4th, 5th, 6th and 9th floor when the top displacements are 8 mm, 24 mm, 56 mm, 88 mm and 136 mm respectively. During the experiment, the dynamic characteristics were measured and the details were introduced in reference [6]. Table 3 shows the test value of frequencies of the first five vibration modes at the initial and failure states.

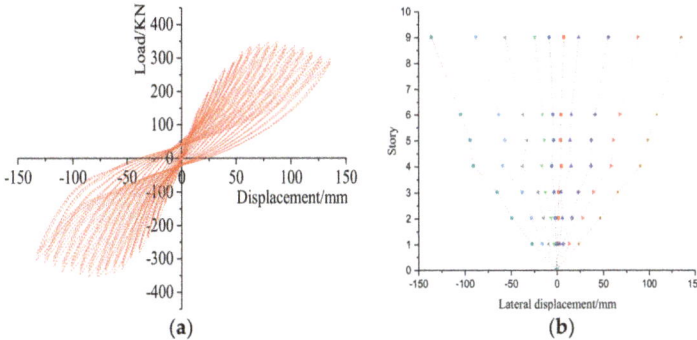

Figure 6. Test results: (a) The hysteretic curve and (b) lateral displacement.

Table 3. Test value of frequencies.

Vibration Modes	Initial State	Failure State
	Test Value/Hz	Test Value/Hz
1	5.94	3.26
2	18.98	6.22
3	25.13	15.96
4	37.93	31.82
5	50.03	42.01

3. Finite Element Model

3.1. Constitutive Model of Materials

In general, bilinear stiffness degeneration behavior is employed to simulate the cumulative damage of structure [36,37]. In this paper, stiffness degeneration of materials are defined in the finite element (FE) model to simulate the cumulative damage of structure. Constitutive model curves of steel and concrete are shown in Figure 7a,b [6], respectively. In Figure 7a, the stress–strain curve exhibits two stages, including elastic and hardening stages. E_s is modulus of elasticity, f_y is yield strength, and E_p is hardening modulus. In Figure 7b, the constitutive model of concrete is based on the Kent-Park model [38], both unconfined and confined concrete fibers in this constitutive model consider tensile strength with linear degeneration. f_c, E_t, f_t, and E_u are ultimate compressive strength, rigidity after cracking, ultimate tensile strength, unloading rigidity. The calculating formula is expressed as below:

$$\sigma = \begin{cases} f_c \left[\frac{2\varepsilon}{0.002} - \left(\frac{\varepsilon}{0.002} \right)^2 \right] & \varepsilon \leq 0.002 \\ f_c [1 - Z(\varepsilon - 0.002)] & 0.002 \leq \varepsilon \leq \varepsilon_{20} \\ 0.2 f_c & \varepsilon \geq \varepsilon_{20} \end{cases} \tag{1}$$

$$Z = \frac{0.5}{\varepsilon_{50u} + \varepsilon_{50h} - 0.002} \tag{2}$$

$$\varepsilon_{50u} = \frac{3 + 0.002 f_c}{0.002 f_c - 1000} \tag{3}$$

$$\varepsilon_{50h} = \frac{3}{4} \rho_s \sqrt{\frac{B}{S_h}} \tag{4}$$

where, ε, Z, ε_{50u}, ε_{50h}, ρ_s, B and S_h are the concrete strain, the slope of the descending branch of the stain-stress curve, the stain at $0.5f_c$, the strain increase of confined concrete over unconfined concrete at $0.5f_c$, the ratio of stirrup, width of concrete core area and the spacing of stirrup, respectively.

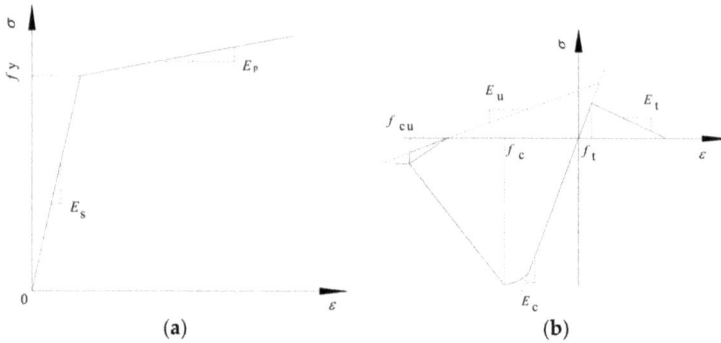

(a) (b)

Figure 7. Constitutive model of materials: (**a**) constitutive model of steel and (**b**) constitutive model of concrete.

3.2. Elements and Boundary Condition

The element types "dispBeamColumn" and "LayeredShell" are employed to divide the cross section of members into units with a certain number. According to the variation positions of concrete and steel in the specimen, the fibers at corresponding positions can be defined with different constitutive models. The cross sections of different members are shown in Figure 8. And the view of the FE model is shown in Figure 9.

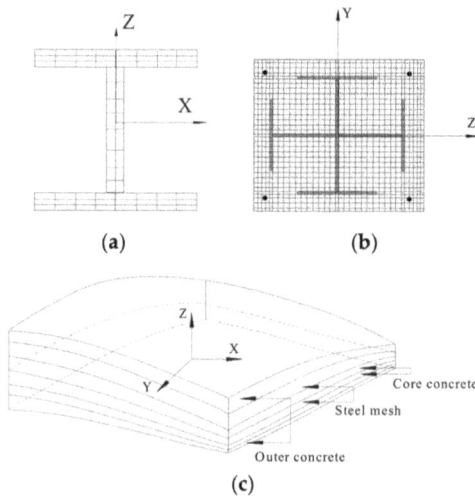

(a) (b)

(c)

Figure 8. Fiber sections: (**a**) beam, (**b**) column and (**c**) wall and floor.

Figure 9. The finite element (FE) model.

Vertical load was applied to the corresponding node. Horizontal load was controlled by displacement referring to the experiment. Based on the experimental conditions, the bottom of the tube and column were defined as fixed, and displacement of nodes on the other same floor was defined as coupled. The mass proportional damping was defined as 0.05.

3.3. Validation of the FE Model

The model size, section dimensions, loading and boundary conditions are exactly replicated in the experiment [6]. The test results are provided in this section. To validate the proposed FE model, simulation and test results are compared focusing on hysteretic curve, skeleton curve, bearing capacity, deformation and dynamic characteristics.

In Figure 10a, the calculating and the test curves are both full and in spindle shaped, the yielding, limiting and failure stages are obvious. In Figure 10b, the calculating and the test curves are consistent in elastic-plastic segment and degenerate segment. Table 4 shows the characteristic values at yielding, limiting, and failure stages. It can be seen that the calculation results are consistent with the test.

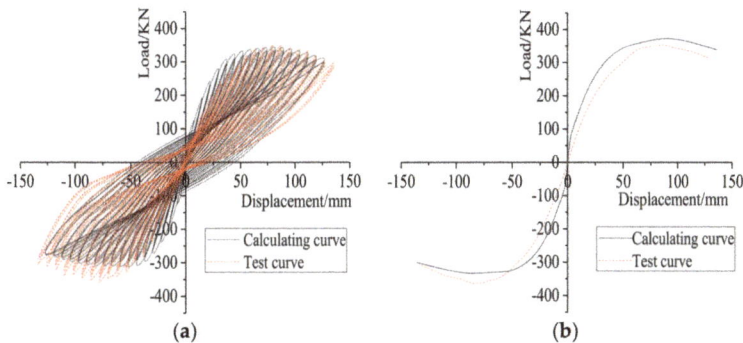

Figure 10. Comparison of hysteretic and skeleton curves: (**a**) Hysteretic curve and (**b**) Skeleton curve.

Table 4. Comparison of characteristic value.

Load	Direction	Test Value/kN	Calculating Value/kN	Absolute Error/%
P_y	Positive	238.49	256.72	7.64
	Negative	−243.52	−251.89	3.43
P_{max}	Positive	351.72	373.69	6.24
	Negative	−356.12	−333.91	6.23
P_u	Positive	308.47	338.44	9.72
	Negative	−304.96	−300.55	1.44

where, P_y, P_{max} and P_u are yield strength, ultimate strength and failure strength respectively.

Figure 11 shows a comparison of lateral displacement at the 1st, 2nd, 3rd, 4th, 5th, 6th and 9th floors resulting from test and simulation. The dotted line is the test value and the solid line is the simulation value. It can be seen that structural deformation resulting from the two approaches are in a good agreement when the top displacements are 8 mm, 24 mm, 56 mm, 88 mm and 136 mm respectively.

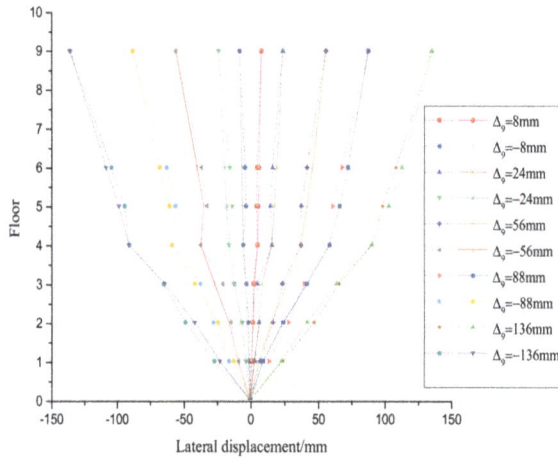

Figure 11. Comparison of lateral displacement.

Table 5 shows the frequencies of the first five vibration modes at the initial and failure state. It can be seen that the natural frequency of structure decreases along with damage accumulation, and that the absolute errors between calculating results and test results is within 5%.

Table 5. Comparison of frequencies.

Vibration Modes	Initial State			Failure State		
	Test Value/Hz	Calculating Value/Hz	Absolute Error/%	Test Value/Hz	Calculating Value/Hz	Absolute Error/%
1	5.94	5.86	1.31	3.26	3.35	2.73
2	18.98	18.13	4.50	6.22	6.01	3.23
3	25.13	23.97	4.60	15.96	15.25	4.47
4	37.93	36.47	3.87	31.82	32.47	2.06
5	50.03	48.43	3.20	42.01	40.85	2.76

4. Progressive Collapse Analysis

4.1. Basic Principle

As stated above, the alternate path method (APM) proposed by General Services Administration (GSA) [10] is applied to assess the potential for progressive collapse by a certain acceptance criteria. As stated in GSA, the demand-capacity ratio (DCR) of adjacent components is applied to evaluate whether progressive collapse occurs as after the failure of critical component.

The component removal is conducted using element-killing technology. The DCR under various cases is assessed using nonlinear static and dynamic analysis. The nonlinear static analysis can be used to simulate the removal scenario induced by triggering events, such as explosions and vehicle accidents. The nonlinear dynamic analysis can be used to simulate the building subjected to an earthquake with failure of critical components. The maximum forces and displacements for each member are recorded. In static analysis, the critical component is removed primarily, and then the analysis goes on to evaluate the structural performance. If the structure is stable, the dynamic analysis will continue based on the static analysis, a seismic wave will be input to the model with critical component removal. The dynamic effects are considered through load-increase factors. For the ground key components loss, load combination was employed as follows:

$$\text{Load} = 2(\text{DL} + 0.25\,\text{LL}) \tag{5}$$

$$\text{Load} = \text{DL} + 0.25\,\text{LL} \tag{6}$$

where Equation (5) is for nonlinear static analysis, Equation (6) is for nonlinear dynamic analysis. DL and LL are dead load and live load respectively.

Based on GSA, DCR is a ratio defined as internal force Q_{UD} of a component after the removal of key component to the ultimate internal force Q_{CE} of the component. The DCR values are calculated as Equation (7). As specified by GSA, the progressive collapse would not occur, as the DCR is smaller than 2; besides the progressive collapse of the structure will occur.

$$\text{DCR} = \frac{Q_{UD}}{Q_{CE}} \tag{7}$$

4.2. Seismic Input for Nonlinear Dynamic Analysis

To assure the accuracy of numerical analysis, the HOLLYWOOD, ELCENTRO and NRIGDE seismic waves are selected to represent the various level of earthquake intensity, including frequent earthquake, fortification earthquake and rare earthquake. Moreover, an artificial wave is also selected. The peak ground acceleration (PGA) of each seismic wave is listed in Table 6, and acceleration-time curves are shown in Figure 12.

Table 6. PGA of seismic waves.

Number	Seismic Wave	PGA/g
1	HOLLYWOOD	0.041
2	ELCENTRO	0.278
3	NRIGDE	0.603
4	Artificial wave	0.540

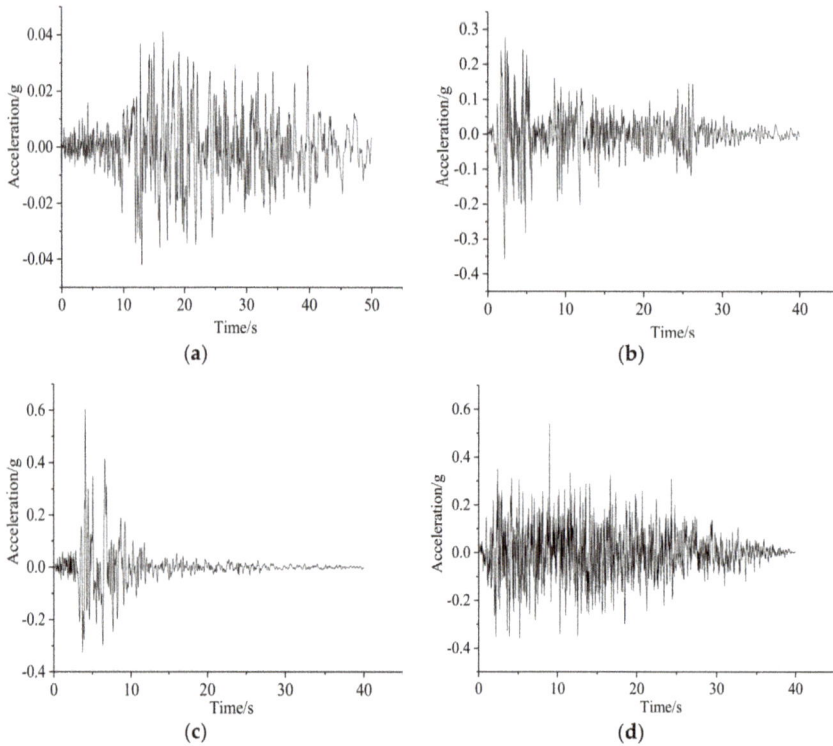

Figure 12. Acceleration-time curves: (**a**) HOLLYWOOD, (**b**) ELCENTRO, (**c**) NRIGDE and (**d**) artificial wave.

Figure 13 shows response spectra of these seismic waves and design response spectrum specified by Chinese code (GB 50011-2001) [33]. When the PGA does not meet the specification, modification should be performed by the equation given below:

$$a'(t) = \frac{A'_{max}}{|A|_{max}} a(t)$$ (8)

where, $a(t)$ is initial time-history relationship of acceleration, $|A|_{max}$ is PGA of seismic wave, $a'(t)$ is modified time-history relationship of acceleration, A'_{max} is the modified result, taken as Table 7.

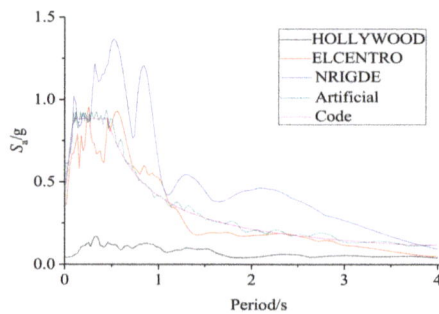

Figure 13. Response spectra of seismic waves.

Table 7. A'_{max}/Gal of seismic waves.

Earthquake Intensity	7	8	9
Frequently occurring	35	70	140
Moderate	107	215	429
Rarely expected	220	400	620

4.3. Progressive Collapse Cases

Table 8 shows the list of analysis cases considered together with the components that are forcibly removed. Due to the symmetry of the cross section of the structure, 5 key components were assumed to be removed. These critical components are side columns C1 and C3, corner column C2, shear walls W1 and W2 at the first floor. Figure 14a,b show the location of these critical components and adjacent beams. To illustrate the analysis method clearly and facilitate analysis of data, case 1 is discussed in detail while discussion of the other cases is relatively concise.

Table 8. Progressive analysis cases.

Case	1	2	3	4	5
Removed component	C1	C2	C3	W1	W2

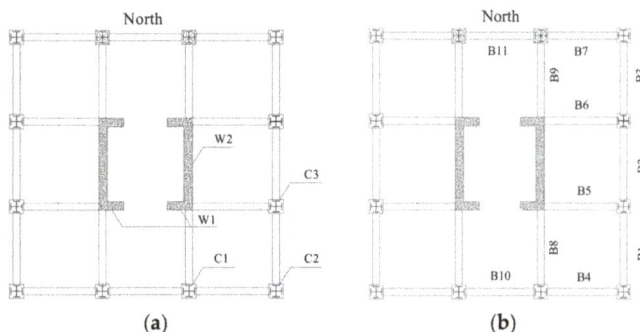

Figure 14. Location of members at the first floor: (**a**) location of key components and (**b**) location of the adjacent beams.

4.4. Nonlinear Static Analysis

4.4.1. Case 1—Column C1 at Ground Floor Removed

For case 1, the ground columns C1 as shown in Figure 14 were suddenly removed (Case 1 in Table 8). Table 9 shows force condition of beams adjacent to C1. It can be seen that the maximum DCR value is 0.86, far less than 2.0. The DCRs of B8 and B10 are larger than the DCR of B4. This result may be caused by the different locations. Compared with the locations of B4, B8 and B10 are closer to the core tube. When column C1 is removed, the load is transferred to B4 by the portion of 20.06%, to B8 by the portion of 45.23% and to B10 by the portion of 34.71%.

Figure 15 shows the lateral displacement of the structure. X-axis is west-east direction, Y-axis is south-north direction. It can be seen that the lateral displacement along the X-axis is larger, and the maximum displacement is 24.5 mm. Moreover, the maximum inter-story displacement angle θ_{max} is far smaller than the limited value 1/800 in Chinese specification (GB50009-2001) [35]. These results indicate that the progressive collapse will not occur under the loss of C1 scenario.

Table 9. Force condition of beams in case 1.

Case	Beam Number	Direction	Bending Moment/kN·m	Portion (%)	Average Value (%)	DCR
1 (C1)	B4	Positive	110.00	15.69	20.06	0.23
		Negative	−202.50	24.43		0.42
	B8	Positive	280.00	39.92	45.23	0.58
		Negative	−418.75	50.53		0.86
	B10	Positive	311.25	44.39	34.71	0.64
		Negative	−207.50	25.04		0.43

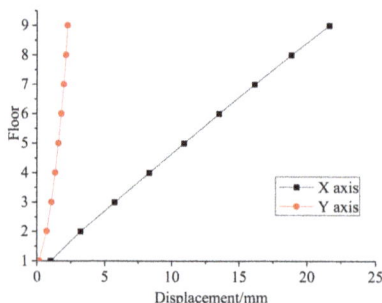

Figure 15. Displacement of structure in case 1.

4.4.2. The Results from Case 2 to Case 5

Table 10 shows the force condition of beams under the other cases. It can be seen that the DCRs are less than 2.0. Due to the structural symmetry, the difference between the DCRs of B1 and B4 is small in case 2. In case 3, the DCR of B5 is larger than the DCRs of B1 and B2. Similar to the case 1, B5 is closer to the core tube than B1 and B2, and the more load is transferred to B5 by the portion of 48.77% while the load is transferred to B1 and B2 by the portion of 20.02% and 32.21% respectively. In case 4 and case 5, B5 and B8 are directly linked to the core tube. Due to the structural symmetry, there is a small difference between the DCRs of B5 and B8, and the DCR of B5 is larger than the DCR of B8 in case 4, but the relationship is reversed in case 5. From the results of the above cases, it can be seen that the closer to the core tube, the more affected the beam gets.

Figure 16 shows the lateral displacement of the structure in different case. The maximum displacements are 22.5 mm, 12.50 mm, 7.80, 5.70 mm respectively. Moreover, the θ_{max} is still smaller than the specified limitation. From case 1 to case 5, in terms of the number, the displacements of 9th floor are the maximum, the maximum displacements and the difference between the maximum value in X-axis and Y-axis showing a decreasing trend. In cases 1, 2 and 4, the maximum values on the X-axis are larger than that on the Y-axis. This means that the removal of C1, C2 and W1 make the ability of structure to resist deformation weaker mostly in X direction. On the contrary, in cases 3 and 5, the removal of C3 and W2 make the ability of structure to resist deformation weaker mostly in Y direction.

Table 10. Force condition of beams in cases 2, 3, 4 and 5.

Case	Beam Number	Direction	Bending Moment/kN·m	Portion (%)	Average Value (%)	DCR
2 (C2)	B1	Positive	213.75	33.79	50	0.44
		Negative	−428.75	66.21		0.88
	B4	Positive	418.75	67.39	50	0.86
		Negative	−207.50	32.61		0.43

Table 10. *Cont.*

Case	Beam Number	Direction	Bending Moment/kN·m	Portion (%)	Average Value (%)	DCR
3 (C3)	B1	Positive	206.25	22.39	20.02	0.42
		Negative	−88.75	17.66		0.18
	B2	Positive	313.75	34.06	32.21	0.64
		Negative	−142.50	28.36		0.29
	B5	Positive	401.25	43.55	48.77	0.82
		Negative	−271.25	53.98		0.55
4 (W1)	B5	Positive	347.50	46.41	47.10	0.71
		Negative	−380.00	47.80		0.78
	B8	Positive	401.25	53.59	52.90	0.82
		Negative	−415.00	52.20		0.85
5 (W2)	B5	Positive	497.50	55.43	56.59	1.02
		Negative	−440.00	57.74		0.90
	B8	Positive	400.00	44.57	43.41	0.82
		Negative	−322.50	42.26		0.66

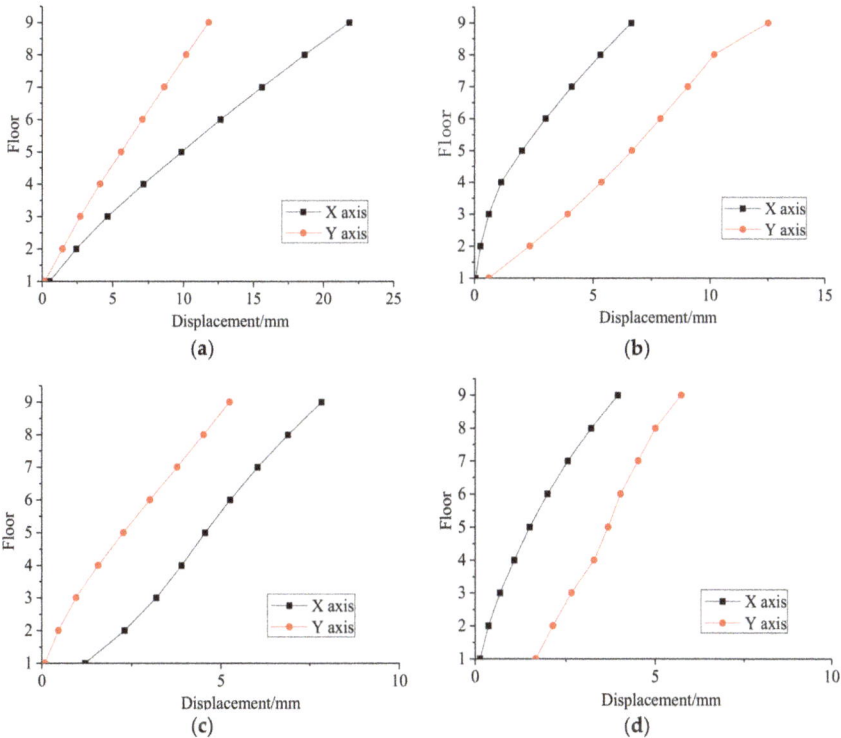

Figure 16. Displacement of structure in (**a**) case 2, (**b**) case 3, (**c**) case 4 and (**d**) case 5.

4.5. Nonlinear Dynamic Analysis

4.5.1. Case 1—Column C1 at Ground Floor Removed

The nonlinear dynamic analysis is to simulate the structure under earthquake action with critical component removal. Under the influences of different seismic waves, the displacement-time curves of the top floor are shown in Figure 17, the inter-story displacement angles θ of structure are shown in

Figure 18. It can be seen the maximum inter-story displacement angle θ_{max} is located at the 6th floor when subjected to HOLLYWOOD and ELCENTRO waves, and the value is within the code limitation. When subjected to NRIDGE and artificial waves, the value reached 0.0016 and 0.0022, which exceed the specified limitation.

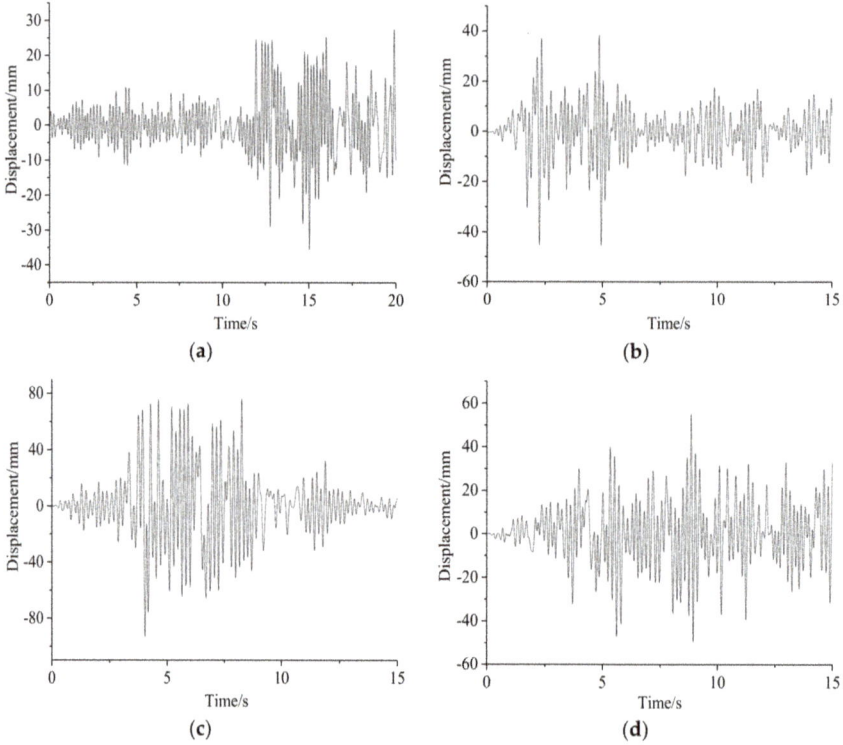

Figure 17. Displacement-time curve of top under different seismic wave in case 1: (**a**) HOLLYWOOD, (**b**) ELCENTRO, (**c**) NRIGDE and (**d**) Artificial wave.

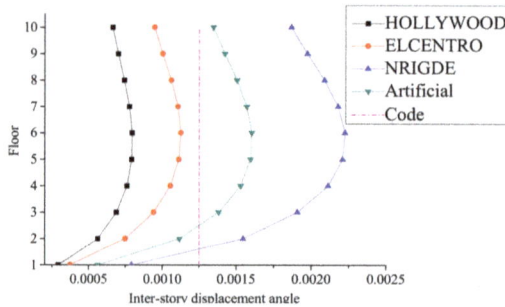

Figure 18. Inter-story displacement angle in case 1.

Figure 19 shows shear distribution between the ground frame and core tube. At the initial stage, the shear ratios of frame to core tube under HOLLYWOOD, ELCENTRO, NRIGDE and artificial seismic waves are 1:3.29, 1:3.17, 1:3.01, and 1:2.96 respectively. The core tube is responsible for the majority of the lateral load. With the stiffness degradation of the core tube, the shear ratios of frame to core tube change to 1: 2.35, 1:2.28, 1:2.18, and 1:2.14 respectively. It means that the core tube plays a role as the first line of defense against lateral load, and the frame plays a role as the second line of defense against lateral load.

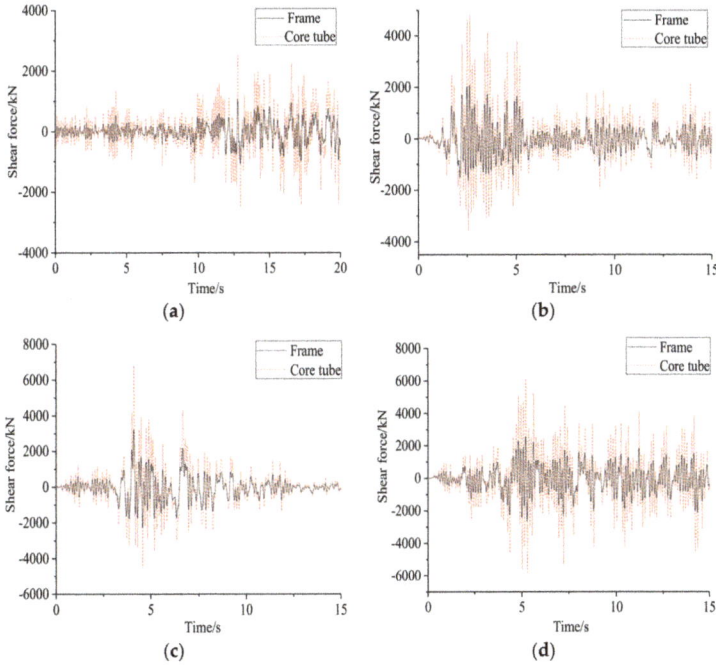

Figure 19. Shear distribution of bottom under different seismic wave in case 1: (**a**) HOLLYWOOD, (**b**) ELCENTRO, (**c**) NRIGDE and (**d**) Artificial wave.

Table 11 shows force condition of beams adjacent to C1 after the sudden failure of C1 under earthquake action. The maximum DCR value of the beams is smaller than 2.0 in the cases of HOLLYWOOD and ELCENTRO seismic waves. In the cases of NRIGDE and Artificial seismic wave, the maximum DCR value of B8 and B10 are 2.92 and 3.04, greater than 2.0. That is to say, the progressive collapse will not occur under the failure of the C1 scenario when subjected to the frequent earthquake or fortification earthquake, but it will occur when subjected to the rare earthquake. Compared to Table 9 in nonlinear static analysis, the loads transferred to B4 and B10 are increased while the load transferred to B8 is reduced. The reason for these changes is the frame has played a greater role of defense against the earthquake load.

Table 11. Force condition of beams in case 1.

Case	Beam Number	Seismic Wave	Average Bending Moment/kN·m	Portion (%)	DCR
1 (C1)	B4	HOLLYWOOD	457.50	27.98	0.94
		ELCENTRO	558.75	23.99	1.15
		NRIGDE	885.00	23.30	1.82
		Artificial wave	763.75	22.36	1.57
	B8	HOLLYWOOD	555.00	33.95	1.14
		ELCENTRO	900.00	38.65	1.79
		NRIGDE	1440.00	37.91	2.92
		Artificial wave	1347.50	39.46	2.77
	B10	HOLLYWOOD	622.50	38.07	1.28
		ELCENTRO	870.00	37.36	1.87
		NRIGDE	1473.75	38.79	3.04
		Artificial wave	1303.75	38.18	2.68

4.5.2. The Results from Cases 2 to Case 5

Table 12 shows the θ_{max} of structure. It can be seen, when the structure is subjected to the same seismic wave, the θ_{max} in case 3 is smaller than other cases. Compared to case 4, the horizontal displacement of structure becomes obviously smaller in case 5.

Table 13 shows the shear distribution between the frame and core tube. It is obvious that the core tube is responsible for more shear than the frame. It obviously increases the shear the frame bears in the cases of shear wall removal. That means the shear is redistributed, the core tube plays a role as the first line of defense against lateral load, and the frame plays a role as the second line of defense against lateral load. Compared to the shear distribution between the ground frame and core tube in case 1, it can be found that the portion of shear assumed by the frame in case 5 is more than double of it in case 1.

Table 14 shows force condition of beams. Compared to Table 10 in nonlinear static analysis, DCRs of the corresponding component in nonlinear dynamic analysis become obviously larger. With increasing earthquake intensity, the DCR under the failure of W2 is larger than 2.0, even in the frequent earthquake. DCR under the failure of W1 and W2 is larger than DCR under the failure of columns. Compared to the failure of C1, the DCR of adjacent beams becomes obviously larger under the failure of C2. That is to say, the failure of the corner column is more likely to cause progressive collapse of the structure than failure of side column. Moreover, the DCRs of adjacent beams under the failure of C1 are obviously larger than that under the failure of C3. That is to say, the failure of side columns is more likely to cause progressive collapse of structure. Compared to the failure of W1, DCRs of adjacent beams become obviously larger under the failure of W2. The failure of shear wall without opening is more likely to cause progressive collapse. Compared to Table 10 in nonlinear static analysis, in terms of cases 2, 4 and 5, due to the structural symmetry, the change of load transfer is not obvious. However, it is obvious for case 3, similar to case 1, the loads transferred to B1 and B2 are increased while the load transferred to B5 is reduced.

Table 12. The maximum inter-story displacement angle in cases 2, 3, 4 and 5.

Case	Seismic Wave	θ_{max}
2 (C2)	HOLLYWOOD	0.00078
	ELCENTRO	0.00108
	NRIGDE	0.00206
	Artificial wave	0.00149
3 (C3)	HOLLYWOOD	0.00076
	ELCENTRO	0.00105
	NRIGDE	0.00199
	Artificial wave	0.00142
4 (W1)	HOLLYWOOD	0.00077
	ELCENTRO	0.00101
	NRIGDE	0.00209
	Artificial wave	0.00151
5 (W2)	HOLLYWOOD	0.00074
	ELCENTRO	0.00093
	NRIGDE	0.00192
	Artificial wave	0.00136

Table 13. Shear distribution in cases 2, 3, 4 and 5.

Case	Seismic Wave	Shear Distribution
2 (C2)	HOLLYWOOD	1:1.76
	ELCENTRO	1:2.19
	NRIGDE	1:2.01
	Artificial wave	1:2.17
3 (C3)	HOLLYWOOD	1:1.81
	ELCENTRO	1:2.03
	NRIGDE	1:2.05
	Artificial wave	1:2.19
4 (W1)	HOLLYWOOD	1:1.37
	ELCENTRO	1:1.61
	NRIGDE	1:1.47
	Artificial wave	1:1.59
5 (W2)	HOLLYWOOD	1:1.23
	ELCENTRO	1:1.58
	NRIGDE	1:1.44
	Artificial wave	1:1.56

Table 14. Force condition of beams in cases 2, 3, 4 and 5.

Case	Beam Number	Seismic Wave	Average Bending Moment/kN·m	Portion (%)	DCR
2 (C2)	B1	HOLLYWOOD	748.75	48.74	1.54
		ELCENTRO	885.00	51.12	1.82
		NRIGDE	1576.25	49.03	3.24
		Artificial wave	1445.00	48.31	2.97
	B4	HOLLYWOOD	787.50	51.26	1.62
		ELCENTRO	846.25	48.88	1.74
		NRIGDE	1638.75	50.97	3.37
		Artificial wave	1546.25	51.69	3.18
3 (C3)	B1	HOLLYWOOD	365.00	25.44	0.75
		ELCENTRO	447.50	22.13	0.92
		NRIGDE	730.00	23.86	1.5
		Artificial wave	627.50	22.70	1.29
	B2	HOLLYWOOD	622.50	43.38	1.28
		ELCENTRO	870.00	43.02	1.79
		NRIGDE	1245.00	40.69	2.56
		Artificial wave	1090.00	39.44	2.24
	B5	HOLLYWOOD	447.50	31.18	0.92
		ELCENTRO	705.00	34.86	1.45
		NRIGDE	1085.00	35.46	2.23
		Artificial wave	1046.25	37.86	2.15
4 (W1)	B5	HOLLYWOOD	602.50	48.20	1.24
		ELCENTRO	782.50	40.94	1.61
		NRIGDE	1426.25	47.90	2.93
		Artificial wave	1196.25	46.68	2.46
	B8	HOLLYWOOD	647.50	51.80	1.33
		ELCENTRO	1128.75	59.06	2.32
		NRIGDE	1551.25	52.10	3.19
		Artificial wave	1366.25	53.32	2.81

Table 14. *Cont.*

Case	Beam Number	Seismic Wave	Average Bending Moment/kN·m	Portion (%)	DCR
5 (W2)	B5	HOLLYWOOD	715.00	47.59	1.47
		ELCENTRO	1026.25	44.52	2.11
		NRIGDE	1502.50	46.33	3.09
		Artificial wave	1430.00	50.87	2.94
	B8	HOLLYWOOD	787.50	52.41	1.62
		ELCENTRO	1278.75	55.48	2.63
		NRIGDE	1740.00	53.67	3.58
		Artificial wave	1381.25	49.13	2.84

5. Conclusions

In this paper, a 3-D finite element model was first built with the OpenSees software to simulate the behavior of SRC frame-RC core tube hybrid buildings under sudden component removal. The method and principle for the modeling techniques and progressive analysis are described in detail. The model also incorporates non-linear material characteristics and non-linear geometric behavior. A 1/5 scaled, 10-story 3-bay model was built for the validation of the proposed modeling method. The numerical results are presented and compared to experimental data, a good agreement is obtained. Using the proposed model, the progressive collapse analysis under loss of key components and seismic waves input were conducted. The following conclusions can be drawn within the limitation of the current study presented in this paper.

Even though the progressive collapse is a rare event when subjected to only the loss of a column or shear wall, the possibility of progressive collapse should not be neglected. When the seismic wave was entered into the component removed model, especially the model under the loss of shear wall scenario, DCR of the adjacent beams significantly increased. For the SRC frame-RC core tube hybrid structure, the possibility of progressive collapse increases with the increase of earthquake intensity, especially under the loss of shear wall scenario.

In this paper, the portion of shear assumed by the frame in case 5 is more than double of that in case 1 under a seismic wave. A loss of the W5 seriously weakened the carrying capacity of the core tube. Under the same general conditions, removal of a shear wall is the most likely to cause progressive collapse, and then next likely is removal of a corner column, and lastly removal of a side column. In addition, removal of a shear wall without opening is more likely to cause progressive collapse than the shear wall with opening.

When subjected to the loss of component and earthquake, the internal forces will be redistributed between the frame and core tube and the frame is responsible for more load. The concept of multi-lines of seismic defense is reflected in SRC frame-RC core tube hybrid buildings. The core tube plays a role as the first line of defense against progressive collapse, the second line of collapse resistance is provided by the frame.

Author Contributions: All authors substantially contributed to this work. W.X. and X.C. were the scientific coordinator of the research, designed the experimental campaign and analyzed the results. Y.X. supervised all the research and revised the results. Y.C. and X.L. contributed on the experimental tests. All authors helped with the writing of the paper and give final approval of the version to be submitted and any revised versions.

Funding: The research presented in this paper was funded by Natural Science Foundation of Hubei Province of China (Grant No. 2016CFB604), Natural Science Foundation of China (Grant Nos. 51108041, 51378077), Science Foundation of the Education Department of Hubei Province of China (Grant No. D20161305), and it is grateful for their support.

Acknowledgments: It is very grateful for the supports from Lei Zeng.

Conflicts of Interest: The authors declare no conflict of interest.

References

1. Zeng, L.; Parvasi, S.M.; Kong, Q.; Huo, L.; Lim, I.; Li, M.; Song, G. Bond slip detection of concrete-encased composite structure using shear wave based active sensing approach. *Smart. Mater. Struct.* **2015**, *24*, 125026. [CrossRef]
2. Xu, C.H.; Zeng, L.; Zhou, Q.; Tu, X.; Wu, Y. Cyclic performance of concrete-encased composite columns with t-shaped steel sections. *Int. J. Civ. Eng.* **2015**, *13*, 456–467.
3. Xiao, Y.F.; Zeng, L.; Cui, Z.K.; Jin, S.Q.; Chen, Y.G. Experimental and analytical performance evaluation of steel beam to concrete-encased composite column with unsymmetrical steel section joints. *Steel Compos. Struct.* **2017**, *23*, 17–29. [CrossRef]
4. Xue, J.; Lavorato, D.; Bergami, A.; Nuti, C.; Briseghella, B.; Marano, G.; Ji, T.; Vanzi, I.; Tarantino, A.; Santini, S. Severely Damaged Reinforced Concrete Circular Columns Repaired by Turned Steel Rebar and High-Performance Concrete Jacketing with Steel or Polymer Fibers. *Appl. Sci.* **2018**, *8*, 1671. [CrossRef]
5. Fiore, A.; Marano, G.C.; Laucelli, D.; Monaco, P. Evolutionary modeling to evaluate the shear behavior of circular reinforced concrete columns. *Adv. Civ. Eng.* **2014**, *2014*, 169–182. [CrossRef]
6. Zeng, L.; Xiao, Y.; Chen, Y.; Jin, S.; Xie, W.; Li, X. Seismic Damage Evaluation of Concrete-Encased Steel Frame-Reinforced Concrete Core Tube Buildings Based on Dynamic Characteristics. *Appl. Sci.* **2017**, *7*, 314. [CrossRef]
7. Tavakoli, H.R.; Afrapoli, M.M.; Tavakoli, H.R. Robustness analysis of steel structures with various lateral load resisting systems under the seismic progressive collapse. *Eng. Fail. Anal.* **2018**, *83*, 88–101. [CrossRef]
8. Shi, Y.C.; Li, Z.X.; Hao, H. A new method for progressive collapse analysis of RC frames under blast loading. *Eng. Struct.* **2010**, *32*, 1691–1703. [CrossRef]
9. Kodur, V.K.; Naser, M.; Pakala, P.; Varma, A. Modeling the response of composite beam–slab assemblies exposed to fire. *J. Constr. Steel Res.* **2013**, *80*, 163–173. [CrossRef]
10. General Services Administration. *Progressive Collapse Analysis and Design Guidelines for New Federal Office Buildings and Major Modernization Projects*; General Services Administration: Washington, DC, USA, 2003.
11. Department of Defense. *Unified Facilities Criteria: Design of Buildings to Resist Progressive Collapse*; UFC 4-023-03; Department of Defense: Washington, DC, USA, 2013.
12. Li, H.; Cai, X.; Zhang, L.; Zhang, B.; Wang, W.; Li, H. Progressive collapse of steel moment-resisting frame subjected to loss of interior column: Experimental tests. *Eng. Struct.* **2017**, *150*, 203–220. [CrossRef]
13. Forquin, P.; Chen, W. An experimental investigation of the progressive collapse resistance of beam-column RC sub-assemblages. *Constr. Build. Mater.* **2017**, *152*, 1068–1084. [CrossRef]
14. Song, B.I.; Sezen, H. Experimental and analytical progressive collapse assessment of a steel frame building. *Eng. Struct.* **2013**, *56*, 664–672. [CrossRef]
15. Guo, L.; Gao, S.; Fu, F.; Wang, Y. Experimental study and numerical analysis of progressive collapse resistance of composite frames. *J. Constr. Steel Res.* **2013**, *89*, 236–251. [CrossRef]
16. Lu, X.; Lin, K.; Li, Y.; Guan, H.; Ren, P.; Zhou, Y. Experimental investigation of RC beam-slab substructures against progressive collapse subject to an edge-column-removal scenario. *Eng. Struct.* **2017**, *149*, 91–103. [CrossRef]
17. Wang, T.; Chen, Q.; Zhao, H.; Zhang, L. Experimental study on progressive collapse performance of frame with specially shaped columns subjected to middle column removal. *Shock Vib.* **2016**, *2016*, 7956189. [CrossRef]
18. Cosgun, T.; Sayin, B. Damage assessment of RC flat slabs partially collapsed due to punching shear. *Int. J. Civ. Eng.* **2017**, *16*, 725–737. [CrossRef]
19. Lima, C.; Martinelli, E.; Macorini, L.; Izzuddin, B.A. Modelling beam-to-column joints in seismic analysis of RC frames. *Earthq. Struct.* **2017**, *12*, 119–133. [CrossRef]
20. Boonmee, C.; Rodsin, K.; Sriboonma, K. Gravity load collapse behavior of nonengineered reinforced concrete columns. *Adv. Civ. Eng.* **2018**, *2018*, 9450978. [CrossRef]
21. Zhang, Q.; Li, Y. The performance of resistance progressive collapse analysis for high-rise frame-shear structure based on opensees. *Shock Vib.* **2017**, *2017*, 3518232. [CrossRef]
22. Li, S.; Shan, S.; Zhai, C.; Xie, L. Experimental and numerical study on progressive collapse process of rc frames with full-height infill walls. *Eng. Fail. Anal.* **2016**, *59*, 57–68. [CrossRef]
23. Tavakoli, H.R.; Hasani, A.H. Effect of Earthquake characteristics on seismic progressive collapse potential in steel moment resisting frame. *Earthq. Struct.* **2017**, *12*, 529–541.

24. Eskandari, R.; Vafaei, D.; Vafaei, J.; Shemshadian, M.E. Nonlinear static and dynamic behavior of reinforced concrete steel-braced frames. *Earthq. Struct.* **2017**, *12*, 191–200. [CrossRef]
25. Elshaer, A.; Mostafa, H.; Salem, H. Progressive collapse assessment of multistory reinforced concrete structures subjected to seismic actions. *KSCE J. Civ. Eng.* **2017**, *21*, 184–194. [CrossRef]
26. Ferraioli, M. Dynamic increase factor for nonlinear static analysis of rc frame buildings against progressive collapse. *Int. J. Civ. Eng.* **2017**, 1–23. [CrossRef]
27. Ellobody, E.; Young, B. Numerical simulation of concrete encased steel composite columns. *J. Constr. Steel Res.* **2011**, *67*, 211–222. [CrossRef]
28. Weng, J.; Tan, K.H.; Lee, C.K. Adaptive superelement modeling for progressive collapse analysis of reinforced concrete frames. *Eng. Struct.* **2017**, *151*, 136–152. [CrossRef]
29. Adom-Asamoah, M.; Banahene, J.O. Nonlinear seismic analysis of a super 13-element reinforced concrete beam-column joint model. *Earthq. Struct.* **2016**, *11*, 905–924. [CrossRef]
30. Psyrras, N.K.; Sextos, A.G. Build-x: Expert system for seismic analysis and assessment of 3d buildings using opensees. *Adv. Eng. Softw.* **2018**, *116*, 23–35. [CrossRef]
31. Lu, X.; Xie, L.; Guan, H.; Huang, Y.; Lu, X. A shear wall element for nonlinear seismic analysis of super-tall buildings using opensees. *Finite Elem. Anal. Des.* **2015**, *98*, 14–25. [CrossRef]
32. China Academy of Building Research. *Code for Design of Steel Structure*; China Architecture & Building Press: Beijing, China, 2010.
33. China Academy of Building Research. *Code for Seismic Design of Buildings*; China Architecture & Building Press: Beijing, China, 2010.
34. China Academy of Building Research. *Technical Specification for Concrete Structures of Tall Building*; China Architecture & Building Press: Beijing, China, 2010.
35. China Academy of Building Research. *Load Code for the Design of Building Structures*; China Architecture & Building Press: Beijing, China, 2012.
36. Luco, N.; Bazzurro, P.; Cornell, C.A. Dynamic versus static computation of the residual capacity of a mainshock-damaged building to withstand an aftershock. In Proceedings of the 13th World Conference on Earthquake Engineering, Vancouver, BC, Canada, 1–6 August 2004.
37. Li, Q.; Ellingwood, B.R. Performance evaluation and damage assessment of steel frame buildings under main shock–aftershock earthquake sequences. *Earthq. Eng. Struct. D* **2010**, *36*, 405–427. [CrossRef]
38. Kent, D. Flexural members with confined concrete. *J. Struct. Div.* **1971**, *97*, 1969–1990.

applied
sciences

MDPI

Review

Continuum Damage-Healing and Super Healing Mechanics in Brittle Materials: A State-of-the-Art Review

Chahmi Oucif [1,*] and Luthfi Muhammad Mauludin [1,2]

[1]   Institute of Structural Mechanics (ISM), Bauhaus-Universität Weimar, Marienstraße 15,
     D-99423 Weimar, Germany
[2]   Teknik Sipil, Politeknik Negeri Bandung, Gegerkalong Hilir Ds.Ciwaruga, Bandung 40012, Indonesia;
     luthfi.muhammad.mauludin@uni-weimar.de
*   Correspondence: chahmi.oucif@uni-weimar.de

Received: 21 October 2018; Accepted: 13 November 2018; Published: 22 November 2018

Abstract: Over the last several years, self-healing materials have become more and more popular in terms of damage reparation. Moreover, a recent theoretical investigation of super healing materials that aims at repairing and strengthening itself was also developed. This research area is well known by the rich experimental studies compared to the numerical investigations. This paper provides a review of the literature of continuum damage-healing and super healing mechanics of brittle materials based on continuum damage and healing mechanics. This review includes various damage-healing models, methodologies, hypotheses and advances in continuum damage and healing mechanics. The anisotropic formulations of damage and healing mechanics are also highlighted. The objective of this paper is also to review the super healing theory based on continuum damage-healing mechanics and its role in material and structure strengthening. Finally, a conclusion of the reviewed damage-healing models is pointed out and future perspectives are given.

Keywords: self-healing; damage-healing mechanics; super healing; anisotropic; brittle material

---

1. Introduction

Brittle materials are subjected to microstructural degradations that lead to their failure. The material degradation is the result of the nucleation and growth of microvoids and microcracks. This phenomenon is expressed and termed by damage. In recent years, self healing materials have been used to repair damage in materials. Therefore, many investigations are conducted on self-healing materials. French Academy of Science discovered the self-healing theory in 1836. They found that calcium carbonate results from cement hydration on exposure to atmosphere that concerns the autogenous self-healing mechanism [1,2]. This is due to hydration of cement or carbonation of calcium hydroxide [3]. The second category, called autonomous self-healing, was first proposed in [4]. The self-healing concept aims at automatically repairing the damages occurring in the material. Inspired from this idea, Barbero et al. [5] developed the continuum damage-healing mechanics (CDHM) for composited materials. CDHM represents the extension of the continuum damage mechanics (CDM) in which the healing effect is introduced into the constitutive equations. Furthermore, many investigations based on fracture mechanics and advanced finite element methods such as discrete particle/element method [6] and smooth particle hydrodynamics (SPH) [7] were also carried out on self-healing materials. For more details on these methods, the reader can refer to Refs. [8–13] for the discrete particle/element method, Refs. [14–17] for the SPH method, Refs. [18–24] for meshfree method, [25–30] for the multiscale method, Refs. [31–35] for the phase field method, and Refs. [36–42] for advances in fracture mechanics.

The presentation of CDM was first given by Kachanov [43] in which the continuum damage mechanics framework was originally applied to handle the response of the creep failure of metal alloys. This framework was further developed by Rabotnov [44] in which the damage factor concept was introduced. CDM framework was further extended by many researchers who aimed at describing the process of damage [45,46] where it was assumed that the material starts to rupture once the damage variable reaches a critical level. At the beginning of the application of CDM, much attention was given to the analysis of damage due to creep [47–50]. Later on, further developments were carried out using the principles of continuum damage mechanics [51–66]. In general, damage mechanics interests in the study of the material in different scales, namely, microscopic, mesoscopic, macroscopic, and mixed scale (statistical method), in which the damage models are applied to describe the variation of the material properties and material failure due to crack initiation and propagation. The basic issue of CDM is to quantify the damage in the material. Many researchers defined the damage variable as the ratio of the number of damaged and total cross-section [67,68], while other researchers used the concept of the effective stress to define the damage variable [69,70]. Another method used to calculate the damage variable which is based on the elastic stiffness reduction was also proposed by Lemaitre [71], and investigated further by many researchers [72,73]. The damage variable can be expressed as a scalar variable in the case of isotropic material and as a tensor in the case of anisotropic material [74,75].

The description of the quasi-static behavior of ductile and brittle materials came subsequently on [76–80]. Application of conventional local damage models results in ill-posedness problems and strain localization due to the softening behavior of brittle materials which can be avoided using particular a simulation. This simulation can be performed using the developed nonlocal damage models of integral and gradient types [81–83]. A large range of applicability of the nonlocal theories can be found in literature, which can simulate the crack initiation and propagation based on continuum mechanics. The nonlocal damage models raised some limitations which are still not completely resolved [84,85]. The characteristic length is a parameter intrinsic to the material whose characterization as well as the physical sense strongly depends on the material model chosen. Whatever the non-local method chosen, it enriches the description of the classical continuum mechanics.

Comparing to the investigations that carried out on continuum damage mechanics, the focus on the continuum damage healing mechanics is still in its infancy. In the present work, an overview of continuum damage-healing and super healing mechanics and their applicability is provided. Review of different aspects of damage and healing measures based on cross-sectional area and elastic stiffness reduction is given. Afterwards, advances of damage-healing models applied on brittle materials are discussed. The anisotropic formulation of damage and healing and some advances are also reviewed. Finally, the super healing theory that aims at strengthening of materials and structures is given and discussed in detail.

2. Damage and Healing Configurations

In this section, review of damage and healing variables is presented based on CDHM. According to CDM, the undamaged, damaged and effective material states defined by the undamaged cross-section S_0, damaged cross-section S_φ and effective cross-section $\bar{S}$ are respectively illustrated in Figure 1. The initial, damaged and effective configurations are also represented by their elastic modulus E_0, E_φ and $\bar{E}$, respectively. The damage variable can mostly be defined based on either cross-sectional or elastic stiffness reduction. The expressions of the damage variable based on cross-sectional reduction and elastic stiffness reduction are expressed as follows [86]:

$$\varphi = \frac{S_\varphi}{S_0} \quad with \;\; 0 \leq \varphi \leq 1, \tag{1}$$

$$\varphi = \frac{\bar{E} - E_\varphi}{E}, \tag{2}$$

where φ is the damage variable and it takes the value $\varphi = 0$ when the material is undamaged and the value $\varphi = 1$ when the material is totally damaged. In [87], the authors defined the damage variable based on shear modulus G, Poisson's ratio ν and the bulk modulus K reduction respectively as follows:

$$\varphi = \frac{\bar{G} - G}{G}, \tag{3}$$

$$\varphi = \frac{\bar{\nu} - \nu}{\nu}, \tag{4}$$

$$\varphi = \frac{\bar{K} - K}{K}, \tag{5}$$

where $\bar{G}$, $\bar{\nu}$ and $\bar{K}$ are, respectively, the effective shear modulus, effective Poisson's ratio and the effective bulk modulus. As the damage is well-known to reduce the cross-section and material stiffness, the healing takes an opposite role of the recovering of the cross-section and material stiffness (see Figure 2). Then, the healing variable can be defined respectively as a function of the cross-section and material stiffness recovery follows [86]:

$$h = \frac{S_h}{S_\varphi} \quad with \quad 0 \le h \le 1, \tag{6}$$

$$h = 1 - \frac{\bar{E} - E_h}{\varphi E}, \tag{7}$$

where h, S_h and E_h represent the healing variable, healed cross-section and elastic stiffness, respectively. The values of $h = 0$, $0 < h < 1$, and $h = 1$ represent the unhealed, partially healed, and fully healed material states, respectively. In [88], the authors defined two healing variables that reflect the healing effect in the case of coupled and uncoupled self-healing mechanics. The former mechanism assumes that the healing and damage occur simultaneously, while the latter one assumes that the healing is introduced when damage is constant (more details are given in Section 3.6). The expression of the healing variable in the case of coupled self-healing mechanism (h_c) is similar to Equation (7), while, for the uncoupled self-healing mechanism (h_u), it is written as follows [88]:

$$h_u = \frac{S_h - S_\varphi}{S_h} \quad with \quad 0 \le h_u \le 1. \tag{8}$$

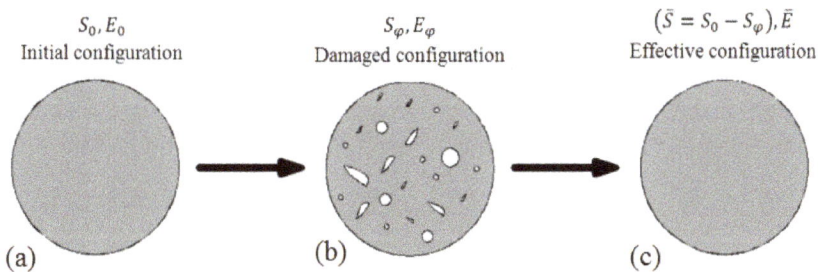

S_0, E_0
Initial configuration

S_φ, E_φ
Damaged configuration

$(\bar{S} = S_0 - S_\varphi), \bar{E}$
Effective configuration

(a)   (b)   (c)

Figure 1. Undamaged, damaged and effective material states [89]. (Copyright, 2018, Journal of engineering mechanics).

When damages are removed from (Figure 1c), the relationship between the nominal stress σ and effective stress $\bar{\sigma}$ becomes as follows [43,66,76]:

$$\bar{\sigma} = \frac{\sigma}{1 - \varphi}. \tag{9}$$

$$S_\varphi, E_\varphi$$
Damaged configuration

$$(S_0 - S_\varphi + S_h), E_h$$
Healed configuration

$$(S_0 - S_\varphi - S_h), \bar{E}$$
Effective configuration

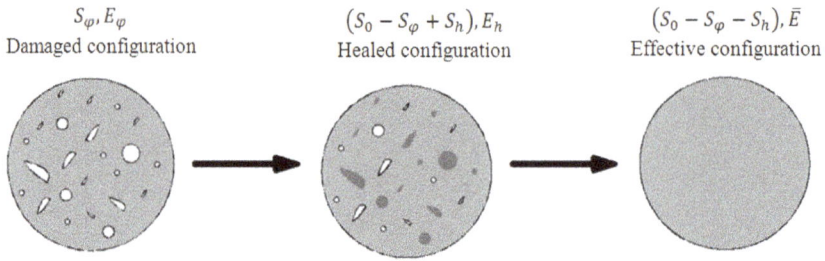

Figure 2. Damaged, partially healed, and effective material states [89]. (Copyright, 2018, Journal of engineering mechanics).

According to CDHM and following the configurations in Figure 2, Equation (9) becomes in the case of self-healing materials as follows [88,90]:

$$\bar{\sigma} = \frac{\sigma}{1 - \varphi(1 - h)}. \tag{10}$$

Another relation of the nominal stress and effective stress was proposed in [91]. The authors assume that the material is totally healed when $h = 0$ and it is totally damaged when $h = 1$. This proposition takes the following expression:

$$\bar{\sigma} = \frac{\sigma}{((1 - \varphi) + \varphi(1 - h))}. \tag{11}$$

It is clear from Equation (10) that the effective stress is equal to the nominal stress when the material is fully healed ($h = 1$ & $\varphi = 0$), while it approaches infinity when the material is totally damaged ($h = 0$ & $\varphi = 1$).

3. Damage-Healing Formulations

In this section, review of proposed formulations of healing laws applied on brittle materials are discussed. Some of them are not based on CDHM, but they are highlighted in this section in order to provide the reader an overall idea of mathematical, mechanical and phenomenological propositions of healing laws. Next, the damage-healing models that have been developed since 2005 and applied on different materials based on CDHM are also reviewed. The main differences and limitations between these models are also discussed.

3.1. Healing Model Based on Parameter Recovery

A theory of crack healing of polymers was developed by Wool and O'Connor [92] based on a recovery parameter R which is defined as a convolution product. According to this theory, the healing is defined in terms of five stages of healing: (a) surface rearrangement, (b) surface approach, (c) wetting, (d) diffusion and (e) randomization (Figure 3). Different mechanical properties of the material were considered in the intact state of the material such as fracture stress σ_∞, strain at failure ϵ_∞, tensile modulus Y_∞ and fracture energy E_∞ when the healing history is subjected. The healing history was measured such that the five stages of healing occur simultaneously and the mechanical properties of the material represent the sum of wetting and diffusion process initiated at different times. Based on this assumption, the healing variable was defined as follows:

$$R = \int_{\tau=\infty}^{\tau=t} R_h(t - \tau) \frac{d\phi(\tau, X)}{d\tau} d\tau, \tag{12}$$

where $R_h(t)$ is the intrinsic healing function, $\phi(\tau, X)$ is the wetting diffusion function and τ is the nucleation time and represents the running variable of the time axis. The intrinsic healing function was related to the wetting and diffusion for the measure of recovery based on stress or energy consideration. The wetting is obtained when two free surfaces touch each other in which the time is controlled by self-diffusion of the overlapping free surfaces. The diffusion is controlled by the stage of surface rearrangement. When the material is damaged and the cracks appear, the molecular ends start to be able to move on the surface following the wetting stage. When two surfaces start to come into contact, their diffusion across the interface results in the healing and recovery of part of the initial strength. Two cases of wetting diffusion function are considered, namely instant wetting and constant rate wetting. In the case of instant wetting, the two surfaces wet instantaneously at time $t = 0$ and the wetting diffusion function is expressed as follows:

$$\frac{d\phi}{dt} = \delta(t),$$
(13)

where $\delta(t)$ is the Dirac-delta function. Consequently, the healing variable in Equation (12) and the intrinsic healing function become similar, as follows:

$$R = R_h(t) = R_0 + Kt^{1/4}/\sigma_\infty,$$
(14)

where K and σ_∞ are the material constant and the fracture strength of the intact material, respectively. On the other hand, in the case of constant rate wetting, the wetting diffusion function is written as:

$$\frac{d\phi(t)}{dt} = k_d U(t),$$
(15)

where k_d and $U(t)$ are the wetting rate and the Heaviside step function, respectively. Thus, the healing variable is expressed as follows:

$$R = R_0 k_d t + 4k_d K t^{5/4}/5\sigma_\infty.$$
(16)

According to Equations (14) and (16), it is observed that the wetting components of the healing variable is time-independent in the case of instant wetting, while it is time-dependent in the case of constant wetting rate. In addition, it is concluded that Equations (14) and (16) are defined based on empirical assumption using large number of material parameters. Figure 4 shows the plot $R - R_0$ with respect to the crack healing.

3.2. Fracture Mechanics Based Healing Model

A crack closing model applied on linear and isotropic viscoelastic materials was developed by Schapery [93]. Time-dependent constitutive equations based on continuum mechanics were proposed in which the crack length and contact size are predicted, and the whole healing process is considered. The crack healing model was based on crack area reduction $\dot{a}_b$, which is related to the Poisson's ratio, fracture process zone, effective bond energy and the tensile bond force. $\dot{a}_b$ is expressed as follows:

$$\dot{a}_b = \pi \left[4\Gamma_b' \right]^{(2+1)/m} \left[\left(1 - v^2 D_1^+ \gamma_m / C_m \right) \right]^{1/m} \left[E_R^+ / \left(1 - v^2 \right) K_I^R \right]^{2(1+1/m)} /8\sigma_b^2 C_m^2,$$
(17)

where Γ_b', E_R^+, σ_b, v and K_I^R are the effective bond energy, elastic modulus, tensile bond force, Poisson's ratio and mode I stress intensity factor. γ_m, C_m, D_1^+ and m are material constants. The crack closing model is known by the fact that there is no difference between the crack closing based on wetting and on diffusion. Moreover, the model is formulated based on different materials that make it difficult to be realized in the case of anisotropic material.

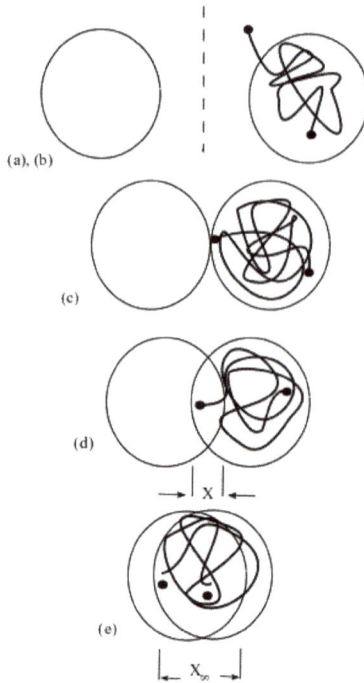

Figure 3. Five stages of healing of two random-coil chains on opposite crack surfaces [92]. (Copyright, 1981, Journal of Applied Physics)

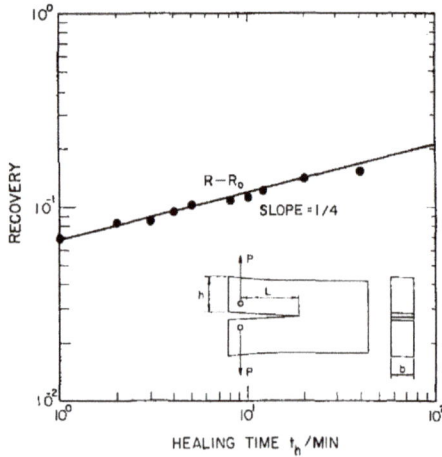

Figure 4. Log of healing based on fracture load recovery [92]. (Copyright, 1981, Journal of Applied Physics).

Based on the healing variable proposed in [92] and the crack area reduction in Equation (17) [93], a new formulation was proposed in [94,95] in which they used the variable R to simulate the healing effect in bituminous materials as follows:

$$\frac{d\phi(t, X)}{dt} = \dot{a}_b = \beta \left[\frac{1}{D_1 k_m} \left\{ \frac{\pi W_c}{4(1 - v^2)\sigma_b^2 \beta} \right\} \right]^{-1/m}, \tag{18}$$

where W_c, β and k_m are the work of cohesion, the healing process zone and material constant, respectively. The rest of the parameters are defined previously in Equation (17). It should be noted that the parameters related to the proposed Equation (18) are difficult to be identified due to the lack of enough experimental data.

3.3. Creep Damage-Healing Model for Salt Rock

Rock salt is generally subjected to creep damage and cracks that result in the increase of the permeability of the material. Because damage results in inelastic flow in rock salt under hydrostatic compression, an extension of continuum damage approach to the healing of creep damage was developed in [96]. It was assumed that the macroscopic strain rate is influenced by the healing mechanism along with damage and creep. Anisotropic healing was also considered such that the conjugate stress measure for healing can be expressed as

$$\sigma_{eq}^h = \frac{1}{3}(I_1 - x_{10}\sigma_1), \tag{19}$$

where I_1, x_{10} and σ_1 are the first invariant of the Cauchy stress, material constant and the maximum principal stress, respectively. It is considered that the healing is isotropic when $x_{10} = 0$ and anisotropic when $x_{10} \neq 0$. The kinetic equations of the healing were formulated based on an experimental observation that suggests two healing mechanisms can be activated in (Waste Isolation Pilot Plant) WIPP salt. The first mechanism assumes that the healing is present in a much smaller time period which results in unchanged damage variable, while the second mechanism assumes that the healing is present in a larger time period which reduces the damage variable. The healing variable proposed in [96] is considered to be the first-order kinetic equation expressed as

$$h = \frac{\omega \sigma_{eq}^{h_2} H\left(\sigma_{eq}^{h_2}\right)}{\tau_2 \mu} \tag{20}$$

where ω, H, τ_2, μ are the damage variable, Heaviside function, time characteristic constant and the shear modulus, respectively. h_2 describes the removal of damage. The healing variable defined in Equation (20) is characterized by the fact that only an individual healing mechanism can be simulated. This leads to the difficulty of finite element implementation of the healing model due to the challenging identification of the healing time involving non-relative loading histories. Thus, the starting time and the period of the healing become ambiguous.

A simplified version of the healing model proposed in [96] was further developed in [97]. Instead of defining two separate healing mechanisms, a single healing mechanism based on damage, kinetic equation and equivalent stress was proposed. The simplified healing variable was defined with only one kinetic equation for one healing mechanism in contrast to the previous investigation in which one kinetic equation for each healing mechanism was defined. After modification of the kinetic equation, the following healing variable was obtained:

$$h = \frac{\omega \left(\sigma_{eq}^h - \sigma_b\right) H\left(\sigma_{eq}^h - \sigma_b\right)}{\tau \mu} \tag{21}$$

with

$$\sigma_b = x_7 \left| \frac{\sigma_1 - \sigma_3}{x_2 x_7} \right|^{1/x_6},$$ (22)

where σ_1, σ_3 and σ_{eq} are principal stresses and equivalent stress, respectively. x_2, x_6 and x_7 are material constants. According to Equation (21), healing can be activated only when $\sigma_{eq} > \sigma_b$. In addition, the healing reduces the volumetric strain to zero. The axial and lateral strains are also recovered under hydrostatic compression. In [98], a thermodynamic framework of CDHM was proposed in which the concept of healing surface and loading-unloading conditions were used. Rate-dependent and rate-independent formulations were also given and applied in the case of isotropic healing. The general thermodynamic framework was applied to study the healing crushed rock salt. The surface-based healing function takes the following expression:

$$F = s : s - c_B \left[(B^{sp})^2 + \left(B^d \right)^2 \right] + c_S (\omega)^2 - F_0,$$ (23)

where s, B, ω and F_0 are the Cauchy stress tensor, constrained modulus, surface energy per unit area, and the material positive constant, respectively. c_B and c_S are positive material parameters related to the changes in material parameters and surface area. If the healing surface $F < 0$, the material is supposed unhealed, while the healing is occurring when $F = 0$. The simulation of the densification of crushed rock salt revealed that the healing model is able to describe the healing mechanism in terms of Young's modulus and inelastic strain recovery, even though the formulation was limited to isotropic material behavior.

Further investigations of damage-healing of salt rock were undertaken. Based on the formulation in [97], the authors in [99] proposed an anisotropic damage-healing formulation for the modeling of creep process in salt rock. The healing process was defined with respect to a viscoplastic scalar healing variable, and the healing strain component was used to account for the reduction of the deformation. However, it was assumed that the healing compensates deformation only in the lateral directions. In addition, the crack healing was assumed to result in an instant reduction of deformation in which τ represents the characteristic time needed to close the cracks of the material subject to salt creep. The simplified healing variable that accounts only for diffusion subject to compressive mean stress was expressed as

$$\dot{h} = \frac{tr(A) \, pH(p)}{\tau G},$$ (24)

where A, p, H and G are the new damage variable, first stress invariant, Heaviside function, and the shear modulus, respectively. For more details of the formulation and application of Equation (24), the reader can refer to [100–104]. Xu et al. [105] implemented an elastoplastic damage healing model of mudstone and defined the healing variable as function of the non-associated dissipation criterion. Unlike the expression of the healing variables defined in Equations (6) and (8), a new healing variable equation is expressed as

$$h = \frac{A_h}{A_{ud}},$$ (25)

where A_h and A_{ud} are the healed and undamaged cross-section areas, respectively. It was assumed that the undamaged cross-section area of the healed cross-section area carries the loads. Following this assumption, the effective stress was expressed as follows:

$$\bar{\sigma} = \frac{\sigma}{(1-d)(1+h)}.$$ (26)

According to this formulation, the cross-section area of the material is divided into three regions: undamaged A_{ud}, unhealed A_{uh} and healed A_h cross-section areas. The damages in the mudstone cannot be fully healed, which results in the unhealed cross-section area being greater than zero. Therefore, it was assumed for simplicity that the healed cross-section area exhibits similar mechanical behavior to that of the original material. However, the boundary conditions of the healing and damage of Equation (26) was unclear. For further self-healing investigations on geomaterials, the reader can refer to [106–114].

3.4. Micro-Damage Healing Models for Asphalt Mixtures

In [115], an elastic-viscoelastic model with healing for asphalt concrete subjected to fatigue loading was proposed. The model was extended from the work presented in [116] in which the pseudo-strain variables are given in [117] and adopted to eliminate the dependency of the stress–strain material behavior to time. The irreversible thermodynamic framework was used to simulate the healing of micro-damage. Growing damage was simulated using uniaxial viscoelastic constitutive equations that are extended to account for the micro-damage healing. Uniaxial tensile tests under cyclic loading were conducted under controlled-strain and controlled-stress models with rest periods. In order to induce damage in the specimens, two stress–strain levels were used in the tests. The rest periods introduced during each test vary from 0.5 to 32 min. The variation of the material stiffness before and after the rest period was studied as a function of the number of cycles (Figure 5). Region I in Figure 5 depicts the reduction of the stiffness due to the damage evolution without a rest period, while region II depicts the reduction of the stiffness due to the damage evolution after rest period. After the introduction of the rest period, it was shown that the stiffness increases from point B to point A due to micro-damage-healing and decreases after damage of the healed material. Based on the experimental results and stiffness variation in regions I and II, the following healing function was proposed:

$$H = \left[S_{B,i}^R + C_2 \left(S_{2,i} \right) \right] C_3 \left(S_{3,i} \right) - C_1 \left(S_{1n} \right) - \sum_{j=1}^{i-1} \left(S_{B,j}^R - S_{C,j}^R \right), \tag{27}$$

when $S^R > S_{B,i}^R$ (region I),

$$H = \sum_{j=1}^{i} \left(S_{B,j}^R - S_{C,j}^R \right), \tag{28}$$

when $S^R < S_{B,i}^R$,

where $(S_{2,i})$ represents the healing evolution during the ith rest period and $(S_{3,i})$ represents the damage evolution after the ith rest period. $C_1(S_{1n})$ and (S_{1n}) are the material function and the normalized damage variable. Several experimental investigations were carried out for the study of the micro-damage healing of asphalt mixtures (e.g., [118–124]). Although the implemented micro-damage-healing model was able to describe the hysteretic behavior under controlled-strain and controlled-stress modes, the identification of the experimental data to simulate the healing behavior is not an easy task.

Another micro-damage-healing model applied to asphalt mixtures subjected to fatigue loads was proposed in [125]. The damage healing model was extended from the viscoelastic, viscoplastic and viscodamage model. The authors defined a healing variable which is a function of the healing time and history, damage level and temperature. The proposed healing variable takes the following expression:

$$\dot{h} = \Gamma^h \left(T \right) \left(1 - \bar{\phi} \right)^{b_1} \left(1 - h \right)^{b_2}, \tag{29}$$

where $\dot{h}$ is the rate of the healing variable, Γ^h is the healing viscosity parameter and b_1 and b_2 are material constants. $\Gamma^h(T)$ is a function of temperature and is the parameter that determines the speed of the healing. It is expressed as

$$7\Gamma^h(T) = \Gamma_0^h exp[-\delta_3(1 - \frac{T}{T_0})], \tag{30}$$

where Γ_0^h and δ_3 are the healing viscosity parameter at temperature T_0 and the healing-temperature coupling parameter, respectively. According to Equation (29), it was assumed that the healing starts to evolve once the temperature reaches a certain reference level (temperature threshold) and decreases when the temperature is less than the reference level. The model was applied to predict the behavior of creep-recovery tests in compression and in tension. An example of the results of the evolution of the creep strain and the effective damage density as function of time in compression is shown in Figure 6. From Figure 6a, it is shown that the introduction of the healing improves the material behavior compared to the model without healing. On the other hand, one can see from Figure 6b that the effective damage increases during loading and decreases during the rest period, while it remains stable during unloading.

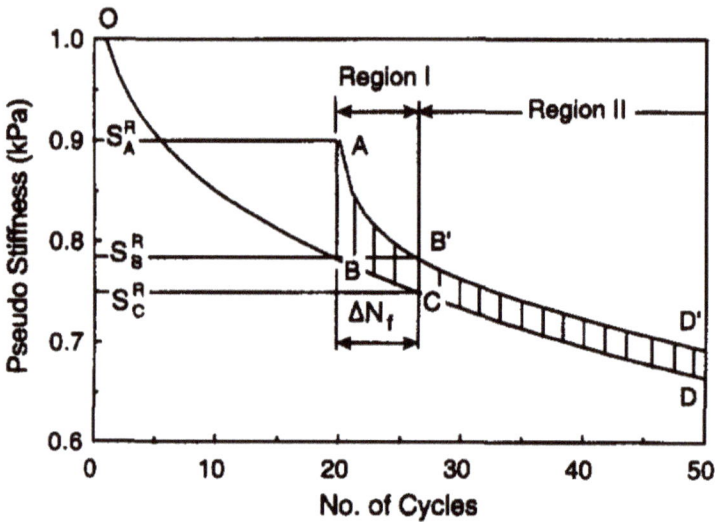

Figure 5. Variation of the stiffness before and after the rest period [115]. (Copyright, 1998, Journal of engineering mechanics).

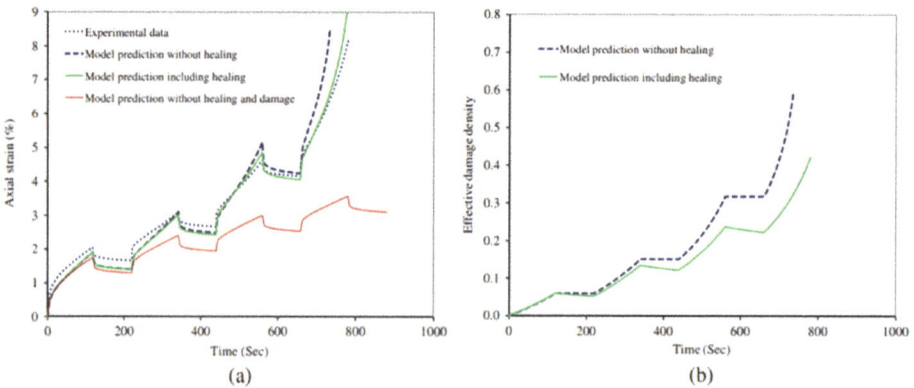

Figure 6. Results of repeated recovery test in compression with 120 s loading time and 100 s of rest period. (**a**) compared creep strain; (**b**) evolution of the effective damage density [125]. (Copyright, 2010, International Journal of Engineering Science).

The micro-damage healing law proposed in [125] was further investigated in [90,126–128]. In [90], a continuum damage mechanics framework was proposed to simulate the micro-damage-healing of materials subjected to cyclic loading. The hypotheses' strain, elastic energy and power equivalence were used to relate the strain tensor and tangent stiffness in the damage and healing configurations. The authors worked on the update of the current stress tensors in the damaged and healing configurations. Examples of the uniaxial constant strain and stress rates were applied. The results revealed that the hypotheses of strain equivalence and power equivalences overestimate the elastic strain energy in the healing configuration compared to the one in the damaged configuration. On the other hand, the strain equivalence hypothesis overestimates the expanded power in the healing configuration compared to the one in the damaged configuration, while the elastic energy equivalence hypothesis underestimates the expanded power in the healing configuration compared to the one in the damaged configuration. It should be noted that these results apply to both strain-controlled and stress-controlled uniaxial tests. The same authors used the micro-damage healing model to simulate fatigue damage of asphalt concrete [126]. They also studied the effect of compressive stresses on the crack closure. This phenomenon is called the "unilateral effect" and is discussed in more detail in Section 3.7. Based on the formulation presented in [125], a theoretical framework of cohesive zone healing model was proposed in [127] and implemented into a finite element code in [128]. The effect of different parameters such as damage history, healing history and resting time were studies. For further investigations on the visco-damage-healing models, the reader can refer to [129–133].

3.5. Curing-Based Damage Healing Law

In [134], the authors proposed a new phenomenological damage-healing model applied to polymers. The proposed healing variable concerned the autonomous self-healing concept and was associated with the curing mechanism of the healing agent and the catalyst. In a microcapsules-based self-healing concept, the propagated cracks break the microcapsules, which results in the release of the healing agent. This latter fills the crack, reacts with the catalyst and they form a solid material in the crack area (Figure 7). It was assumed in this work that the process of cure leads to the mechanical properties variation and stiffness recovery [135]. The formulation was not limited to the healed material, re-damage of the healed material was also considered. The same equations of healing, damage and effective stress in Equations (1), (6) and (10), respectively, were used. As previously shown in Equation (12) [92], the convolution integral was used to define the healing as follows:

$$h(t) = \int_{s=t^c}^{t} d(s)\eta_h exp(-\eta_h[t-s])ds, \tag{31}$$

where $d(s)$, η_h are the damage variable during the healing period and the parameter that determines the speed of the healing process, respectively. t^c and t are the initial time of the healing and the healing time, respectively. It was assumed that the damage threshold decreases by the introduction of the healing as it is increased due to damage. Therefore, a damage threshold equation was defined which assumes that the behavior of the fully healed material is similar to the original material and the evolution of the healing variable does not affect the increase of the damage variable at constant deformations. The healing was introduced in three cases. The first one concerns the introduction of the healing during the rest period. When the material is loaded and unloaded, the healing variable evolves during a rest period. Afterwards, the material is reloaded and comparison of the stress–strain response of the healed and original materials is carried out. Figure 8 elucidates the stress–strain response of the healed material when the healing is introduced during different rest periods.

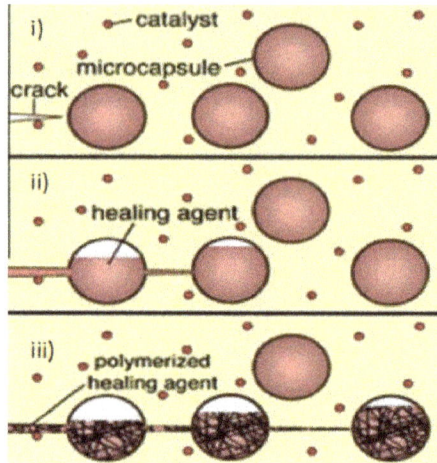

Figure 7. Microcapsules-based self-healing mechanism: (**i**) crack formation; (**ii**) the crack breaks the microcapsules and the healing agent releases into the crack area; (**iii**) solidification of healing agent in contact with the catalyst [136]. (Copyright, 2012, Construction and Building Materials).

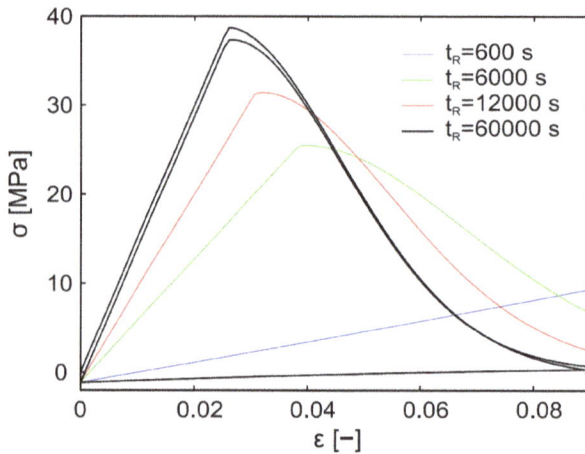

Figure 8. Stress–strain response of the healed material with different rest periods [134]. (Copyright, 2013, Computational Mechanics).

In the second example, the healing is introduced when the material is partially damaged while the assumption of its evolution during a required recovery time is kept. It should be noted that, in this example, the strain is released before the evolution of the healing, which means that damage is constant in this phase. The third example concerns the introduction of the healing during a rest period while the strain is assumed to be constant. For further works on modeling of self-healing polymers and microcapsules-based self-healing, the reader can refer to [137–141].

3.6. Damage-Healing Law for Concrete

In the previous work [88], a continuum damage-healing model for autonomous and autogenous self-healing concrete was proposed. In this model, a time-dependent healing variable representing the opposite of the damage variable was proposed. Isotropic material was considered at macroscale subjected to tensile load. The concept of coupled and uncoupled self-healing mechanism was introduced. The uncoupled self-healing mechanism represents the autogenous self-healing, and the coupled mechanism represents the the autonomous self-healing. In addition, both mechanisms were applied in the case of the so-called nonlinear self-healing theory, and comparison with the linear self-healing theory was performed. The authors in [142] proposed the nonlinear self-healing theory. It concerns the generalized nonlinear and quadratic self-healing theories. It should be noted that the classical (linear) self-healing theory is represented in Equation (10). It is called linear because the equation is linear in h. It was revealed that, in the case of small damage, the linear healing theory is a special case of the nonlinear healing theory. The configuration of the nonlinear healing theory is illustrated in Figure 9. Classical damage variable φ is used to describe the damaged material state in Figure 9a, while the healing variable h is used to describe the partially healed material state in Figure 9b. According to the nonlinear healing theory, a partial area of damage is subjected to healing as clearly shown in Figure 9b in which the healed area S_h is less than the damaged area S_φ. The theory of decomposition of the damage variable was used to obtain the combined healing/damage variable φ_{hd}. For more details on the decomposition theory, the reader can refer to [75,143]. It should be noted that the combined healing/damage variable of the classical self-healing theory takes the following expression:

$$\varphi_{hd} = \varphi\,(1 - h)\,. \tag{32}$$

$$S_\varphi$$
Damaged area

$$S_\varphi, S_h$$
Damaged and healed areas

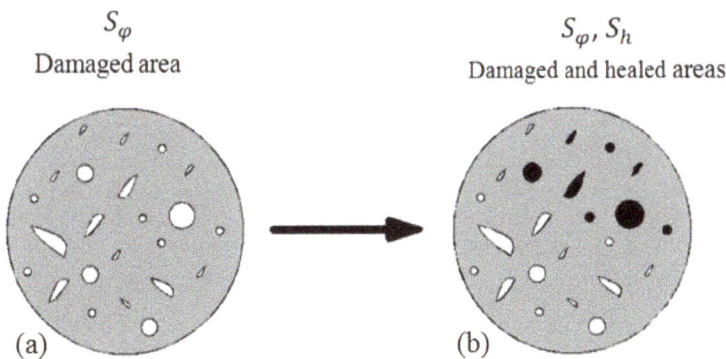

(a)   (b)

Figure 9. Configuration of the nonlinear self-healing theory [89].

According to the nonlinear self-healing theory, Equation (32) becomes respectively in the case of generalized nonlinear and quadratic self-healing theories as follows:

$$\varphi_{hd} = \frac{\varphi\,(1 - h)}{1 - h\varphi}, \tag{33}$$

$$\varphi_{hd} = \varphi\,(1 - h)\,(1 + \varphi h)\,. \tag{34}$$

Equations (33) and (34) represent the expression of the nonlinear healing the generalization of Equation (32) in the case of generalized nonlinear and quadratic self-healing theories, respectively. Following the expressions in Equations (33) and (34), the equations of the effective stress of the generalized nonlinear and quadratic self-healing models are respectively expressed as follows:

$$\sigma = \frac{\sigma\,(1 - h\varphi)}{1 - \varphi}, \tag{35}$$

$$\sigma = \frac{\sigma}{1 - (\varphi - h\varphi)\,(1 + \varphi h)}\,. \tag{36}$$

The damage healing model proposed in [88] was formulated based on the introduction of the healing variable into the damage model. When Mazars damage model for concrete material was adapted, the damage variable was expressed as [143]

$$
\begin{cases}
\varphi = 1 - \frac{(1-A)k_0}{\varepsilon_u} - A \, exp[\beta(\varepsilon_u - k_0)], & if \ \varepsilon_u > k_0, \\
\varphi = 0, & if \ \varepsilon_u < k_0,
\end{cases}
\tag{37}
$$

where A and β are material parameters and k_0 is the strain threshold of damage. ε_u is the unidirectional strain. It was assumed that the healing is introduced during loading, unloading, and rest period phases; deformed and undeformed material states. The deformed state represents the coupled self-healing mechanism, while the undeformed state represents the uncoupled self-healing mechanism. In the first case, damage and healing evolve simultaneously, and, in the second case, the healing is introduced during unloading or rest periods. In addition, the generalized nonlinear self-healing formulation was applied in the case of coupled self-healing mechanism, and the quadratic self-healing theory was applied in the case of uncoupled self-healing mechanism. The influence of several parameters on the healing efficiency was studied. It concerns the damage history, rest period and material characteristics. The material characteristics were defined mathematically using the parameter γ. The proposed healing variable was expressed in two cases; uncoupled and coupled self-healing mechanisms as follows:

$$
\begin{cases}
h_u(t) = 1 - exp\left[-\gamma\varphi(t_h)\left(t_{hf} - t_{hi}\right)\right], & if \ \dot{\varphi} = 0, \\
h_u(t) = 0, & if \ \dot{\varphi} > 0,
\end{cases}
\tag{38}
$$

$$
\begin{cases}
h_c(t) = 1 - exp\left[-\gamma\varphi(t_h)\left(t_{hf} - t_{hi}\right)\right], & if \ \dot{\varphi} > 0 \ \& \ \varphi \geq \varphi_{cr}, \\
h_c(t) = 0, & if \ \dot{\varphi} > 0 \ \& \ \varphi < \varphi_{cr},
\end{cases}
\tag{39}
$$

where h_u, h_c γ are the uncoupled healing variable, uncoupled healing variable and the material parameter, respectively. $\varphi(t_h)$ is the damage variable during the healing period t_h and φ_{cr} represents the critical damage that induces the healing process. $\varphi(t_h)$ is constant after unloading phase ($\dot{\varphi} = 0$). In this model, the healing efficiency is described during loading in which damage and healing evolve simultaneously. In this case, damage evolves until failure. The healing is assumed to start at time t_{hi} and stops at time t_{hf}. The healing period is defined by t_h. The material paramaters influencing the healing efficiency considered in this work were the following:

- History of loading and damage;
- Rest period;
- Material characteristics that were reflected mathematically in this present work represented by the parameter γ.

Figure 10 shows the stress–time response of the damage healing model in the case of uncoupled self-healing mechanism according to the classical self-healing theory. It is clear that the stiffness recovery is partially recovered for a short period of healing, while it is fully recovered with 30,000 s of rest period. Figure 11 shows the stress–time response of the model in the case of coupled self-healing mechanism. In this example, different values of the material parameter γ were considered. It is shown that small value of the parameter γ results in partial stiffness recovery of the material, and $\gamma = 0.02$ results in complete stiffness recovery. The coupled and uncoupled self-healing mechanisms were also applied using the nonlinear self-healing theory. It was found that the healing efficiency is underestimated using both coupled and uncoupled nonlinear self-healing compared to the linear self-healing theory. Further investigations on self-healing concrete and cementitious materials can be found in [144–150].

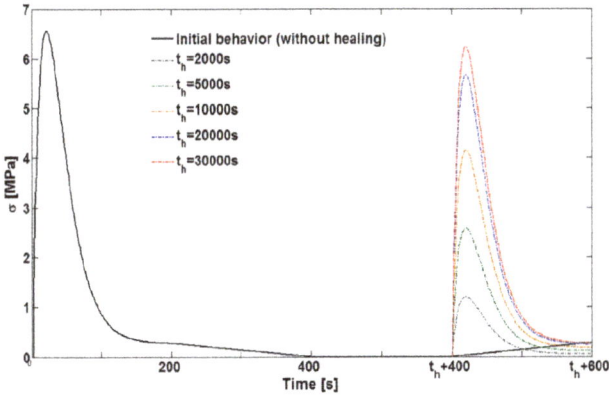

Figure 10. Stress–time response of the uncoupled damage healing model with various rest periods [88]. (Copyright, 2018, Theoretical and Applied Fracture Mechanics).

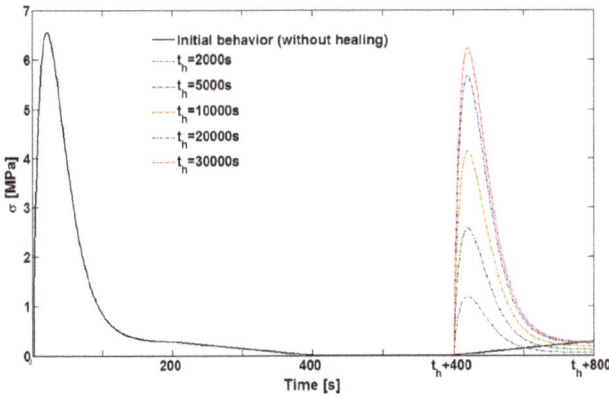

Figure 11. Stress–time response of the coupled damage healing model with different values of the material parameter [88]. (Copyright, 2018, Theoretical and Applied Fracture Mechanics).

3.7. Unilateral-Effects-Based Models

Damage material weakens the mechanical properties of the material. These properties can be recovered if the cracks close again. When the material is subjected to tensile loads and followed by compressive loads in the same direction, the cracks close in the compression domain and the material recovers its stiffness. This is known by the unilateral effects. The unilateral effect is simulated with the distinguishment between damage in tension and damage in compression using two damage variables. Using damage variable in tension and damage variable in compression, the loading mode for a diffuse network of identical microcracks is defined. The unilateral effect can be classified as a healing process due to its effect of crack closure. Many authors pointed out that taking into account the unilateral effect often leads to ambiguity in the computational analysis [151,152]. In [153], the authors developed an isotropic 3D damage model for quasi-brittle materials that accounts for the microcracks closure. An anisotropic continuum damage framework accounting for the unilateral effect was proposed in [154]. In [155], the crack closure was simulated through the decomposition of the stress and strain tensors into positive and negative projection operators. Zhu and Arson [156] proposed a thermodynamic framework to study the effect of the mechanical stress and temperature on the crack opening and closing in rocks, and crack closure was simulated through unilateral effect. The authors in [157] proposed a micro-macro chemo-mechanical damage-healing model to simulate the evolution of the salt stiffness due to microcracks opening, closing and propagation. A unilateral effect was taken into account to simulate the crack closure and stiffness recovery under compression. The proposed model was found to be able to predict the stiffness recovery by the unilateral effect of crack closure. In [158], a nonlocal formulation of concrete damage model with unilateral effects. Unlike the use of spectral decomposition of stress or strain, the unilateral effect was simulated using the trace of the strain tensor. Matallah and La Borderie [159] developed an inelastic-damage model that simulates the crack opening due to inelasticity and elastic modulus recovery due to crack closure. The crack closure was described using a scalar damage variable that is coupled with the Unitary Crack Opening (UCO). UCO is the internal variable that describes the inelastic strain. A function S called Cracks Opening Indicator was introduced in order to control the vanishing of the inelastic strain effect in the material when it is loaded under compression. It was assumed that the function S takes the value zero when the cracks are completely closed and takes the value of 1 when the cracks are completely opened. The expression of the function S was defined as

$$S = 1 - \frac{F_t^{ac}}{F_t^{\sigma_c}},\tag{40}$$

where F_t^{ac} is the actual tension yield function value and $F_t^{\sigma_c}$ is the tension yield function value corresponding to the crack closure stress σ_c. It was also assumed that the material recovers its initial stiffness when $F(\sigma_{ij}) = F_t^{\sigma_c}$. It should be noted that the proposed model was not able to simulate complex problems that occur during complex unloading phase because the function S represents a scalar variable.

4. Anisotropic Damage-Healing Formulations

Damage and healing of brittle materials are generally simulated by conventional continuum damage-healing mechanics in which scalar damage and healing variables are used to describe the relationship between nominal stress and effective stress; isotropic damage models. In addition, anisotropic damage-healing formulations were also recently proposed and studied using second-order and fourth-order damage and healing tensors. Murakami [160] was the first who generalized the multi-axial anisotropic formulation of the description of the material degradation. Although most of the works conducted on CDHM are based on isotropic presentation, an anisotropic CDHM was also investigated by the introduction of a healing tensor. According to the formulation [161], the damage variable tensor is expressed as

$$\phi_{ij}n_i = \frac{(dAn_j - d\bar{A}\bar{n}_j)}{dA}; 0 \leq (\phi_{ij}\phi_{ij})^{1/2} \leq 1,\tag{41}$$

where dAn_j and $d\bar{A}\bar{n}_j$ are the damage and effective fictitious area vectors, respectively. When the healing is introduced into the material, the effective area increases. Figure 12 shows the anisotropic damage and healing configurations [161]. According to the presentation in Figure 12, the authors in [161] proposed a second rank anisotropic healing variable tensor as follows:

$$h_{ij}n_i^d = \frac{\left(\phi_{jk}dAn_k - dA^h n_j^h\right)}{dA^d}; 0 \leq (h_{ij}h)^{1/2} \leq 1,\tag{42}$$

where h_{ij} describes the relationship between the damaged area vector dAn_i and the effective healed area vector. In the same paper, k_{ijkl} is denoted the fourth-order anisotropic damage variable tensor and describes the elastic modulus degradation as follows:

$$\begin{cases} k_{ijkl}^{(1)} = \left(\bar{E}_{ijmn} - E_{ijmn}^{d} \right) \bar{E}_{mnkl'}^{-1} \\ k_{ijkl}^{(2)} = \left(\bar{E}_{mnkl} - E_{mnkl}^{d} \right) \bar{E}_{ijmn'}^{-1} \end{cases} \tag{43}$$

where $\bar{E}_{ijkl}$ and E_{ijkl}^{d} are the undamaged and damaged elastic tensors. The subscripts in Equation (43) represent the two different mathematical tensorial expressions of the damage tensor. In addition, a new fourth rank healing tensor h'_{ijkl} was also defined to measure the elastic modulus recovery as follows:

$$\begin{cases} h_{ijkl}^{'(1)} = \left(E_{ijmn}^{h} - E_{ijmn}^{d} \right) E_{mnkl'}^{d-1} \\ h_{ijkl}^{'(2)} = \left(E_{mnkl}^{h} - E_{mnkl}^{d} \right) E_{ijmn'}^{d-1} \end{cases} \tag{44}$$

where E_{ijkl}^{h} is the healed elastic modulus. It was assumed that the material is undamaged when $h'_{ijkl} = 0_{ijkl}$ is the fourth rank zero tensor) and is fully healed when $h_{ijkl}^{'max} = k_{ijkl}^{max}$.

The generalization of the relational between the effective stress and nominal stress of Equation (9) is expressed in the case of anisotropic materials as follows [64,74]:

$$\bar{\sigma}_{ij} = M_{ijkl} \sigma_{kl}, \tag{45}$$

where M_{ijkl} represents the fourth-rank damage effect tensor. $\bar{\sigma}_{ij}$ and σ_{kl} are the effective and Cauchy stress tensors, respectively. The relationship between M and φ was investigated in the literature [66,87,162] and expressed as

$$M = \frac{1}{1 - \varphi}. \tag{46}$$

In the case of anisotropic damage-healing mechanics, Equations (35) and (36) become

$$\bar{\sigma}_{ij} = \left[M_{ijkl}^{-1} + \left(I_{ijmn} - M_{ijmn}^{-1} \right) : H_{mnkl}^{-1} \right]^{-1} \sigma_{kl}, \tag{47}$$

$$\bar{\sigma}_{ij} = \left[I_{ijmn} - \left(I_{ijmn} - M_{ijmn}^{-1} \right) \left(I_{mnlp} - (n+1) H_{mnlp}^{-1} \right) \left(I_{lpkl} + (n+1) \left(I_{mnsf} - M_{mnsf}^{-1} \right) H_{sfkl}^{-1} \right]^{-1} \sigma_{kl}, \tag{48}$$

where I_{ijmn} is the fourth-rank identity tensor and H is the fourth-rank healing tensor. This equation was obtained by assuming that the tensor H corresponds to 1/h [66,87,162]. Based on CDHM, it can be observed from Equation (47) that the parameter $\varphi(1 - h)$ is generalized to become $\left(I_{ijmn} - M_{ijmn}^{-1} \right)\left(I_{mnkl} - H_{mnkl}^{-1} \right)$.

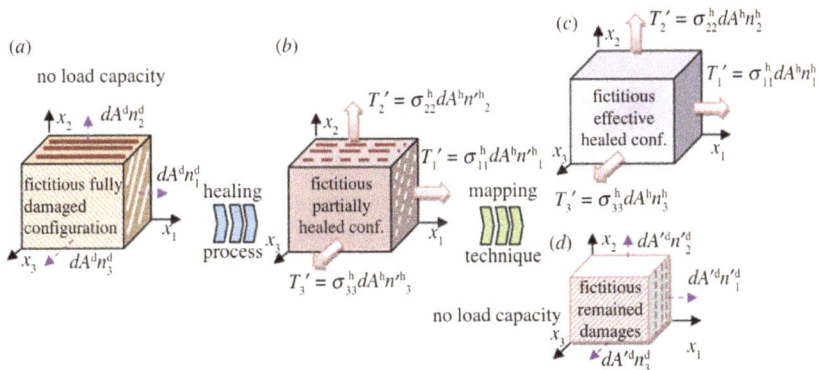

Figure 12. (a) fictitious damaged state; (b) fictitious healed state; (c) fictitious effective healed material state; and (d) fictitious damaged state [161]. (Copyright, 2012, The Royal Society).

The proposed healing tensor was later decomposed to healing tensor for cracks and healing tensor for voids in [163] along the lines of the decomposition theory applied on scalar based healing definition in [73,163]. For more details on the definition of the anisotropic damage variable tensors that were defined based on cross-section area reduction and elastic stiffness degradation, the reader can refer to [75,143]. The same authors proposed an anisotropic presentation of new damage variables that are called Fabric Tensors [74,87,162]. Later on, the same authors of [161] extended their work and proposed a coupled viscoplastic-viscodamage-viscohealing model to study the irregular behavior of glassy polymers [129]. Power function was added to the Frederick–Armstong–Philips–Chaboche (FAPC) model in the expression of the dynamic recovery of the hardening function. This latter results in increase of back stress evolution that cannot describe the irregular responses associated with the inelastic responses of glassy polymers. A thermodynamic viscoplastic-viscodamage-viscohealing framework was presented where the healing was assumed to be coupled or uncoupled. The same coupled and uncoupled healing systems were investigated previously using a thermodynamic framework of elasto-plastic-damage-healing problems [164].

Asphalt concrete is a multiphased material and exhibits complicated mechanical behavior and multiple modes of degradation. In [165], the authors developed a viscoelastic-viscoplastic model coupled to anisotropic damage in which a second-order tensor damage tensor was introduced in order to relate the nominal and effective stresses. The damage tensor was divided into permanent and non-permanent parts. The first part represents the classical damage process, while the second part represents the self-healing during unloading and rest period. A creep recovery test was simulated using the healing model. The rest concerns the application of a pressure of 1 MPa during 800 s. Afterwards, unloading period of 50 s and rest period of 3000 s were imposed. A reduction of the degradation was observed during the unloading and rest period (Figure 13). Some investigations were also carried out in which anisotropic damage is coupled with a scalar healing variable. In this regard, the authors in [99,156] developed a thermodynamic damage healing model applied to salt rock with alternative fabric descriptors. Later on, the anisotropy induced by the healing was also modeled in [103]. The effect of crack opening, closure and healing on the stiffness evolution was described by means of a multiscale model. Fabric tensors are used to relate the microcrack evolution with the macroscopic deformation rate. In [25,145], the anisotropic Cosserat continuum model was used to simulate the damage, healing and plastic of granular materials. Combination of damage and healing was defined in terms of undamaged and damaged elastic moduli tensors. Other investigations on the anisotropic definition of damage variable based on elastic modulus tensor degradation were undertaken. This concerns the decomposition of the stiffness [73], definition of anisotropic damage tensors based on Poisson's ratio, shear modulus and bulk modulus degradation [87], and description of damage in series and damage in parallel [166].

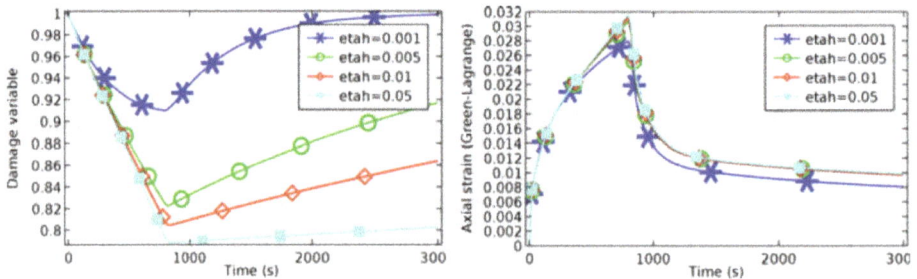

Figure 13. Effect of creep recovery load on evolution of degradation (**left**); evolution of axial strain due to healing parameter effect resulting from the creep-recovery test (**right**) [165]. (Copyright, 2016, Springer Nature).

5. Super Healing Theory

Recent research investigation reveals that self-healing presents a crucial solution for the strengthening of the materials. This solution is termed as Super Healing. Super healing theory was first proposed in [167]. Once the stiffness of the material is recovered due to self-healing, further healing can result as a strengthening material. In this section, we present the theory of the super healing model within the framework of continuum damage mechanics.

The super healing process comes into play after complete healing of the material ($h = 1$) in which the healing mechanism continues beyond its limit $h = 1$. After this limit, the strengthening and enhancing of the material properties takes place instead of healing, and the material will be able to heal and strengthen itself. A refined theory of super healing was proposed in [89]. According to the theory, the same healing material is assumed to be used as super healing material (Figure 14). In this case, the value of the healing variable can increase beyond what is necessary to recover the initial stiffness of material.

From Equation (10) of the self-healing theory, it can be observed that, when the material is fully healed, the healing variable h takes the value of 1. In the theory of super healing, once the material recovers its initial stiffness (E_0), the healing is supposed to be continued ($h > 1$). In this case, the healing will act as a strengthening material. In Figure 13, the super healing configuration is illustrated. The super healed material is characterized by its higher elastic modulus E_{sh} which is higher than the elastic modulus of the healed and original material ($E_{sh} > \bar{E}$). In Figure 15, the variation of the elastic modulus of the material in different configurations is illustrated. The material is undamaged in the initial state and its stiffness is represented by the initial elastic modulus E_0. When the material is subjected to external loading and after the energy exceeds the material threshold, damage accumulates via the variable φ. In this case, the material is damaged and its stiffness is represented by the elastic modulus E_φ, which is inferior to the initial stiffness. The material can be partially or fully healed. Thus, the elastic moduli E_{ph} and E_{fh} represent the stiffness of the partially and fully healed material, respectively. Introducing the super healing material leads to the enhancing and strengthening of the material stiffness in which the elastic modulus of the super-healed/strengthened material is higher than the elastic modulus of the fully healed and original material.

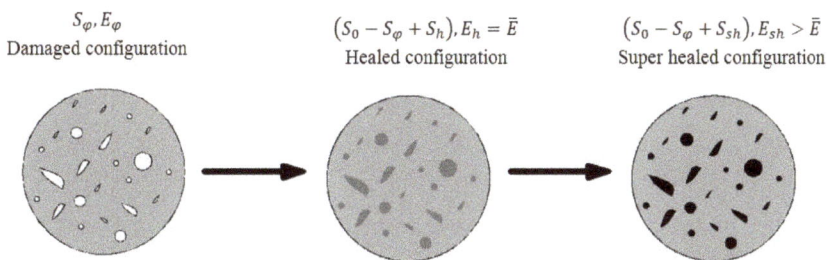

S_φ, E_φ
Damaged configuration

$(S_0 - S_\varphi + S_h), E_h = \bar{E}$
Healed configuration

$(S_0 - S_\varphi + S_{sh}), E_{sh} > \bar{E}$
Super healed configuration

Figure 14. Damaged, healed, and super healed material states [89]. (Copyright, 2018, Journal of engineering mechanics).

According to the super healing theory, it is supposed that the healing continues beyond its limit after the material recovers its initial stiffness. In this phase, the healing variable will reach large values such as 2, 3, 4, ..., x. x represents the maximal value of super healing h_s that can be applied. The super healing theory is categorized into two mechanisms: single and multiple super healing mechanisms. In the first mechanism, the variable h_s is defined by a large value at one single point of the material. This mechanism can be found in reality for example in the case of microcapsules-based self-healing concrete [4,168,169]. In this case, when only one single crack appears in the material and is healed further due to self-healing, the strengthening of this material due to super healing material should take one large single value of h_s that enhances the stiffness of the material in the area of the single crack. In the second mechanism, the value h_s ($h_s \geq 1$) is defined by small values different points of the material. This case can be found for example when the concrete is damaged in different points (multiple cracks) in which the super healing acts with small values of h_s different points. The values of h_s depends in this case on the number of healed cracks. The number of super healing variables is called n. The limitation of the second mechanism is that the variable h_s is able to take only one value that is constant at every point of the material. Unlike the super healing theory proposed in [167], in the refined super healing theory [89], the super healing variable h_s is not restricted to only integer values. It can also take non-integer values. According to the refined super healing theory, the relation of the nominal and effective stresses is expressed as

$$\bar{\sigma} = \frac{\sigma}{1 + [h_s\,(n+1) - 1]\,\varphi}. \tag{49}$$

Equation (49) represents the main result governing the super healing theory. From Equation (49), when the number of super healing parameter n approaches infinity, the effective stress vanishes, which is irrespective of the damage and super healing variables. In addition, when the damage variable $\varphi = 1$, the effective stress retains a finite value. This is explained by the fact that the material will not rupture even though the damage is high. Equation (49) of the super healing theory was also generalized to anisotropic formulation as follows:

$$\bar{\sigma}_{ij} = \left[M_{ijkl}^{-1} + (n+1)\,Hs_{ijmn}^{-1} : \left(I_{mnkl} - M_{mnkl}^{-1} \right) \right]^{-1} \sigma_{kl}, \tag{50}$$

where Hs is the fourth-rank super healing tensor corresponding to the super healing variables h_s defined in Equation (49). In addition, examples of one-dimensional and plane stress were applied. It was shown that the proposed super healing theory is applicable in the case of plane stress. Figure 15 shows the effects of the self-healing and super healing mechanisms. From Figure 16b, it is seen that the material enhances its stiffness when the super healing effect is introduced. Generalized nonlinear and quadratic super healing formulation was also presented along the lines of the nonlinear self-healing theory previously presented, and comparison of super healing models was given (Table 1).

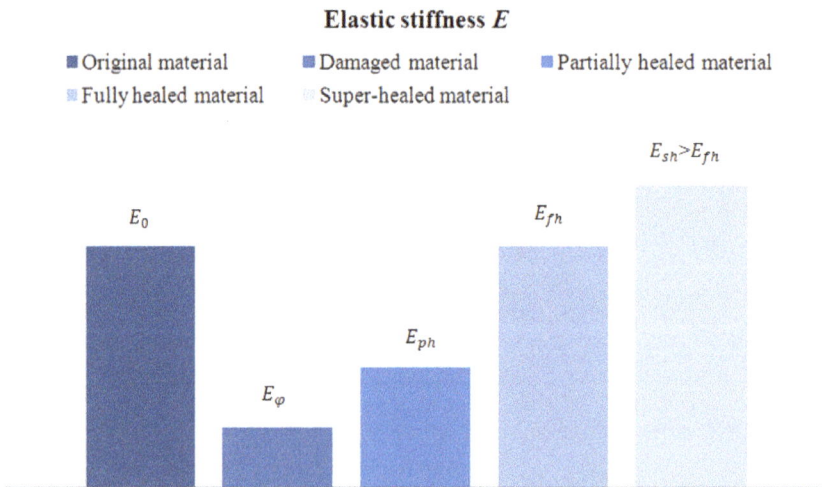

Figure 15. Variation of material stiffness from initial to super healed state [89]. (Copyright, 2018, Journal of engineering mechanics).

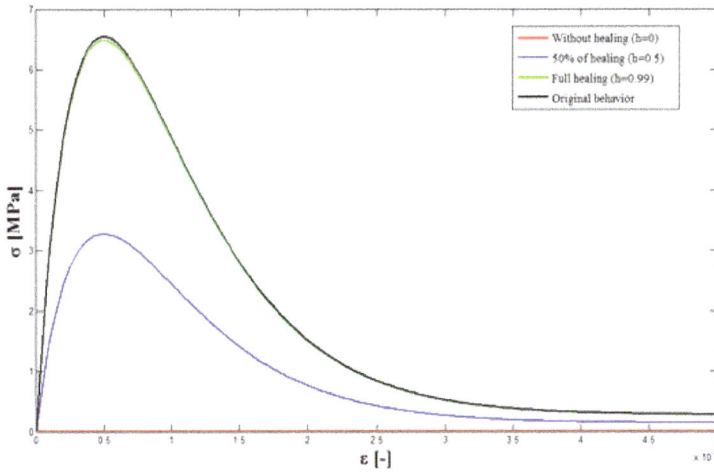

(a)

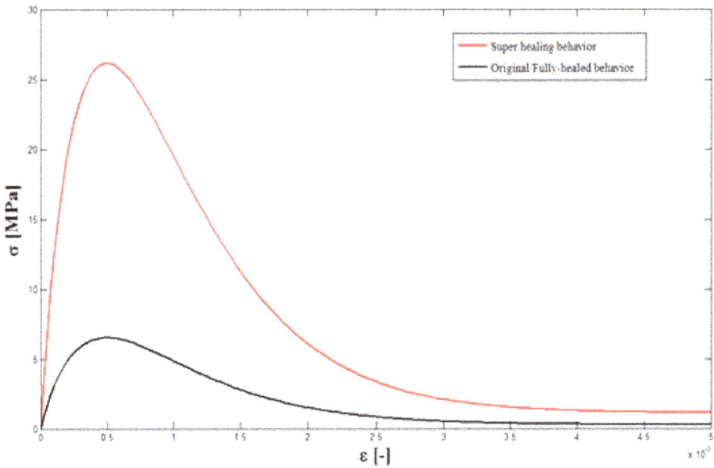

(b)

Figure 16. Effects of (**a**) healing and (**b**) super healing mechanisms [89]. (Copyright, 2018, Journal of engineering mechanics).

Table 1. Comparison between super healing models.

Super Healing Model	Equation of the Ratio φ_{sd}/φ	Equation of the Ratio $\tilde{\sigma}/\sigma$
Linear super healing (LSH)	$\frac{\varphi_{sd}}{\varphi} = 1 - h_s\,(n+1)$	$\frac{\tilde{\sigma}}{\sigma} = \frac{1}{1-[1-h_s(n+1)]\varphi}$
Generalized nonlinear super healing (NSH)	$\frac{\varphi_{sd}}{\varphi} = \frac{1-h_s(n+1)}{1-\varphi h_s(n+1)}$	$\frac{\tilde{\sigma}}{\sigma} = \frac{1-h_s(n+1)\varphi}{1-\varphi}$
Quadratic super healing (QSH)	$\frac{\varphi_{sd}}{\varphi} = 1 - h_s\,(n+1) + \varphi h_s\,(n+1) - \varphi h_s^2\,(n+1)^2$	$\frac{\tilde{\sigma}}{\sigma} = \frac{1}{1-[1-h_s(n+1)+\varphi h_s(n+1)-\varphi h_s^2(n+1)^2]\varphi}$

Figure 17 shows the comparison of the super healing models. The results revealed that the generalized nonlinear super healing model is the most appropriate to describe the super healing concept. In addition, the link between the proposed theory and the theory of undamageable materials [162,170–172] has been studied. It was found that both theories lead to a material that undergoes zero damage during the deformation process. Later on, an investigation of the super healing theory in terms of the elastic stiffness variation was performed in which the hypotheses of elastic strain and elastic energy equivalence were used [86]. Using the hypothesis of elastic strain equivalence, the following expressions of damage, healing and super healing elastic stiffness are respectively expressed as

$$E_\varphi = \bar{E} \left(1 - \varphi\right), \tag{51}$$

$$E_h = \bar{E} \left[1 - \varphi \left(1 - h\right)\right], \tag{52}$$

$$E_{sh} = \bar{E} \left[1 + (R - 1) \varphi\right]. \tag{53}$$

From Equation (53), it can be seen that, with $R > 1$, the healed elastic modulus E_{sh} is greater than the effective elastic modulus $\bar{E}$, while they become equal when $R = h_s$. On the other hand, when $R = h_s = 0$, it becomes equal to the damaged elastic modulus. Using the hypothesis of elastic energy equivalence, the following expressions of damage, healing and super healing elastic stiffness are respectively expressed as

$$E_\varphi = \bar{E} \left(1 - \varphi\right)^2, \tag{54}$$

$$E_h = \bar{E} \left[1 - \varphi \left(1 - h\right)\right]^2, \tag{55}$$

$$E_{sh} = \bar{E} \left[1 + (R - 1) \varphi\right]^2. \tag{56}$$

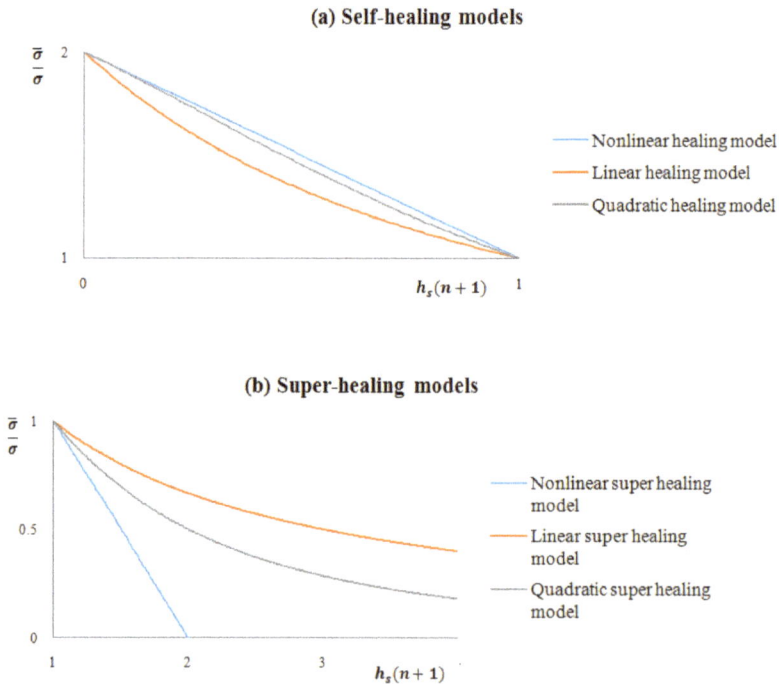

Figure 17. Comparison between linear, generalized nonlinear and quadratic models: (**a**) self-healing models; (**b**) super healing models [89]. (Copyright, 2018, Journal of engineering mechanics).

Equation (56) of the super healing can represent Equation (55) when the healing is introduced and can represent Equation (54) when the material is only damaged. If the material is undamaged and unhealed, the super healing elastic modulus becomes similar to the initial elastic modulus, while, when the material is fully healed, the super healing modulus becomes similar to the effective elastic modulus. On the other hand, when super healing is introduced, the super healing elastic modulus becomes greater than the effective elastic modulus. Table 2 presents a summary of the elastic moduli, damage, healing and super healing variables in the case of elastic strain equivalence and elastic energy equivalence. It should be noted that R represents the super healing variable ($R = h_s(n+1)$).

Table 2. Elastic modulus in the damaged, healed, and super healed material states using the hypotheses of elastic strain equivalence and elastic energy equivalence.

Phase	Elastic Strain Equivalence		Elastic Energy Equivalence	
	Elastic Modulus	Variable	Elastic Modulus	Variable
Damage	$E_\varphi = \bar{E}(1-\varphi)$	$\varphi = \frac{\bar{E}-E_d}{\bar{E}}$	$E_\varphi = \bar{E}(1-\varphi)^2$	$\varphi = 1 - \sqrt{\frac{E_\varphi}{\bar{E}}}$
Healing	$E_h = \bar{E}[1 - \varphi(1-h)]$	$h = 1 - \frac{\bar{E}-E_h}{\varphi\bar{E}}$	$E_h = \bar{E}[1 - \varphi(1-h)]^2$	$h = 1 - \frac{\sqrt{\bar{E}}-\sqrt{E_h}}{\varphi\sqrt{\bar{E}}}$
Super healing	$E_{sh} = \bar{E}[1 + (R-1)\varphi]$	$R = \frac{E_{sh}-\bar{E}}{\varphi\bar{E}}+1$	$E_{sh} = \bar{E}[1 + (R-1)\varphi]^2$	$R = \frac{\sqrt{E_h}-\sqrt{\bar{E}}}{\varphi\sqrt{\bar{E}}}+1$

6. Self-Healing Metals

In this section, a brief review of different mechanisms of self-healing of ductile materials such as metals is presented. The self-healing concept was widely exploited on polymer, concrete, and ceramic materials; however, few investigations were carried out on self-healing of metals due to the nature of the healing of each material. Metals are known by their high melting temperature, which leads to a challenging process of the healing. There are two mechanisms of self-healing metals: liquid state mechanism and solid state mechanism. The first one is based on adding shape memory alloys (SMA) to the metal matrix that represents a liquid at high temperature. The second one is based on diffusion of solute into the cracks and voids.

6.1. Liquid State Healing Mechanism

The most commonly method for self-healing of metals is the embedding of healing agent into the metal matrix [173]. When the metal is subjected to heating, the matrix becomes liquid, and thus the healing agent becomes able to heal the damage. In addition, damage can be healed in different lifetimes of metals due to the availability of the liquid healing agent. Many investigations were carried out on self-healing mechanisms using SMA [174–176]. Figure 18 illustrates the liquid state healing mechanism. When the metal composite is subjected to tensile stress resulting in crack formation, interfacial debonding will occur by crack due to the low strength at the interface of and high strength of SMA. The crack in the metal composite is supposed to heal when the SMA is subjected to high temperature.

Several numerical models were developed to predict the behavior of liquid state healing mechanism. In [177], the authors developed a numerical model to describe the thermomechanical behavior in the interface SMA-matrix. Two-dimensional elasto-plastic model was applied on the matrix and one-dimensional material model was applied on the SMA wires using material subroutine implemented in the software package Abaqus (Version 6.3, 2002, Pawtucket, RI, USA, Hibbitt, Karlsson & Sorensen, Inc., 7.9.3–3). The model shows its ability to describe the behavior of SMA wires at different temperature levels. The authors in [175,176] developed a model that analyzes the relationship between the strength of matrix, stress of SMA wires, and volume fraction of reinforcement. In [178], Zhu et al. developed a three-dimensional model of metal-matrix composites reinforced by SMA. The self-healing mechanism was modelled based on pre-strained SMA wires. In addition, micromechanical approaches were also proposed by many researchers in order to demonstrate the healing efficiency of SMA-based composite structures [179–181]. The effectiveness of the description of the microstructure behavior is high. Nonetheless, it is not an easy task to be applied at the specimen level. For more information on the different mechanisms of self-healing metals in fine scale and structural scale, the reader can refer to [182].

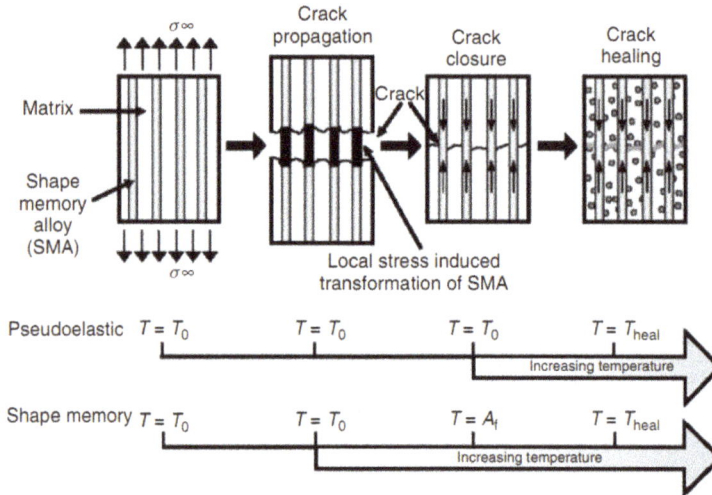

Figure 18. Schematic of liquid state healing mechanism [173]. (Copyright, 2009, John Wiley & Sons).

6.2. Solid State Healing Mechanism (Precipitation Healing)

This method of healing is based on the minimization of the energy system and decreasing the solubility of the element when the material is subjected to a decreasing temperature. In this case, the alloy changes its phase from liquid to solid upon solidification. The nucleation of precipitation occurs at unstable phases, high energy defect in grain boundaries and free surfaces. Figure 19 illustrates the solute migration along a high diffusion path and precipitation on high energy surfaces. In [183], the authors revealed that aluminium alloys are subjected to healing mechanisms by solute precipitation during creep and fatigue loading. Moreover, in [184], the authors revealed that this mechanism is similar to a precipitation mechanism in powder alloys. Steel material is found to be the material that most demonstrates the efficiency of self-healing cavities. However, creep strength and ductility of the steels can graduate when subjected to high temperatures. In [185,186], it was found that sulfur accelerates creep cavitation. For more information on the healing mechanism of solid state healing, the reader can refer to [173].

Several numerical models were developed to predict the behavior of solid state healing mechanisms. In [187–189], molecular dynamic modelings were developed to describe the behavior of solid state healing of microcracks in aluminium and copper. It was shown that dislocation around the microcrack induces healing. In [190], Wei et al. used the same concept to study the crack healing in iron along the lines of the theory applied on aluminium and copper. Based on a finite element method, Huang et al. [191] showed that there are two stages of healing of cracks in the form of ellipsoid subjected to high pressure. The first stage concerns the shrinkage of microcracks and the second one concerns the splitting microcracks. Later on, in [192,193], the authors presented a thermodynamic approach to study the void shrinkage rate considering the void surface, grain boundary and elastic energy. In [194], the impact of high energy electromagnetic field on the elasto-plastic damage material was modelled. The influence of different parameters such as melting and evaporation of metal was considered. The authors in [195] developed a numerical model to describe the creep cavity growth and strain rates in metals through self-healing. It was found that Fe-W alloys represents a good alternative to be used for self-healing at high temperature.

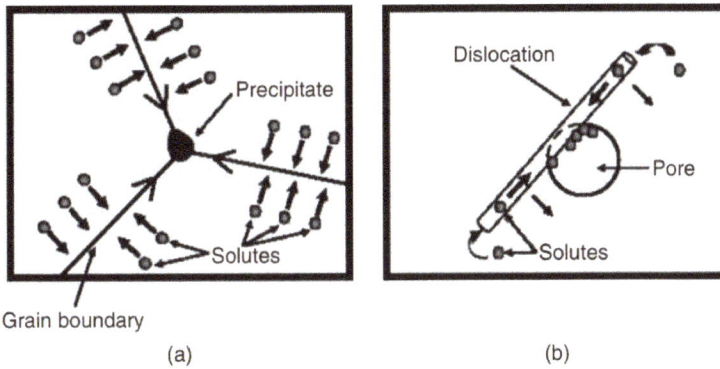

Figure 19. (a) solute diffusion along grain boundary and (b) mobile dislocation moving solutes to a pore [173]. (Copyright, 2009, John Wiley & Sons).

7. Conclusions and Perspectives

A state-of-the-art review of continuum damage-healing and super healing mechanics applied on brittle materials was presented in the present paper. The main features of damage-healing and super healing mechanics considered are as follows:

- The measure and presentation of the healing variable in both autonomous and autogenous self-healing mechanisms.
- The evolution equations of the healing models based on CDHM.
- The influence of different mechanical and environmental parameters on the healing efficiency.
- The effect of the self-healing and super healing on the mechanical behavior of the material.
- The anisotropic presentation of damage-healing and super healing with tensorial formulation.
- The effect of the new strengthening theory based on the super healing and CDHM.

The CDHM represents an extension of the CDM. Based on it, the initial stiffness recovery and enhancing of the mechanical properties of the materials while taking into account many parameters (e.g., microcapsule percentage, temperature, healing time, damage history, ... etc.) is described. In addition, two self-healing mechanisms are mechanically studied: autonomous and autogenous. They are also termed respectively by coupled and uncoupled healing mechanisms. Each damage-healing formulation proposed in literature is based on the experimental data of self-healing materials while considering some assumptions for simplicity of the computational analysis (e.g., isotropy of the material instead of anisotropy). This is due to the heterogeneity of the brittle materials. For instance, concrete is an heterogeneous material that has a complex fracture behavior. Taking into account the complexity nature of concrete material in the healing analysis is not an easy task. Moreover, the softening behavior of brittle materials leads to mesh-dependence of their responses due to strain localization when local damage-healing models are used. This issue was thoroughly addressed in CDM using non-conventional damage models (e.g., gradient and nonlocal damage models), while non-conventional damage-healing model is not yet addressed except for a short explanation given in [88].

To the best knowledge of the authors, the healing can regularize the problem of strain localization and mesh-dependency provided damage/healing in the time range it is applied if it is rate-dependent. This regularization happens in the suitable range of time that recovers the original stiffness of the material; minimization and elimination of damage, especially in the case coupled self-healing mechanism in which damage and healing evolve simultaneously. In this case, when one point of material is damaged, the microcapsules (or hollow fibers) are broken and the healing agents are released from the microcapsules. The damage evolves first and the healing agent evolves after it is released, which leads to the deactivation and elimination of the damage evolution. Therefore, it is highly necessary to develop some non-conventional damage-healing models using nonlocal healing variables coupled to local or nonlocal damage variables. In addition, further investigation of anisotropic damage-healing mechanics is needed in which new healing tensors can be proposed. Finally, further studies on the super healing theory will be an interesting task in terms of focusing on some limitations of the theoretical framework (e.g., plasticity, assumption that h_s takes only one constant value at every point of the material). It is hoped that future studies will be carried out in the manufacturing technology along the lines of the super healing theory.

Author Contributions: C.O. conducted the literature review and wrote the paper. L.M.M. contributed in the revision of the paper.

Funding: The first author would like to acknowledge the Deutscher Akademischer Austauschdienst (DAAD) for the financial support of this work. The second author would like to acknowledge the RISTEK-DIKTI (Directorate General of Resources for Science, Technology and Higher Education. Ministry of Research, Technology and Higher Education of Indonesia) under funding agreement No: 153.39/E4.4/2014.

Conflicts of Interest: The authors declare that there is no conflict of interest.

References

1. Hilloulin, B.; Grondin, F.; Matallah, M.; Loukili, A. Modelling of autogenous healing in ultra high performance concrete. *Cem. Concr. Res.* **2014**, *61*, 64–70. [CrossRef]
2. Yang, Y.; Lepech, M.D.; Yang, E.H.; Li, V.C. Autogenous healing of engineered cementitious composites under wet–dry cycles. *Cem. Concr. Res.* **2009**, *39*, 382–390. [CrossRef]
3. Giannaros, P.; Kanellopoulos, A.; Al-Tabbaa, A. Sealing of cracks in cement using microencapsulated sodium silicate. *Smart Mater. Struct.* **2016**, *25*, 084005. [CrossRef]
4. White, S.R.; Sottos, N.; Geubelle, P.; Moore, J.; Kessler, M.; Sriram, S.; Brown, E.; Viswanathan, S. Autonomic healing of polymer composites. *Nature* **2001**, *409*, 794–797. [CrossRef] [PubMed]
5. Barbero, E.J.; Greco, F.; Lonetti, P. Continuum damage-healing mechanics with application to self-healing composites. *Int. J. Damage Mech.* **2005**, *14*, 51–81. [CrossRef]
6. Herbst, O.; Luding, S. Modeling particulate self-healing materials and application to uni-axial compression. *Int. J. Fract.* **2008**, *154*, 87–103. [CrossRef]
7. Hall, J.; Qamar, I.; Rendall, T.; Trask, R. A computational model for the flow of resin in self-healing composites. *Smart Mater. Struct.* **2015**, *24*, 037002. [CrossRef]
8. Rabczuk, T.; Belytschko, T. Cracking particles: A simplified meshfree method for arbitrary evolving cracks. *Int. J. Numer. Methods Eng.* **2004**, *61*, 2316–2343. [CrossRef]
9. Rabczuk, T.; Gracie, R.; Song, J.H.; Belytschko, T. Immersed particle method for fluid–structure interaction. *Int. J. Numer. Methods Eng.* **2010**, *81*, 48–71. [CrossRef]
10. Rabczuk, T.; Zi, G.; Bordas, S.; Nguyen-Xuan, H. A simple and robust three-dimensional cracking-particle method without enrichment. *Comput. Methods Appl. Mech. Eng.* **2010**, *199*, 2437–2455. [CrossRef]
11. Zhou, S.; Zhu, H.; Ju, J.W.; Yan, Z.; Chen, Q. Modeling microcapsule-enabled self-healing cementitious composite materials using discrete element method. *Int. J. Damage Mech.* **2017**, *26*, 340–357. [CrossRef]
12. Areias, P.; Rabczuk, T.D.; de Sá, J.C. A novel two-stage discrete crack method based on the screened Poisson equation and local mesh refinement. *Comput. Mech.* **2016**, *58*, 1003–1018. [CrossRef]
13. Gui, Y.L.; Bui, H.H.; Kodikara, J.; Zhang, Q.B.; Zhao, J.; Rabczuk, T. Modelling the dynamic failure of brittle rocks using a hybrid continuum-discrete element method with a mixed-mode cohesive fracture model. *Int. J. Impact Eng.* **2016**, *87*, 146–155. [CrossRef]
14. Rabczuk, T.; Eibl, J. Simulation of high velocity concrete fragmentation using SPH/MLSPH. *Int. J. Numer. Methods Eng.* **2003**, *56*, 1421–1444. [CrossRef]

15. Kalameh, H.A.; Karamali, A.; Anitescu, C.; Rabczuk, T. High velocity impact of metal sphere on thin metallic plate using smooth particle hydrodynamics (SPH) method. *Front. Struct. Civ. Eng.* **2012**, *6*, 101–110.

16. Rabczuk, T.; Xiao, S.P.; Sauer, M. Coupling of mesh-free methods with finite elements: Basic concepts and test results. *Commun. Numer. Methods Eng.* **2006**, *22*, 1031–1065. [CrossRef]

17. Rabczuk, T.; Eibl, J.; Stempniewski, L. Numerical analysis of high speed concrete fragmentation using a meshfree Lagrangian method. *Eng. Fract. Mech.* **2004**, *71*, 547–556. [CrossRef]

18. Triantafyllou, S.P.; Chatzis, M.N. A new damage-healing smooth hysteretic formulation for the modeling of self-healing materials. In Proceedings of the 8th GRACM International Congress on Computational Mechanics, Volos, Greece, 12–15 July 2015.

19. Rabczuk, T.; Areias, P.; Belytschko, T. A meshfree thin shell method for non-linear dynamic fracture. *Int. J. Numer. Methods Eng.* **2007**, *72*, 524–548. [CrossRef]

20. Rabczuk, T.; Zi, G.; Bordas, S. Enriched Finite Element and Meshfree Methods for Dynamic Crack Propagation Problems. In Proceedings of the 5th Australasian Congress on Applied Mechanics, Brisbane, Australia, 10–12 December 2007; p. 570.

21. Zi, G.; Rabczuk, T.; Wall, W. Extended meshfree methods without branch enrichment for cohesive cracks. *Comput. Mech.* **2007**, *40*, 367–382. [CrossRef]

22. Rabczuk, T.; Bordas, S.; Zi, G. A three-dimensional meshfree method for continuous multiple-crack initiation, propagation and junction in statics and dynamics. *Comput. Mech.* **2007**, *40*, 473–495. [CrossRef]

23. Rabczuk, T.; Areias, P. A meshfree thin shell for arbitrary evolving cracks based on an extrinsic basis. *Comput. Model. Eng. Sci.* **2006**, *16*, 115–130.

24. Rabczuk, T.; Samaniego, E. Discontinuous modelling of shear bands using adaptive meshfree methods. *Comput. Methods Appl. Mech. Eng.* **2008**, *197*, 641–658. [CrossRef]

25. Li, X.; Wang, Z.; Zhang, S.; Duan, Q. Multiscale modeling and characterization of coupled damage-healing-plasticity for granular materials in concurrent computational homogenization approach. *Comput. Methods Appl. Mech. Eng.* **2018**, *342*, 354–383. [CrossRef]

26. Talebi, H.; Silani, M.; Rabczuk, T. Concurrent multiscale modeling of three dimensional crack and dislocation propagation. *Adv. Eng. Softw.* **2015**, *80*, 82–92. [CrossRef]

27. Budarapu, P.R.; Gracie, R.; Bordas, S.P.; Rabczuk, T. An adaptive multiscale method for quasi-static crack growth. *Comput. Mech.* **2014**, *53*, 1129–1148. [CrossRef]

28. Talebi, H.; Silani, M.; Bordas, S.P.; Kerfriden, P.; Rabczuk, T. A computational library for multiscale modeling of material failure. *Comput. Mech.* **2014**, *53*, 1047–1071. [CrossRef]

29. Budarapu, P.R.; Gracie, R.; Yang, S.W.; Zhuang, X.; Rabczuk, T. Efficient coarse graining in multiscale modeling of fracture. *Theor. Appl. Fract. Mech.* **2014**, *69*, 126–143. [CrossRef]

30. Budarapu, P.; Javvaji, B.; Reinoso, J.; Paggi, M.; Rabczuk, T. A three dimensional adaptive multiscale method for crack growth in Silicon. *Theor. Appl. Fract. Mech.* **2018**, *96*, 576–603. [CrossRef]

31. Pan, Y.; Tian, F.; Zhong, Z. A continuum damage-healing model of healing agents based self-healing materials. *Int. J. Damage Mech.* **2018**, *27*, 754–778. [CrossRef]

32. Amiri, F.; Millán, D.; Shen, Y.; Rabczuk, T.; Arroyo, M. Phase-field modeling of fracture in linear thin shells. *Theor. Appl. Fract. Mech.* **2014**, *69*, 102–109. [CrossRef]

33. Msekh, M.A.; Sargado, J.M.; Jamshidian, M.; Areias, P.M.; Rabczuk, T. Abaqus implementation of phase-field model for brittle fracture. *Comput. Mater. Sci.* **2015**, *96*, 472–484. [CrossRef]

34. Msekh, M.A.; Silani, M.; Jamshidian, M.; Areias, P.; Zhuang, X.; Zi, G.; He, P.; Rabczuk, T. Predictions of J integral and tensile strength of clay/epoxy nanocomposites material using phase field model. *Compos. Part B Eng.* **2016**, *93*, 97–114. [CrossRef]

35. Msekh, M.A.; Cuong, N.; Zi, G.; Areias, P.; Zhuang, X.; Rabczuk, T. Fracture properties prediction of clay/epoxy nanocomposites with interphase zones using a phase field model. *Eng. Fract. Mech.* **2018**, *188*, 287–299. [CrossRef]

36. Areias, P.; Rabczuk, T. Steiner-point free edge cutting of tetrahedral meshes with applications in fracture. *Finite Elem. Anal. Des.* **2017**, *132*, 27–41. [CrossRef]

37. Ren, H.; Zhuang, X.; Rabczuk, T. Dual-horizon peridynamics: A stable solution to varying horizons. *Comput. Methods Appl. Mech. Eng.* **2017**, *318*, 762–782. [CrossRef]

38. Nguyen-Thanh, N.; Zhou, K.; Zhuang, X.; Areias, P.; Nguyen-Xuan, H.; Bazilevs, Y.; Rabczuk, T. Isogeometric analysis of large-deformation thin shells using RHT-splines for multiple-patch coupling. *Comput. Methods Appl. Mech. Eng.* **2017**, *316*, 1157–1178. [CrossRef]

39. Ren, H.; Zhuang, X.; Cai, Y.; Rabczuk, T. Dual-horizon peridynamics. *Int. J. Numer. Methods Eng.* **2016**, *108*, 1451–1476. [CrossRef]

40. Areias, P.; Rabczuk, T.; Msekh, M. Phase-field analysis of finite-strain plates and shells including element subdivision. *Comput. Methods Appl. Mech. Eng.* **2016**, *312*, 322–350. [CrossRef]

41. Nguyen, B.; Tran, H.; Anitescu, C.; Zhuang, X.; Rabczuk, T. An isogeometric symmetric Galerkin boundary element method for two-dimensional crack problems. *Comput. Methods Appl. Mech. Eng.* **2016**, *306*, 252–275. [CrossRef]

42. Areias, P.; Msekh, M.; Rabczuk, T. Damage and fracture algorithm using the screened Poisson equation and local remeshing. *Eng. Fract. Mech.* **2016**, *158*, 116–143. [CrossRef]

43. Kachanov, L. On the creep fracture time. *IZV AKAD* **1958**, *8*, 26–31.

44. Rabotnov, Y.N. Creep rupture in applied mechanics. In Proceedings of the 12th International Congress on Applied Mechanics, Stanford, CA, USA, 26–31 August 1968; pp. 342–349.

45. Rabotnov, Y.N. *Creep Problems in Structural Members*; North-Holland Publishing Company: Amsterdam, The Netherlands, 1969.

46. Lemaitre, J.; Chaboche, J. *A Non-Linear Model of Creep Fatigue Damage Cumulation*; ONERA: Palaiseau, France, 1394; p. 174.

47. Hayhurst, D. Creep rupture under multi-axial states of stress. *J. Mech. Phys. Solids* **1972**, *20*, 381–382. [CrossRef]

48. Leckie, F.A.; Hayhurst, D. Constitutive equations for creep rupture. *Acta Metall.* **1977**, *25*, 1059–1070. [CrossRef]

49. Chaboche, J. Continuum damage mechanics: Present state and future trends. *Nucl. Eng. Des.* **1987**, *105*, 19–33. [CrossRef]

50. Lin, J.; Dunne, F.; Hayhurst, D. Aspects of testpiece design responsible for errors in cyclic plasticity experiments. *Int. J. Damage Mech.* **1999**, *8*, 109–137. [CrossRef]

51. Murakami, S.; Ohno, N. Creep damage analysis in thin-walled tubes. *Inelast. Behav. Press. Vessel Pip. Compon.* **1978**, 55–69.

52. Murakami, S. Effect of cavity distribution in constitutive equations of creep and creep damage. In Proceedings of the EUROMECH Colloquium on Damage Mechanics, Cachan, France, 7–11 September 1981.

53. Murakami, S.; Ohno, N. A continuum theory of creep and creep damage. In *Creep in Structures*; Springer: Berlin, Germany, 1981; pp. 422–444.

54. Murakami, S. Damage mechanics approach to damage and fracture of materials. *Rairo* **1982**, *3*, 1–13.

55. Chaboche, J.L. Continuous damage mechanics—A tool to describe phenomena before crack initiation. *Nucl. Eng. Des.* **1981**, *64*, 233–247. [CrossRef]

56. Chaboche, J. *Une loi Différentielle D'Endommagement de Fatigue Avec Cumulation Non Linéaire*; Office Nationale d'Etudes et de Recherches Aérospatiales: Palaiseau, France, 1974.

57. Chaboche, J.L. Continuum damage mechanics: Part I. general concepts. *J. Appl. Mech.* **1988**, *55*, 59–72. [CrossRef]

58. Chaboche, J.L. Continuum damage mechanics: Part II—Damage growth, crack initiation, and crack growth. *J. Appl. Mech.* **1988**, *55*, 65–72. [CrossRef]

59. Simo, J.; Ju, J. Strain-and stress-based continuum damage models—II. Computational aspects. *Int. J. Solids Struct.* **1987**, *23*, 841–869. [CrossRef]

60. Simo, J.; Ju, J. Strain-and stress-based continuum damage models—I. Formulation. *Math. Comput. Model.* **1989**, *12*, 378. [CrossRef]

61. Simo, J.; Ju, J.; Taylor, R.; Pister, K. On strain-based continuum damage models: Formulation and computational aspects. *Const. Laws Eng. Mater.* **1987**, *1*, 233–245.

62. Simo, J.; Ju, J. On continuum damage-elastoplasticity at finite strains. *Comput. Mech.* **1989**, *5*, 375–400. [CrossRef]

63. Voyiadjis, G.Z.; Kattan, P.I. A plasticity-damage theory for large deformation of solids? I. Theoretical formulation. *Int. J. Eng. Sci.* **1992**, *30*, 1089–1108. [CrossRef]

64. Voyiadjis, G.; Park, T. Anisotropic damage effect tensors for the symmetrization of the effective stress tensor. *J. Appl. Mech.* **1997**, *64*, 106–110. [CrossRef]

65. Voyiadjis, G.Z.; Park, T. Local and interfacial damage analysis of metal matrix composites using the finite element method. *Eng. Fract. Mech.* **1997**, *56*, 483–511. [CrossRef]

66. Voyiadjis, G.Z. *Advances in Damage Mechanics: Metals and Metal Matrix Composites*; Elsevier: Amsterdam, The Netherlands, 2012.

67. Cordier, G.; Van, K.D. Strain hardening effects and damage in plastic fatigue. In *Physical Non-Linearities in Structural Analysis*; Springer: Berlin, Germany, 1981; pp. 52–55.

68. Bodner, S. A procedure for including damage in constitutive equations for elastic-viscoplastic work-hardening materials. In *Physical Non-Linearities in Structural Analysis*; Springer: Berlin, Germany, 1981; pp. 21–28.

69. Kachanov, L. *Introduction to Continuum Damage Mechanics*; Springer Science & Business Media: Berlin, Germany, 2013; Volume 10.

70. Lemaitre, J.; Chaboche, J.L. Aspect phénoménologique de la rupture par endommagement. *J. Méc. Appl.* **1978**, *2*, 317–365.

71. Lemaitre, J.; Dufailly, J. Modelization and identification of endommagement plasticity of material. In Proceedings of the 3rd French Congress of Mechanics, Grenoble, France, 1977; pp. 17–21.

72. Voyiadjis, G.Z. Degradation of elastic modulus in elastoplastic coupling with finite strains. *Int. J. Plast.* **1988**, *4*, 335–353. [CrossRef]

73. Voyiadjis, G.Z.; Kattan, P.I. Decomposition of elastic stiffness degradation in continuum damage mechanics. *J. Eng. Mater. Technol.* **2017**, *139*, 021005. [CrossRef]

74. Voyiadjis, G.Z.; Kattan, P.I. Damage mechanics with fabric tensors. *Mech. Adv. Mater. Struct.* **2006**, *13*, 285–301. [CrossRef]

75. Kattan, P.I.; Voyiadjis, G.Z. Decomposition of damage tensor in continuum damage mechanics. *J. Eng. Mech.* **2001**, *127*, 940–944. [CrossRef]

76. Rabczuk, T.; Akkermann, J.; Eibl, J. A numerical model for reinforced concrete structures. *Int. J. Solids Struct.* **2005**, *42*, 1327–1354. [CrossRef]

77. Dunant, C.F.; Bordas, S.P.; Kerfriden, P.; Scrivener, K.L.; Rabczuk, T. An algorithm to compute damage from load in composites. *Front. Archit. Civ. Eng. China* **2011**, *5*, 180–193. [CrossRef]

78. Silani, M.; Ziaei-Rad, S.; Talebi, H.; Rabczuk, T. A semi-concurrent multiscale approach for modeling damage in nanocomposites. *Theor. Appl. Fract. Mech.* **2014**, *74*, 30–38. [CrossRef]

79. Silani, M.; Talebi, H.; Hamouda, A.M.; Rabczuk, T. Nonlocal damage modelling in clay/epoxy nanocomposites using a multiscale approach. *J. Comput. Sci.* **2016**, *15*, 18–23. [CrossRef]

80. Thai, T.Q.; Rabczuk, T.; Bazilevs, Y.; Meschke, G. A higher-order stress-based gradient-enhanced damage model based on isogeometric analysis. *Comput. Methods Appl. Mech. Eng.* **2016**, *304*, 584–604. [CrossRef]

81. Abiri, O.; Lindgren, L.E. Non-local damage models in manufacturing simulations. *Eur. J. Mech. A/Solids* **2015**, *49*, 548–560. [CrossRef]

82. Geers, M.; Peerlings, R.; Brekelmans, W.; de Borst, R. Phenomenological nonlocal approaches based on implicit gradient-enhanced damage. *Acta Mech.* **2000**, *144*, 1–15. [CrossRef]

83. Rojas-Solano, L.B.; Grégoire, D.; Pijaudier-Cabot, G. Interaction-based non-local damage model for failure in quasi-brittle materials. *Mech. Res. Commun.* **2013**, *54*, 56–62. [CrossRef]

84. Simone, A.; Askes, H.; Sluys, L.J. Incorrect initiation and propagation of failure in non-local and gradient-enhanced media. *Int. J. Solids Struct.* **2004**, *41*, 351–363. [CrossRef]

85. Peerlings, R.; Geers, M.; De Borst, R.; Brekelmans, W. A critical comparison of nonlocal and gradient-enhanced softening continua. *Int. J. Solids Struct.* **2001**, *38*, 7723–7746. [CrossRef]

86. Oucif, C.; Voyiadjis, G.Z.; Kattan, P.I.; Rabczuk, T. Investigation of the super healing theory in continuum damage and healing mechanics. *Int. J. Damage Mech.* **2018**. [CrossRef]

87. Voyiadjis, G.Z.; Kattan, P.I. A comparative study of damage variables in continuum damage mechanics. *Int. J. Damage Mech.* **2009**, *18*, 315–340. [CrossRef]

88. Oucif, C.; Voyiadjis, G.Z.; Rabczuk, T. Modeling of damage-healing and nonlinear self-healing concrete behavior: Application to coupled and uncoupled self-healing mechanisms. *Theor. Appl. Fract. Mech.* **2018**, *96*, 216–230. [CrossRef]

89. Oucif, C.; Voyiadjis, G.Z.; Kattan, P.I.; Rabczuk, T. Nonlinear Superhealing and Contribution to the Design of a New Strengthening Theory. *J. Eng. Mech.* **2018**, *144*, 04018055. [CrossRef]

90. Darabi, M.K.; Al-Rub, R.K.A.; Little, D.N. A continuum damage mechanics framework for modeling micro-damage healing. *Int. J. Solids Struct.* **2012**, *49*, 492–513. [CrossRef]

91. Voyiadjis, G.Z.; Shojaei, A.; Li, G.; Kattan, P. Continuum damage-healing mechanics with introduction to new healing variables. *Int. J. Damage Mech.* **2012**, *21*, 391–414. [CrossRef]

92. Wool, R.; O'connor, K. A theory crack healing in polymers. *J. Appl. Phys.* **1981**, *52*, 5953–5963. [CrossRef]

93. Schapery, R. On the mechanics of crack closing and bonding in linear viscoelastic media. *Int. J. Fract.* **1989**, *39*, 163–189. [CrossRef]

94. Little, D.N.; Bhasin, A. Exploring Mechanism of H ealing in Asphalt Mixtures and Quantifying its Impact. In *Self Healing Materials*; Springer: Berlin, Germany, 2007; pp. 205–218.

95. Bhasin, A.; Little, D.N.; Bommavaram, R.; Vasconcelos, K. A framework to quantify the effect of healing in bituminous materials using material properties. *Road Mater. Pavement Des.* **2008**, *9*, 219–242. [CrossRef]

96. Chan, K.; Bodner, S.; Fossum, A.; Munson, D. *Constitutive Representation of Damage Development and Healing in WIPP Salt*; Technical report; Sandia National Labs.: Livermore, CA, USA, 1994.

97. Chan, K.; Bodner, S.; Munson, D. Recovery and healing of damage in WIPP salt. *Int. J. Damage Mech.* **1998**, *7*, 143–166. [CrossRef]

98. Miao, S.; Wang, M.L.; Schreyer, H.L. Constitutive models for healing of materials with application to compaction of crushed rock salt. *J. Eng. Mech.* **1995**, *121*, 1122–1129. [CrossRef]

99. Xu, H.; Arson, C.; Chester, F. Stiffness and Deformation of Salt Rock Subject to Anisotropic Damage and Temperature-Dependent Healing. In Proceedings of the 46th US Rock Mechanics/Geomechanics Symposium, Chicago, IL, USA, 24–27 June 2012. American Rock Mechanics Association.

100. Arson, C.; Xu, H.; Chester, F.M. On the definition of damage in time-dependent healing models for salt rock. *Géotech. Lett.* **2012**, *2*, 67–71. [CrossRef]

101. Zhu, C.; Arson, C. Theoretical Bases of Thermomechanical Damage and DMT-Healing Model for Rock. In Proceedings of the Geo-Congress 2014: Geo-Characterization and Modeling for Sustainability, Atlanta, GA, USA, 23–26 February 2014; pp. 2785–2794.

102. Zhu, C.; Arson, C. *Using Microstructure Descriptors to Model Thermo-Mechanical Damage and Healing in Salt Rock*; Georgia Institute of Technology: Atlanta, GA, USA, 2014.

103. Zhu, C.; Arson, C. A model of damage and healing coupling halite thermo-mechanical behavior to microstructure evolution. *Geotech. Geol. Eng.* **2015**, *33*, 389–410. [CrossRef]

104. Zhu, C.; Arson, C. *Fabric-Enriched Modeling of Anisotropic Healing Induced by Diffusion in Granular Salt*; Georgia Institute of Technology: Atlanta, GA, USA, 2015.

105. Xu, J.; Qu, J.; Gao, Y.; Xu, N. Study on the Elastoplastic Damage-Healing Coupled Constitutive Model of Mudstone. *Math. Probl. Eng.* **2017**, *2017*, 6431607. [CrossRef]

106. Ju, J.; Yuan, K. New strain-energy-based coupled elastoplastic two-parameter damage and healing models for earth-moving processes. *Int. J. Damage Mech.* **2012**, *21*, 989–1019. [CrossRef]

107. Ju, J.; Yuan, K.; Kuo, A. Novel strain energy based coupled elastoplastic damage and healing models for geomaterials—Part I: Formulations. *Int. J. Damage Mech.* **2012**, *21*, 525–549. [CrossRef]

108. J. W. Ju.; K. Y. Yuan.; A. W. Kuo.; J. S. Chen Novel Strain Energy Based Coupled Elastoplastic Damage and Healing Models for Geomaterials—Part II: Computational Aspects. *Int. J. Damage Mech.* **2012**, *21*, 551–576. [CrossRef]

109. Yuan, K.; Ju, J. New strain energy–based coupled elastoplastic damage-healing formulations accounting for effect of matric suction during earth-moving processes. *J. Eng. Mech.* **2012**, *139*, 188–199. [CrossRef]

110. Hampel, A. Description of damage reduction and healing with the CDM constitutive model for the thermo-mechanical behavior of rock salt. In *Mechanical Behavior of Salt VIII*; Taylor & Francis Group: London, UK, 2012; pp. 1–10.

111. Hong, S.; Yuan, K.; Ju, J. New strain energy-based thermo-elastoviscoplastic isotropic damage–self-healing model for bituminous composites—Part I: Formulations. *Int. J. Damage Mech.* **2017**, *26*, 651–671. [CrossRef]

112. Hong, S.; Yuan, K.; Ju, J. New strain energy-based thermo-elastoviscoplastic isotropic damage–self-healing model for bituminous composites—Part II: Computational aspects. *Int. J. Damage Mech.* **2017**, *26*, 672–696. [CrossRef]

113. Hong, S.; Yuan, K.; Ju, J. Initial strain energy-based thermo-elastoviscoplastic two-parameter damage–self-healing models for bituminous composites—Part I: Formulations. *Int. J. Damage Mech.* **2016**, *25*, 1082–1102. [CrossRef]

114. Hong, S.; Yuan, K.; Ju, J. Initial strain energy-based thermo-elastoviscoplastic two-parameter damage self-healing model for bituminous composites—Part II: Computational aspects. *Int. J. Damage Mech.* **2016**, *25*, 1103–1129. [CrossRef]

115. Lee, H.J.; Kim, Y.R. Viscoelastic continuum damage model of asphalt concrete with healing. *J. Eng. Mech.* **1998**, *124*, 1224–1232. [CrossRef]

116. Lee, H.J.; Kim, Y.R. Viscoelastic constitutive model for asphalt concrete under cyclic loading. *J. Eng. Mech.* **1998**, *124*, 32–40. [CrossRef]

117. Schapery, R.A. Correspondence principles and a generalizedJ integral for large deformation and fracture analysis of viscoelastic media. *Int. J. Fract.* **1984**, *25*, 195–223. [CrossRef]

118. Carpenter, S.; Shen, S. A Dissipated Energy Approach to Study HMA Healing 36 in Fatigue. *Transp. Res. Rec.* **2006**, *1970*, 178–185. [CrossRef]

119. Kim, B.; Roque, R. Evaluation of healing property of asphalt mixtures. *Transp. Res. Rec.* **2006**, *1970*, 84–91. [CrossRef]

120. Prager, S.; Tirrell, M. The healing process at polymer–polymer interfaces. *J. Chem. Phys.* **1981**, *75*, 5194–5198. [CrossRef]

121. Shen, S.; Carpenter, S. Application of the dissipated energy concept in fatigue endurance limit testing. *Transp. Res. Rec.* **2005**, *1929*, 165–173. [CrossRef]

122. Shen, S.; Airey, G.D.; Carpenter, S.H.; Huang, H. A dissipated energy approach to fatigue evaluation. *Road Mater. Pavement Des.* **2006**, *7*, 47–69. [CrossRef]

123. Menozzi, A.; Garcia, A.; Partl, M.N.; Tebaldi, G.; Schuetz, P. Induction healing of fatigue damage in asphalt test samples. *Constr. Build. Mater.* **2015**, *74*, 162–168. [CrossRef]

124. Riara, M.; Tang, P.; Mo, L.; Javilla, B.; Chen, M.; Wu, S. Systematic evaluation of fracture-based healing indexes of asphalt mixtures. *J. Mater. Civ. Eng.* **2018**, *30*, 04018264. [CrossRef]

125. Al-Rub, R.K.A.; Darabi, M.K.; Little, D.N.; Masad, E.A. A micro-damage healing model that improves prediction of fatigue life in asphalt mixes. *Int. J. Eng. Sci.* **2010**, *48*, 966–990. [CrossRef]

126. Darabi, M.K.; Al-Rub, R.K.A.; Masad, E.A.; Little, D.N. Constitutive modeling of fatigue damage response of asphalt concrete materials with consideration of micro-damage healing. *Int. J. Solids Struct.* **2013**, *50*, 2901–2913. [CrossRef]

127. Alsheghri, A.A.; Al-Rub, R.K.A. Thermodynamic-based cohesive zone healing model for self-healing materials. *Mech. Res. Commun.* **2015**, *70*, 102–113. [CrossRef]

128. Alsheghri, A.A.; Al-Rub, R.K.A. Finite element implementation and application of a cohesive zone damage-healing model for self-healing materials. *Eng. Fract. Mech.* **2016**, *163*, 1–22. [CrossRef]

129. Voyiadjis, G.Z.; Shojaei, A.; Li, G. A generalized coupled viscoplastic–viscodamage–viscohealing theory for glassy polymers. *Int. J. Plast.* **2012**, *28*, 21–45. [CrossRef]

130. Darabi, M.K.; Al-Rub, R.K.A.; Masad, E.A.; Huang, C.W.; Little, D.N. A thermo-viscoelastic–viscoplastic–viscodamage constitutive model for asphaltic materials. *Int. J. Solids Struct.* **2011**, *48*, 191–207. [CrossRef]

131. Shojaei, A.; Li, G.; Voyiadjis, G.Z. Cyclic viscoplastic-viscodamage analysis of shape memory polymers fibers with application to self-healing smart materials. *J. Appl. Mech.* **2013**, *80*, 011014. [CrossRef]

132. Underwood, B.; Zeiada, W. Characterization of microdamage healing in asphalt concrete with a smeared continuum damage approach. *Transp. Res. Rec.* **2014**, *2447*, 126–135. [CrossRef]

133. Karki, P.; Li, R.; Bhasin, A. Quantifying overall damage and healing behaviour of asphalt materials using continuum damage approach. *Int. J. Pavement Eng.* **2015**, *16*, 350–362. [CrossRef]

134. Mergheim, J.; Steinmann, P. Phenomenological modelling of self-healing polymers based on integrated healing agents. *Comput. Mech.* **2013**, *52*, 681–692. [CrossRef]

135. Mergheim, J.; Possart, G.; Steinmann, P. Modelling and computation of curing and damage of thermosets. *Comput. Mater. Sci.* **2012**, *53*, 359–367. [CrossRef]

136. Wu, M.; Johannesson, B.; Geiker, M. A review: Self-healing in cementitious materials and engineered cementitious composite as a self-healing material. *Constr. Build. Mater.* **2012**, *28*, 571–583. [CrossRef]

137. Sanada, K.; Mizuno, Y.; Shindo, Y. Damage progression and notched strength recovery of fiber-reinforced polymers encompassing self-healing of interfacial debonding. *J. Compos. Mater.* **2015**, *49*, 1765–1776. [CrossRef]

138. Ahmed, A.; Sanada, K.; Fanni, M.; El-Moneim, A.A. A practical methodology for modeling and verification of self-healing microcapsules-based composites elasticity. *Compos. Struct.* **2018**, *184*, 1092–1098. [CrossRef]

139. Mauludin, L.M.; Zhuang, X.; Rabczuk, T. Computational modeling of fracture in encapsulation-based self-healing concrete using cohesive elements. *Compos. Struct.* **2018**, *196*, 63–75. [CrossRef]

140. Mauludin, L.M.; Oucif, C. Interaction between matrix crack and circular capsule under uniaxial tension in encapsulation-based self-healing concrete. *Undergr. Space* **2018**, *3*, 181–189. [CrossRef]

141. Mauludin, L.M.; Oucif, C. The effects of interfacial strength on fractured microcapsule. *Front. Struct. Civ. Eng.* **2018**, 1–11. [CrossRef]

142. Voyiadjis, G.Z.; Kattan, P.I. Mechanics of damage, healing, damageability, and integrity of materials: A conceptual framework. *Int. J. Damage Mech.* **2017**, *26*, 50–103. [CrossRef]

143. Voyiadjis, G.Z.; Kattan, P.I. Decomposition of healing tensor: In continuum damage and healing mechanics. *Int. J. Damage Mech.* **2018**, *27*, 1020–1057. [CrossRef]

144. Zhu, H.; Zhou, S.; Yan, Z.; Ju, J.W.; Chen, Q. A two-dimensional micromechanical damage-healing model on microcrack-induced damage for microcapsule-enabled self-healing cementitious composites under tensile loading. *Int. J. Damage Mech.* **2015**, *24*, 95–115. [CrossRef]

145. Li, X.; Du, Y.; Duan, Q.; Ju, J.W. Thermodynamic framework for damage-healing-plasticity of granular materials and net damage variable. *Int. J. Damage Mech.* **2016**, *25*, 153–177. [CrossRef]

146. Zhu, H.; Zhou, S.; Yan, Z.; Ju, J.W.; Chen, Q. A two-dimensional micromechanical damage–healing model on microcrack-induced damage for microcapsule-enabled self-healing cementitious composites under compressive loading. *Int. J. Damage Mech.* **2016**, *25*, 727–749. [CrossRef]

147. Shahsavari, H.; Baghani, M.; Sohrabpour, S.; Naghdabadi, R. Continuum damage-healing constitutive modeling for concrete materials through stress spectral decomposition. *Int. J. Damage Mech.* **2016**, *25*, 900–918. [CrossRef]

148. Ozaki, S.; Osada, T.; Nakao, W. Finite element analysis of the damage and healing behavior of self-healing ceramic materials. *Int. J. Solids Struct.* **2016**, *100*, 307–318. [CrossRef]

149. Davies, R.; Jefferson, A. Micromechanical modelling of self-healing cementitious materials. *Int. J. Solids Struct.* **2017**, *113*, 180–191. [CrossRef]

150. Kazemi, A.; Baghani, M.; Shahsavari, H.; Abrinia, K.; Baniassadi, M. Application of elastic-damage-heal model for self-healing concrete thick-walled cylinders through thermodynamics of irreversible processes. *Int. J. Appl. Mech.* **2017**, *9*, 1750082. [CrossRef]

151. Chaboche, J.L. Damage induced anisotropy: On the difficulties associated with the active/passive unilateral condition. *Int. J. Damage Mech.* **1992**, *1*, 148–171. [CrossRef]

152. Cormery, F.; Welemane, H. A critical review of some damage models with unilateral effect. *Mech. Res. Commun.* **2002**, *29*, 391–395. [CrossRef]

153. Welemane, H.; Goidescu, C. Isotropic brittle damage and unilateral effect. *C. R. Méc.* **2010**, *338*, 271–276. [CrossRef]

154. Alliche, A. A continuum anisotropic damage model with unilateral effect. *Mech. Sci.* **2016**, *7*, 61–68. [CrossRef]

155. Bielski, J.; Skrzypek, J.; Kuna-Ciskal, H. Implementation of a model of coupled elastic-plastic unilateral damage material to finite element code. *Int. J. Damage Mech.* **2006**, *15*, 5–39. [CrossRef]

156. Zhu, C.; Arson, C. A thermo-mechanical damage model for rock stiffness during anisotropic crack opening and closure. *Acta Geotech.* **2014**, *9*, 847–867. [CrossRef]

157. Xianda, S.; Zhu, C.; Arson, C. *Chemo-Mechanical Damage and Healing of Granular Salt: Micro-Macro Modeling*; Georgia Institute of Technology: Atlanta, GA, USA, 2016.

158. He, W.; Wu, Y.F.; Xu, Y.; Fu, T.T. A thermodynamically consistent nonlocal damage model for concrete materials with unilateral effects. *Comput. Methods Appl. Mech. Eng.* **2015**, *297*, 371–391. [CrossRef]

159. Matallah, M.; La Borderie, C. Inelasticity–damage-based model for numerical modeling of concrete cracking. *Eng. Fract. Mech.* **2009**, *76*, 1087–1108. [CrossRef]

160. Murakami, S. Mechanical modeling of material damage. *J. Appl. Mech.* **1988**, *55*, 280–286. [CrossRef]

161. Voyiadjis, G.Z.; Shojaei, A.; Li, G.; Kattan, P.I. A theory of anisotropic healing and damage mechanics of materials. *Proc. R. Soc. A* **2012**, *468*, 163–183. [CrossRef]

162. Voyiadjis, G.Z.; Yousef, M.A.; Kattan, P.I. New tensors for anisotropic damage in continuum damage mechanics. *J. Eng. Mater. Technol.* **2012**, *134*, 021015. [CrossRef]

163. Voyiadjis, G.Z.; Kattan, P.I. On the decomposition of the damage variable in continuum damage mechanics. *Acta Mech.* **2017**, *228*, 2499–2517. [CrossRef]

164. Voyiadjis, G.Z.; Shojaei, A.; Li, G. A thermodynamic consistent damage and healing model for self healing materials. *Int. J. Plast.* **2011**, *27*, 1025–1044. [CrossRef]

165. Balieu, R.; Kringos, N.; Chen, F.; Córdoba, E. Multiplicative viscoelastic-viscoplastic damage-healing model for asphalt-concrete materials. In Proceedings of the 8th RILEM International Conference on Mechanisms of Cracking and Debonding in Pavements, Nantes, France, 7–9 June 2016; pp. 235–240.

166. Voyiadjis, G.Z.; Kattan, P.I. Mechanics of damage processes in series and in parallel: A conceptual framework. *Acta Mech.* **2012**, *223*, 1863–1878. [CrossRef]

167. Voyiadjis, G.Z.; Kattan, P.I. Healing and super healing in continuum damage mechanics. *Int. J. Damage Mech.* **2014**, *23*, 245–260. [CrossRef]

168. Wang, J.; Soens, H.; Verstraete, W.; De Belie, N. Self-healing concrete by use of microencapsulated bacterial spores. *Cem. Concr. Res.* **2014**, *56*, 139–152. [CrossRef]

169. Dong, B.; Fang, G.; Wang, Y.; Liu, Y.; Hong, S.; Zhang, J.; Lin, S.; Xing, F. Performance recovery concerning the permeability of concrete by means of a microcapsule based self-healing system. *Cem. Concr. Compos.* **2017**, *78*, 84–96. [CrossRef]

170. Voyiadjis, G.Z.; Kattan, P.I. Introduction to the mechanics and design of undamageable materials. *Int. J. Damage Mech.* **2013**, *22*, 323–335. [CrossRef]

171. Voyiadjis, G.Z.; Kattan, P.I. On the theory of elastic undamageable materials. *J. Eng. Mater. Technol.* **2013**, *135*, 021002. [CrossRef]

172. Voyiadjis, G.Z.; Kattan, P.I. Governing differential equations for the mechanics of undamageable materials. *Eng. Trans.* **2014**, *62*, 241–267.

173. Ghosh, S.K. *Self-Healing Materials: Fundamentals, Design Strategies, and Applications*; John Wiley & Sons: Hoboken, NJ, USA, 2009.

174. Files, B.S. Design of a Biomimetic Self-Healing Superalloy Composite. Ph.D. Thesis, Northwestern University, Evanston, IL, USA, 1997.

175. Manuel, M.V. Design of a Biomimetic Self-Healing Alloy Composite. Ph.D. Thesis, Northwestern University, Evanston, IL, USA, 2007.

176. Manuel, M.V.; Olson, G.B. Biomimetic self-healing metals. In Proceedings of the 1st International Conference on Self-Healing Materials, Noordwijik aan Zee, The Netherlands, 18–20 April 2007; pp. 18–20.

177. Burton, D.; Gao, X.; Brinson, L. Finite element simulation of a self-healing shape memory alloy composite. *Mech. Mater.* **2006**, *38*, 525–537. [CrossRef]

178. Zhu, P.; Cui, Z.; Kesler, M.S.; Newman, J.A.; Manuel, M.V.; Wright, M.C.; Brinson, L.C. Characterization and modeling of three-dimensional self-healing shape memory alloy-reinforced metal-matrix composites. *Mech. Mater.* **2016**, *103*, 1–10. [CrossRef]

179. Araki, S.; Ono, H.; Saito, K. Micromechanical analysis of crack closure mechanism for intelligent material containing TiNi fibers. *JSME Int. J. Ser. Solid Mech. Mater. Eng.* **2002**, *45*, 208–216.

180. Bor, T.C.; Warnet, L.; Akkerman, R.; de Boer, A. Modeling of stress development during thermal damage healing in fiber-reinforced composite materials containing embedded shape memory alloy wires. *J. Compos. Mater.* **2010**, *44*, 2547–2572. [CrossRef]

181. Kawai, M.; Ogawa, H.; Baburaj, V.; Koga, T. Micromechamical analysis for hysteretic behavior of unidirectional TiNi SMA fiber composites. *J. Intell. Mater. Syst. Struct.* **1999**, *10*, 14–28. [CrossRef]

182. Grabowski, B.; Tasan, C.C. Self-healing metals. In *Self-Healing Materials*; Springer: Berlin, Germany, 2016; pp. 387–407.

183. Lumley, R.; Polmear, I. Advances in self-healing metals. In Proceedings of the First International Conference on Self Healing Materials, Series in Materials Science, Noordwijk aan Zee, The Netherlands, 18–20 April 2007; Volume 1.

184. Lumley, R.; Morton, A.; Polmear, I. Enhanced creep performance in an Al–Cu–Mg–Ag alloy through underageing. *Acta Mater.* **2002**, *50*, 3597–3608. [CrossRef]

185. Shinya, N.; Kyono, J.; Laha, K.; Masuda, C. Self-healing of creep damage through autonomous boron segregation and boron nitride precipitation during high temperature use of austenitic stainless steels. In Proceedings of the First International Conference on Self-Healing Materials, Noordwijk aan Zee, The Netherlands, 18–20 April 2007.

186. Laha, K.; Kyono, J.; Sasaki, T.; Kishimoto, S.; Shinya, N. Improved creep strength and creep ductility of type 347 austenitic stainless steel through the self-healing effect of boron for creep cavitation. *Metall. Mater. Trans. A* **2005**, *36*, 399–409. [CrossRef]

187. Li, S.; Gao, K.; Qiao, L.; Zhou, F.; Chu, W. Molecular dynamics simulation of microcrack healing in copper. *Comput. Mater. Sci.* **2001**, *20*, 143–150. [CrossRef]

188. Shen, L.; Kewei, G.; Lijie, Q.; Wuyang, C.; Fuxin, Z. Molecular dynamics simulation of the role of dislocations in microcrack healing. *Acta Mech. Sin.* **2000**, *16*, 366–373. [CrossRef]

189. Zhou, G.; Gao, K.; Qiao, L.; Wang, Y.; Chu, W. Atomistic simulation of microcrack healing in aluminium. *Model. Simul. Mater. Sci. Eng.* **2000**, *8*, 603. [CrossRef]

190. Wei, D.; Han, J.; Tieu, K.; Jiang, Z. Simulation of crack healing in BCC Fe. *Scr. Mater.* **2004**, *51*, 583–587. [CrossRef]

191. Huang, P.; Li, Z.; Sun, J. Shrinkage and splitting of microcracks under pressure simulated by the finite-element method. *Metall. Mater. Trans. A* **2002**, *33*, 1117–1124. [CrossRef]

192. Wang, H.; Li, Z. The shrinkage of grain-boundary voids under pressure. *Metall. Mater. Trans. A* **2003**, *34*, 1493–1500. [CrossRef]

193. Wang, H.; Li, Z. Stability and shrinkage of a cavity in stressed grain. *J. Appl. Phys.* **2004**, *95*, 6025–6031. [CrossRef]

194. Kukudzhanov, K. Modeling the healing of damage of metal by high-energy pulsed electromagnetic field. *Lett. Mater. PIS Mater.* **2018**, *8*, 27–32.

195. Versteylen, C.; Sluiter, M.; van Dijk, N. Modelling the formation and self-healing of creep damage in iron-based alloys. *J. Mater. Sci.* **2018**, *53*, 14758–14773. [CrossRef]

applied
sciences

MDPI

Article

Three Dimensional CS-FEM Phase-Field Modeling Technique for Brittle Fracture in Elastic Solids

Sauradeep Bhowmick * and Gui-Rong Liu

Department of Aerospace Engineering and Engineering Mechanics, University of Cincinnati, Cincinnati, OH 45219, USA; liugr@ucmail.uc.edu
* Correspondence: bhowmisp@mail.uc.edu

Received: 30 October 2018; Accepted: 22 November 2018; Published: 4 December 2018

Abstract: The cell based smoothed finite element method (CS-FEM) was integrated with the phase-field technique to model brittle fracture in 3D elastic solids. The CS-FEM was used to model the mechanics behavior and the phase-field method was used for diffuse fracture modeling technique where the damage in a system was quantified by a scalar variable. The integrated CS-FEM phase-field approach provides an efficient technique to model complex crack topologies in three dimensions. The detailed formulation of our combined method is provided. It was implemented in the commercial software ABAQUS using its user-element (UEL) and user-material (UMAT) subroutines. The coupled system of equations were solved in a staggered fashion using the in-built non-linear Newton–Raphson solver in ABAQUS. Eight node hexahedral (H8) elements with eight smoothing domains were coded in CS-FEM. Several representative numerical examples are presented to demonstrate the capability of the method. We also discuss some of its limitations.

Keywords: Brittle Fracture; cell-based smoothed-finite element method (CS-FEM); Phase-field model; ABAQUS UEL

1. Introduction

In the past decade, Liu et al. [1,2] generalized the gradient smoothing approach in meshfree method [3] and proposed the smoothed finite element method(S-FEM) to overcome some of the inherent shortcomings of the classical finite element method(FEM) such as overly stiff behavior, sensitivity to mesh distortions, and stress inaccuracy. The SFEM combines the FEM with the traditional meshfree methods in producing more accurate results with higher efficiency [1,2]. It uses the base mesh of the FEM and reconstructs the strain field using the gradient smoothing technique. Thus, in contrast to the weak formulations of the FEM [4], the weakened weak formulation used here [1,2] further softens the model [5], giving it a closer to exact stiffness [6]. The gradient smoothing operation facilitates creation of various smoothing domains based on elements, nodes, edges (2D), and faces (3D) over which the stress/strain is evaluated [1]. This, in turn, produces a wide variety of results and gives the analyst much needed freedom to design models as per the requirement. For example, the node based smoothing domains (NS-FEM) produce upper bound solutions for force driven problems [7]. The edge/face based smoothing domains (ES/FS-FEM) produce very accurate results using even triangular and tetrahedral mesh [8]. A combination of the NS-FEM and the ES-FEM produces very accurate close to exact solutions in the error norm [6,9]. Amongst other applications [10–14], the S-FEM has been effectively applied to fracture problems for quasi-static crack propagation [15], anisotropic materials [16], dynamic fracture [1] and for singular geometries of arbitrary order [17,18]. Although these studies have proven to be quite accurate and efficient when compared to the FEM, they treat the crack as an discrete entity, a formulation which has its own inherent problems.

Fracture is the primary cause of failure in the majority of engineering structures. An initially existing small crack often leads to catastrophic failures by propagation under external loads. Over the past few decades, several theories and computational techniques have been developed to accurately model fracture and predict crack propagation in engineering systems. Classically, cracks are dealt with discretely in theoretical/computational fracture mechanics. The discontinuities have been separately modeled and special methodologies have been developed to accurately model the singular stress field near the crack tip. The collapsed quadrilateral or triangular element in the FEM [4] is the most widely used, however several other methods such as the boundary element method (BEM) [19], mesh-free methods [3,20], the extended finite element method [21], and SFEM [18,22,23] have proven to be equally efficient. BEM uses a mesh only on the boundaries, thereby reducing the computational complexity by one order, but is unable to treat material non-linearities. S-FEM uses a base mesh of linear elements with an enrichment only around the crack-tip, thereby mitigating the computational cost of using a quadratic mesh throughout the domain [18]. However, in S-FEM, as in the classical FEM, the crack path is mesh dependent; cracks can only propagate along edges of elements and the tracking of crack front for three-dimensional problems is computationally very expensive. The X-FEM overcomes this mesh dependency but uses additional degrees of freedom and the integration in the weak form for the elements containing the crack becomes particularly very complex in three-dimensional fracture problems.

An alternate way of modeling fracture has gained popularity in the computational community in the past decade or two. In this method, the crack is modeled as a diffused entity and represented by a continuous scalar variable called the phase-field [24]. This variable differentiates between the broken and intact material phases and, contrary to the discrete approach, provides a smooth transition between them [25–28]. This method is based on the variational theory of fracture [24] addressing the shortcomings of the original Griffith's theory, which could not predict crack nucleation or complex crack paths. The elegance of this method lies in the fact that it can successfully predict complex crack behaviors such as crack branching, curved crack paths, and crack merging even in three dimensions without any ad-hoc criterion. This gives it an immediate advantage over the traditional methods where the prediction of such phenomenons is quite complex. Apart from the primary applications of brittle fracture [28], the phase-field approach has been developed and applied to many complex fracture mechanics problems including large deformations [29], plasticity [30], multiphysics [31,32], and dynamics [33,34]. It has also been previously implemented in the commercial code ABAQUS by using its user subroutine features [27,28]. The phase-field model, however, is not without its disadvantages, including having a very refined mesh in the expected crack propagation region. Advanced fracture algorithms such as the screened Poisson's equations [35–37] for crack propagation are developed to overcome such shortcomings. However, here we limit ourselves to the application of the phase-field method and discuss its useful features as well as disadvantages.

In this work, we integrate the cell based S-FEM with phase-field model to simulate 3-D crack propagation. This integration is done on the commercial platform of ABAQUS. ABAQUS is one of the most widely used commercial codes and its excellent inbuilt solvers and sophisticated visualization tools are particularly attractive for implementing user developed element formulations. A CS-FEM formulation for the eight-node hexahedral elements (3D-CS-FEM-H8) is used, similar to the one used in Xuan et al. [38]. We consider each node to have an additional degree of freedom (phase-field) in addition to the standard displacements. The phase-field and the displacement variables are solved in a staggered fashion to attenuate instability [27]. The 3D CS-FEM-H8 has already been proven to have faster convergence rate and better accuracy than the standard FEM, thus its assimilation with the phase-field method which computes complex crack topologies without any ad-hoc criterion, produces another efficient technique to model crack propagation. One of the significant disadvantages of implementing this method in ABAQUS is that the software does not let users access data for any surrounding elements, thus we deviate from the classical way of calculating strains/stresses in CS-FEM [1,38], and develop another novel, simple yet quite effective approach. It is noteworthy that this is also the reason for our inability to implement S-FEM models of higher accuracy into the software.

However, with all these shortcomings, as demonstrated by a number of examples, this method proves to be quite an effective tool to predict complex crack behaviors.

This paper is outlined as follows: In Section 2, we outline the phase-field model of fracture and the 3D-CS-FEM-H8, and discuss the implementation details of the combined method into ABAQUS. In Section 3, we provide numerical examples to validate our method. In Section 4, we conclude the paper with a summary of our findings and proposed future works.

2. Methods

2.1. Governing Equations of a 3D Elastic Solid with Discontinuity

Consider a three-dimensional (3D) arbitrary, homogeneous, linear elastic domain Ω bounded by Γ such that $\Gamma = \Gamma_u \cup \Gamma_t$, $\Gamma_u \cap \Gamma_t = 0$ and an internal traction free crack Γ_c, as shown in Figure 1, where Γ_u and Γ_t are the displacement and traction boundary surfaces, respectively. The equilibrium equations are given as [1,4]:

$$\nabla \cdot \sigma + f^b = 0 \tag{1}$$

where ∇ is the divergence operator, σ is the Cauchy stress tensor and f^b is the body force.

The Dirichlet and Neumann boundary conditions are given as:

$$u(x,t) = \bar{u}(x,t) \quad \text{on} \quad \Gamma_u \tag{2}$$

$$\sigma \cdot n = f^t \quad \text{on} \quad \Gamma_t \tag{3}$$

$$\sigma \cdot n = 0 \quad \text{on} \quad \Gamma_c \tag{4}$$

where n is the outward unit normal vector on the boundary area Γ and $\bar{u}$ is the prescribed displacement on the boundary Γ_u.

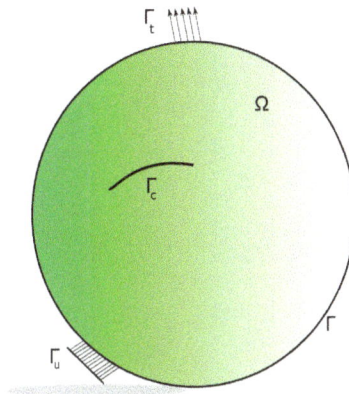

Figure 1. An arbitrary discontinuous three-dimensional body.

The stress–strain relation is given by the constitutive equation

$$\sigma = C \cdot \epsilon \tag{5}$$

where C is the matrix of elastic constants.

Assuming small displacements and strain, we can define the compatibility relation between strain and displacement as:

$$\epsilon = \nabla^s u(x) \tag{6}$$

The elastic strain energy density is given by

$$\psi_e(\epsilon) = \frac{1}{2}\lambda\epsilon_{ii}\epsilon_{jj} + \mu\epsilon_{ij}\epsilon_{ij} \tag{7}$$

where λ and μ are the Lamé constants.

2.2. Review of Phase-Field Model for Brittle Fracture

Following the derivations in [25,27,39], we discuss briefly about the formulations of phase-field approximations for diffuse fracture modeling.

According to the variational theory of fracture, the crack propagates in such a way the total energy of a system is always minimized. Thus, we approach by minimizing the energy functional. The total energy of a discontinuous system is given by

$$\Pi = \psi_e(\epsilon) + \psi_f - \psi_{ext}(u) \tag{8}$$

where $\psi_e(\epsilon)$ is the elastic strain energy as given in Equation (7) integrated over the entire volume Ω, ψ_f is the fracture surface energy and $\psi_{ext}(u)$ is the external potential energy.

The fracture surface energy ψ_f is given by

$$\psi_f = G_c \int_\Omega \gamma(c)d\Omega \tag{9}$$

where G_c is the critical energy release rate proposed by Griffith, c is the phase-field parameter and $\gamma(c)$ is the density function which was calculated to be [28]

$$\gamma(c) = \frac{1}{2}[\frac{1}{l_c}c^2 + \frac{l_c}{2}|\nabla c|^2] \tag{10}$$

where l_c is the length scale parameter.

The external potential energy is given by the summation of the body force b and the surface traction t_Γ:

$$\Psi_{ext}(u) = \int_\Omega b \cdot u dV + \int_\Gamma t_\Gamma \cdot u dA \tag{11}$$

The total internal potential energy of a system can be written as the summation of the bulk energy and the energy required for the formation of crack [24,28,39]

$$\Psi(u,c) = \int_\Omega [(1-c)^2 + d]\psi(\epsilon)d\Omega + \int_\Omega \frac{G_c}{2}[l_c\nabla c \cdot \nabla c + \frac{1}{l_c}c^2]d\Omega \tag{12}$$

where d is a numerical stabilization parameter.

By variation of the external and internal energy potentials (using Equations (8)–(12)) and thereby imposing the principal of virtual displacements we obtain the governing equations of the model:

$$[(1-c)^2 + d]\frac{\partial\sigma_{ij}}{\partial x_i} + b_j = 0 \quad in \quad \Omega \tag{13}$$

$$[(1-c)^2 + d]n_i\sigma_{ij} = t_j \quad in \quad \Gamma_t \tag{14}$$

$$u_j = \bar{u}_j \quad in \quad \Gamma_u \tag{15}$$

$$-G_c l_c \frac{\partial^2 c}{\partial x_i \partial x_i} + [\frac{G_c}{l_c} + 2\psi(\epsilon)]c = 2\psi(\epsilon) \quad in \quad \Omega \tag{16}$$

$$\frac{\partial c}{\partial x_i}n_i = 0 \quad in \quad \Gamma \tag{17}$$

2.3. Three-Dimensional Cell Based Smoothed-Finite Elements

In S-FEM, the idea of strain smoothing is combined with the standard underlying mesh of FEA by dividing each element into a number of sub-cells [1]. We use the eight-noded hexahedral elements with eight smoothing cells (Figure 2). They have already been proven to have a better accuracy than the standard eight-noded hexahedral element used in FEA with eight Gaussian points [38] using the same set of elements. Classical FEM involves isoparametric transformation for every element, making the problem very much mesh dependent because it involves inverting the Jacobian matrix. The solutions tend to vary with minor mesh distortions and for better modeling of certain topological features one needs mesh refinement or higher order elements, which makes the problem computationally expensive. The S-FEM can solve majority of computational problems using a base of linear triangular/tetrahedral mesh, which can be generated automatically. Since the shape functions are created on the basis of radial point interpolation, creating special elements to treat special topological features [18] is of minor hassle and does not need any transition or patch elements. In S-FEM, a smoothing operation is performed on the gradient of the displacement field for each smoothing cell in an element. Subsequently, the interior integration on each of the smoothing cell is transferred to the boundary surface area using the divergence theorem. Herein lies one of the biggest advantages of the S-FEM. This feature makes the solution pretty impervious to mesh distortion since we can avoid the isoparametric transformation and also contributes in saving computational cost. In this case, the gradient smoothing is the spatial average of the strain, over a smoothing hexahedral cell. A smoothing operation is performed to the gradient of displacement for each smoothing cell in an element:

$$\nabla u^h(x_C) = \int_\Omega \nabla u^h(x)\Phi(x - x_C)d\Omega \tag{18}$$

where x_C and Ω represent the center and volume of the smoothing domain, respectively, and Φ is a smoothing function.

Integration by parts on Equation (18) yields:

$$\nabla u^h(x_C) = \int_\Gamma u^h(x)n(x)\Phi(x - x_C)d\Gamma - \int_\Omega u^h(x)\nabla\Phi(x - x_C)d\Omega \tag{19}$$

For simplicity, a piecewise constant smoothing function is applied, which is assumed to be constant within Ω_C and vanish everywhere else,

$$\Phi(x - x_C) = \begin{cases} \dfrac{1}{V_c} & \text{for } x \in \Omega_c \\ 0 & \text{for } x \notin \Omega_c \end{cases} \tag{20}$$

where $V_C = \int_{\Omega_c}^d \Omega$ and Ω_c is the smoothing cell.

Substituting Equation (20) into Equation (18), we can get the smoothed gradient or strain field

$$\tilde{\nabla} u^h(x_C) = \int_{\Gamma_C} u^h(x)n(x)\Phi(x - x_C)d\Gamma = \frac{1}{V_c}\int_{\Gamma_c} u^h(x)n(x)d\Gamma \tag{21}$$

with $d\Gamma_c$ denoting the boundary of the smoothing cell. It is noteworthy that the choice of the smoothing function makes the second term of Equation (19) vanish and the area integration is thus converted to line integration along the boundary of the smoothing cell.

The smoothed strain can be obtained as

$$\tilde{\varepsilon}^h(x_C) = \sum_{I=1}^n \tilde{B}_I(x_C)d_I \tag{22}$$

where $\tilde{B}_I$ is the smoothed strain gradient matrix given by

$$
\tilde{B}_I = \begin{bmatrix} \tilde{b}_{I1}(x_C) & 0 & 0 \\ 0 & \tilde{b}_{I2}(x_C) & 0 \\ 0 & 0 & \tilde{b}_{I3}(x_C) \\ \tilde{b}_{I2}(x_C) & \tilde{b}_{I1}(x_C) & 0 \\ 0 & \tilde{b}_{I2}(x_C) & \tilde{b}_{I3}(x_C) \\ \tilde{b}_{I1}(x_C) & 0 & \tilde{b}_{I3}(x_C) \end{bmatrix} \tag{23}
$$

where

$$
\tilde{b}_{I1}(x_C) = \frac{1}{V_C} \int_{\Gamma_C} N_I(x) n_k(x) d\Gamma \tag{24}
$$

N_I is the regular shape functions for a eight-node hexahedral finite element. Since we are using eight-noded hexahedral elements, one Gauss point at the center is sufficient to integrate along each surface boundary. The above equation then reduces to

$$
\tilde{b}_{I1}(x_C) = \frac{1}{V_C} \sum_{I=1}^{M} N_I(x_i^{GP}) n_{ki}^C A_i^C \tag{25}
$$

where x_i is the midpoint or the intersection of the diagonals of the boundary segment and n_i and A_i are the corresponding outward unit normal and surface area, respectively.

The smoothed stiffness matrix of an element is thus given by the assembly of the stiffness of each smoothing cell

$$
\tilde{K}_e = \sum_C \tilde{B}_C^T C \tilde{B}_C V_C \tag{26}
$$

The overall smoothed stiffness matrix of the domain is calculated by the summation of sub-stiffness matrices for nodes I in relation to node J as in FEM, except that the summation here is performed over the smoothing domains, not elements:

$$
\tilde{K}_{IJ}^{CS-FEM} = \sum_{i=1}^{N_e} \sum_{m=1}^{N_c} \int_{\Omega_{i,m}^s} \tilde{B}_I^T C \tilde{B}_J d\Omega = \sum_{i=1}^{N_e} \sum_{m=1}^{N_c} \tilde{B}_I^T C \tilde{B}_J V_{i,m}^s = \sum_{k=1}^{N_e} \tilde{B}_I^T C \tilde{B}_J V_k^s \tag{27}
$$

Thus, it is an algebraic summation over the stiffness of each element and the equations required to calculate the stiffness matrix of an element are the Equations (23) and (24) where we do not need the spatial derivative of the shape functions, thereby eliminating the Jacobian inversion problem. Only the values of shape function at certain points (Gauss point/center of each surface of a smoothing cell) are needed (Figure 2).

The calculation of displacements or force vector, based on the nature of the problem is similar to classical FEM, by solving the following algebraic equation

$$
\tilde{K}^{CS-FEM} \tilde{d} = \tilde{f} \tag{28}
$$

Once the displacements are calculated, we deviate from the standard stress/strain calculations used in [1,2,38]. This is because ABAQUS does not let users obtain information from surrounding elements; the user element (UEL) is written such that it only specifies the formulation of a single element. In a typical CS-FEM setting [1,2], the stress/strains at a node is calculated by the weighted average of the values of the surrounding smoothing domains of the node. In this case, due to our inability to access data from surrounding elements and because of the way the UEL is formulated, we calculate the elemental strains as

$$
\epsilon = \tilde{B} U \tag{29}
$$

where $\tilde{B}$ is the smoothed strain displacement matrix obtained at the center of the smoothing cell. The corresponding elemental stress is given by

$$\sigma_{ij} = C_{ijkl}\epsilon_{kl} \tag{30}$$

ABAQUS then allows its internal algorithms to calculate the nodal stress/strain values. Similarly, the elemental displacement values are given as

$$u_i = N_{ij}U_j \tag{31}$$

where U_j is the 24 × 1 nodal displacement vector and u_i is the 3 × 1 elemental displacement vector.

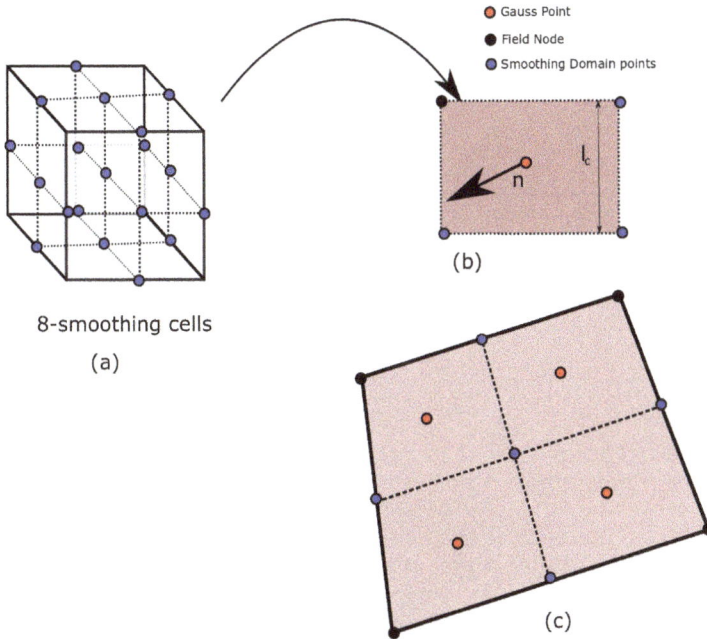

Figure 2. A 3D H8 element subdivided into eight smoothing cells: (**a**) subdivision into smoothing cells by joining the midpoints of the edges and the faces; (**b**) face area of a smoothing cell with its outward normal and length; and (**c**) entire face of the element containing surfaces of the four smoothing cells.

2.4. Implementation in ABAQUS UEL

We implemented the PhaseField-CSFEM in ABAQUS (ABAQUS 2017, Dassault Systemes, Providence, RI, USA) and solved it using a staggered scheme. Equations (13)–(17) contain the coupled phase and displacement fields which can be solved using a monolithic approach [26,28], however we decoupled and solved them separately using the SFEM in ABAQUS UEL. In the literature, the staggered scheme has presented quite a few stability advantages over the monolithic solution [25,27].

The decoupled governing equations are:

$$r_c = \int_\Omega ([\frac{g_c}{l_c}c - 2(1-c)H](N^c)^T + g_c l_c (B^c)^T \nabla c)d\Omega \tag{32}$$

where r_c is the residual force vector and N^c and B^c are the shape function and the strain gradient matrix for 3D eight-node hexahedral elements, respectively.

The phase-field component in the decoupled stiffness matrix is given by:

$$K_c = \int_\Omega ([\frac{G_c}{l_c} + 2H](N^c)^T(N^c) + G_c l_c (B^c)^T(B^c))d\Omega \tag{33}$$

Similarly, the displacement-field contribution in the stiffness matrix is given by B^u matrix calculated from Equation (23),

$$K_u = \int_\Omega [(1-c)^2 + d](B^u)^T C(B^u)d\Omega \tag{34}$$

and the internal force(residual) vector is

$$r_u = \int_\Omega [(1-c)^2 + d](B^u)^T \sigma_0 d\Omega \tag{35}$$

where σ_0 is the stress calculated by Equation (30).

Thus, the following equation is solved using the modified Newton–Raphson scheme

$$\begin{bmatrix} u \\ c \end{bmatrix}_{t+\delta t} = \begin{bmatrix} u \\ c \end{bmatrix}_t - \begin{bmatrix} K_u & 0 \\ 0 & K_c \end{bmatrix}_t \begin{bmatrix} r_u \\ r_c \end{bmatrix}_t \tag{36}$$

The term H in Equations (32) and (33) is a history field which is equal to the strain energy from the previous step. In the staggered scheme, in the first iteration of every load step, the history variable is updated via $H_{n+1} = \psi_n$ and the displacement field (Equations (34) and (35)) is solved by updating the value of the phase-field from the previous load step (c_n). The history variable satisfies certain properties [25] and ensures that no penalty term is necessary to enforce the irreversibility of the crack field.

In this formulation, the B^u matrix in Equations (34) and (35) is calculated based on the smoothing cells, using CS-FEM-H8.

The non-linear system of equations (Equation (36)) is solved via an incremental iterative approach using a Newton–Raphson scheme. The continuum mechanics equations are solved by the more accurate CS-FEM [2] and the phase-field equations are solved by the standard FEM. In the UEL, the user needs to specify the tangent stiffness matrix $AMATRX$ and the internal force vector RHS, which the solver calls for each element. The properties are calculated at the center of the smoothing domain for each subcell. In the staggered scheme of solutions [27], we use the *common* block in the UEL to facilitate the transfer of variables. Through the *common* block, the UEL allows a user to write formulations of multiple elements in a single code. We define three elements, the first two being the phase-field and the displacement elements and the third a dummy element written to facilitate post-processing in the ABAQUS viewer. ABAQUS is unaware of the inherent shape functions used in the element formulation(UEL) and thus is unable to extrapolate the results on its own. We write a user material (UMAT), where the results stored as solution variables $STATEV$ are transferred from the UEL. This is possible because the element connectivity and the shape function of the elements are exactly the same as the C3D8 elements in ABAQUS library. Thus, we imply that there is a dummy mesh of H8 elements whose material properties, chosen such that there is no resistance to strain, are provided to the UMAT. The corresponding variables are stored as $SVARS$ in the UEL and transferred to the UMAT via the *common* statement.

3. Numerical Modeling

We present several numerical examples in this section to validate the accuracy of the 3D-CS-FEM-phase field method for linear elastic brittle fracture and also to discuss some shortcomings of the implementation in ABAQUS. First, three benchmark examples were tested and subsequently we simulated comparatively complex models. All models were meshed using eight-node hexahedral elements and eight smoothing domains were used for CS-FEM formulations. Since the phase-field represents a diffused fracture representation, no special crack-tip element was necessary for capturing the crack path. The mesh size was significantly reduced to successfully resolve the length scale parameter near the expected crack propagation zone. As discussed by Borden et al. [34], the approximate element size should be less than half of the corresponding length scale parameter. Contour plots representing the phase-field variable which signifies the crack path are presented for all examples.

3.1. Single Edge Notched Tensile Sample

We simulated crack propagation in a rectangular bar, with a finite opening, under far-field tension. The bottom surface was fixed and a tensile pull was applied on the top surface of the rectangular block (Figure 3). The crack was located at an edge and the material parameters were: $E = 210$ kN/mm^2, $v = 0.3$, and $G_c = 5 \times 10^{-4}$ kN/mm. The length scale parameter was $l_c = 0.08$ mm. We used a uniform mesh of 45,000 hexahedral elements, which was refined near the expected crack propagation zone (edge side of 0.02 mm) to successfully resolve the length scale parameter for better reproduction of the ultimate strength of the sample [25,39]. Since this was a diffused fracture representation, we did not need to use any special crack-tip elements; the length scale parameter provided the transition from the damaged to intact zone. As expected, we saw the contour plot of crack evolution; it propagated along the straight path (Figure 4). We also studied the influence of the Griffith's energy release rate parameter G_c on the crack pattern. We tested the model for $G_c = 5 \times 10^{-4}$, 4×10^{-4}, 2×10^{-4}, and 1×10^{-4} kN/mm, and observd that this parameter had no influence on the ultimate crack path. The resultant load–displacement curve is presented in Figure 5. It shows that the slope of the curve is same for small displacements for different energy release rates. However, as displacement increased, the force reduced before it reached the ultimate load, which itself increased with G_c. There was a sudden drop of the load after the crack initiation, which signified decrease in strength and ultimate failure. We also performed experiments with further reduction of mesh size near the expected crack path, however that did not affect crack pattern and the force–displacement curve, thereby proving that the damage pattern was convergent.

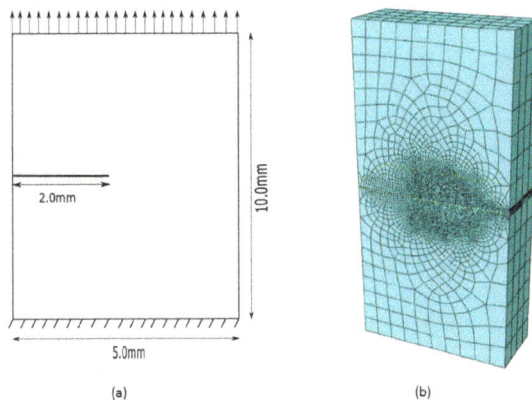

Figure 3. A single edge notched tensile specimen: (**a**) representational 2D geometry; thickness = 2.0 mm; and (**b**) hexahedral mesh with smaller element size near the expected propagation zone.

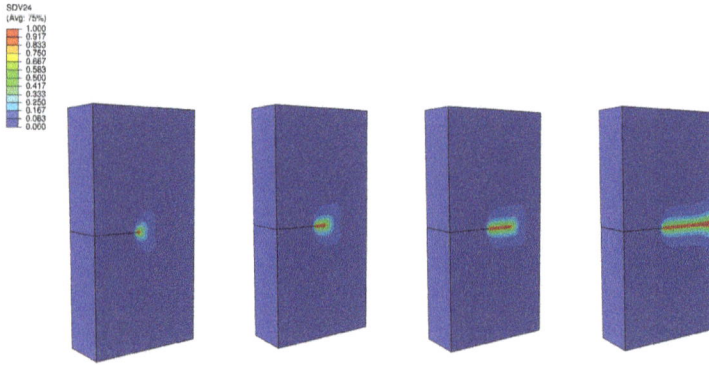

Figure 4. Crack path with increasing time steps.

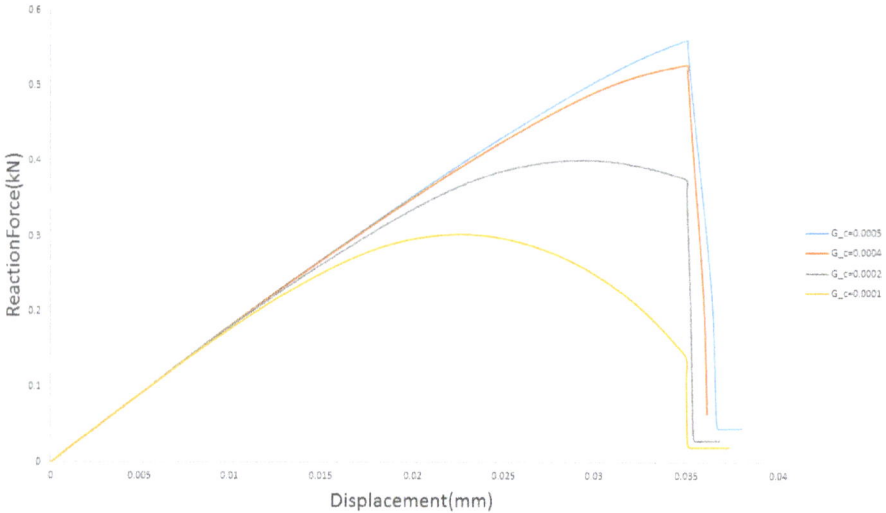

Figure 5. Load–displacement curves for varying G_c.

3.2. Single Edge Notched Shear Sample

We next considered a similar example, to simulate crack propagation in a rectangular bar, with a finite opening, under shear loading. The crack was located at an edge and the material and length scale parameters were: $E = 210$ kN/mm^2, $\nu = 0.2$, $l_c = 0.2$ mm, and $G_c = 2.7 \times 10^{-3}$ kN/mm. We used a hexahedral mesh of 49,000 elements, refined with elements of edge size about 0.03 mm, near the expected crack path (Figure 6). The simulation used displacement controlled boundary condition, with the bottom surface fixed and a horizontal displacement acting on the top surface. We noticed a hint of damage originating from the side, after the deformation due to the crack beginning to propagate. The observed crack path was similar to the results obtained in Zeng et al. [22] who used ES-FEM-T3 to simulate crack propagation. We performed a similar mesh size reduction study as in Section 3.1, where our findings corroborated that the predicted crack patterns were convergent.

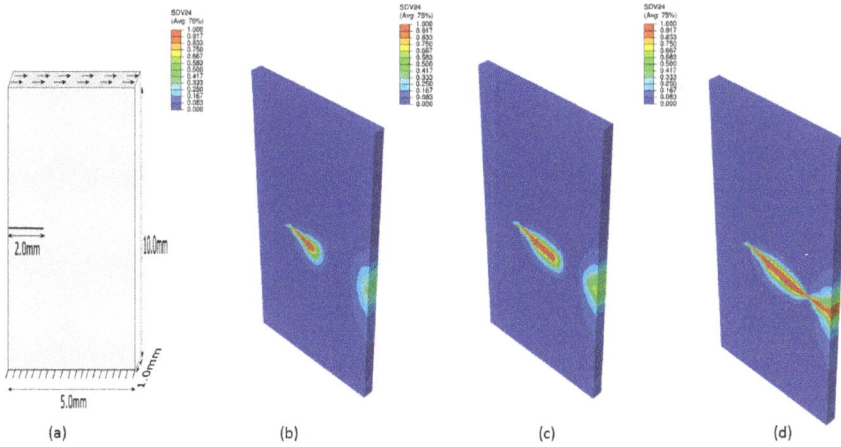

Figure 6. A single edge notched sample under shear loading: (**a**) panel geometry; (**b**) crack initiation; (**c**) crack path after 300 step-times; (**d**) crack path after 500 step-times; and (**e**) final crack path.

3.3. Double Edge Notched Tensile Sample

A doubly notched symmetric rectangular bar, subjected to uniaxial tension was tested in this example (Figure 7). The material and length scale parameters were: $E = 210$ kN/mm^2, $\nu = 0.3$, $l_c = 0.1$ mm, and $G_c = 2.7 \times 10^{-3}$ kN/mm. The model had approximately 35,000 H8 elements with reduced element size near the expected crack path and the boundary conditions were the same as presented in Section 3.1. The crack paths obtained by this code are presented in Figure 8, and were in excellent agreement with Msekh et al. [40].

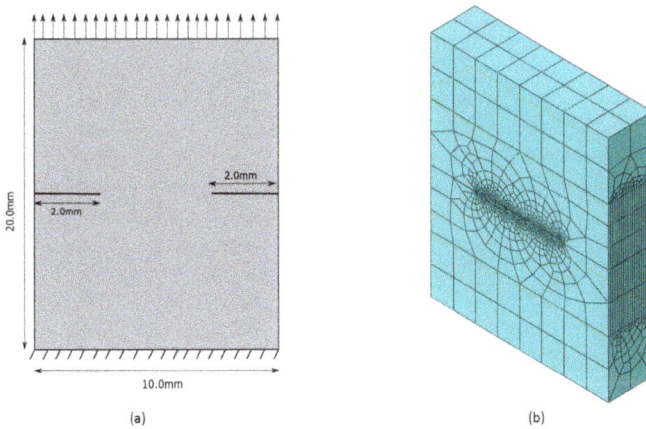

Figure 7. A double edge notched tensile sample: (**a**) 2D Geometric representation of the structure; thickness = 2.0 mm; and (**b**) biased hexahedral mesh.

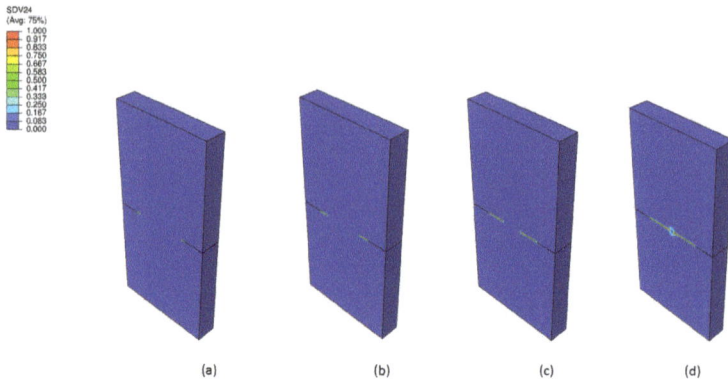

Figure 8. Crack path at different time steps: (**a**) initial damage after 100 time steps; (**b**) initial damage after 100 time steps; (**c**) initial damage after 100 time steps; and (**d**) final crack path.

3.4. 3D Specimen with Notch and Three Openings

In this example, we tested a relatively difficult problem to solve with traditional discrete fracture modeling techniques. In coordination with [40], we had a geometry with three openings and an initial notch under tensile pull (Figure 9). Initially, an incremental displacement of $\Delta u = 10^{-3}$ mm was applied. The relative positions of the openings and the notch were altered to see the difference in fracture pattern. The material and length scale parameters were: $E = 210$ kN/mm^2, $v = 0.2$, $l_c = 0.02$ mm, and $G_c = 5.0 \times 10^{-3}$ kN/mm. Here, we observed two different crack paths: when the openings were aligned, in the same line, the fracture zone from the notch merged with the one originating from the nearest opening and the crack then propagated through the openings. This is basically a phenomenon of crack arrest which comes into existence due to the merging of damaged zones. Although the stress field near the notch was singular, the field near the openings also suffered from the effect of stress concentration [4], which further led to damage. The beauty of this method lies in the fact that it can also elegantly capture the damage initiation from the openings without any ad-hoc criterion. Our results conformed with the findings of Msekh [40] (Figure 10). However, when the openings are not aligned symmetrically, we saw a difference in crack path behavior. The usual crack arrest phenomenon occurred as in the previous case, but, unlike in [40] (Figure 11), where the crack propagates through the first two holes and merges with the third hole and then continues, we saw that the crack propagation continued along the same line, with a slight deflection towards the damaged zone of the third opening. Additionally, another crack propagation started from the damaged zone of the third opening. We believe this is just a by-product of using different displacement increments for the simulation. When the step size of our displacement increments were reduced, $\Delta u = 10^{-5}$ mm, we observed a different crack path, as shown, in Figure 12 which was much closer to the reference results. This indicates a drawback of the method, where the size of each displacement increment has to be suitably chosen to simulate the correct phenomenon.

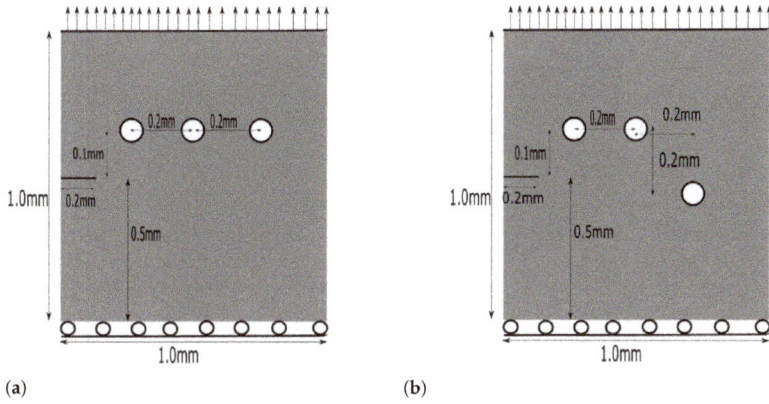

(a)   (b)

Figure 9. Geometry with notch and openings (in *x-y* plane): (**a**) openings are aligned along a straight-line; and (**b**) openings are aligned haphazardly.

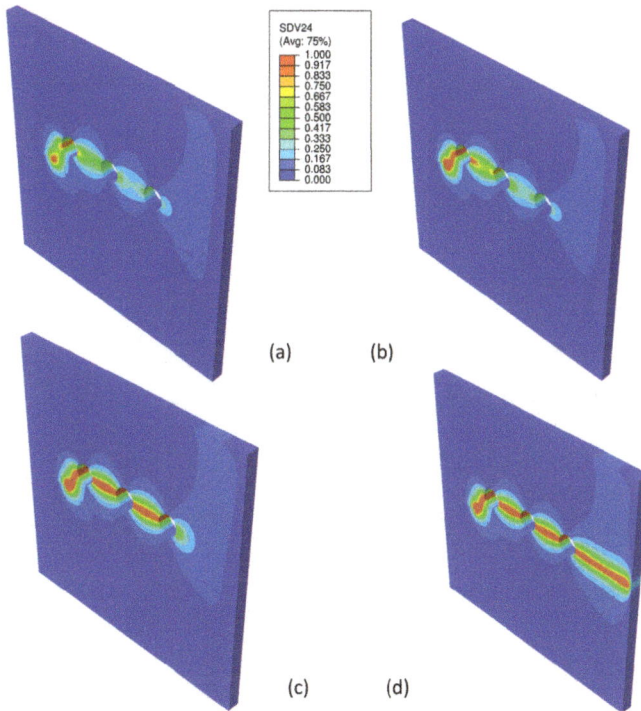

(a)   (b)

(c)   (d)

Figure 10. Crack propagation steps: (**a**) crack propagation begins (**b**) crack arrest due to damage from opening; (**c**) crack propagates through the openings; and (**d**) crack continues in the previous path.

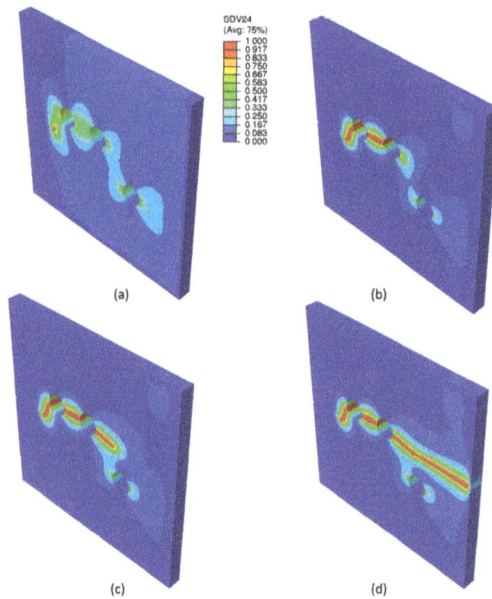

Figure 11. Crack propagation steps for higher incremental step size: (**a**) crack propagation after 100 steps (**b**) crack propagation through holes; (**c**) slight bend in path due to damage zone influence of hole below; and (**d**) propagates along straight path.

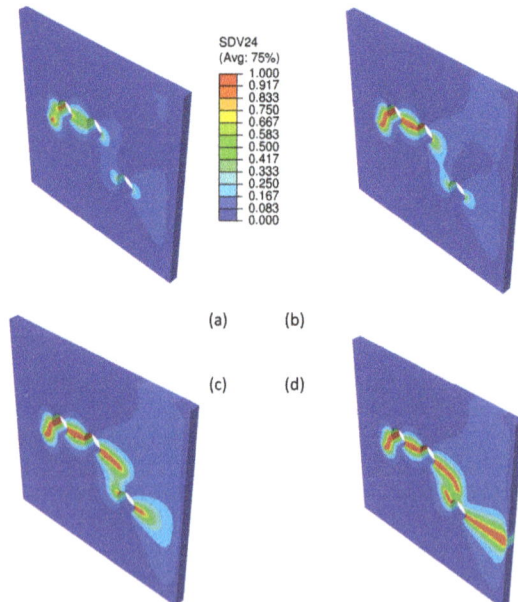

Figure 12. Crack propagation steps for lower incremental step size: (**a**) crack propagation after 100 steps (**b**) crack propagation through holes; (**c**) deflected path due to damage zone influence of hole below; and (**d**) crack initiates at the stress concentration zone and propagates.

3.5. 3D Bi-Material Notched Specimen

This example highlighted one of the advantages of the phase-field method, where we reproduced the phenomenon of crack branching. The geometry and material parameters were chosen similar to those in [27], where we had a notch in the the softer material region with specifications given in Figure 13. The surface on the right was fixed while an incremental displacement pull was applied on the left. The plate had a thickness of 5 mm. We observed that the damage initiation and crack propagation occurred as usual in any edge-notched specimen but, as soon as the crack reached the stiffer material, it branched. This is because the energy required to propagate through the stiff material is higher than the energy required to branch and continue along the material boundaries. The resultant crack path was in pretty good agreement with the available result in [27].

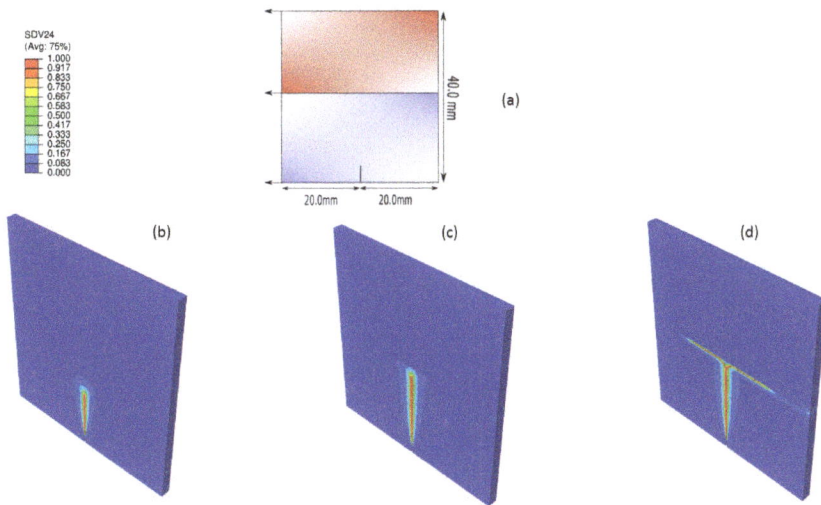

Figure 13. Crack propagation in a bi-material specimen: (**a**) representational 2-D geometry; thickness = 5 mm; (**b**) crack initiation; (**c**) crack propagation through the soft material; (**d**) crack reaches the material discontinuity junction; and (**e**) crack branching.

3.6. Crack Propagation in Thick Walled Cylinder

We performed another numerical example to demonstrate the advantages of using the phase-field method to simulate crack propagation. We modeled two parallel cracks at different elevations in a thick walled cylinder under uni-axial tension. The dimensions were: $t = 0.1$ mm, $r = 1$ mm, $a = 0.2$ mm, and $h = 1$ mm (Figure 14). The material and length scale parameters were: $E = 210$ kN/mm^2, $v = 0.2$, $l_c = 0.075$ mm, and $G_c = 2.7 \times 10^{-3}$ kN/mm. We observed the crack merging phenomenon in this case. The cracks initiated from the respective notches but, when the damage zones were influenced by each other, the two cracks merged (Figure 15). This phenomenon was simulated based on the physics of the system without any ad-hoc criterion set up for the crack propagation path, thereby restating that the phase-field technique can be used to simulate many such cases that would traditionally be a bit more complex using discrete treatment.

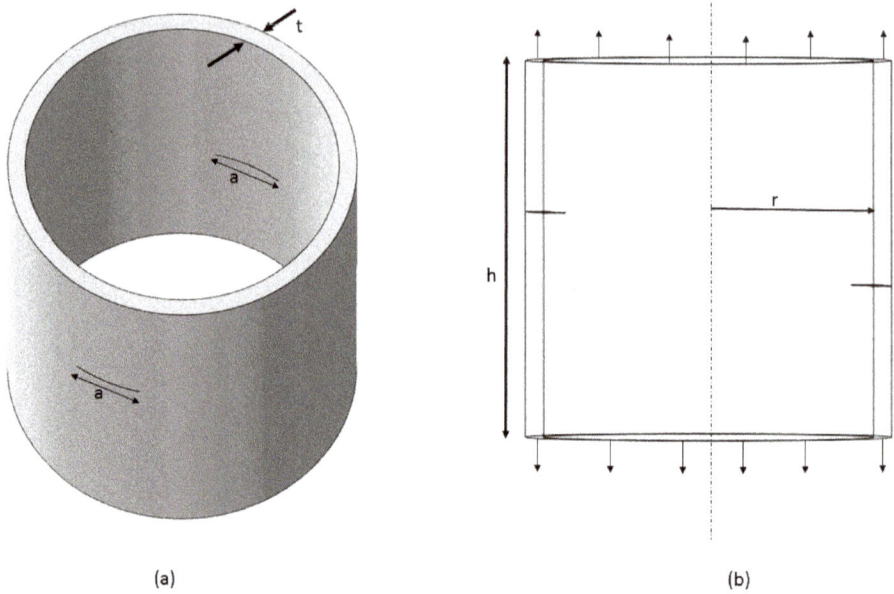

Figure 14. Crack propagation in a thick-walled cylinder: (a) isometric view of the cylinder; and (b) relative position of the two Cracks.

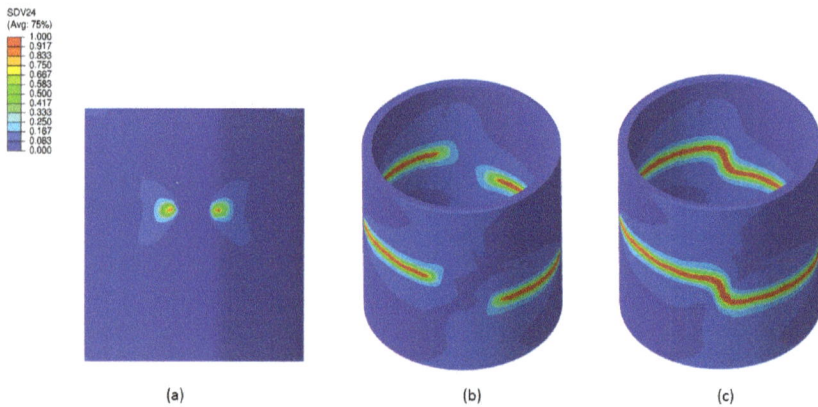

Figure 15. Crack propagation in a thick-walled cylinder: (a) crack initiation; (b) pattern of the two cracks propagating; and (c) final crack merging.

4. Discussion

The primary objective of this study was to predict realistic crack paths in three-dimensional solids. The CS-FEM-H8 element is implemented in the commercial platform of ABAQUS. The element was already proven to have a very good accuracy in displacement norm [38]; here, we combined it with the phase-field model for diffuse fracture to simulate crack propagation in 3D linear elastic brittle solids. ABAQUS user elements (UEL) has its inherent limitations in not letting the user access data from surrounding elements, thus we developed a much simpler technique to calculate stress/strain for smoothing domains instead of the traditional weighted average approach. We sacrificed the stress convergence rate due to this new formulation, however still obtained accurate results that predict complex crack topologies such as crack branching and curved crack paths. Several drawbacks of the method were observed during the simulations: the phase-field variable propagation is very much mesh dependent, thus one needs to have a very fine mesh (approximate characteristic element size of $(1/2)$ the length scale parameter, or less [24,25,28]) near the expected crack propagation zone to efficiently resolve the length-scale parameter. This is not an issue for geometries where experimentally obtained crack paths are already known, however, for complex geometries and unknown crack paths, we needed to have a very fine mesh throughout the domain to facilitate independent crack propagation which increases the computational cost. Once the length scale parameter has been successfully resolved, further reduction of the mesh size does not alter the crack path. We also observed that the selection of displacement increment step size affects the resultant crack path. The final fracture pattern is not dependent on the Griffith's energy release rate parameter, however the ultimate strength of the sample decreases with decreasing G_c. Implementation in ABAQUS, as already discussed, has its own drawbacks in not letting the user implement more accurate S-FEM models such as ES-FEM/α-FEM [18]. This formulation was developed based on the H8 element with eight smoothing cells, which was not generated automatically for complex geometries. Analysts probably need to resort to the use of tetrahedral (T4) elements in cases of complex topologies. This in turn affects the accuracy because the formulations of the T4 elements are the same for both CS-FEM and FEM [1]. However, this opens up an avenue for future research in this domain, where in-house codes can be used to solve the same problem using a much more accurate ES-FEM/FS-FEM and a base mesh of T3/T4 elements, which are very easily generated for complex geometries. Moreover, adaptive meshing schemes designed based on values of the phase-field parameter can be utilized to reduce the computational cost of having very fine mesh. In 2D cases, CS-FEM is slightly more efficient than FEM [39]. However, in this 3D setting, the computational cost is similar (quantitative difference of around 1%) for both FEM and CS-FEM formulations. This is due to the assembly process of the stiffness matrix for each smoothing domain which involves integration over each surface of the cell. However, the better accuracy in the displacement norm and the comparative "soft" property in stiffness matrix with respect to FEM gives CS-FEM an advantage.

Thus, it can be concluded that this method is another efficient and novel computational technique to model crack propagation in 3D structures.

Author Contributions: Formal analysis, S.B.; Investigation, S.B.; Methodology, S.B.; Supervision, G.-R.L.; Validation, S.B.; Writing—original draft, S.B.; and Writing—review and editing, S.B. and G.-R.L.

Funding: This research received no external funding.

Conflicts of Interest: The authors declare no conflict of interest.

References

1. Liu, G.; Trung, N.T. *Smoothed Finite Element Methods*; CRC Press: Boca Raton, FL, USA, 2010.
2. Liu, G.R.; Dai, K.Y.; Nguyen, T.T. A smoothed finite element method for mechanics problems. *Comput. Mech.* **2007**, *39*, 859–877. [CrossRef]
3. Liu, G.R. *Meshfree Methods—Moving Beyond the Finite Element Method*; CRC Press: Boca Raton, FL, USA, 2009.
4. Liu, G.R.; Quek, S.S. *The Finite Element Method—A Practical Course*; CRC Press: Boca Raton, FL, USA, 2003.
5. Liu, G.R.; Nguyen, T.T.; Dai, K.Y.; Lam, K.Y. Theoretical aspects of the smoothed finite element method (SFEM). *Int. J. Numer. Methods Eng.* **2007**, *71*, 902–930. [CrossRef]
6. Liu, G.; Nguyen-Thoi, T.; Lam, K. A novel alpha finite element method (αFEM) for exact solution to mechanics problems using triangular and tetrahedral elements. *Comput. Methods Appl. Mech. Eng.* **2008**, *197*, 3883–3897. [CrossRef]

7.   Liu, G.; Nguyen-Thoi, T.; Nguyen-Xuan, H.; Lam, K.   A node-based smoothed finite element method (NS-FEM) for upper bound solutions to solid mechanics problems.   *Comput. Struct.* **2009**, *87*, 14–26. [CrossRef]

8.   Liu, G.; Nguyen-Thoi, T.; Lam, K. An edge-based smoothed finite element method (ES-FEM) for static, free and forced vibration analyses of solids. *J. Sound Vib.* **2009**, *320*, 1100–1130. [CrossRef]

9.   Zeng, W.; Liu, G.; Jiang, C.; Nguyen-Thoi, T.; Jiang, Y. A generalized beta finite element method with coupled smoothing techniques for solid mechanics. *Eng. Anal. Bound. Elem.* **2016**, *73*, 103–119. [CrossRef]

10.   Nguyen-Xuan, H.; Rabczuk, T.; Bordas, S.; Debongnie, J. A smoothed finite element method for plate analysis. *Comput. Methods Appl. Mech. Eng.* **2008**, *197*, 1184–1203. [CrossRef]

11.   Nguyen-Thoi, T.; Liu, G.; Vu-Do, H.; Nguyen-Xuan, H. A face-based smoothed finite element method (FS-FEM) for visco-elastoplastic analyses of 3D solids using tetrahedral mesh. *Comput. Methods Appl. Mech. Eng.* **2009**, *198*, 3479–3498. [CrossRef]

12.   Nguyen-Thoi, T.; Vu-Do, H.; Rabczuk, T.; Nguyen-Xuan, H. A node-based smoothed finite element method (NS-FEM) for upper bound solution to visco-elastoplastic analyses of solids using triangular and tetrahedral meshes. *Comput. Methods Appl. Mech. Eng.* **2010**, *199*, 3005–3027. [CrossRef]

13.   Nguyen-Xuan, H.; Liu, G.; Thai-Hoang, C.; Nguyen-Thoi, T. An edge-based smoothed finite element method (ES-FEM) with stabilized discrete shear gap technique for analysis of Reissner–Mindlin plates. *Comput. Methods Appl. Mech. Eng.* **2010**, *199*, 471–489. [CrossRef]

14.   Nguyen-Thoi, T.; Liu, G.R.; Vu-Do, H.C.; Nguyen-Xuan, H. An edge-based smoothed finite element method for visco-elastoplastic analyses of 2D solids using triangular mesh. *Comput. Mech.* **2009**, *45*, 23–44. [CrossRef]

15.   Nourbakhshnia, N.; Liu, G.R. A quasi-static crack growth simulation based on the singular ES-FEM. *Int. J. Numer. Methods Eng.* **2001**, *88*, 473–492. [CrossRef]

16.   Jiun-Shyan, C.; Cheng-Tang, W.; Sangpil, Y.; Yang, Y. A stabilized conforming nodal integration for Galerkin mesh-free methods. *Int. J. Numer. Methods Eng.* **2000**, *50*, 435–466. [CrossRef]

17.   Nguyen-Xuan, H.; Liu, G.; Bordas, S.; Natarajan, S.; Rabczuk, T. An adaptive singular ES-FEM for mechanics problems with singular field of arbitrary order. *Comput. Methods Appl. Mech. Eng.* **2013**, *253*, 252–273. [CrossRef]

18.   Bhowmick, S.; Liu, G. On singular ES-FEM for fracture analysis of solids with singular stress fields of arbitrary order. *Eng. Anal. Bound. Elem.* **2018**, *86*, 64–81. [CrossRef]

19.   Guo, Z.; Liu, Y.; Ma, H.; Huang, S. A fast multipole boundary element method for modeling 2-D multiple crack problems with constant elements. *Eng. Anal. Bound. Elem.* **2014**, *47*, 1–9. [CrossRef]

20.   Belytschko, T.; Lu, Y.; Gu, L. Crack propagation by element-free Galerkin methods. *Eng. Fract. Mech.* **1995**, *51*, 295–315. [CrossRef]

21.   Belytschko, T.; Black, T. Elastic crack growth in finite elements with minimal remeshing. *Int. J. Numer. Methods Eng.* **1999**, *45*, 601–620. [CrossRef]

22.   Zeng, W.; Liu, G.; Kitamura, Y.; Nguyen-Xuan, H. A three-dimensional ES-FEM for fracture mechanics problems in elastic solids. *Eng. Fract. Mech.* **2013**, *114*, 127–150. [CrossRef]

23.   Liu, G.; Nourbakhshnia, N.; Zhang, Y. A novel singular ES-FEM method for simulating singular stress fields near the crack tips for linear fracture problems. *Eng. Fract. Mech.* **2011**, *78*, 863–876. [CrossRef]

24.   Bourdin, B.; Francfort, G.; Marigo, J.J. Numerical experiments in revisited brittle fracture. *J. Mech. Phys. Solids* **2000**, *48*, 797–826. [CrossRef]

25.   Miehe, C.; Hofacker, M.; Welschinger, F. A phase field model for rate-independent crack propagation: Robust algorithmic implementation based on operator splits. *Methods Appl. Mech. Eng.* **2010**, *199*, 2765–2778. [CrossRef]

26.   Miehe, C.; Welschinger, F.; Hofacker, M. Thermodynamically consistent phase-field models of fracture: Variational principles and multi-field FE implementations. *Int. J. Numer. Methods Eng.* **2010**, *83*, 1273–1311. [CrossRef]

27.   Molnár, G.; Gravouil, A. 2D and 3D ABAQUS implementation of a robust staggered phase-field solution for modeling brittle fracture. *Finite Elem. Anal. Des.* **2017**, *130*, 27–38. [CrossRef]

28.   Msekh, M.A.; Sargado, J.M.; Jamshidian, M.; Areias, P.M.; Rabczuk, T. ABAQUS implementation of phase-field model for brittle fracture. *Comput. Mater. Sci.* **2015**, *96*, 472–484. [CrossRef]

29.   Miehe, C.; Schänzel, L.M. Phase field modeling of fracture in rubbery polymers. Part I: Finite elasticity coupled with brittle failure. *J. Mech. Phys. Solids* **2014**, *65*, 93–113. [CrossRef]

30. Miehe, C.; Aldakheel, F.; Raina, A. Phase field modeling of ductile fracture at finite strains: A variational gradient-extended plasticity-damage theory. *Int. J. Plast.* **2016**, *84*, 1–32. [CrossRef]

31. Miehe, C.; Schänzel, L.M.; Ulmer, H. Phase field modeling of fracture in multi-physics problems. Part I. Balance of crack surface and failure criteria for brittle crack propagation in thermo-elastic solids. *Methods Appl. Mech. Eng.* **2015**, *294*, 449–485. [CrossRef]

32. Miehe, C.; Hofacker, M.; Schänzel, L.M.; Aldakheel, F. Phase field modeling of fracture in multi-physics problems. Part II. Coupled brittle-to-ductile failure criteria and crack propagation in thermo-elastic–plastic solids. *Methods Appl. Mech. Eng.* **2015**, *294*, 486–522. [CrossRef]

33. Hofacker, M.; Miehe, C. A phase field model of dynamic fracture: Robust field updates for the analysis of complex crack patterns. *Int. J. Numer. Methods Eng.* **2012**, *93*, 276–301. [CrossRef]

34. Borden, M.J.; Verhoosel, C.V.; Scott, M.A.; Hughes, T.J.; Landis, C.M. A phase-field description of dynamic brittle fracture. *Methods Appl. Mech. Eng.* **2012**, *217–220*, 77–95. [CrossRef]

35. Rabczuk, T.; Belytschko, T. Cracking particles: a simplified meshfree method for arbitrary evolving cracks. *Int. J. Numer. Methods Eng.* **2004**, *61*, 2316–2343. [CrossRef]

36. Areias, P.; Msekh, M.; Rabczuk, T. Damage and fracture algorithm using the screened Poisson equation and local remeshing. *Eng. Fract. Mech.* **2016**, *158*, 116–143. [CrossRef]

37. Areias, P.; Rabczuk, T.; de Sá, J.C. A novel two-stage discrete crack method based on the screened Poisson equation and local mesh refinement. *Comput. Mech.* **2016**, *58*, 1003–1018. [CrossRef]

38. Nguyen-Xuan, H.; Nguyen, H.V.; Bordas, S.; Rabczuk, T.; Duflot, M. A cell-based smoothed finite element method for three-dimensional solid structures. *KSCE J. Civ. Eng.* **2012**, *16*, 1230–1242. [CrossRef]

39. Bhowmick, S.; Liu, G.R. A phase-field modeling for brittle fracture and crack propagation based on the cell-based smoothed finite element method. *Eng. Fract. Mech.* **2018**, *204*, 369–387. [CrossRef]

40. Msekh, M.A. Phase Field Modeling for Fracture with Applications to Homogeneous and Heterogeneous Materials. Ph.D. Thesis, Bauhaus-Universität Weimar, Weimar, Germany, 2017, doi:10.25643/bauhaus-universitaet.3229.

applied
sciences

MDPI

Article

Cracking Risk and Overall Stability Analysis of Xulong High Arch Dam: A Case Study

Peng Lin [1,*], Pengcheng Wei [1], Weihao Wang [2] and Hongfei Huang [2]

[1]   Department of Hydraulic Engineering, Tsinghua University, Beijing 100084, China;
      wei-pc18@mails.tsinghua.edu.cn
[2]   Changjiang Institute of Survey, Planning, Design and Research, Wuhan 430010, China;
      wangweihao@cjwsjy.com.cn (W.W.); huanghongfei@cjwsjy.com.cn (H.H.)
*   Correspondence: celinpe@tsinghua.edu.cn; Tel.: +86-139-1050-5719

Received: 31 October 2018; Accepted: 3 December 2018; Published: 10 December 2018

Featured Application: The cracking risk, the overall stability, and the reinforcement measures are directly related to the long-term stability of arch dams. The three safety factors and five stress zones have a great significance for arch dam-foundation design of cracking control and overall stability evaluation.

Abstract: It is of great significance to study the cracking risk, the overall stability, and the reinforcement measures of arch dams for ensuring long-term safety. In this study, the cracking types and factors of arch dams are summarized. By employing a nonlinear constitutive model relating to the yielding region, a fine three-dimensional finite element simulation of the Xulong arch dam is conducted. The results show that the dam cracking risk is localized around the outlets, the dam heel, and the left abutment. Five dam stress zones are proposed to analysis dam cracking state base of numerical results. It is recommended to use a shearing-resistance wall in the fault f57, replace the biotite enrichment zone with concrete and perform consolidation grouting or anchoring on the excavated exposed weak structural zone. Three safety factors of the Xulong arch dam are obtained, $K\_1 = 2\~2.5$; $K\_2 = 5$; $K\_3 = 8.5$, and the overall stability of the Xulong arch dam is guaranteed. This study demonstrates the significance of the cracking control of similar high arch dams.

Keywords: the Xulong arch dam; yielding region; cracking risk; overall stability; dam stress zones

1. Introduction

A series of super-high arch dams (height over 200 m) have been constructed or are being planned in China. Most of them are distributed in the mountainous areas in southwest China (Figure 1), and therefore are subject to complex engineering challenges, such as high seismic intensity, high slope, huge water thrust, and so forth. The geological conditions are also very complex, for example, the deep-cutting valley, high ground stress, and some unfavorable geological conditions, such as atypical faults, dislocation interfaces, altered rock masses, and weak rock masses [1–3]. The complex geological conditions may lead to the crack of dam concrete or foundation, which eventually leads to dam failure [4]. Therefore, the construction of super-high arch dams still faces many challenges. Cracks may initiate in the outlets [5], heel [6], surface, and interior of dam concrete blocks [2], then propagate and coalescence along horizontal or vertical directions in concrete blocks. The main cracking factors include temperature variations, heat from concrete hydration, shrinkage and creep, dam foundation uncoordinated deformation, earthquake, and seepage effect [4].

Some research studies have been done related to the cracking mechanism on concrete blocks of dams experimentally and theoretically. Study on thermal mechanics of the concrete dam includes the temperature variation of the external environment and heat from concrete hydration. When the temperature gradient changes dramatically inside the dam, the thermal stresses will concentrate and cause the cracks to initiate under cold wave conditions [7], in cold areas [8], under unfavorable solar radiation [9], and in dry, hot valley regions [10]. Temperature load is the main cracking factor of the Karaj arch dam [11]. The nonlinear analysis of concrete arch dams is necessary to check the stability of cracks in high tensile stress areas. Maken et al. [12] investigated the mechanical properties and showed that stress relaxation is affected by the concrete temperature. They used a finite-element modeling procedure for assessing the thermal mechanical behaviors of concrete dams, and successfully reproduced the oblique cracks present on the downstream face of Daniel Johnson dam. The distribution of stresses in roller-compacted concrete dams is greatly affected by the starting date of the roller-compacted concrete placement schedule [13]. Self-weight and weak foundation [14,15], uneven settlement of arch dam foundation, and earthquake [16,17] can also lead to the cracking of arch dams. Hariri-Ardebili et al. [18] assessed seismic cracks in three types of concrete dams, namely gravity, buttress, and arch, using an improved 3D coaxial rotating smeared crack model. The cracking factors of arch dams often interact with each other in a nonlinear manner, and therefore cannot be considered separately.

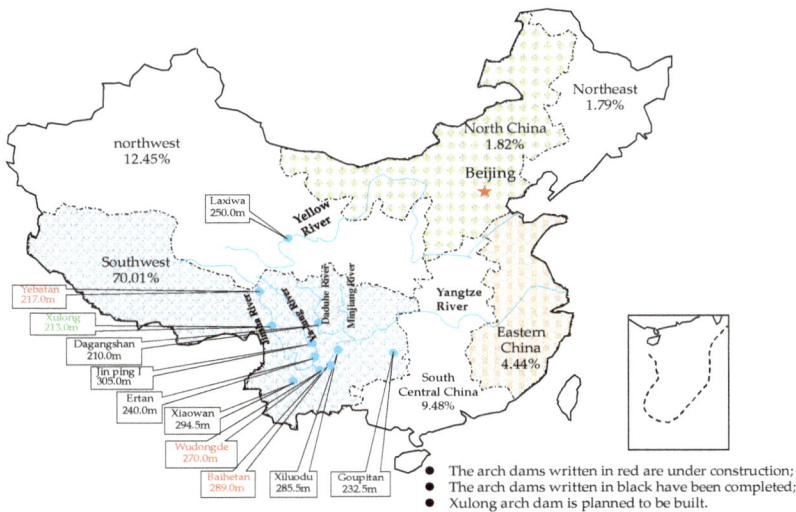

Figure 1. Distribution of hydropower resources and super-high arch dams in China.

The concrete dam cracking problems have been studied by different methods in the past 30 years as follows. (1) The finite element method (FEM) is widely used in the numerical simulation of dam cracking, including the FEM based on elastoplastic mechanics, the FEM based on fracture mechanics [19,20], and the FEM based on damage mechanics [21,22]. Based on linear elastic crack mechanics and three-dimensional boundary element modeling, Feng et al. [23] presented a procedure to analyze the cracks in arch dams. Chen et al. [24] introduced the existing constitutive model of large, light, reinforced concrete structures and the deficiency of the design procedure, and they also presented a three-dimensional nonlinear cracking response simulation procedure for outlet structures. The overall stability of arch dams is analyzed by the three-dimensional numerical simulation [25]. Sato et al. [26] simulated the thermal stress of a concrete dam by three-dimensional linear elastic FEM, and the autogenous shrinkage strain was added to the thermal strain. (2) Dam geomechanical model tests were widely carried out in the United States, Switzerland, Yugoslavia, Russia, Germany, Italy, Japan, and Sweden in the 1970s and 1980s [27]. The geomechanical model mainly refers to the model that reflects the specific engineering geological structure in a small range, such as the faults, fractures, and weak zones in the dam foundation, and it follows the similarity theory [28]. Lin et al. [2]

analyzed the cracking characteristics of the Xiaowan arch dam surfaces and rock mass failure process of the abutments. They judged the alteration zones, weak rock masses, and other faults in the abutments that caused the arch dam to crack and proposed the method of foundation reinforcement. (3) Many scholars also focus on different numerical methods to simulate dam cracking processes, for example, element-free method [29], interface stress element method, and boundary element method [30]. Prototype monitoring is also widely used in arch dam cracking analysis. However, there are still many areas for improvement in the analysis of cracking of arch dams. Linear finite element is not capable of revealing the actual state of the structure. As a popular simulation method, the nonlinear numerical method has no uniform standard for the selection of material constitutive models and parameters and the setting of boundary conditions. Although the geomechanical model of rupture test is straightforward, the loading control, boundary condition simulation, and error analysis of measurement data need to be further investigated. The cracking theory of arch dams has not been fully studied, especially for the location and propagation of cracks.

This study first summarizes the main analysis methods, cracking types, and factors of arch dams according to the cracking cases. In order to analyze the cracking and overall stability of the Xulong high arch dam, a nonlinear constitutive model and overall stability criterion are employed, and numerical simulation on the overall stability, cracking analysis of dam outlets, and arch abutments are performed. This study aims at proposing effective reinforcement methods and prevention methods for arch dam cracking. Through the analysis of the yielding region and stress before and after the reinforcement, the reinforcement methods of the Xulong arch dam are determined. Based on the analysis of the first and third principal stresses and yielding region of the arch dam, a method for crack prevention based on five stress zones of arch dams is proposed.

2. Summary of Cracking Types and Effect Factors of High Arch Dams

2.1. Cracking Types

Concrete arch dams are generally constructed of massive plain concrete with almost no tensile resistance. Tensile stress can occur due to concrete shrinkage, temperature variations, rigidity weakening due to seepage effects, and other factors like earthquake and large deformation of the dam abutments due to a weak foundation. The tensile stress often leads to cracking of arch dams, which affects the safety and stable operation of arch dams.

Table 1 summarizes the completed time, dam height, cracking position, and cracking reason of the main arch dams in the world. The crack types of arch dams are illustrated in Figure 2 according to the common cracking positions of arch dams.

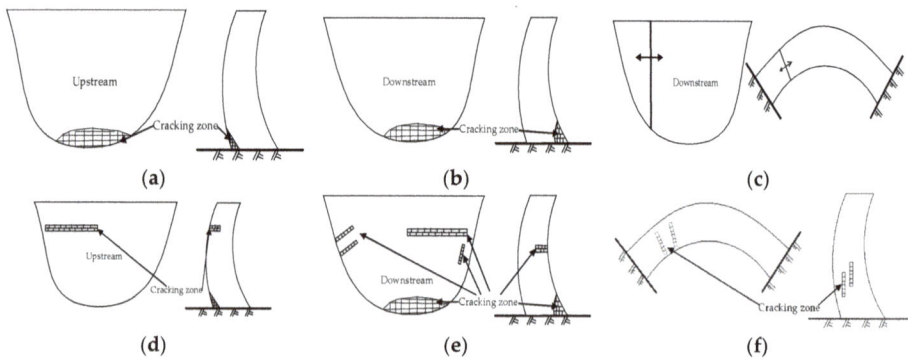

Figure 2. Cracking type of arch dam: (**a**) dam heel crack; (**b**) dam toe crack; (**c**) transverse joint open; (**d**) horizontal cracks at the upstream surface; (**e**) cracks at the downstream surface; (**f**) internal cracks.

Table 1. Summary of cracking types in high arch dams.

Dam Name	Country	Operation Year	Height (m)	Cracking Description	Main Cracking Causes
Buffalo Bill Arch Dam	America	1910	107.0	Vertical cracks at downstream surface	Temperature (extreme thermal gradients)
Packard Sama Dam	America	1928	113.0	Different settlement	Earthquake
Stewart Mountain Dam [31]	America	1930	64.6	Visible surface, mainly upstream surface	Alkali–silica reactions and expansions
Zeuzier Dam	Switzerland	1956	156.0	Transverse joints open at upstream Peripheral joints form downstream	Foundation
Sardine Dam	Italy	1957	115.0	Horizontal cracks at upstream surface	Temperature
Santa Maria Dam	Switzerland	1968	117.0	Leakage in dam foundation upstream	Foundation
Daniel Johnson Dam [12]	Canada	1968	214.0	Oblique cracks at downstream surface	Seasonal temperature
				Plunging cracks at the heel of the dam	Geometric discontinuities
Kolnbrein Dam [6,14]	Austria	1977	200.0	Horizontal construction joints open Cracking at dam heel	Foundation
Zillergrundl Dam [14]	Austria	1985	186	Horizontal cracks at heelVertical cracks in the elevator shaft	Concrete hydration heat
Sayano-Shushenskaya Dam [14]	Former Soviet Union	1989	242.0	Vertical cracks in the gallery Horizontal cracks at upstream surface	Concrete hydration heat and temperature
Shuanghe Arch Dam [15]	China	1991	82.3	Seven vertical cracks at downstream surface	Self-weight and weak foundation
Ertan Dam	China	2000	240.0	Cracking at downstream surface	Foundation Temperature
Xiaowan Dam [2]	China	2010	294.5	Internal cracking	Temperature
Goupitan Dam	China	2011	232.5	Cracking around the bottom outlets	Concrete hydration heat and temperature

2.2. Cracking Factors

Many factors lead to the cracking of super-high arch dams, including different concrete materials, concrete temperature control measures, geological condition of dam foundation, seepage, geological exploration, arch dam profile, and so forth. The cracking factors of arch dams are summarized as follows.

(1) Concrete materials. Different concrete materials have different properties such as hydration heat and tensile strength. High-strength concrete generally has a large content of cement, leading to high hydration heat. When the external temperature changes sharply or the temperature control measures are not appropriate, high-strength concrete can easily crack. Concrete materials should be selected according to different dam structures and high-strength concrete should not be used blindly.

(2) Site selection of the arch dam. The complex geological conditions of the dam foundation directly affect the stress and deformation distribution of the arch dam. The uneven deformation of both abutments and different stiffness between arch dam and foundation can easily lead to arch dam cracking. Appropriate reinforcement methods are important for reducing the cracking risk of arch dams.

(3) Temperature control and maintenance. Concrete temperature control measures are directly related to concrete thermal stress. The sharp increase or decrease of the external temperature has more influence on the dam abutment, heel, and outlets. The temperature control methods should be designed and implemented before the arch dam is built. For the special structures such as outlets, it is necessary to consider the cracking caused by the cavern drafts flowing and the outlets should be closed.

(4)   Dam profile design. Profile design needs to consider specific geological conditions. Arch dam profile is directly related to the stress distribution of the dam body. Outlets and dam heel should be considered especially. The effects of different profiles on the stress, deformation, overall stability, and cracking of the arch dam can be comprehensively compared by using the method of dividing load of the arch beam, FEM, and geomechanical model.

3. Numerical Modeling of the Xulong High Arch Dam

3.1. Numerical Method

Using a 3D nonlinear finite element analysis, the convergence of elastoplastic analysis solution shows the stability of the structure. The yielding condition of the ideal elastoplastic model adopts Drucker–Prager (D-P) yielding criterion [5].

(1) Safety factors of the arch dam

Both in the geomechanical model experiment and 3D nonlinear numerical simulation, overloading, strength reduction, and comprehensive method are main methods to analyze the ultimate state and safety factors of arch dams [2,32], as illustrated in Figure 3. According to different overloading ways, overloading method includes increasing the bulk density of the upstream water and upstream water level (Figure 3a,b). The comprehensive method combines the overloading and strength reduction method.

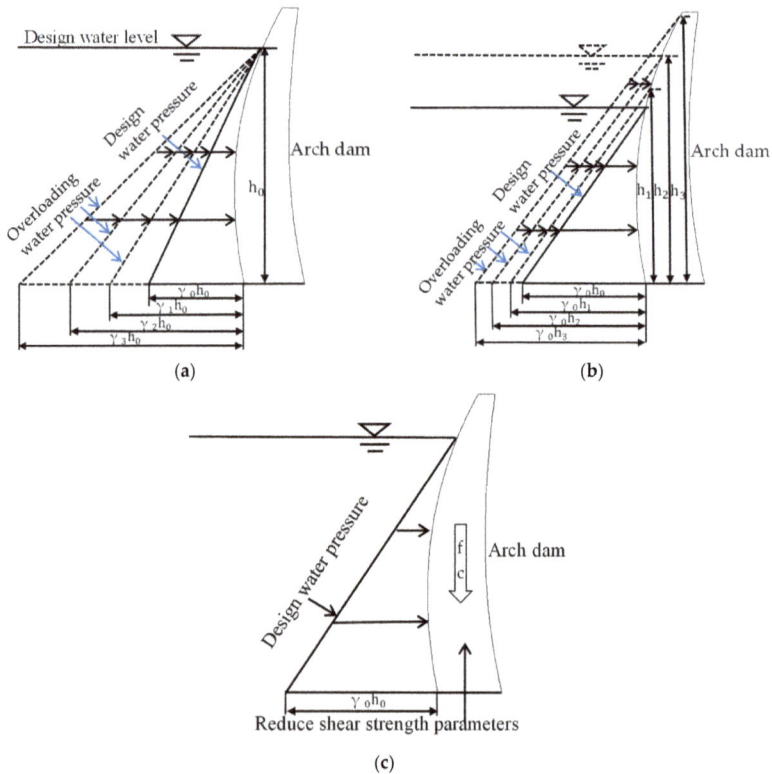

Figure 3. Overloading and strength reduction method. (**a**) Increasing the bulk density of the upstream water; (**b**) increasing the upstream water level; (**c**) strength reduction method.

The arch dam is a high-order statically indeterminate structure. In the process of overloading or strength reduction, the deformation of arch dams can be divided into three stages, namely the elastic deformation stage, plastic deformation stage, and total failure stage. Corresponding to three deformation stages, three safety factors, K_1, K_2, K_3, are employed to evaluate the overall stability of the dam [2].

K_1 represents the safety factor of crack initiation of the arch dam. A crack is generally initiated at the dam heel. As shown in Figure 3a, $K_1 = \frac{\gamma_1}{\gamma_0}$; in Figure 3b, $K_1 = \frac{h_1}{h_0}$.

K_2 represents the safety factor of structural nonlinear behavior initiation. The dam has a large displacement due to nonlinear deformation. Dam cracks rapidly propagate and coalesce with each other. As shown in Figure 3a, $K_2 = \frac{\gamma_2}{\gamma_0}$; in Figure 3b, $K_2 = \frac{h_2}{h_0}$.

K_3 represents the safety factor of the maximum loading of the dam–foundation system. The yielding regions connect. The dam foundation is totally destroyed and the undertaking capacity is lost. As shown in Figure 3a, $K_3 = \frac{\gamma_3}{\gamma_0}$; in Figure 3b, $K_3 = \frac{h_3}{h_0}$.

(2) Evaluation of connection of yielding region

Adopting to increase the bulk density of the upstream water (Figure 3a), the yielding region of dam and foundation is simulated under overloading condition. The connection of the yielding region means that it connects piece by piece and forms movement mechanism so that the integrity stiffness of arch dam–foundation is weakened. The cracking or yielding caused by excessive local stress reduces the constraint to adjust the internal stress. If the yielding region is not fully connected, the structure can still provide support. This method can fully reflect the adjustment process of nonlinear stresses and exert the bearing capacity of the high-order statically indeterminate arch dam.

Failure criterion can be defined as Equation (1), that is, iteration does not converge.

$$\varnothing(A_i) \geq 0. \tag{1}$$

where $\varnothing$ is yielding surface. A_i is a mechanism composed of local yielding regions, that is, formed by the connection of the local yielding regions. A_i can be expressed as: $A_i = \varphi_j \cup \varphi_m \cup \ldots \cup \varphi_n$. When any A_i is formed, the structure loses its integral stability. In addition, $\det[K] \leq 0$, where $[K]$ is the integral stiffness matrix in FEM analysis.

3.2. Brief Introduction of Xulong Super-High Arch Dam

The Xulong hydropower station is located on the main stream of Jinsha River, juncture of the Deqin county, Yunnan Province and Derong county, Sichuan Province. The total storage capacity of it is 829 million m^3, and it has an installed capacity of 2220 MW. The principal structures consist of a double-curvature arch dam with a height of 213 m, underground powerhouse, diversion tunnel, and plunge pool. There are 3 upper outlets and 4 middle outlets placed in the numbers 9~12 dam monoliths. The entrances of the upper outlets are at elevation level (EL) 2286 m. The size of the middle outlets is 8 m × 6 m (height × width) at the entrances and 5 m × 7.2 m (height × width) at the exit. The entrances of the middle outlets are at EL 2222 m. The ratio of thickness and height of the arch dam is 0.217 and the length of the crest on the upper surface is 482.9 m.

The dam site is deep canyon topography with an aspect ratio of 1.8. The Triassic Indosinian granite dike at the dam site slopes into the riverbed. The left bank is Mesoproterozoic Xiongsong Group plagioclase amphibole schist and the right bank is Mesoproterozoic Xiongsong Group migmatite. According to the double-hole acoustic testing results, the fresh granite, migmatite, and plagioclase amphibole schist have longitudinal wave velocities of 4300~5900 m/s, 4000~5600 m/s, and 3000~5800 m/s, respectively. The average longitudinal wave velocities of the fresh granite, migmatite, and plagioclase amphibole schist are 5100 m/s, 5000 m/s, and 4600 m/s, respectively, which can mainly be classified as the type II rock mass. The type II rock masses have an acoustic velocity of 4800~5500 m/s, weak permeability, and good uniformity. The width of strongly unloading zones due to the dam construction is generally less than 30 m. The strongly unloading rocks are only at the EL

2300 and 2308 m, which can be classified as the type IV rock mass. The width of weakly unloading zones is between 15 and 35 m with the type III1 rock masses and III2 rock masses. A total of 70 faults are found on the surface of the dam site, of which the statistics are shown in Figure 4. Ten of them are low-angle faults. The width of the faults is 0.20~0.50 m, and the length of the faults is usually less than 100 m. The fault strikes can be divided into 4 groups: NE~NEE group, NWW group, NNE group, and NNW~NW group. F1, f3, f10, f11, f26, f57, f74, and f75 are relatively large faults. Figure 5 illustrates the bedrock distribution of the dam–foundation interface.

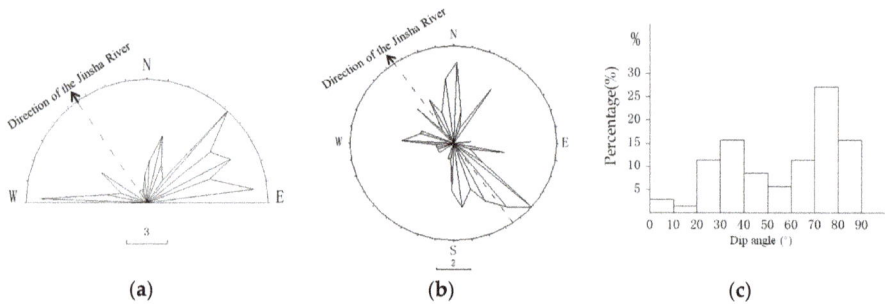

Figure 4. Surface fault statistics in dam site area. (a) Rose of the strike; (b) rose of dip; (c) dip angle histogram.

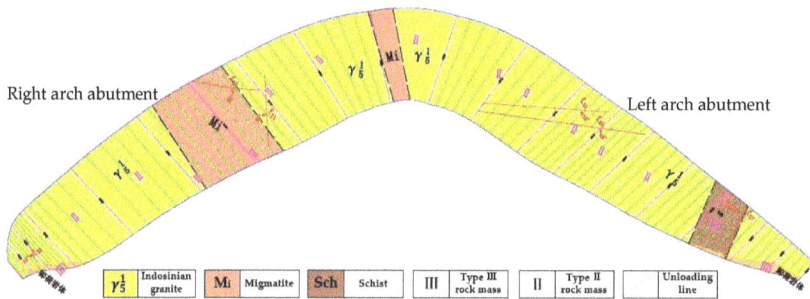

Figure 5. Bedrock distribution of the foundation surface.

3.3. Numerical Model and Analysis Cases

Figure 6 illustrates the 3D numerical model of the Xulong high arch dam and foundation, and the distribution of main faults including F1, f3, f10, f11, f26, f57, f74, and f75. In this 3D model, the simulation range is 840 m × 800 m × 553 m (length × width × height). The numerical model adopts 8-node hexahedral elements, with the total number of 129,241 elements and 147,331 nodes. There are 34,284 elements and 41,204 nodes for the dam.

Based on laboratory testing, the main physical–mechanical parameters of the rock masses and dam concrete are listed in Table 2. Considering the influence of temperature and complex geological conditions on the cracking of arch dams, this study uses the overloading method to judge the overall stability of the arch dam and foundation. Temperature load, self-weights of the dam and foundation, water pressure, and silt pressure are considered in the ten analysis cases as follows.

In order to compare the effect of temperature rise and drop loading on the stress and displacement of the arch dam during the construction period after arch closure, temperature drop loading is applied in cases 1 and 10, and the other loads in cases 1 and 10 are the same. In order to obtain the three safety factors to evaluate the dam overall stability, cases 1–9 correspond to 1–9 times of overloading, respectively, and the other loads in cases 1 and 9 are the same.

The influence of temperature loading on cracking of the arch dam during operation period is analyzed. Under the long-term external temperature variation, the temperature loading can be divided into mean and linear temperature difference. The mean and linear temperature difference under normal water level is illustrated in Table 3.

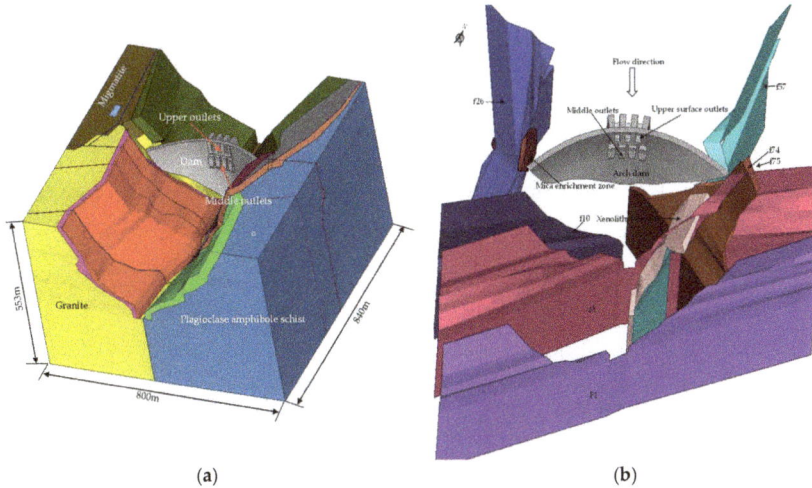

(a)   (b)

Figure 6. The 3D numerical model. (**a**) dam-foundation overall model; (**b**) main faults distribution.

Table 2. Physical–mechanical parameters of the rock masses and dam materials.

Materials	Bulk Density (t/m^3)	Deformation Modulus (GPa)	Poisson's Ratio	Shear Strength	
				C' (MPa)	F'
Dam concrete	2.40	25.0	0.167	5.0	1.7
Rock of type II	2.70	24.0	0.22	1.2	1.1
Rock of type III1	2.60	17.5	0.24	1.05	1.0
Rock of type III2	2.55	12.5	0.26	1.0	0.95
Rock of type IV	2.50	6.0	0.30	0.65	0.60

Table 3. The mean and linear temperature difference under normal water level.

EL (m)	Normal Water Level + Temperature Rise		Normal Water Level + Temperature Drop	
	Mean Temperature Difference	Linear Temperature Difference	Mean Temperature Difference	Linear Temperature Difference
2308	9.40	0.00	2.65	0.00
2302	7.67	3.09	3.22	−0.49
2290	4.75	8.65	2.06	2.06
2270	2.42	12.54	0.68	5.67
2245	2.24	14.18	0.92	7.99
2220	2.79	14.66	1.67	8.98
2195	3.60	14.85	2.62	9.29
2170	3.58	14.72	2.65	9.45
2145	1.43	9.96	0.83	6.53
2120	−0.97	4.74	−1.35	2.60
2095	−1.88	2.66	−2.12	1.32

4. Cracking Analysis of the Xulong High Arch Dam

4.1. Effect of Temperature Load on Stress and Displacement of the Xulong Arch Dam

For the analysis cases 1 and 10, the displacement and stress distribution of the dam (Figure 7), characteristic stresses (Table 4), and maximum displacement along river direction at different key locations (Table 5) are obtained. The maximum displacement along river direction is 32.9 mm near the EL 2189–2226 m (case 1) and 20.2 mm at the dam crest (case 10). The dam tensile stress of case 1 is slightly greater than that of case 10. The maximum tensile stress near the left arch abutment is bigger than that of the right arch abutment. This is related to different geological conditions on the left and right bank of the arch dam.

The sudden drop in temperature has a greater impact on the tensile stress and the displacement of the arch dam, which increases the possibility of dam cracking. This is why the temperature drop loads are applied to the arch dam in cases 1 to 9. The insulation work of the arch dam should be done in February and March, especially at the crest and outlets of the dam.

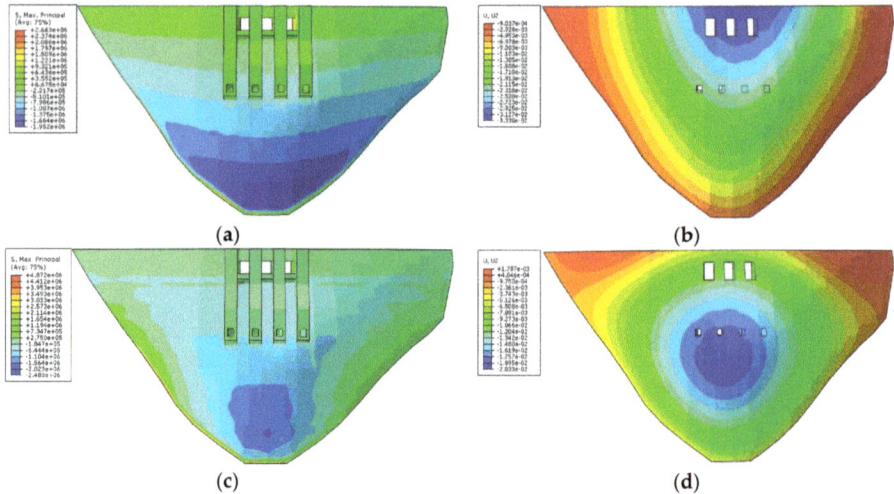

(a)

(b)

(c)

(d)

Figure 7. The first principal stress and the displacement along the river direction distribution under analysis cases 1 & 10. (**a**) The first principal stress distribution under analysis case 1 (Unit: Pa); (**b**) the displacement along the river distribution under analysis case 1 (Unit: mm); (**c**) the first principal stress distribution under analysis case 10 (Unit: Pa); (**d**) the displacement along the river distribution under analysis case 10 (Unit: mm).

Table 4. Characteristic stresses at different key locations under analysis cases 1 & 10 (unit: MPa).

Location	Content	Case 1	Case 2
Upstream surface	Maximum tensile stress of dam heel	0.9	0.89
	Maximum tensile stress near left arch abutment	1.18	1.17
	Maximum tensile stress near right arch abutment	0.97	0.94
Downstream surface	Maximum compression stress of dam toe	6.93	7.35
	Maximum compression stress near left arch abutment	8.76	8.88
	Maximum compression stress near right arch abutment	8.53	8.49

Table 5. Maximum displacement along river direction at different key locations under analysis cases 1 & 10.

	Case 1			Case 10		
	Left Arch Abutment	Arch Crown	Right Arch Abutment	Left Arch Abutment	Arch Crown	Right Arch Abutment
Maximum (mm)	6.55	32.9	5.04	8.29	20.2	5.98
EL (m)	2167.3	2308	2153	2167.3	2263	2153

4.2. Cracking Analysis of Dam Outlets

The outlets affect the stress continuity of the dam. The large tensile stress near the upstream surface may be the main cause of the outlets cracking. The maximum tensile stress of the upper and middle outlets is about 0.9 and 0.48 MPa, respectively (Figure 8a,e). Therefore, the upper outlets should have a larger cracking risk than the middle outlets. In particular, the tensile stress of the left and right upper outlets are relatively large due to the pier. Figure 8c,d illustrates the possible cracking positions of the outlets.

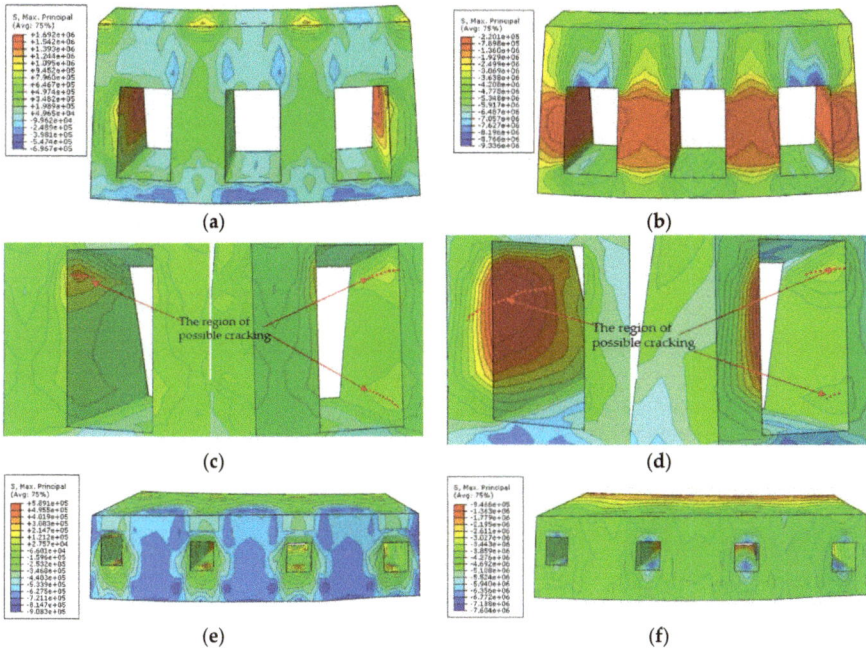

Figure 8. The first and third principal stress distribution and possible cracking positions of outlets under analysis case (Unit: Pa). (**a**) The first principal stress distribution of upper outlets; (**b**) the third principal stress distribution of upper outlets; (**c**) possible crack positions of the middle upper outlet; (**d**) possible crack positions of the side upper outlet; (**e**) the first principal stress distribution of middle outlets; (**f**) the third principal stress distribution of middle outlets.

Cracks may continue to propagate if the pore water pressure in the crack reaches 0.5 MPa [5]. Therefore, it is necessary to strictly control the cracks at the outlets, especially the possible cracking positions predicted in Figure 8c,d. More attention should be paid to the reinforcement bars of the pier and outlets to prevent tension cracks. Appropriate concrete materials which have the abrasion-resistance capacity may be used around the outlets. The concrete strength should be selected according to the discharge flow and velocity.

4.3. Cracking Analysis of the Dam Heel and the Dam Abutments

There are always stress concentrations near the upper dam heel. The maximum tensile stress of the Xulong arch dam heel is 0.9 MPa under the analysis case 1 (Figure 7). The yielding region and crack usually first appear at the dam heel and abutments as the load gradually increases. It is related to the discontinuous geometric shape and stiffness.

The stress change law of the dam heel is analyzed without considering the seepage pressure. It is assumed that the crack depth is 7.7 m, 15.4 m, and 23.1 m, respectively, that is, 1/6, 1/3, and 1/2 of the dam bottom thickness. The maximum tensile stress of the dam heel is 0.875 MPa, 0.796 MPa, and 0.874 MPa. With the increase of crack depth, the tensile stress decreases first due to the increase of gravity stress at the crack and then increases due to the increase of shear stress.

The geological condition of both abutments are complicated (Figure 9). In particular, the fault f57 and xenolith of the left abutment, the fault f26 and biotite enrichment zone of the right bank have a great influence on the stress distribution of the arch dam abutments.

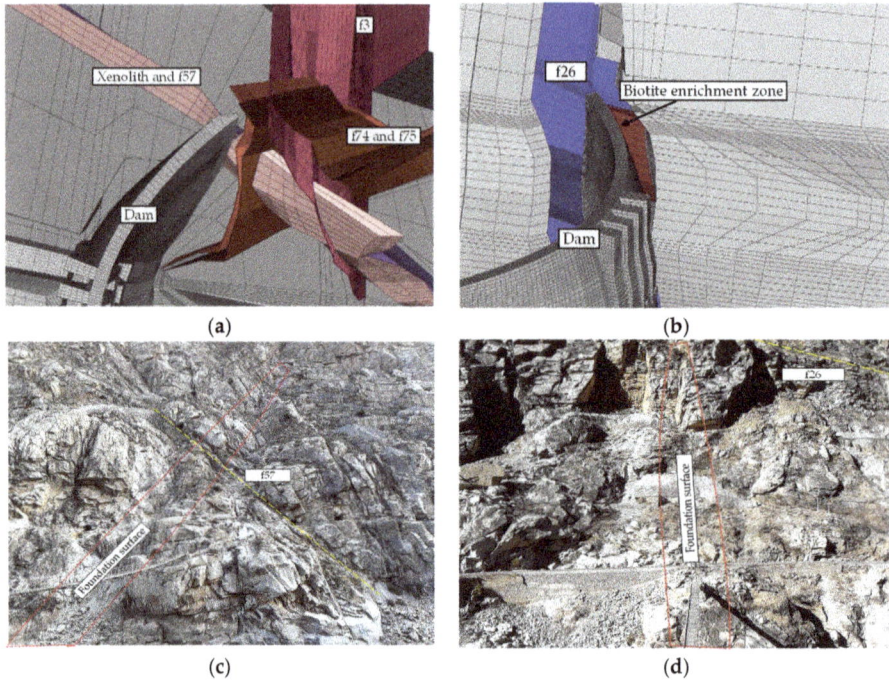

Figure 9. Complex geological condition of the Xulong arch abutments. (**a**) Left arch abutment of the model; (**b**) right arch abutment of the model; (**c**) left arch abutment of the site (the foundation face has not been excavated); (**d**) right arch abutment of the site (the foundation face has not been excavated).

5. Overall Stability and Reinforcement Analysis of Xulong Arch Dam

5.1. Overall Stability Analysis

The overall stability analysis of the Xulong arch dam adopts the methods in Section 3.1 and obtains three safety factors, $K_1 = 2 \sim 2.5$; $K_2 = 5$; $K_3 = 8.5$. The capacity curve of the maximum displacement along river direction of the arch crown is illustrated in Figure 10. With 2 to 2.5 times overloading, cracks initiate at the dam heel. At the bottom of the dam to EL 2220 m, local yielding occurs on the upstream of the left and right abutments. Therefore, the safety factor of crack initiation is estimated to be 2~2.5.

When five times overloading, cracks initiate at the foundation surface and propagate from the upstream to the downstream between EL 2095 m and EL 2258 m. The maximum crack depth is about 0.5 times the thickness of the dam. The local region of the dam toe and the outlets of the downstream begin to yield and gradually propagate to the surrounding region. The yield region of the foundation surface between the dam heel and toe tends to coalesce and the capacity curve starts to be nonlinear at five times overloading (Figure 10). Therefore, the safety factor of structural nonlinear behavior initiation of the dam is judged as 5.

When eight times overloading, the yielding region is not fully connected to form a movement mechanism, so the structure can still provide support (Figure 11a,b). When nine times overloading, the foundation surface forms two connected yielding regions and a movement mechanism (Figure 11c,d). The displacement along river direction of the arch crown increases faster at eight and nine times overloading (Figure 10). The overall stability of arch dam–foundation is lost. Therefore, the ultimate undertaking coefficient of the arch dam is judged as 8.5.

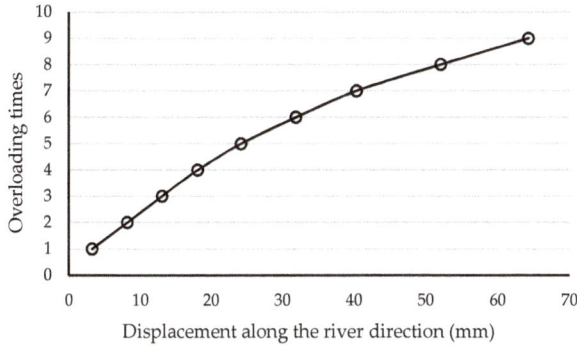

Figure 10. The capacity curve of the maximum displacement along river direction of the arch crown.

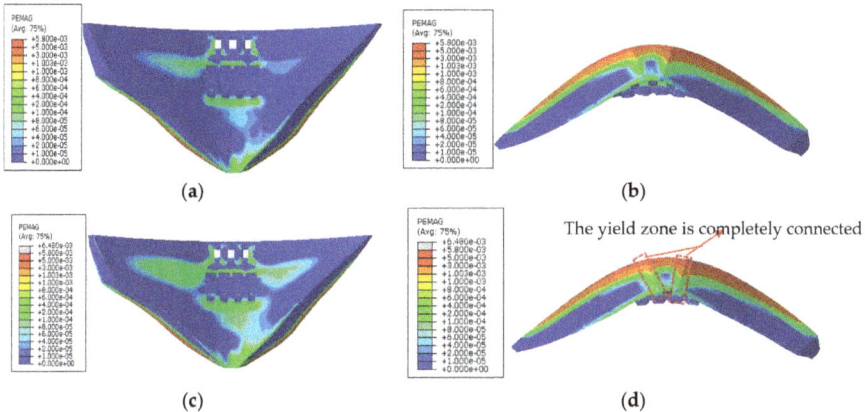

(a)

(b)

(c)

The yield zone is completely connected

(d)

Figure 11. The yielding region under different analysis cases (PEMAG: plastic strain magnitude). (a) Downstream surface under case 8; (b) foundation surface under case 8; (c) downstream surface under case 9; (d) foundation surface under case 9.

The displacement distributions of the dam are basically consistent in different overloading times (Figure 12). The maximum displacement along the river direction of the arch crown is around the dam crest and increases with the increase of overloading.

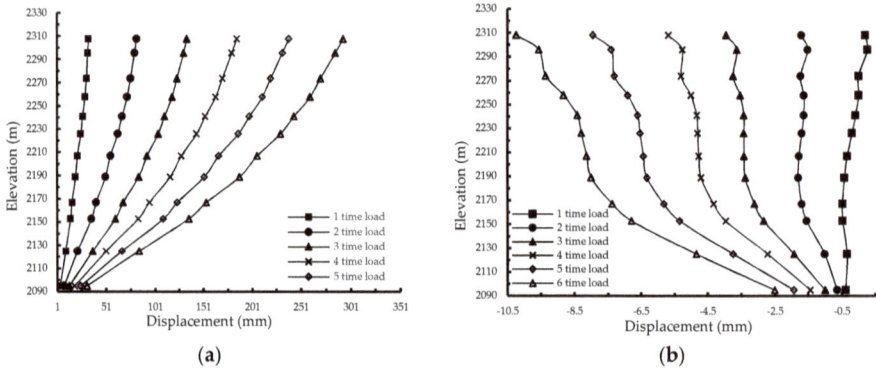

Figure 12. Crown displacement under various analysis cases 1 to 6. (**a**) Crown displacement along river direction; (**b**) crown displacement cross river direction.

The arch thrust distribution characteristics of several high arch dams are compared in Figure 13. The middle and lower elevation arch thrusts of the dam are huge and the upper elevation arch thrust is small. The Xulong and other arch dams have the same thrust distribution characteristic. The large arch thrust region is consistent with the large yielding region. The distribution characteristics of the yielding region and arch thrust can be used as a validation of the five stress zones in Section 5.2.

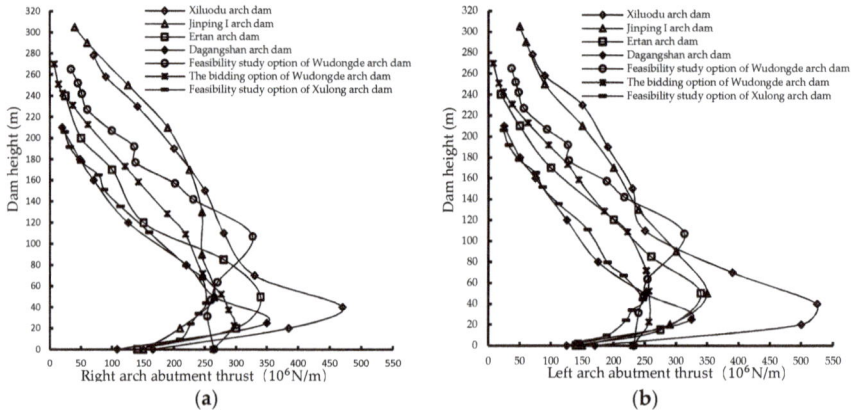

Figure 13. The distribution characteristics of arch thrust of several high arch dams. (**a**) Right arch abutment thrust; (**b**) left arch abutment thrust.

5.2. Discussion on Dam Stress Zones

Based on past analytical experience and the analysis of the stress, displacement, and yielding region of the Xulong arch dam, five stress zones of the arch dam are proposed as follows. Figure 14 is a schematic diagram of the five stress zones. The five stress zones can better guide the crack prevention of the arch dam.

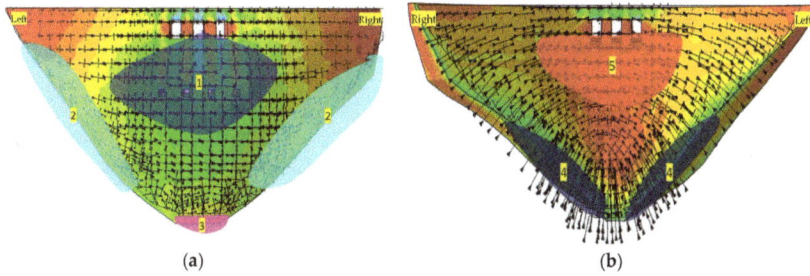

Figure 14. Five stress zones of high arch dam. (**a**) Upstream surface; (**b**) Downstream surface. Note: The number 1–5 represents compression zone of upstream surface, tensile and compressive zone of upstream arch abutment, tensile stress zone of upstream dam heel, compression stress zone of downstream arch abutment, and tensile stress zone of downstream surface, respectively.

(1) Three-way compression zone of upstream surface

This zone ranges from about 1/5 to 4/5 dam height, and the zone width is close to the height. The stress state in this zone indicates the structural state of the arch dam and it is important to control the compression stress in this zone. In general, the maximum compression stress is around the arch crown beam at 1/3 elevation of the arch dam. The compression stress results of the finite element analysis are around 6.2~8.0 MPa.

(2) Tensile and compressive zone of upstream arch abutment

The stress state may be tensile stress in the direction of both beam and arch or one of the directions is tensile stress. When upstream water pressure is considered, it is the state of double-tension single-compression or double-compression single-tension. More attention should be paid to control the tensile stress of this area to prevent cracking. The calculation results show that the tensile stress of the left arch abutment of the Xulong dam reaches 1.18 MPa of case 1. It is suggested to control the tensile stress of this area to less than 1.5 MPa when the FEM is adopted.

(3) Tensile stress zone of upstream dam heel

Based on the analysis of several super-high arch dams in China, it is suggested that the tensile stress should be strictly controlled within 1.4 MPa if the tensile stress of arch dams is based on FEM simulation. The dam heel tensile stress of the Xulong arch dam is 0.9 MPa. Although the upstream bottom joint can reduce the tensile stress of the dam heel, attention should be paid to the effect of hydraulic fracturing. The upstream bottom joint cannot affect the construction of the curtain grouting. The high tensile stress is related to the discontinuous geometric shape of the arch dam heel. The cracking of the dam heel should be paid more attention to.

(4) Compression stress zone of downstream arch abutment

This zone ranges from the bottom to the middle height of the dam. Normally, the largest compression stress is in this zone and it is important to control it. The compression stress of the left arch abutment of the Xulong dam reaches 8.88 MPa of case 10. It is suggested to control the compression stress of this zone to less than 14 MPa when the FEM is adopted.

(5) Tensile stress zone of downstream surface

The arch dam's downstream surface between the upper to middle elevation is a tensile zone, and the tensile stress is in the direction of the beam. The tensile stress may be large here due to the pier. When the upstream water level is low, this tensile stress zone will shift to the left and right arch abutments. The results of the geomechanical model test also show that the cracking of the downstream arch abutment basically extends to the center of the dam along the normal of the foundation surface, which is the failure of the tension and shear [2].

5.3. Abutment Reinforcement Suggestion

Based on the analysis of the overall stability, stress, and displacement of the arch dam, it is considered that the fault f57, xenolith, fault f26, and biotite enrichment zone have a great effect on the stress distribution of the arch abutments.

In order to improve the stress state of the dam abutments and decrease the cracking risk during long-term operation, it is recommended to use a shearing-resistance wall in the fault f57, to replace the biotite enrichment zone with concrete, and to perform consolidation grouting or anchoring on the excavated exposed weak structural zone. Figure 15 illustrates the shearing-resistance tunnel for the left arch abutment and the concrete replacement for the right arch abutment.

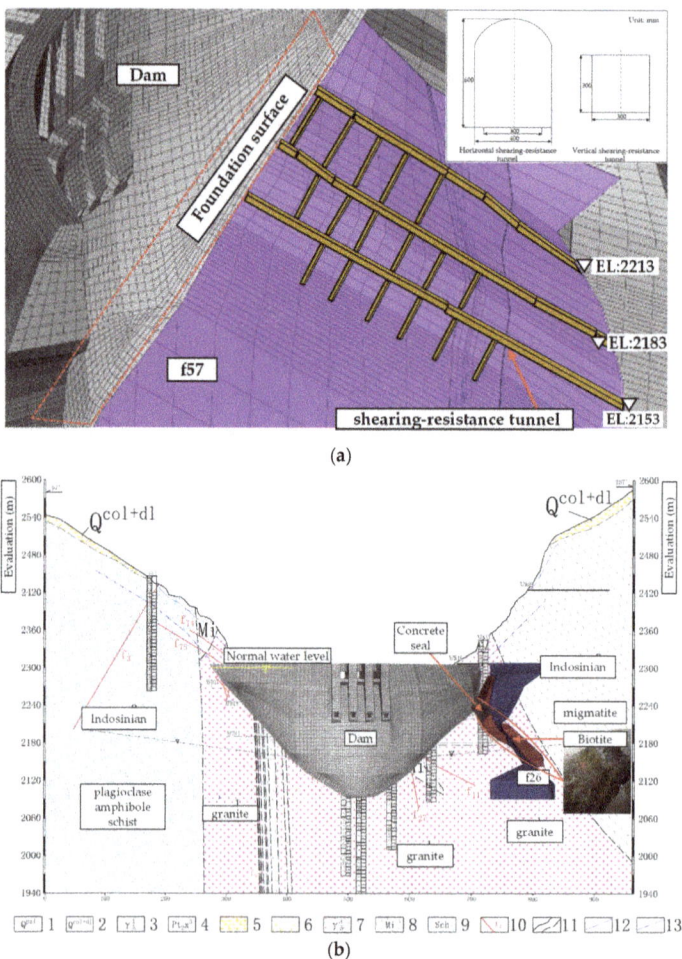

(a)

(b)

Figure 15. The abutments reinforcement method of the Xulong arch dam. (**a**) Shearing-resistance tunnel for the left abutment; (**b**) concrete replacement for the right abutment.

Through numerical simulation, the tensile stress and yielding zone changes of the arch abutments are obtained before and after reinforcement (Figures 16 and 17). The first principal stresses of the left and right arch abutments decrease by about 0.13 and 0.17 MPa, respectively. The reinforcement of the abutments reduces the first principal stress and improves the stress state of the arch abutments, thereby reducing the cracking risk. The reinforcement method also improves the comprehensive shear strength of the side-slip surface and ensures a certain safety margin for the anti-sliding of the arch abutments.

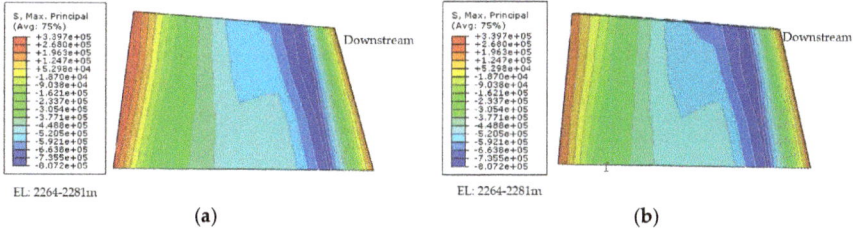

Figure 16. The first principal stress at the right arch abutment from EL 2264 m to 2281 m (Unit: Pa). (**a**) Before reinforcement; (**b**) after reinforcement.

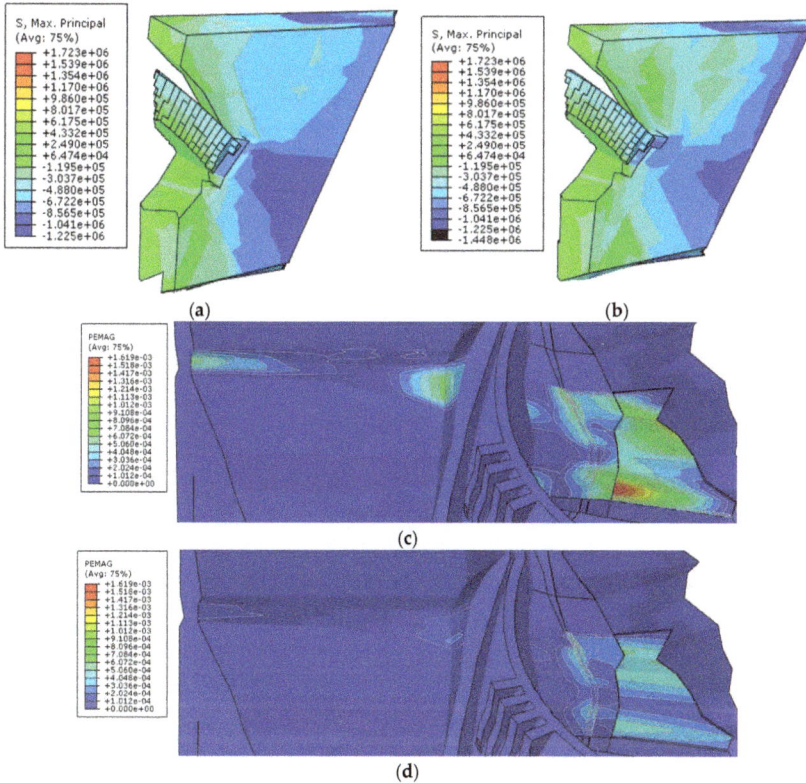

Figure 17. The first principal stress and yielding region distribution of the left abutment (Stress unit: Pa. PEMAG: plastic strain magnitude). (**a**) before concrete replacement; (**b**) after concrete replacement; (**c**) before concrete replacement; (**d**) after concrete replacement.

6. Conclusions

In this paper, the different cracking types and effect factors are summarized. The cracking risk, overall stability, and abutment reinforcement of the Xulong arch dam are analyzed through numerical simulation. The following conclusions can be drawn:

(1) A nonlinear constitutive model relating to the yielding region is proposed to evaluate dam cracking risk and overall stability. The temperature gradient change has a greater impact on the tensile stress and displacement of the arch dam, which increases dam cracking risk. In particular, the tensile stress of the left and right upper outlets are relatively large due to the pier.
(2) The three safety factors of the Xulong arch dam are obtained, $K_1 = 2{\sim}2.5$; $K_2 = 5$; $K_3 = 8.5$, and the dam overall stability is guaranteed.
(3) The five dam stress zones are proposed to analyze the dam cracking base of numerical results. It is recommended to use a shearing-resistance wall in the fault f57, replace the biotite enrichment zone with concrete, and perform consolidation grouting or anchoring on the excavated exposed weak structural zone. With optimal design of the dam structure according to the different stress characteristics of the five stress zones, the cracking risk and overall stability of the Xulong arch dam can be better controlled.

Author Contributions: P.L. and P.W. performed the numerical analysis on crack risk and overall stability evaluation; W.W. and H.H. provided original dam design parameters and scheme.

Funding: This research was funded by Changjiang Institute of survey, planning, design and research, China.

Conflicts of Interest: The authors declare no conflict of interest.

References

1. Lin, P.; Liu, X.-L.; Hu, S.-Y.; Li, P.-J. Large deformation analysis of a high steep slope relating to the Laxiwa reservoir, China. *Rock Mech. Rock Eng.* **2016**, *49*, 2253–2276. [CrossRef]
2. Lin, P.; Zhou, W.-Y.; Liu, H.-Y. Experimental study on cracking, reinforcement, and overall stability of the Xiaowan super-high arch dam. *Rock Mech. Rock Eng.* **2015**, *48*, 819–841. [CrossRef]
3. Lin, P.; Shi, J.; Zhou, W.-Y.; Wang, R.-K. 3D geomechanical model tests on asymmetric reinforcement and overall stability relating to the Jinping I super-high arch dam. *Int. J. Rock Mech. Min. Sci.* **2018**, *102*, 28–41. [CrossRef]
4. Duffaut, P. The traps behind the failure of Malpasset arch dam, France, in 1959. *J. Rock Mech. Geotech. Eng.* **2013**, *5*, 335–341. [CrossRef]
5. Lin, P.; Liu, H.-Y.; Li, Q.-B.; Hu, H. Effects of outlets on cracking risk and integral stability of super-high arch dams. *Sci. World J.* **2014**, *2014*, 312827. [CrossRef] [PubMed]
6. Xia, S.-Y.; Lu, S.-W. Approach to cracking mechanism of Kolnbrein arch dam heel. *Des. Hydroelectr. Power Stn.* **1999**, *15*, 26–33. (In Chinese)
7. Zhang, X.-F.; Wang, X.-P.; Huang, Y.; Li, S.-Y. Simulation study on temperature stress of RCC arch dam under cold wave conditions. *J. Water Resour. Water Eng.* **2018**, *29*, 192–197. (In Chinese) [CrossRef]
8. Zhang, X.; Liu, X.-H.; Jing, X.-Y.; Wang, Q.; Chang, X.-L. Study on effect of thermal stress compensation for MgO concrete of high arch dam in cold area. *Water Resour. Power* **2013**, *31*, 82–85. (In Chinese)
9. Mirzabozorg, H.; Hariri-Ardebili, M.A.; Shirkhan, M. Impact of solar radiation on the uncoupled transient thermo-structural response of an arch dam. *Sci. Iran.* **2015**, *22*, 1435–1448.
10. Liang, R.-Q. Study of thermal control and crack prevention for high arch dam in dry-hot valley region. *Yangtze River* **2014**, *45*, 42–45. (In Chinese) [CrossRef]
11. Sheibany, F.; Ghaemian, M. Effects of environmental action on thermal stress analysis of Karaj concrete arch dam. *J. Eng. Mech.* **2006**, *132*, 532–544. [CrossRef]
12. Maken, D.D.; Léger, P.; Roth, S.N. Seasonal thermal cracking of concrete dams in northern regions. *J. Perform. Constr. Facil.* **2014**, *28*, 04014014. [CrossRef]

13. Waleed, A.M.; Jaafar, M.S.; Noorzaei, J.; Bayagoob, K.H.; Amini, R. Effect of placement schedule on the thermal and structural response of RCC dams, using finite element analysis. In Proceedings of the Geo Jordan Conference 2004, Irbid, Jordan, 12–15 July 2004; pp. 94–104. [CrossRef]

14. Jia, J.-S.; Li, X.-Y. Dam heel cracking problem in high arch dams and new measure for solution. *J Hydraul. Eng.* **2008**, *39*, 1183–1188. (In Chinese) [CrossRef]

15. Lin, P.; Chen, X.; Zhou, W.-Y.; Yang, R.-Q.; Wang, R.-K. Simulation on back analysis of Shuanghe arch dam cracking. *Rock Soil Mech.* **2003**, *24*, 53–56. (In Chinese) [CrossRef]

16. Câmara, R.J. A method for coupled arch dam-foundation-reservoir seismic behaviour analysis. *Earthq. Eng. Struct. Dyn.* **2000**, *29*, 441–460. [CrossRef]

17. Lotfi, V.; Espandar, R. Seismic analysis of concrete arch dams by combined discrete crack and non-orthogonal smeared crack technique. *Eng. Struct.* **2004**, *26*, 27–37. [CrossRef]

18. Hariri-Ardebili, M.A.; Seyed-Kolbadi, S.M. Seismic cracking and instability of concrete dams: Smeared crack approach. *Eng. Fail. Anal.* **2015**, *52*, 45–60. [CrossRef]

19. Mi, Y.; Aliabadi, M.H. Dual boundary element method for three-dimensional fracture mechanics analysis. *Eng. Anal. Bound. Elem.* **1992**, *10*, 161–171. [CrossRef]

20. Gerstle, W.H.; Ingraffea, A.R.; Perucchio, R. Three-dimensional fatigue crack propagation analysis using the boundary element method. *Int. J. Fatigue* **1988**, *10*, 187–192. [CrossRef]

21. Lubliner, J.; Oliver, J.; Oller, S.; Onate, E. A plastic-damage model for concrete. *Int. J. Solids Struct.* **1989**, *25*, 299–326. [CrossRef]

22. Lee, J.; Fenves, L.G. Plastic-damage model for cyclic loading of concrete structures. *J. Eng. Mech.* **1998**, *124*, 892–900. [CrossRef]

23. Feng, L.M.; Pekau, O.A.; Zhang, C.H. Cracking analysis of arch dams by 3D boundary element method. *J. Struct. Eng.* **1996**, *122*, 691–699. [CrossRef]

24. Chen, J.; Soltani, M.; An, X. Experimental and numerical study of cracking behavior of openings in concrete dams. *Comput. Struct.* **2005**, *83*, 525–535. [CrossRef]

25. Sharan, S.K. Efficient finite element analysis of hydrodynamic pressure on dams. *Comput. Struct.* **1992**, *42*, 713–723. [CrossRef]

26. Sato, H.; Miyazawa, S.; Yatagai, A. Thermal crack estimation of dam concrete considering the influence of autogenous shrinkage. In Proceedings of the 10th International Conference on Mechanics and Physics of Creep, Shrinkage, and Durability of Concrete and Concrete Structures (CONCREEP), Vienna, Austria, 21–23 September 2015; pp. 1289–1298. [CrossRef]

27. Chow, W.-Y.; Yang, R.-Q. Determination of stability of arch dam abutment using finite element method and geomechanical models. In Proceedings of the 4th Australia—New Zealand Conference on Geomechanics, Perth, Australia, 14–18 May 1984; Volume 2, pp. 595–600.

28. Zhou, W.-Y.; Yang, R.-Q.; Liu, Y.-R.; Lin, P. Research on geomechanical model of rupture tests of arch dams for their stability. *J. Hydroelectr. Eng.* **2005**, *24*, 53–58. (In Chinese) [CrossRef]

29. Kou, X.-D.; Zhou, W.-Y. The application of element-free method to approximate calculation of arch dam crack propagation. *J. Hydraul. Eng.* **2000**, *31*, 28–35. (In Chinese)

30. Portela, A.; Aliabadi, M.H.; Rooke, D.P. The dual boundary element method: Effective implementation for crack problems. *Int. J. Numer. Methods Eng.* **2010**, *33*, 1269–1287. [CrossRef]

31. Singhal, A.C.; Nuss, L.K. Cable anchoring of deteriorated arch dam. *J. Perform. Constr. Facil.* **1991**, *5*, 19–36. [CrossRef]

32. Hariri-Ardebili, M.A.; Saouma, V.E. Single and multi-hazard capacity functions for concrete dams. *Soil Dyn. Earthq. Eng.* **2017**, *101*, 234–249. [CrossRef]

applied
sciences

MDPI

Article

Long-Term Behaviour of Precast Concrete Deck Using Longitudinal Prestressed Tendons in Composite I-Girder Bridges

Haiying Ma, Xuefei Shi * and Yin Zhang

Department of Bridge Engineering, Tongji University, Shanghai 200092, China; mahaiying@tongji.edu.cn (H.M.); zhangyinzian@126.com (Y.Z.)
* Correspondence: shixf@tongji.edu.cn; Tel.: +86-216-598-2956

Received: 12 November 2018; Accepted: 5 December 2018; Published: 13 December 2018

Abstract: Twin-I girder bridge systems composite with precast concrete deck have advantages including construction simplification and improved concrete strength compared with traditional multi-I girder bridge systems with cast-in-place concrete deck. But the cracking is still a big issue at interior support for continuous span bridges using twin-I girders. To reduce cracks occurrence in the hogging regions subject to negative moments and to guarantee the durability of bridges, the most essential way is to reduce the tensile stress of concrete deck within the hogging regions. In this paper, the prestressed tendons are arranged to prestress the precast concrete deck before it is connected with the steel girders. In this way, the initial compressive stress induced by the prestressed tendons in the concrete deck within the hogging region is much higher than that in regular concrete deck without prestressed tendons. A finite element analysis is developed to study the long-term behaviour of prestressed concrete deck for a twin-I girder bridge. The results show that the prestressed tendons induce large compressive stresses in the concrete deck but the compressive stresses are reduced due to concrete creep. The final compressive stresses in the concrete deck are about half of the initial compressive stresses. Additionally, parametric study is conducted to find the effect to the long-term behaviour of concrete deck including girder depth, deck size, prestressing stress and additional imposed load. The results show that the prestressing compressive stress in precast concrete deck is transferred to steel girders due to concrete creep. The prestressed forces transfer between the concrete deck and steel girder cause the loss of compressive stresses in precast concrete deck. The prestressed tendons can introduce some compressive stress in the concrete deck to overcome the tensile stress induced by the live load but the force transfer due to concrete creep needs be considered. The concrete creep makes the compressive stress loss and the force redistribution in the hogging regions, which should be considered in the design the twin-I girder bridge composite with prestressed precast concrete deck.

Keywords: concrete creep; prestressing stress; compressive stress; FE analysis; force transfer

---

1. Introduction

A two or multiple-I girder system has two or more steel I girders connected with diaphragms and composite with a concrete deck using shear studs. The steel girders are in tension and the concrete deck is in compression in the regions of positive moments (i.e., the sagging region) under vertical loads, which makes good use of material advantages of steel and concrete. While within the regions of negative moments (i.e., the hogging regions), the concrete deck is in tension under vertical loads and the tensile stress may increase due to concrete shrinkage and creep. Concrete cracking is a big issue for the hogging regions [1–5]. High performance concrete with larger tension strength can be used [6] but the cost is substantial. One common way to make concrete deck in compression with initial compressive stress to overcome the tensile stress induced by live load.

One way is to arrange prestressed tendons in the concrete deck in the hogging regions. The prestressing compressive stress in the concrete is to overcome any additional tensile stress induced by vertical loads and additional second order effect of shrinkage and creep. Miyamoto et al. found that using external tendons could be considered an effective method of strengthening bridges deteriorating due to overloading [7]. Deng and Morcous proposed a new prestressed concrete-steel composite girder, which uses pretensioned concrete bottom flange to provide initial compressive stress in the concrete deck [8,9]. Wang et al. investigated the behaviour of reinforced concrete strengthened with externally prestressed tendons and they found that the basalt fibre reinforced polymer (BFRP) was feasible to strengthen the beam behaviour [10].

Except using tendons to introduce compressive stress in concrete deck, some construction strategies are used. Temporary loads are sometimes applied to the sagging regions (the regions of positive moment) before the concrete cast in the hogging regions. In this way, the compressive stress is induced in the hogging regions after the hogging region concrete is hardened and the temporary loads are removed. Marí et al. and Dezi et al. studied the behaviour of composite bridges considering different construction processes and they found that the construction sequence could affect the tension stress in concrete deck in the hogging regions [11,12]. Liu et al. analysed the jacking-up method to prestress the concrete deck and they found that jacking-up the interior support could efficiently introduce compressive stress in the concrete deck in the hogging regions to overcome the tensile stresses induced by shrinkage and live load [13].

Either prestressed tendons or construction strategies can introduce initial compressive stress in the concrete deck to avoid or reduce concrete cracking in the hogging regions. However, for the preconnected composite girder systems, the prestressed forces are applied to the whole composite section and mostly are applied to the steel section. Kwon et al. and Hällmark et al. studied the behaviour of steel-concrete composite girders with prestressing tendons before concrete deck connected to steel girders [14,15]. Su et al. studied the behaviour of a continuous composite box girder with prefabricated prestressed concrete slab in the hogging region [16]. Tong et al. studied the long-term behaviour of the composite box girders with post connected prestressed concrete deck and the research shows the prestressed concrete deck before connected with steel box girders can improve concrete shrinkage [17].

In recent years, a significant amount of continuous twin-I girders with precast concrete deck are built in China. The cracks in the hogging regions are usually controlled by the crack width control [18–20]. However, crack width control is not an efficient way to improve the behaviour in the hogging regions. The way to arrange prestressed tendons in concrete deck in the hogging regions is used for twin-I girders with post connected prestressed precast concrete deck. In the paper, the long-term behaviour of a continuous twin-I girder bridge is investigated to find the creep effect on the compressive stresses induced by the prestressed tendons. Additionally, parametric study is developed to find the creep effect on the prestressing the concrete deck and the force transfer between steel girders and concrete deck.

2. Case of a Twin-I Girder Composite with Precast Deck

The bridge is composed of steel two-I girders with precast concrete deck and the steel girders and precast concrete deck are composite with shear studs within the voids. The span arrangement is 4×35 m. The girder spacing is 8.95 m. The steel girder depth is 1.7 m. Cross beams are arranged with a spacing of 7 m. Interior cross beams are not connected with the concrete deck and end cross beams at the ends of bridge are connected with the concrete deck through the shear studs. The width of the precast concrete deck is 16.75 m. For the concrete deck, the prestressed tendons are arranged in the hogging regions. Figure 1 presents the structure components in the hogging regions of the bridge.

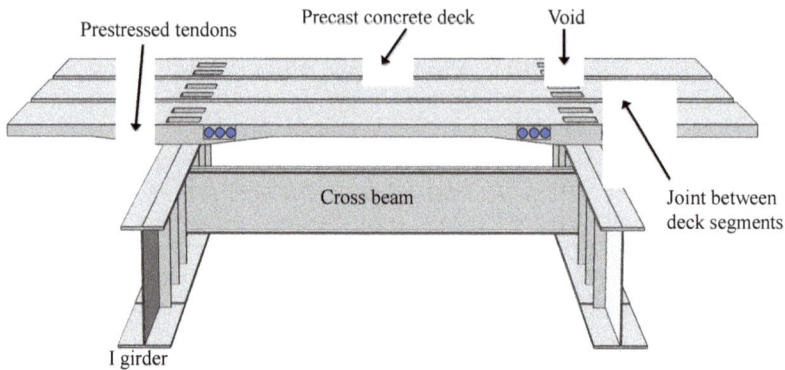

Figure 1. A continuous twin-I girder with prestressed precast concrete deck in the hogging region.

As shown in Figure 2, the construction procedure including five steps:

(1) The steel girders are lifted and connected to be a four-span continuous system.
(2) All the precast concrete segments are lifted to the steel girders and the concrete deck and the steel girders are not composite (the concrete in the voids are not casted).
(3) Only the precast concrete deck segments within the hogging regions (e.g., within the regions at interior supports) are prestressed and the steel girders are not composite with the concrete deck at this time.
(4) The joints and voids are casted with concrete to make the concrete deck composite with the steel girders.
(5) The transverse tendons are prestressed and the bridge is constructed with the wearing surface and the attached appurtenances (barriers, railings, lights, etc.).

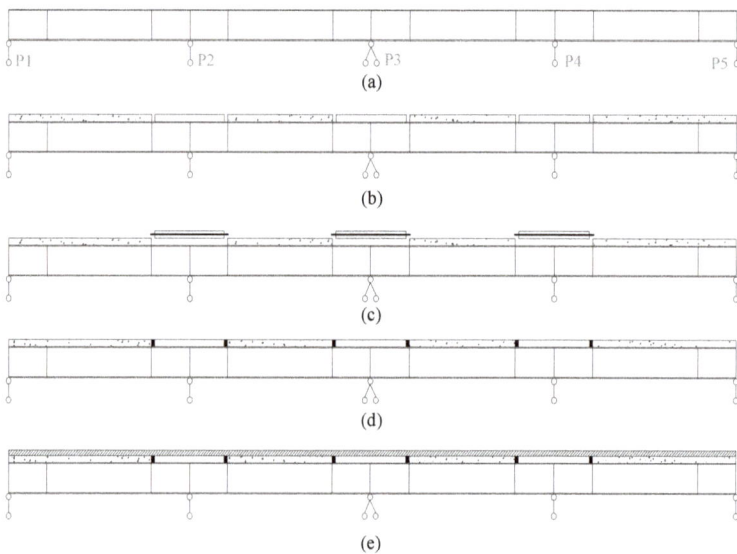

Figure 2. Construction sequence of the twin-I girder bridge: (**a**) Steel girder erection; (**b**) Precast concrete deck segments erection; (**c**) Prestressing concrete deck in the hogging regions; (**d**) Cast-in-place joints and void to connect girders and deck; (**e**) Constructed condition with the wearing surface and the attached appurtenance.

3. Finite Element Model

3.1. Elements and Meshes

The software ANSYS is used to develop analysis in the paper [21,22]. FE analyses can predict and analyse the behaviour of steel-composite bridges [1,9,16,23]. Solid elements (element type of solid 45) are used to model a concrete deck and shell elements (element type of shell 43) are used to model steel girders and stiffeners. Spring element (element type of combine14) is used to model shear studs to connect the concrete deck and steel girders. The concrete deck and the steel girders are assumed fully connected by the shear studs. Link elements (element type of link 8) are used to model prestressed tendons. The prestress forces in the tendons are applied with temperature. Figure 3 shows the finite element model of the bridge. The default convergence criteria are used in the analyses.

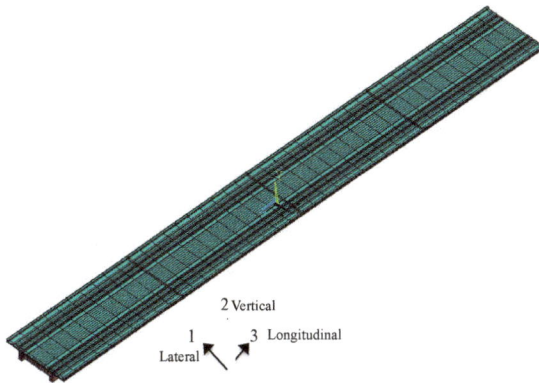

Figure 3. Finite element model of a twin-I girder bridge with precast concrete deck.

3.2. Material Models

The steel material of the girders is modelled using an elastic isotropic material in the elastic range with an elastic modulus of 200 GPa and Poisson's ratio of 0.3 and a perfectly plastic isotropic material in the inelastic range. The yield strength of the steel material is 345 MPa. The deck concrete has 23.1 MPa compressive strength (Ministry of Transport of the People's Republic of China) [19,22,23]. An empirical stress-strain model for unconfined concrete proposed by Oh and Sause is used for the uniaxial stress-strain relationship of concrete [24].

The creep and shrinkage are included in the model based on the equations from Ministry of Transport of the People's Republic of China [25]. The shrinkage effect is applied to the models through temperature decrease in the concrete material. ANSYS does not have direct method to calculate the creep effect. It gives metal creep model to model the creep in concrete. There are 13 creep equations in ANSYS and one used often is as follows:

$$\dot{\varepsilon}_{cr} = C_1 \sigma^{C_2} \varepsilon_{cr}^{C_3} e^{-C_4/T} \tag{1}$$

where, ε_{cr} is creep strain; $\dot{\varepsilon}_{cr}$ is creep variance ratio of time; σ is stress; T is absolute temperature; C1 through C4 is parameters to be calculated. Usually, there are two ways to simplify the equation. One is to assume that creep variance ratio of time is only related to stress with C2 = 1 and C3 = C4 = 0 (Method A). Thus the equations is simplified as follows:

$$\Delta \varepsilon_{cr} = C_1 \sigma \Delta t \tag{2}$$

For concrete with constant stress, the creep strains satisfy:

$$\varepsilon_{cr} = \varepsilon_0 \phi(t, t_0) = \frac{\sigma_0}{E} \phi(t, t_0) \tag{3}$$

Equation (3) is changed with time of Δt:

$$\frac{\Delta \varepsilon_{cr}}{\Delta t} = \varepsilon_0 \phi(t, t_0) = \frac{\sigma_0}{E} \frac{\Delta \phi(t, t_0)}{\Delta t} \tag{4}$$

Within the time of Δt, C1 is calculated as follows:

$$C_i = \frac{\Delta \phi(t_i, t_{i-1})}{\Delta t} \frac{1}{E} \tag{5}$$

Another way is assuming there is linear relationship between creep variation rate and strain with C2 = C4 = 0, C3 = 1 (Method B), which is denoted as follows:

$$\Delta \varepsilon_{cr} = C_1 \varepsilon \Delta t \tag{6}$$

For concrete with constant stress, the creep strains satisfy:

$$\varepsilon(t) = \varepsilon_e + \varepsilon_c(t) = (1 + \phi(t, t_0)) \varepsilon_e \tag{7}$$

Thus within the time of Δt, C1 is calculated as follows:

$$C1 = \frac{\Delta \phi(t_i, t_{i-1})}{\Delta t (1 + \phi(t_i, t_{i-1}))} \tag{8}$$

Figure 4 presents the validation of these two ways to analyse the creep effect for a column applied with a vertical constant force (denoted as *P*) and shows that the ways agree with the results using theoretical analysis.

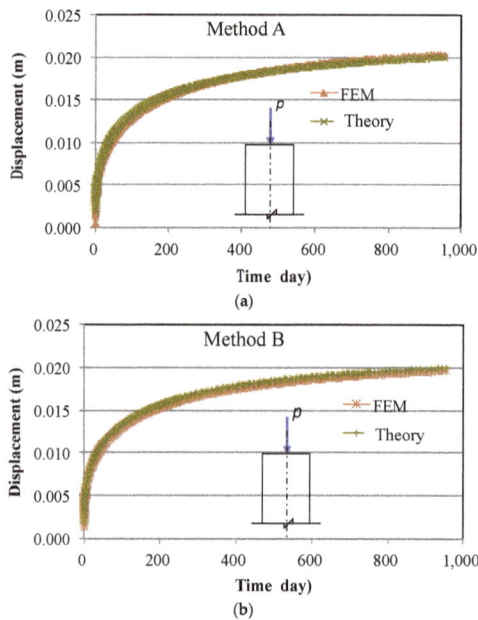

Figure 4. Finite element analysis using metal creep model for creep effect: (a) Method A; (b) Method B.

3.3. Boundary Conditions

Continuously supported boundary conditions are used for the bridge model. The vertical displacements ($U2$) at the bottom of the flange nodes are restrained at each support. At each support, the lateral displacements ($U1$) of the bottom flange nodes at the bottom of the flange nodes are restrained. The longitudinal displacements ($U3$) of the bottom flange nodes at the middle support are only restrained.

4. Long-Term Behaviour Analysis

4.1. Prestressed Concrete Deck Condition

Table 1 gives the induced compressive stresses in the precast concrete deck within the hogging regions. After the tendons prestressed, the compressive stresses on the deck top surface are from −6 Mpa to −7 Mpa (negative value denotes compression) and are from −7 Mpa to −8 Mpa on the deck bottom surface. The tendons are not located at the neutral axis of the deck cross-section and cause the difference between the top surface and the bottom surface.

Table 1. Stress and deformation analysis results.

Analysis Result		Prestressed Concrete Deck Condition	Constructed Condition	10,000-Day Creep
Concrete deck	Bottom surface	−7~−8 MPa	−4.4~−6.2 MPa	−2.3 MPa~−3.3 MPa
	Top surface	−6~−7 MPa	−4.7~−5.6 MPa	−2.6 MPa~−3.5 MPa
Steel girder	Top flange	135 Mpa	165 MPa	10–20 MPa
	Bottom flange	143 MPa	167 MPa	155–165 MPa
Tendons		-	1135–1175 MPa	1115–1180 MPa
Deflection		0.066 m	0.074 m	0.040 m

4.2. Constructed Condition

Figures 5 and 6 present the normal stress variation in the concrete deck and the steel girders under the bridge constructed condition. Within the hogging regions, the compressive stresses vary from −5.6 MPa to −6.4 MPa at middle interior support (P3) and vary from −4.7~−5.6 MPa at the other interior supports (P2 and P4). The maximum tension stresses in the steel girders are about 165 MPa at P2. After the bridge constructed, the compressive stresses in the concrete deck within the hogging regions are not small.

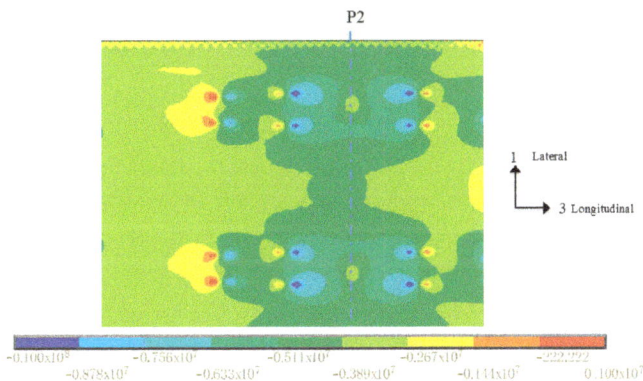

Figure 5. Stress at the top surface of the concrete deck under the constructed condition in the hogging region (Pa).

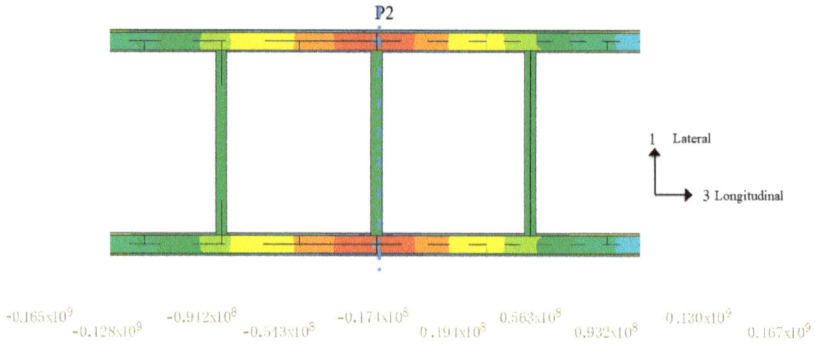

Figure 6. Stress at the top flange of the steel girders under the constructed condition in the hogging region (Pa).

4.3. Long-Term Behaviour

Figure 7 shows the normal stress variation on the concrete deck top surface after 10,000 days creep near P2. Compared with the stresses under the constructed condition, the stresses in the concrete deck change due to the concrete creep, especially in the hogging regions. The compressive stresses decrease to −2.6~−3.5 MPa, with a decrease of about 3 MPa from the constructed condition. Figure 8 presents the stress variation in the steel girder. The flange stresses in the girder change a lot compared with the constructed condition that top flange stress is in compression with stress from −10 to −20 MPa. The stresses in the tendons are checked and it is found that no changes occur for the tendons. The results show that the concrete creep reduces the initial compressive stress in the concrete deck and causes force transfer between the concrete deck and the steel girders.

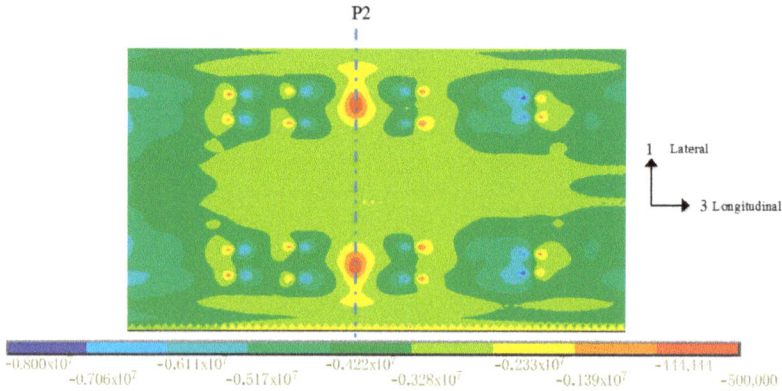

Figure 7. Stress at the top surface of the concrete deck in the hogging region after 10,000-day creep (Pa).

Table 1 summarizes the stress change in the bridge due to the concrete creep. The results show that the stresses in the concrete deck and the steel girders change due to the concrete creep. The force transfer occurs between the concrete deck and the steel girders and it mostly occurs within the hogging regions.

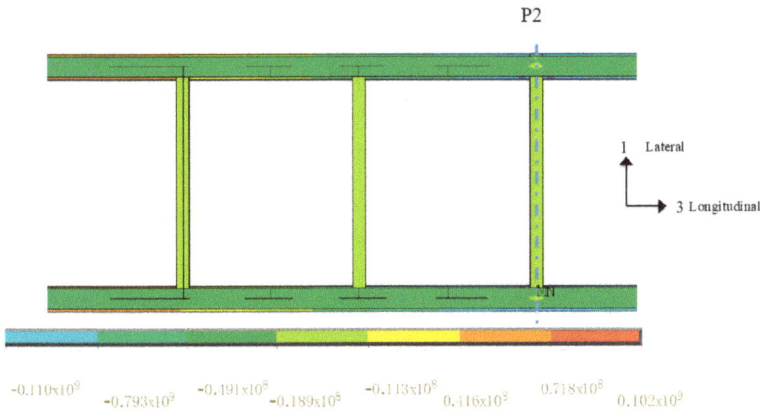

P2

1 Lateral

3 Longitudinal

-0.110x10⁹
-0.793x10⁹
-0.191x10⁸
-0.189x10⁸
-0.113x10⁸
0.416x10⁸
0.718x10⁸
0.102x10⁹

Figure 8. Stress at the top flange of the steel girders in the hogging region after 10000-day creep (Pa).

5. Parametric Studies

To study the effects of different parameters on the prestressing the concrete deck and the force transfer between the concrete deck and steel girders, a simplified two-span I girder is conducted. The continuous girder has two spans of 3 m + 3 m. The precast concrete deck has width of 0.5 m. The parameters include girder depth, concrete deck thickness, prestressed compressive stress in concrete deck and additional imposed vertical load. The additional imposed load is to model the condition that long-term load applied on the bridge system. Note that 10,000 days creep is considered in the analyses.

5.1. Girder Depth

Different girder depths are analysed and discussed to find the effect of girder stiffness on the creep effect. Figure 9 gives the stress variation in the hogging region. The initial prestressing stress is 10 MPa. Along with the increase of girder depth, the stress variations due to shrinkage and creep decrease slightly, which indicates that the girder stiffness has little effect on the creep effect. The results also show that the stress loss due to creep is over than 50% of the initial prestressing stress. The stresses in the tendons do not change and the stress loss in the concrete deck is transferred to steel girders.

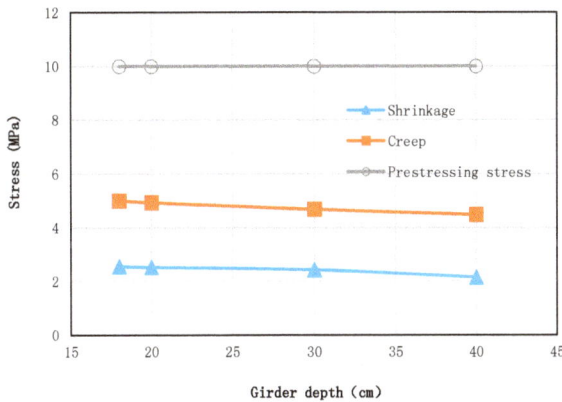

Figure 9. Stress variation in the concrete deck in the hogging region due to different effects for different girder depths.

5.2. Deck Thickness

Different deck thickness are analysed and discussed to find the effect of deck size on the creep effect. Figure 10 presents the stress variation in the hogging region. The initial prestressing stress is 10 MPa. The results show that the stresses due to shrinkage have no change with the increase of deck thickness. Along with the increase of deck thickness, the stress variations due to creep varies but the variation is not linear. The stress variation induced by creep increases with the increase the deck thickness up to 180 mm thickness and then decreases. With the thickness of 180 mm, the stress loss is the biggest one. But the difference is not big and the difference between the thickness of 180 mm and 160 mm is 6%. The results also show that the stress loss due to creep is over than 50% of the initial prestressing stress.

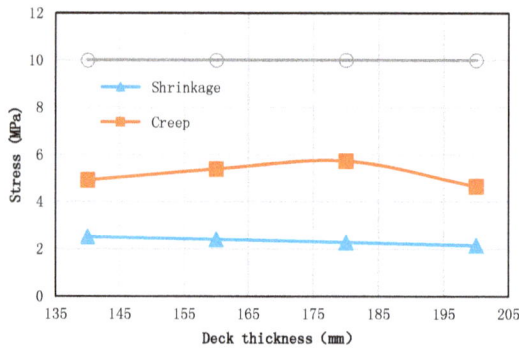

Figure 10. Stress variation in the concrete deck in the hogging region due to different effects for different deck thicknesses.

5.3. Prestressing Stress

Figure 11 presents the results to analyse the initial prestressing stress applied to concrete deck. The stress due to shrinkage has no change with the increase of prestressing stress. Along with the increase of the initial prestressing stress, the stress loss due to creep increases. The ratio between the stress loss due to creep to the initial prestressing stress is larger for smaller prestressing stress (e.g., 67% for 5 MPa prestressing stress and 49% for 10 MPa prestressing stress). The results show that the prestressing stress is decreased in the concrete deck due to creep and the loss mostly is larger than 50% of the initial prestressing stress.

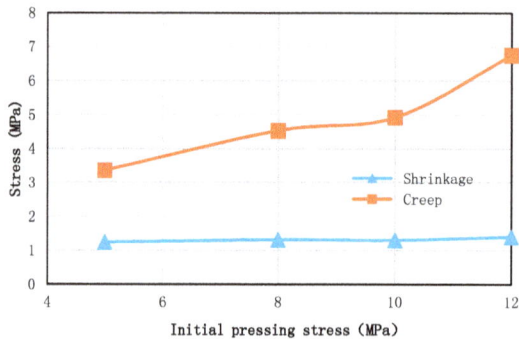

Figure 11. Stress variation in the concrete deck in the hogging region due to different effects for different prestressing stress by tendons.

5.4. Additional Imposed Load

Additional imposed load is to model the condition under long-term dead load and live load, which induces tension in the concrete deck in the hogging regions. The tension stress induced by the additional imposed load is used to denote the value of the imposed load. Figure 12 presents the results to analyse the effect of additional imposed load to the creep effect. In the figure, "shrinkage" denotes the stress variation induced by the concrete shrinkage, "creep" denotes the stress variation induced by the concrete creep considering 10,000 days and "prestressing stress" denotes the initial compressive stress introduced by the prestressed tendons. The initial prestressing stress is 10 MPa. The stress due to shrinkage has no change with the increase of additional imposed load. With the increase of imposed load, the stress loss due to creep increase, which shows that the additional imposed load have a big effect on the creep effect.

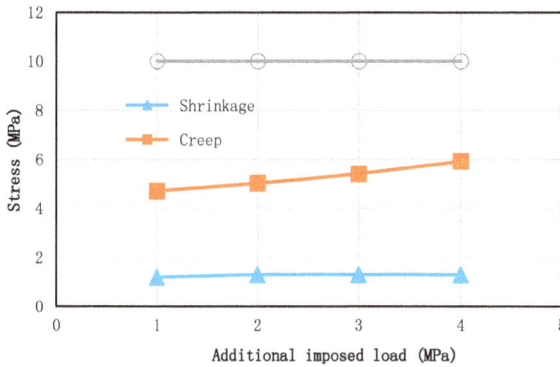

Figure 12. Stress variation in the concrete deck in the hogging region due to different effects for different additional imposed load.

6. Conclusions

The paper investigated the behaviour of a four-span continuous twin-I girder bridge using prestressed precast concrete deck due to concrete creep. Simplified two-span I girder models are analysed to find the effect on the prestressed compression and force transfer between concrete deck and steel girders. Major findings are summarized as follows:

(1) For the continuous twin-I girder bridge, the prestressed tendons introduce compressive stress in the concrete deck and the compressive stress under constructed condition is big and it can overcome the tensile stress induced by shrinkage and live load.
(2) In the hogging regions, the prestressed stresses in the concrete deck are reduced due to the concrete creep effect and the decrease is up to 50% of the initial prestressing stress.
(3) The stresses in the steel girders in the hogging regions vary big, especially for girder flange in tension and the changes are due to force transfer from compressive stress in concrete deck.
(4) The stresses in the tendons have almost no change and the prestressed force transfers from concrete deck to steel girders in the hogging regions.
(5) The steel girder stiffness has no effect on the prestressing stress loss in the concrete deck.
(6) The concrete deck, initial prestressing stress and additional imposed load have an effect on the initial prestressing stress loss in the concrete deck due to concrete creep.
(7) The prestressing stress loss in the concrete due to creep mostly is over 50% and it is transferred to steel girders to change the stress distribution of composite section in the hogging regions.

Author Contributions: Conceptualization, H.M. and X.S.; numerical analysis, H.M., X.S. and Y.Z.; writing: H.M. and Y.Z.

Funding: This research was funded by National Key R&D Program of China, grant number 2018YFC0809606, National Natural Science Foundation of China, grant number 51608378 and Science and Technology Commission of Shanghai Municipality (18DZ1201203, 17DZ1204300) and the Fundamental Research Funds for the Central Universities.

Conflicts of Interest: The authors declare no conflict of interest.

References

1. Gara, F.; Leoni, G.; Dezi, L. Slab cracking control in continuous steel-concrete bridge decks. *J. Bridge Eng.* **2013**, 1319–1327. [CrossRef]
2. Macorini, L.; Fragiacomo, M.; Amadio, C.; Izzuddin, B.A. Long-term analysis of steel–concrete composite beams: FE modeling for effective width evaluation. *Eng. Struct.* **2006**, *28*, 1110–1121. [CrossRef]
3. Oehlers, D.J.; Bradford, M.A. *Composite Steel and Concrete Structures: Fundamental Behavior*; Elsevier: Oxford, UK, 2013.
4. Ryu, H.K.; Chang, S.P.; Kim, Y.J.; Kim, B.S. Crack control of a steel and concrete composite plate girder with prefabricated slabs under hogging moments. *Eng. Struct.* **2005**, *27*, 1613–1624. [CrossRef]
5. Xia, Y.; Wang, P.; Sun, L. Neutral axis position based health monitoring and condition assessment techniques for concrete box girder bridges. *Int. J. Struct. Stab. Dyn.* **2019**, *19*. [CrossRef]
6. Xia, Y.; Nassif, H.; Su, D. Early-age cracking in high performance concrete decks of typical Curved steel girder bridges. *J. Aerosp. Eng. (ASCE)* **2017**, *30*, B4016003. [CrossRef]
7. Miyamoto, A.; Tei, K.; Nakamura, H.; Bull, J. Behavior of prestressed beam strengthened with external tendons. *J. Struct. Eng.* **2000**, 1033–1044. [CrossRef]
8. Deng, Y.; Morcous, G. Efficient prestressed concrete-steel composite girder for medium-span bridges. I: System description and design. *J. Bridge Eng.* **2013**, 1347–1357. [CrossRef]
9. Deng, Y.; Morcous, G. Efficient prestressed concrete-steel composite girder for medium-span bridges. II: Finite-element analysis and experimental investigation. *J. Bridge Eng.* **2013**, 1358–1372. [CrossRef]
10. Wang, X.; Shi, J.; Wu, G.; Yang, L.; Wu, Z. Effectiveness of basalt FRP tendons for strengthening of RC beams through the external prestressing technique. *Eng. Struct.* **2015**, *101*, 34–44. [CrossRef]
11. Marí, A.; Mirambell, E.; Estrada, I. Effects of construction process and slab prestressing on the serviceability behaviour of composite bridges. *J. Constr. Steel Res.* **2003**, *59*, 135–163. [CrossRef]
12. Dezi, L.; Gara, F.; Leoni, G. Construction sequence modeling of continuous steel-concrete composite bridge decks. *Steel Compos. Struct.* **2006**, *6*, 123–138. [CrossRef]
13. Liu, X.; Liu, Y.; Luo, J.X.H. Behavior of Continuous Composite Bridge during Construction with Jacking up Interior Support, National Bridge Conference 2012, pp. 658–662, 2012. (In Chinese). China Highway and Transportation Society. Available online: http://caj.d.cnki.net//KDoc/docdown/pubdownload.aspx?dk=kdoc%3apdfdown%3aa31df52a5345b1bdc7ff1d792e853f28 (accessed on 12 November 2018).
14. Kwon, G.; Engelhardt, M.D.; Klingner, R.E. Behavior of post-installed shear connectors under static and fatigue loading. *J. Constr. Steel Res.* **2010**, *66*, 532–541. [CrossRef]
15. Hällmark, R.; Collin, P.; Möller, M. The behaviour of a prefabricated composite bridge with dry deck joints. *Struct. Eng. Int.* **2013**, *23*, 47–54. [CrossRef]
16. Su, Q.; Yang, G.; Bradford, M. Behavior of a continuous composite box girder with a prefabricated prestressed-concrete slab in its hogging-moment region. *J. Bridge Eng.* **2015**, B4014004. [CrossRef]
17. Tong, T.; Yu, X.; Su, Q. Coupled effects of concrete shrinkage, creep, and cracking on the performance of postconnected prestressed steel-concrete composite girders. *J. Bridge Eng.* **2017**. [CrossRef]
18. Ministry of Construction of the People's Republic of China. *Code for Design of Concrete Structures*; GB50010-2010; Ministry of Construction of the People's Republic of China: Beijing, China, 2015.
19. Ministry of Transport of the People's Republic of China. *Chinese Code for Design of Highway Reinforced Concrete and Pre-Stressed Concrete Bridge and Culverts Beijing*; JTG D62-2004; Ministry of Transport of the People's Republic of China: Beijing, China, 2004.
20. Ministry of Transport of the People's Republic of China. *Code for Design of Highway Reinforced Concrete and Prestressed Concrete Bridges and Culverts*; JTG D62-2012; Ministry of Transport of the People's Republic of China: Beijing, China, 2012.
21. *ANSYS Release 12.0 [Computer Software, version 12.0]*; ANSYS, Inc.: Canonsburg, PA, USA, 2009.

22. ANSYS, Inc. *ANSYS Manual*; ANSYS, Inc.: Canonsburg, PA, USA, 2009.
23. Pan, H.; Azimi, M.; Yan, F. Time-Frequency-Based Data-Driven Structural Diagnosis and Damage Detection for Cable-Stayed Bridges. *J. Bridge Eng.* **2018**, *23*, 04018033. [CrossRef]
24. Ministry of Construction of the People's Republic of China. *Code for Design of Steel Structures*; GB50017-2003; Ministry of Construction of the People's Republic of China: Beijing, China, 2003.
25. Ministry of Construction of the People's Republic of China. *Code for Design of Steel and Concrete Composite Bridges*; GB50917-2013; Ministry of Construction of the People's Republic of China: Beijing, China, 2013.
26. Oh, B.; Sause, R. Empirical Models for Confined Concrete under Uniaxial Loading. In *International Symposium on Confined Concrete*; ACI SP-238; American Concrete Institute: Farmington Hills, MI, USA, 2006; pp. 141–156.

applied
sciences

MDPI

Article
Grouting Process Simulation Based on 3D Fracture Network Considering Fluid–Structure Interaction

Yushan Zhu [1], Xiaoling Wang [1,*], Shaohui Deng [2], Wenlong Chen [1], Zuzhi Shi [1], Linli Xue [1] and Mingming Lv [1]

[1] State Key Laboratory of Hydraulic Engineering Simulation and Safety, Tianjin University, Tianjin 300072, China; zhuyushan3@tju.edu.cn (Y.Z.); chenwl@tju.edu.cn (W.C.); Zuzhi_Shi@163.com (Z.S.); xuelinli@tju.edu.cn (L.X.); lxm_02@tju.edu.cn (M.L.)
[2] Yalong River Hydropower Development Company, Ltd., Chengdu 610051, China; dengshaohui@tju.edu.cn
* Correspondence: wangxl@tju.edu.cn; Tel.: +86-022-2789-0911

Received: 29 January 2019; Accepted: 12 February 2019; Published: 15 February 2019

Abstract: Grouting has always been the main engineering measure of ground improvement and foundation remediation of hydraulic structures. Due to complex geological conditions and the interactions between the grout and the fractured rock mass, which poses a serious challenge to the grouting diffusion mechanism analysis, fracture grouting has been a research hotspot for a long time. In order to throw light on the grout diffusion process in the fractured rock mass and the influence of grout on the fracture network, and to achieve more realistic grouting numerical simulation, in this paper a grouting process simulation approach considering fluid–structure interaction is developed based on the 3D fractured network model. Firstly, the relationship between fracture apertures and trace lengths is used to obtain a more realistic value of fracture aperture; then a more reliable model is established; subsequently, based on the 3D fracture network model, different numerical models are established to calculate fluid dynamics (grout) and structure deformation (fractured rock mass), and the results are exchanged at the fluid–structure interface to realize the grouting process simulation using two-way fluid-structure interaction method. Finally, the approach is applied to analyze the grouting performance of a hydropower station X, and the results show that the grouting simulation considering fluid–structure interaction are more realistic and can simultaneously reveal the diffusion of grout and the deformation of fracture, which indicates that it is necessary to consider the effect of fluid–structure interaction in grouting simulation. The results can provide more valuable information for grouting construction.

Keywords: grouting; fracture network modeling; numerical simulation; fluid–structure interaction

1. Introduction

Many high dams are built in areas with complicated geological conditions, and the numerous fractures and voids will increase the permeability and decrease the strength of the rock mass, which affect the stability and safety of the dam's foundations. As a common and effective measure to improve the geological conditions of dam foundations, grouting is used to fill up the joints and fractures in the rock mass so as to prevent seepage and improve the bearing capacity and deformation resistance [1–4]. However, fracture grouting is still a difficult issue due to complicated fracture distribution, complex fluid–structure interaction effects, and incomplete information on grout diffusion behavior and corresponding fracture deformation. In order to reveal the grouting mechanisms in the fractured rock mass, it is necessary to select an effective tool for studying the grouting process, especially for considering the fluid–structure interaction.

Computational fluid dynamics (CFD) is an effective tool used to simulate fracture grouting, which can partly overcome some limitations of experiments. In recent years, various researchers have carried out an abundance of research. Saeidi et al. [5] established a numerical model to study the effect of fracture properties on grout flow and penetration length in fractured rock mass using Universal Distinct Element Code (UDEC). Yang et al. [6] simulated the cement grout diffusion process in a single rough fracture by the finite element method. Fu et al. [7] performed numerical simulation on the diffusion process of cement grouting in the fractures of the rock mass to determine reasonable hole spacing and other parameters. Hao et al. [8] developed a numerical simulation of polymer grout diffusion in a single fracture to analyze the pressure distribution. Deng et al. [4] proposed a CFD simulation approach based on 3D fracture network model to study the grouting process of a dam's foundation. Kim et al. [9] used UDEC to simulate the flow of Bingham grout in a single joint with smooth parallel surfaces and considered the hydromechanical coupling to study its effect on grouting performance. Ao et al. [10] simulated the grouting process in underground goaf and analyzed the stability by applying one-way fluid–structure interaction. Liu et al. [11] combined a finite-discrete element method (FDEM) and a grouting flow simulator to consider the hydromechanical coupling effect in the parallel-plate model. The review of previous studies reveals that most of the aforementioned approaches simulated the grouting process in a single fracture, 2D fracture network or simplified rock mass, which cannot reflect the actual diffusion of grout flow in complex fractured rock mass completely. Furthermore, these studies only studied the diffusion of grout in fractures or the effect of grout on the rock mass, without taking the complex fluid–structure interaction between grout flow and rock mass into account, which were different from the actual conditions.

In order to obtain a more realistic simulation of grouting process, the establishment of a precise and reliable three-dimensional fracture network model is an important prerequisite [4]. Since deterministic data of the fracture aperture are not available, in current studies of fracture network modeling and its application the fracture aperture is usually ignored [12–14], reduced to a given value [15], or randomly generated from a given range or geologically conditioned statistical distributions [4,16]. The fracture aperture significant influence the permeability of the fractured rock masses, so it is necessary to establish a fracture network model with a more authentic and accurate fracture aperture to simulate the grouting process. For fractured rock mass, fractures are usually random and complex which are fractal structure with self-similarity; according to this characteristic of a fracture, some scholars have put forward the formula for the relationship between fracture apertures and trace lengths, and this relationship has been widely investigated [17–20]. In this study, this relationship will be introduced into the fracture network modeling to make up for the deficiency of the existing research.

As the pressurized grout penetrates the fractures inside the rock mass, the grout will separate the fracture surfaces from each other, causing an interaction between the grout and the rock mass [21]. Tsang et al. [22] indicated the coupling of processes implies that the both interact in the initiation and progress of each other. On the one hand, the grout pressure induces stresses on the surfaces of the fracture, this will lead to the deformation of the fractures. On the other hand, the change of the fracture aperture will affect the fracture permeability, and then results in the variation of grout performance. So the grouting performance cannot be determined by considering each process independently. Some theoretical studies on the interaction between grout and rock mass have been carried out. GothäLl and Stille [21] analyzed the interaction of two parallel fracture during high pressure grouting and discussed the effect of fracture dilation on the penetrability of fine fractures. Rafi and Stille [23] proposed a procedure for optimizing grouting pressure based on the estimation of grout spread and the identification of jacking of the fracture. Rafi and Stille [24] described the basic mechanism of elastic deformation during grouting and discussed its impact on the spread of grout. In literature [9], the authors strongly recommended that the interaction between the grout flow and fractured rock mass should be included in the grouting analysis in order to have a precise prediction of grout performance. With this in mind, the fluid–structure interaction (FSI) method as a numerical technique is used to solve problems that involve the mutual interaction of fluid and structure. In recent years, with the development of computer performance and increasing interest in more realistic modeling, FSI has attracted extensive attention in the computational field [25]. For instance, this technique has been applied in the problem of internal slip during the operation of progressive cavity pump (PCP)s in oilfield production [26], biomedical problems where blood flow interacts with blood vessel walls [27], and the fluid–structure interaction problem of fracturing structures under impulsive loads [28]. Of interest from a fractured grouting perspective, there is still much work to do in the grouting simulation considering fluid–structure interaction to capture the interaction between grout flow characteristics and deformation of the fractures.

In summary, most of the existing studies simulated the grouting process in a single fracture, 2D fracture network or simplified rock mass, which is inconsistent with actual fractures under complex geological conditions. Moreover, the value of fracture aperture is usually ignored or inaccurate in the fracture network modeling, which will affect the authenticity of grouting simulation. Additionally, rich theoretical research achievements have proposed on the interaction between grout and fracture, and some of the grouting simulations considering fluid-structure interaction are based on single-fracture or just using one-way fluid–solid coupling. Therefore, based on the 3D fractured network model the numerical simulation of the grouting process considering two-way fluid-structure interaction still needs further study.

In this study, a grouting simulation approach considering fluid–structure interaction is developed based on the 3D fractured network model. Firstly, fracture parameters are randomly simulated by the Latin hypercube sampling (LHS) method based on the statistical information from fracture survey and borehole imaging of the exposed surface, the relationship between fracture apertures and trace lengths is used to obtain the value of fracture aperture, then a more reliable 3D fracture network model for dam foundation rock mass is established with VisualGeo software [29]. Next, the CFD simulation model of grout (fluid) and the finite element model of fractured rock mass (structure) are established respectively, and their governing equations are solved in different ways, with the results exchanged through the fluid–structure interface to realize the two-way fluid-structure interaction simulation of the grouting process. Finally, the approach is used in a case study to analyze the dam foundation grouting to investigate the effects of fluid-structure interaction on grouting processes; the results show that the grouting simulation considering fluid–structure interaction are more realistic and can simultaneously reveal the grout diffusion and fracture deformation under the interaction of grout and rock mass, which can provide more valuable information for optimizing the grouting process.

The remaining parts of this paper are organized as follows: the methodology of 3D fracture network modeling and fluid–structure interaction simulation are introduced in Section 2. In Section 3, the approach is applied to analysis of the grouting performance of hydropower station X and the studies on the grouting characteristics are given in this section. Finally, the conclusions are provided in Section 4.

2. Methodology

2.1. Modeling of 3D Fracture Network

2.1.1. Modeling Process of 3D Fracture Network

Due to the large amount of complex fractures in the rock mass of a dam's foundations, it is difficult to determine the exact position and occurrence of each fracture by using a deterministic model. A large number of engineering practices and studies have shown that the fractures have obvious statistical distribution rules and characteristics, so we established the 3D fracture network model which is close to the real fracture conditions in a statistical sense.

The modeling process (Figure 1) mainly includes the following steps: (1) the statistical homogeneous zone is divided firstly, then the fractures in the statistical homogeneous zone are divided into dominant sets and the cracks with similar properties are clustered; (2) the fracture space density and the distributions of the geometry parameters could be obtained based on the statistical analysis; (3) the fracturing parameters are randomly simulated by the LHS method; and (4) a 3D fracture network model is constructed in VisualGeo software.

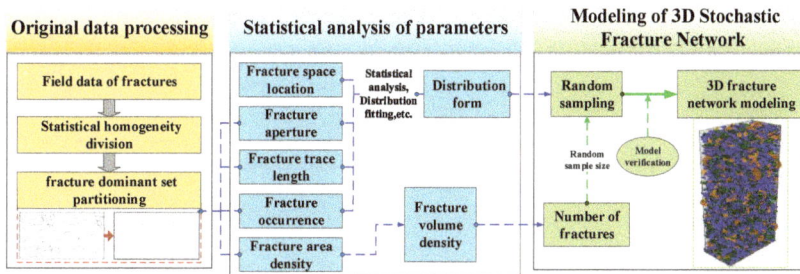

Figure 1. Modeling process of 3D fracture network.

2.1.2. Statistical Analysis of Fracture Geometric Characteristic Parameters

In this study, the fracture was simulated by the Baecher disc model [30] which assumed every fracture as a thin disc (Figure 2). The fracture disc model can be defined by the following formula:

$$C = c(O, V, R, A) \tag{1}$$

the formula defines a fracture disc with center point O, occurrence V, radius R and aperture A. Where $O = (x_0, y_0, z_0)$, $V = (\alpha, \beta)$, α and β are the dip direction and dip angle of the fracture disc respectively, and n is the normal vector of the fracture disc.

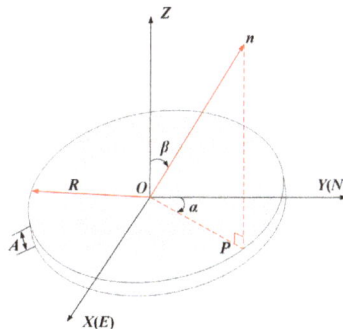

Figure 2. Baecher disc model.

The relationship between the various parameters can be expressed as follows:

$$\begin{cases} a(x-x_0)+b(y-y_0)+c(z-z_0)= 0 \\ (x-x_0)^2 + (y-y_0)^2 + (z-z_0)^2 \le R^2 \\ a = sin\beta sin\alpha, \ b = sin\beta cos\alpha, \ c = cos\beta \\ A = f(R) \end{cases} \tag{2}$$

According to the exposed surface fracture catalog data and digital borehole data, the distributions of fracture geometric characteristic parameters can be determined.

(1) Fracture space location

The Poisson process [31] is widely used to describe fracture location. The fractures are mutually independent and the uniform distribution function are adopted to obtain the coordinates (x_0, y_0, z_0) of the fracture center point.

(2) Fracture density

The Mauldon method [32] is adopted to estimate the fracture volume density. The following equation is used to estimate the trace area density:

$$\lambda_a = \frac{n_1 + 2n_2}{2WH} \tag{3}$$

where λ_a is the trace area density, n_1 is the number of traces which one end can be observed, n_2 is the number of traces which both end can be observed, W is the width of rectangular window, H is the height of rectangular window. Then, Equation (4) is adopted to obtain the fracture volume density:

$$E(\lambda_v) = \frac{E(\lambda_a)}{E(D)E|sinv|} \tag{4}$$

where λ_v is the fracture volume density, D is the fracture diameter, $sinv$ is the sine value of the dip.

(3) Fracture size

To simulate the size of the fracture surface, statistical analysis of the fracture trace length is needed first. Huang et al. [33] put forward the estimation formula of trace length:

$$l = \frac{n_1 + 2n_0}{2N} \frac{\pi WH}{W + H} \tag{5}$$

where l is the fracture trace length in the window, n_0 is the number of traces which neither end can be observed, n_1 is the number of traces which one end can be observed, N is the total number of fracture traces in the window, W is the width of rectangular window, H is the height of rectangular window. When the disc model is used to simulate the fracture, the fracture size is expressed by its diameter. the fracture diameter distribution can be confirmed based on the distribution of trace length.

(4) Fracture occurrence

According to Kemeny and Post [34], the fisher distribution can be used to fit fracture occurrence and obtained relatively better results.

(5) Fracture aperture

Schultz et al. [17,18] conducted a lot of statistical studies and obtained the expression of the relation between fracture aperture and fracture trace length:

$$A = \beta l^n \tag{6}$$

where A is the fracture aperture, l is the fracture trace length, β and n are constants related to the properties of fractured rock mass.

In this study, the value of $n = 1$ is chosen to reflect the self-similarity and fractal of the fracture network [35–37], and then the relationship between fracture aperture and fracture trace length is linear; based on this relationship, the existing survey data are linearly fitted to get the value of γ. Finally, the fracture aperture can be obtained by Equation (6) based on the data of trace length.

2.1.3. Latin Hypercube Sampling (LHS) Random Sampling

After obtaining the determined probability distribution model of each fracture parameter, random sampling of parameters is needed. The essence of the LHS method is to divide the sampling interval according to the sampling times, and then random sampling is carried out from each subinterval. This method avoids the collapse of the sample data and the simulation results are more stable. Therefore, the LHS method is adopted to randomly simulate the fracture parameters in this study.

Then, taking the center point coordinates, diameters, aperture, dip direction and dip angle of fractures simulated by LHS method as input parameters, a three-dimensional model of rock mass fracture network is established by using VisualGeo software.

2.2. Fluid–Structure Interaction Model

The fluid–structure interaction model mainly consists of two parts: computational fluid dynamics (CFD) and computational structure dynamics (CSD). The grouting process is simulated by CFD and the deformation of fractures is calculated by CSD.

The solution of fluid-structure interaction includes a directly coupled solution and partitioned solution. The first one solves the governing equations of fluid and structure simultaneously in the same solver by coupling the governing equations of fluid and structure to the same equation matrix, so its advantage is that there is no time lag problem. However, a direct coupling solution may result in poor convergence and huge computational cost. As a consequence, it is difficult to realize in fact [38]. On the contrary, the second one solves the fluid governing equations and the structure governing equations in different solvers, and the results are exchanged and transmitted through the fluid-structure interface. In this study, we choose the partitioned solution to solve the fluid–structure interaction between grout and fractures.

2.2.1. Computational Fluid Dynamics (CFD) Grouting Numerical Model

The governing equations of the grouting can be described by the continuity equation, momentum equation, two-phase volume of fluid (VOF) equation and Papanastasiou regularized equation.

(1) The two-phase VOF equation

In the process of grouting, the grout drives out air or groundwater, which should be treated as a two-phase flow [4]. The accurate description of the interface between two kinds of incompatible and incompressible fluids is one of the most important issues in multi-fluid flow computations [39], this can be solved by the VOF method which is proposed by Hirt and Nichols [40] to track free fluid surfaces under fixed grid condition. Therefore, the VOF method is used to keep track of the grout-air interface in this paper. In this method, a volume fractional variable $F = F(x, y, z, t)$ for each phase of the model in the computational domain is introduced. $F_g = 1$ indicates that the volume is occupied by grout while $F_g = 0$ indicates that the volume contains no grout and is in the air phase, and $0 < F_g < 1$ stands for the volume that contains both grout and air. Equation (7) is used to describe the motion of the grout-air interface:

$$\frac{\partial F_g}{\partial t} + \rho v \nabla F_g = 0 \tag{7}$$

where F_g is the volume fraction of grout; ρ is the density of fluid in kg/m^3; v is the kinematic viscosity of fluid in m^2/s.

(2) The continuity equation:

$$\frac{\partial \rho}{\partial t} + \nabla(\rho u) = 0 \tag{8}$$

where ρ is the density of fluid in kg/m^3; t is the time in s; u is the velocity of the unit section in m/s.

(3) The momentum equation:

$$\rho \frac{du}{dt} = -\nabla p + \rho g + \nabla(\eta \dot{\gamma}) + F_{st} + S \tag{9}$$

where p is the pressure on the fluid micro-unit in *Pa*, g is the acceleration of gravity in m/s^2; η is the apparent viscosity of fluid in Pa·s; the relationship between v and η is $v = \eta/\rho$; $\dot{\gamma}$ is the shear rate in 1/s, F_{st} is the surface tension force in N/m^3 and is presented as Equation (10) [41]; and S is the momentum resistance source term in N/m^3, including inertia loss term S_i and viscosity loss term Sv; in this paper, S_i can be neglected because of the low velocity of the grout and $S = S_v$. Equation (11) is the expression of the viscosity loss term S_v:

$$F_{st} = -\sigma\nabla\cdot\left(\frac{\nabla F_i}{|\nabla F_i|}\right)\nabla F_i \tag{10}$$

where F_i is the volume fraction of phases; σ is the surface tension coefficient in N/m.

$$S_v = -\frac{\rho v}{\alpha}u \tag{11}$$

where $\frac{1}{\alpha}$ is the viscous drag coefficient and its expression is as follows:

$$\frac{1}{\alpha} = \frac{g}{Kv} \tag{12}$$

where K is the permeability coefficient. In order to obtain single set of equations, ρ and v in Equations (7)–(12) are no longer constants but are variables weighted by the volume fraction of fluid [42]:

$$\rho = F_g\rho_g + (1-F_g)\rho_a \tag{13}$$

$$v = F_g v_g + (1-F_g)v_a \tag{14}$$

where ρ_g, ρ_a, v_g, v_a are the density of grout, the density of air, the kinematic viscosity of grout and the kinematic viscosity of air, respectively.

(4) The Papanastasiou regularized equation

The cement grout with a w/c ratio of less than 1 is usually described by the Bingham model. However, in the Bingham constitutive equation, when the shear rate is close to zero, the apparent viscosity will become infinite, which causes problems in numerical simulation. In order to solve this problem, the Papanastasiou regularized model is used to describe the rheological properties of cement grout [4], as shown in Equation (15):

$$\eta = \begin{cases} m\tau_0 & \dot{\gamma} = 0 \\ \eta_0 + \frac{\tau_0}{\dot{\gamma}}\left[1-e^{-m\dot{\gamma}}\right] & \dot{\gamma} \neq 0 \end{cases} \tag{15}$$

where η is the apparent viscosity of fluid in Pa·s; m is the stress growth parameter in s; τ_0 is the yield stress in Pa; $\dot{\gamma}$ is the shear rate in 1/s; η_0 represents the plastic viscosity in Pa·s; and e is a natural constant. From our previous research [4], it is considered that $m = 100$ can meet the research needs and can make the numerical model effectively express the rheological properties of cement grout.

2.2.2. Computational Structure Dynamics (CSD) Model

The rock mass is elastoplastic materials, since the objective has been to establish a model applicable for grouting problems, the Mohr–Coulomb (M-C) shear yield criterion which is comparable to actual rock was applied to the rock mass. The expressions of the criterion are shown as follows:

$$\tau_n = C + \sigma_n tan\phi \tag{16}$$

if $\tau < \tau_n$, the rock mass is linear elastic material, if $\tau \geq \tau_n$, the rock mass become yield.

The linear elastic model is based on the generalized Hooke law, and the constitutive equation is as follows:

$$\{\varepsilon\} = [D]\{\sigma\} \tag{17}$$

where ε is the strain, D is the elastic matrix, and σ is the stress component.

In the stress space, the form of the yield function is as follows:

$$f = \frac{1}{2}(\sigma_1 - \sigma_3) - \frac{1}{2}(\sigma_1 + \sigma_3)\sin\phi - C\cos\phi \tag{18}$$

where σ_1 and σ_2 are the maximum and minimum principal stresses of material damage, respectively; C is the cohesion; and φ is the internal friction angle. When $f \geq 0$, shear failure will occur in the material.

2.2.3. Fluid–Structure Interaction Analysis Solution

In this paper, Finite Element Analysis (FEA) software solves the structural domain (fractured rock mass) and CFD software solves the fluid domain (grout). The two domains are interconnected on the fluid–structure interface using the SIMULIA Co-Simulation Engine (CSE) [41] which is widely used to couple CFD software and FEA software for fluid–structure interaction simulation [43–45].

As show in Figure 3, in the fluid-structure interaction solving process, CFD software initializes the fluid field and passes loads to the FEA software (pressure + wall shear stress), and the deformations in the structure can be computed by FEA software and passes displacements back to CFD software; this can provide a new deformed geometry for the CFD software to solve the fluid field. This iteration can be repeated until the end of the coupling process.

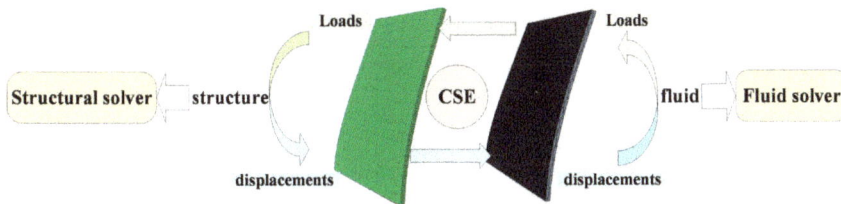

Figure 3. The principle of fluid–structure interaction.

On the fluid–structure interface, the stress (τ) and the displacement (d) of the fluid and structure should be equal:

$$\tau_g n_g = \tau_f n_f \tag{19}$$

$$d_g = d_f \tag{20}$$

where the subscript g and f represent grout and fractured rock mass.

2.2.4. Boundary Conditions

(1) Inlet boundary conditions: according to the data of grouting pressure measured by grouting recorder and taking the mean value of grouting pressure during grouting period, pressure inlet is set at the boundary of grouting borehole interval. The corresponding grout VOF at the inlet is set to 1.

(2) Outlet boundary conditions: the pressure outlet is set at the end boundary of the fracture, and the pressure satisfies the second boundary condition.

(3) Initial conditions: assuming that there is no groundwater during grouting, the fractures are filled with air before grouting, and the initial air VOF in the fracture is set to 1.

(4) Displacement boundary conditions: the bottom boundary of the computational domain is the z-axis constraint, the lateral boundaries are the x- and y-axis constraints.

3. Case Study

Hydropower station X is located in the upper reaches of Lancang River. It is a large-scale hydropower project which mainly generates electricity and takes into account the comprehensive utilization benefits of irrigation and water supply. The hydropower project is mainly composed of a gravelly soil core rockfill dam, left bank slope spillway, water diversion and power generation system, ground workshop, and so on. The installed capacity of the hydropower station is 1400 MW, the maximum dam height is 139.80 m, and total dam crest length is 576.68 m. The dam area exposed strata are mainly from the Middle Jurassic flower group (J2h) and Quaternary strata (Q). The layout of the dam's foundation grouting curtain and the project profile are shown in Figure 4. The dam foundation curtain grouting project is divided into several continuous grouting units. In this study, a typical grouting unit is taken as a case study to simulate a three-dimensional random fracture network. The location of the study area is shown in the red wireframe in Figure 4b. This area is the foundation curtain grouting unit of the river bed dam section and is also the main area of the dam foundation seepage control of this project. In the study area, there are 41 grouting boreholes including 21 in the upstream row and 20 in the downstream row with the hole spacing of 1.5 m, the diameter of the grouting borehole is 75 mm, and the total depth is 60 m.

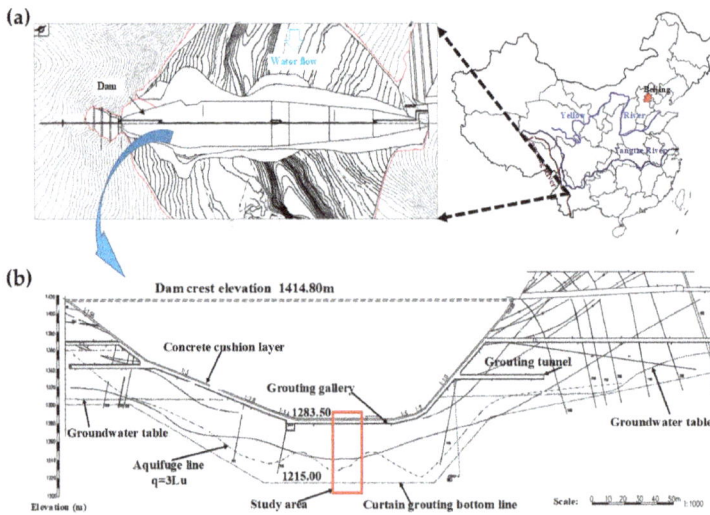

Figure 4. The project profile: (**a**) the layout of project and (**b**) the study area.

3.1. Simulation of 3D Fracture Network

According to the size of the exposed surface and the depth of the grouting hole, the study area is 30 m × 16 m × 60 m (length × width × depth). In the study area, a total of 83 fractures were recorded on the exposure surface, the fracture sketch picture and the fracture dominant sets diagram are shown in Figure 5. The fractures were divided into 3 sets based on the occurrence.

Figure 5. The fracture sketch picture and the fracture dominant sets diagram of the fracture network in the exposed surface: (**a**) Fracture sketch picture and (**b**) Fracture dominant sets diagram.

The distribution of fracture parameters in each set is fitted according to mathematical statistics, the results shows that the trace lengths of the three sets of fractures obey the logarithmic normal distribution and the occurrences obey the Fisher distribution. The statistical results of the fracture parameters of the dam foundation are shown in Table 1 (Taking first set fractures as an example).

Table 1. Statistical results of fracture parameters.

Set	Fracture Number	Regional Volume (m³)	Parameter	Mean	Minimum	Maximum	Distribution
1	2587	28,800	Coordinate X/m	15	0	30	
			Coordinate Y/m	8	0	16	Uniform
			Coordinate Z/m	30	0	60	
			Diameter/m	1.93	0.31	2.89	Lognormal
			Dip direction/degree	345.00	340.00	349.98	
			Dip angle/degree	15	10.00	19.98	Fisher

According to the statistical results of fracture parameters of the dam foundation rock mass, the fracturing parameters are randomly simulated 10 times by the LHS method and the optimal results were obtained. Then, the 3D fracture network models are constructed in VisualGeo software. The effect drawing of the final 3D fracture network model for the study area is shown in Figure 6.

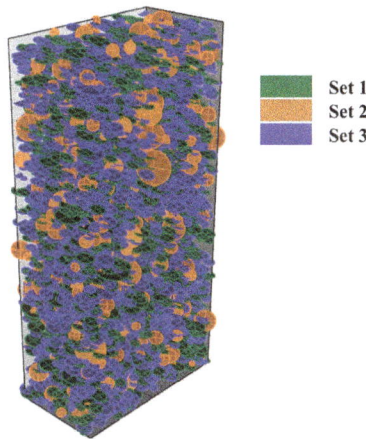

Figure 6. Final 3D fracture network model.

3.2. Grouting Simulation Considering Fluid–Structure Interaction

In this study, the second stage (3–6 m) of the grouting borehole 3-LR1-33 is selected from the established 3D fracture network model as the simulation object, the selected model is a cylindrical area centered on the grouting borehole with the radius of 1.5 m, which contains 20 fractures, including 7 fractures in the first set, 4 fractures in the second set and 9 fractures in the third set (Figure 7). There are 9 independent fractures which are not connected to the existing hydraulic fracture network and removed from the model. Hence, as shown in Figure 8a, the fracture model for fluid computation contains 11 fractures, of which 5 fractures are the primary fractures intersecting with the grouting borehole and the parameters of each fractures are arranged in Table 2. Figure 8b shows the fracture grid model, and the number of cell meshes is 2,319,281. Figure 8c is the rock mass geometrical model and Figure 8d is its grid model with 34,346 elements. The conditions of the simulated calculation are as follows: the water–cement ratio of cement grout is 1:1, the grouting pressure is 1.0 MPa, and the permeability rate is 7.23 Lu. and the physical mechanical parameters of rock masses is shown in Table 3.

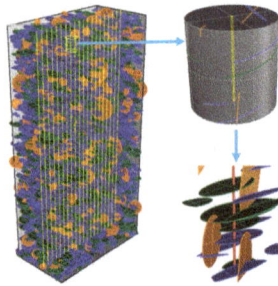

Figure 7. Selected grouting simulation area.

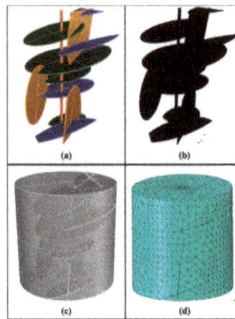

Figure 8. Fluid calculation model: (**a**) fracture geometrical model, (**b**) fracture grid model, and structural calculation model: (**c**) rock mass geometrical model, (**d**) rock mass grid model.

Table 2. Geometric parameters of the computation fracture mode.

Fracture	Coordinate/m			Radius/m	Dip Direction/Deg	Dip Angle/Deg	Aperture/m
	X	Y	Z				
1-1724	19.874	9.182	56.403	0.835	344.432	18.027	0.0089
1-1647	**19.053**	**9.902**	**55.445**	**1.213**	**347.699**	**13.503**	**0.0129**
1-872	19.157	8.701	55.255	0.726	347.079	14.107	0.0077
1-952	17.818	10.036	52.685	0.543	349.080	15.092	0.0058
2-220	**20.193**	**7.555**	**57.400**	**1.246**	**41.687**	**80.996**	**0.0132**
2-562	18.444	14.515	54.789	0.755	41.635	86.409	0.0080
2-121	17.862	13.714	51.170	1.162	39.079	85.291	0.0124
3-1755	18.708	7.543	56.899	1.150	9.167	14.944	0.0122
3-2007	16.190	8.332	56.259	1.099	13.901	11.962	0.0117
3-2881	**19.736**	**7.530**	**55.015**	**0.784**	**7.736**	**10.360**	**0.0083**
3-699	19.515	9.665	54.138	1.035	9.766	19.581	0.0110

[1] The primary fractures are marked in bold in the table.

Table 3. Physical mechanical parameters of rock masses.

Medium	Density (kg/m^3)	Elastic Modulus (MPa)	Poisson Ratio μ	Cohesion C (MPa)	Internal Friction Angle φ (°)
Slate	2690	4000	0.25	9.5	37.7

3.2.1. Analysis of Fluid Calculation

1. Analysis of Grout Diffusion Process

Driven by grouting pressure, the injected grout penetrates into the primary fractures intersecting with grouting borehole and then migrates into the fracture network. As shown in Figure 9, the grout diffusion length increases gradually over time. After 60 s, most of the primary fractures have been filled with grout except primary fracture 1-1647; this is because the radius of fracture 1-1647 is larger and there are two secondary fractures intersecting with it, which hinders the further diffusion of grout in it. After grouting, primary fractures are completely filled with grout, secondary fractures are mostly filled, and other fractures are filled with a little grout.

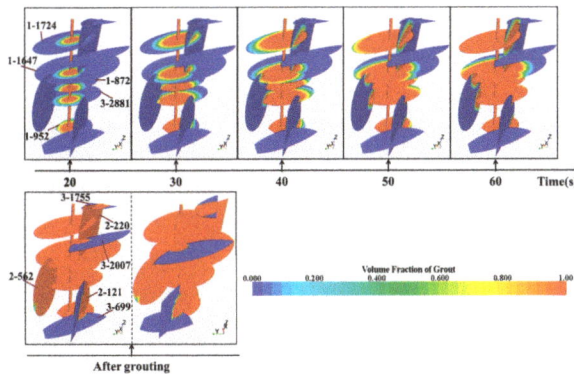

Figure 9. Grouting diffusion process.

For details, five primary fractures (1-1724, 1-1647, 1-872, 3-2881, 1-952) intersecting directly with the grouting borehole are the main passages of the grout, and then the grout diffuses from the primary fracture to the secondary fractures. From Figure 9, it can be seen that there are secondary fractures intersecting with primary fractures, and the grout diffusion at the intersection presents a non-uniform spread shape. Thus, the filling of the primary fractures is influenced by the number of intersecting secondary fractures. As shown in Figure 9, the filling rate of secondary fracture 2-220 is 100%, this is due to the aperture of fracture 2-220 being 13.2 mm, which is bigger than other fractures, and it intersects with two primary fractures, of which the intersecting length with primary fractures 1-1647 is relatively large, reaching 1.6447 m. Secondary fracture 2-562 is basically filled with grout because it intersects with three primary fractures (1-1647, 1-872, 3-2881), but the length of intersection is short, and its aperture is 0.8 mm which is relatively small resulting in incomplete grout filling. Secondary fracture 2-121 is not fully filled with grout due to the hindrance of fracture 3-699 and the small intersections with fracture 3-2881 and 1-952. In summary, the filling of the secondary fractures is related to the fracture aperture and the intersection length between secondary facture and primary fracture. For other fractures (3-1755, 3-699), the grout comes from secondary fractures and they are far from the grouting borehole, so the filling rate is low.

In the actual project, the grout borehole spacing is 1.5 m, and thus the fractures that are not fully filled will be grouted by adjacent grouting boreholes. In addition, the real cement consumption is 244.74 kg, and the simulated cement consumption is 278.41 kg with an error of 13.76% compared to the real cement consumption. Therefore, considering the complexity of geological conditions, there is a good agreement between the simulated and actual values, and this approach can be used effectively to determine the grouting effect.

The maximum radial diffusion length variation of the primary fractures is analyzed and compared in Figure 10. At the beginning of grouting, there is no significant difference in the variation of grout diffusion length with time in each fracture, after 10 s, the difference of diffusion rate in each fracture gradually appeared. The diffusion rate in fracture 1-952 is the lowest because its aperture is the smallest with 5.8 mm and the diffusion rate in fracture 1-1647 is the fastest because its aperture is the biggest with 12.9 mm. As for the other three fractures (1-1724, 1-872, 3-2881), they have similar fracture apertures, but their grout diffusion rate is different, this is due to their different dip angles, and the grout diffusion rate increases with the dip angle. In summary, fracture aperture and dip angle will affect grout diffusion rate and the grout diffusion rate is proportional to them. Besides, fracture 1-1647 with the biggest aperture but smaller dip angle, and the maximum grout diffusion rate indicates that the fracture aperture is the main effect factor.

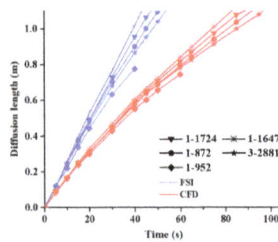

Figure 10. Variation of grouting diffusion length with time.

2. Comparison Analysis

In order to further analyze the simulation results, the effects of fluid–structure interaction in the grouting process are investigated through a comparison with the conventional CFD simulation neglecting these effects.

It can be seen from Figure 10 that the variation trend of the grout diffusion length in each fracture obtained by conventional CFD simulation is same as that obtained by fluid–structure interaction simulation. However, at the same grouting time, the fluid–structure interaction simulation results display a larger grouting diffusion length than the conventional CFD simulation results, that is, the grouting diffusion rate is faster. This can be explained thus: when considering the fluid–structure interaction effect, the fracture deformation occurs due to the grout pressure induces stresses on the surface of the fracture during the grouting process, which will improve the grout penetrability because the grout will diffuse more easily in a fracture with larger aperture. Therefore, ignoring the fluid–structure interaction effects in the simulation of grouting will lead to an underestimation of the grout diffusion ability.

3.2.2. Analysis of Structural Calculation

In this section, the fractured rock mass in the study area is analyzed to explore the effects of fluid–structure interaction on the rock mass. The maximum and minimum principal stresses of rock mass are shown in Figure 11, which indicates that during the grouting process, most of the rock mass is subjected to compressive stress and the maximum compressive stress is 4.936 MPa. The area near the grouting borehole, the local boundary of the rock mass and fractures are partially subjected to tensile stress, and the maximum tensile stress is 3.361 MPa. The maximum tensile stress and the maximum compressive stress of the rock mass during grouting are less than the tensile strength (45.56 MPa) and the compressive strength (6.32 MPa) of the rock mass. Therefore, the rock mass of the grouting area will not generate new fractures or lifting deformation.

In order to understand the influence of grout on the fracture aperture during grouting process, the aperture variation of fracture 1-1724 is analyzed at 50 s and 100 s. As shown in Figure 12, the maximum aperture increment happens at the intersection of grouting holes and fracture, and when at 50 s, the minimum aperture increment occurs at about 0.6 m where the grout front diffuses to. Then at 100 s, the grout diffusion length increases, the minimum aperture increment occurs at about 1.0 m, and the increment of the aperture along the grout diffusion distance is larger than that at 50 s. In summary, the aperture increment decreases non-linearly along the grout diffusion length and as the grouting time increases, the increment of aperture increases and the fracture deformation range expands.

Appl. Sci. **2019**, *9*, 667

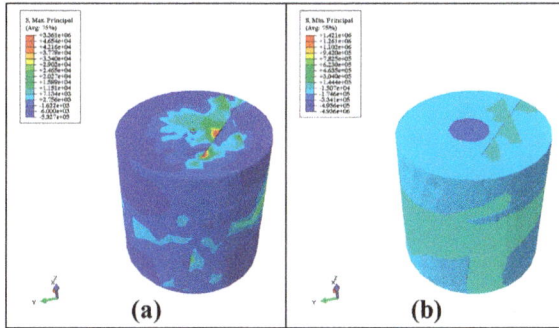

Figure 11. The principal stress contours. (**a**) The first principal stress contours, (**b**) the third principal stress contours.

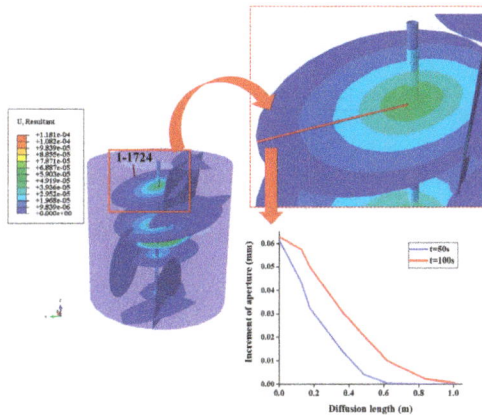

Figure 12. Fracture aperture variation along the grout diffusion length.

Figure 13 shows the displacement vectorgraph of fractured rock mass, it can be seen that the aperture variation trend of primary fractures is approximately the same, for each fracture, the largest displacement occurs at the intersection of the fracture and grouting borehole, and the smallest at the end of the fracture. For secondary fractures, the displacement happens at the intersection of primary and secondary fractures. As shown in Figure 13, the maximum displacement of the whole area occurs at the lower fracture surface of fracture 1-1647, i.e., the region ②. This is because that the distance between fracture 1-1647 and fracture 1-872 is 0.43 m which is relatively close, fracture 2-220 intersects with fracture 1-1647 and inserts into the rock mass between fracture 1-1647 and 1-872, so the rock mass in this region is the most vulnerable and prone to deformation. Similarly, the region ① is also the case where the fracture 2-220 intersects the fracture 1-1724, so that the rock mass in the region ① becomes less rigid, but the fracture 1-1724 is farther from the surrounding fractures, so the deformation of the region ① is smaller than the region ②. For fracture 1-872, we can see that the displacement of its upper surface is small. This is because the grouting process is conducted from top to bottom, so the fracture 1-1647 is grouted and deformation occurred before fracture 1-872, which offsets most deformation of the upper surface of the fracture 1-872. As for region ③, the distance between the fracture 3-2881 and the fracture 1-872 is small with 0.293 m, but the rock mass is still relatively rigid due to only a small portion of fracture 2-562 being inserted in the rock mass of this region, so that region 3 does not have large displacement. Overall, the maximum displacement of the entire area is 0.1181 mm, which is very small, so there will be no generation of new fractures and other adverse conditions in this area under the action of grouting.

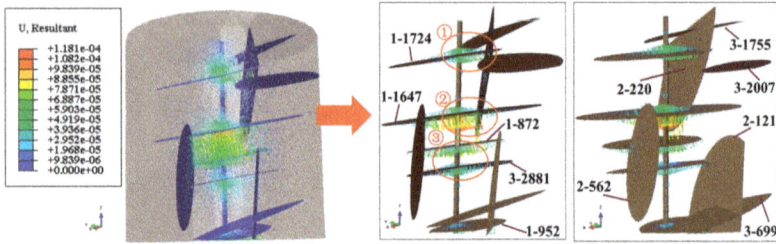

Figure 13. Vectorgraph of fracture displacement.

In summary, the effects of fluid–structure interaction between the grout and the rock mass will affect the grout diffusion process and the rock mass deformation. Therefore, grouting process simulation considering the fluid–structure interaction can better reproduce the grout diffusion and rock deformation process and explore the grouting mechanism under real conditions.

3.3. Parameter Analysis

The parameters affecting the grouting process include grouting pressure, water–cement ratio and elastic modulus of rock mass. In order to discuss the influence of different parameters on the grouting process, the diffusion length of fracture 1-1724 at 20 s is analyzed at different grouting pressures (0.5 MPa, 1.0 MPa, 1.5 MPa) and different water-cement ratio grouts (0.7, 1.0, 2.0), and the maximum displacement of the rock mass is analyzed at different grouting pressure (0.5 MPa, 1.0 MPa, 1.5 MPa) and different elastic modulus (4 GPa, 10 GPa, 20 GPa, 40 GPa). The results are shown in Figures 14 and 15, respectively.

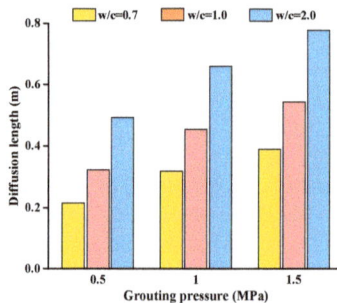

Figure 14. The grout diffusion length at different grouting pressures and different water–cement ratios (20 s).

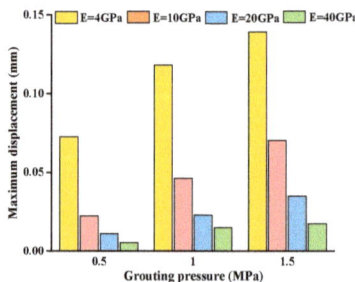

Figure 15. The maximum displacement of rock mass at different grouting pressures and different elastic modulus.

As shown in Figure 14, under the same grouting pressure, the grout diffusion length increases with the water–cement ratio due to the larger the water-cement ratio, the smaller the density, and the yield stress and plastic viscosity of the grout increasing with the decrease of density, which results in a faster grout flow rate. In addition, in the case of the same grout water–cement ratio, the grout diffusion length increases as the grouting pressure increases. Furthermore, as shown in Figure 15, under the same grouting pressure, with the increase in elastic module, the maximum displacement of the rock mass decreases, and with the increase in grouting pressure, the maximum displacement of the rock mass increases accordingly. The maximum displacement under 1.5 MPa is small with 0.1391 mm, indicating that the grouting pressure can be appropriately increased to 1.5 MPa to improve the grouting efficiency without causing adverse conditions.

4. Conclusions

The purpose of this study is to investigate the grouting process in fractured rock mass and the influence of grout on the fracture network; thus, a grouting simulation approach considering fluid–structure interaction based on the 3D fractured network model is developed, and the hydropower station X is taken as a case study to do some in-depth analysis using the proposed approach. The results show that the grouting simulation considering fluid–structure interaction is more realistic and can simultaneously reveal the grout diffusion and fracture deformation, which indicated that it is necessary to consider the effect of fluid–structure interaction in grouting simulation. The main conclusions of this study are as follows:

(1) In fracture network modeling studies, fracture aperture values are often ignored or inaccurate, which will affect the authenticity of grout simulation. Combined with the exposed surface fracture catalog data, we use the relationship between fracture apertures and trace lengths to obtain a more realistic value of fracture aperture and to establish a more reliable model for numerical simulation of grouting.

(2) During the grouting process, the filling of the primary fractures is influenced by the number of intersecting secondary fractures, whilst the filling of the secondary fractures is related to the fracture aperture, and the length of the intersection between the secondary facture and the primary fracture. Fracture aperture and dip angle have a significant effect on the grout diffusion rate, while the fracture aperture is the major influencing factor. Moreover, the effect of fluid–structure interaction between the grout flow and the rock mass has a certain influence on the grout diffusion length and neglecting this effect will cause an underestimation of the grouting performance.

(3) When the fractures in a certain region intersect with each other and are close to other fractures in the surrounding area, the rock mass between such fractures will be the least rigid and prone to deformation during the grouting process.

(4) Grouting pressure, grout water–cement ratio and rock mass elastic module all have effects on the grouting process. Therefore, in the grouting construction process, the appropriate grouting pressure and grout water–cement ratio should be selected according to different geological conditions.

(5) The effects of fluid–structure interaction between the grout and the rock mass will affect the grout diffusion process and the rock mass deformation. Therefore, the grouting process simulation considering the fluid–structure interaction can better analyze grout diffusion and rock deformation, and hence explore the grouting mechanism under real conditions.

These results can provide an important theoretical basis and valuable information for grouting construction, and in future research the grouting performance considering the effect of underground water can be further studied.

Author Contributions: Conceptualization, Y.Z., X.W. and S.D.; Formal analysis, Y.Z. and W.C.; Methodology, Yu.Z. and S.D.; Writing—original draft, Y.Z.; Writing—review and editing, X.W., S.D., W.C., Z.S., L.X. and M.L.

Funding: This research was funded by the National Key R&D Program of China, grant number 2018YFC0406704; the Science Fund for Creative Research Groups of the National Natural Science Foundation of China, grant number 51621092; and the National Natural Science Foundation of China, grant number 51439005.

Conflicts of Interest: The authors declare no conflict of interest.

References

1. Saeidi, O.; Azadmehr, A.; Torabi, S.R. Development of a rock groutability index based on the rock engineering systems (res): A case study. *Indian Geotech. J.* **2014**, *44*, 49–58. [CrossRef]
2. Lin, P.; Zhu, X.X.; Li, Q.B.; Liu, H.Y.; Yu, Y.J. Study on Optimal Grouting Timing for Controlling Uplift Deformation of a Super High Arch Dam. *Rock Mech. Rock Eng.* **2016**, *49*, 115–142. [CrossRef]
3. Mohajerani, S.; Baghbanan, A.; Wang, G.; Forouhandeh, S.F. An Efficient Algorithm for Simulating Grout Propagation in 2D Discrete Fracture Networks. *Int. J. Rock Mech. Min. Sci.* **2017**, *98*, 67–77. [CrossRef]
4. Deng, S.H.; Wang, X.L.; Yu, J.; Zhang, Y.C.; Liu, Z.; Zhu, Y.S. Simulation of grouting process in rock masses under a dam foundation characterized by a 3D fracture network. *Rock Mech. Rock Eng.* **2018**, *51*, 1801–1822. [CrossRef]
5. Saeidi, O.; Håkan, S.; Torabi, S.R. Numerical and analytical analyses of the effects of different joint and grout properties on the rock mass groutability. *Tunn. Undergr. Space Technol.* **2013**, *38*, 11–25. [CrossRef]
6. Yang, P.; Sun, X.Q. Single fracture grouting numerical simulation based on fracture roughness in hydrodynamic environment. *Electron. J. Geotechn. Eng.* **2015**, *20*, 59–67.
7. Fu, P.; Zhang, J.J.; Xing, Z.Q.; Yang, X.D. Numerical Simulation and Optimization of Hole Spacing for Cement Grouting in Rocks. *J. Appl. Math.* **2013**, *2013*, 1–9. [CrossRef]
8. Hao, M.M.; Wang, F.M.; Li, X.L.; Zhang, B.; Zhong, Y.H. Numerical and experimental studies of diffusion law of grouting with expansible polymer. *J. Mater. Civ. Eng.* **2018**, *30*, 04017290. [CrossRef]
9. Kim, H.M.; Lee, J.W.; Yazdani, M.; Tohidi, E.; Nejati, H.R.; Park, E.S. Coupled Viscous Fluid Flow and Joint Deformation Analysis for Grout Injection in a Rock Joint. *Rock Mech. Rock Eng.* **2018**, *51*, 627–638. [CrossRef]
10. Ao, X.F.; Wang, X.L.; Zhu, X.B.; Zhou, Z.Y.; Zhang, X.X. Grouting simulation and stability analysis of coal mine goaf considering hydromechanical coupling. *J. Comput. Civ. Eng.* **2017**, *31*, 04016069. [CrossRef]
11. Liu, Q.S.; Sun, L. Simulation of coupled hydro-mechanical interactions during grouting process in fractured media based on the combined finite-discrete element method. *Tunn. Undergr. Space Technol.* **2019**, *84*, 472–486. [CrossRef]
12. Berrone, S.; Canuto, C.; Pieraccini, S.; Scialò, S. Uncertainty quantification in discrete fracture network models: Stochastic fracture transmissivity. *Comput. Math. Appl.* **2015**, *70*, 603–623. [CrossRef]
13. Li, M.C.; Han, S.; Zhou, S.B.; Zhang, Y. An Improved Computing Method for 3D Mechanical Connectivity Rates Based on a Polyhedral Simulation Model of Discrete Fracture Network in Rock Masses. *Rock Mech. Rock Eng.* **2018**, *51*, 1789–1800. [CrossRef]
14. Han, X.D.; Chen, J.P.; Wang, Q.; Li, Y.Y.; Zhang, W.; Yu, T.W. A 3D Fracture Network Model for the Undisturbed Rock Mass at the Songta Dam Site Based on Small Samples. *Rock Mech. Rock Eng.* **2016**, *49*, 611–619. [CrossRef]
15. Yan, C.Z.; Zheng, H. Three-dimensional hydromechanical model of hydraulic fracturing with arbitrarily discrete fracture networks using finite-discrete element method. *Int. J. Geomech.* **2016**, *17*, 04016133. [CrossRef]
16. Shiriyev, J. Discrete Fracture Network Modeling in a Carbon Dioxide Flooded Heavy Oil Reservoir. Master's Thesis, Middle East Technical University, Ankara, Turkey, 2014.
17. Schultz, R.A.; Soliva, R.; Fossen, H.; Okubo, C.H.; Reeves, D.M. Dependence of displacement–length scaling relations for fractures and deformation bands on the volumetric changes across them. *J. Struct. Geol.* **2008**, *30*, 1405–1411. [CrossRef]
18. Schultz, R.A.; Klimczak, C.; Fossen, H.; Olson, J.E.; Exner, U.; Reeves, D.M.; Soliva, R. Statistical tests of scaling relationships for geologic structures. *J. Struct. Geol.* **2013**, *48*, 85–94. [CrossRef]

19. Liu, R.C.; Li, B.; Jiang, Y.J.; Huang, N. Review: Mathematical expressions for estimating equivalent permeability of rock fracture networks. *Hydrogeol. J.* **2016**, *24*, 1623–1649. [CrossRef]
20. Klimczak, C.; Schultz, R.A.; Parashar, R.; Reeves, D.M. Cubic law with aperture-length correlation: Implications for network scale fluid flow. *Hydrogeol. J.* **2010**, *18*, 851–862. [CrossRef]
21. GothäLl, R.; Stille, H. Fracture–fracture interaction during grouting. *Tunn. Undergr. Space Technol.* **2010**, *25*, 199–204. [CrossRef]
22. Tsang, C.F. Coupled hydromechanical-thermochemical processes in rock fractures. *Rev. Geophys.* **1991**, *29*, 537–551. [CrossRef]
23. Rafi, J.Y.; Stille, H. Control of rock jacking considering spread of grout and grouting pressure. *Tunn. Undergr. Space Technol.* **2014**, *40*, 1–15. [CrossRef]
24. Rafi, J.Y.; Stille, H. Basic mechanism of elastic jacking and impact of fracture aperture change on grout spread, transmissivity and penetrability. *Tunn. Undergr. Space Technol.* **2015**, *49*, 174–187. [CrossRef]
25. Cerroni, D.; Fancellu, L.; Manservisi, S.; Menghini, F. Fluid structure interaction solver coupled with volume of fluid method for two-phase flow simulations. *AIP Conf. Proc.* **2016**, *1738*, 030024. [CrossRef]
26. Chen, J.; Liu, H.; Wang, F.S.; Shi, G.C.; Cao, G.; Wu, H.G. Numerical prediction on volumetric efficiency of progressive cavity pump with fluid–solid interaction model. *J. Pet. Sci. Eng.* **2013**, *109*, 12–17. [CrossRef]
27. Villiers, A.M.D.; Mcbride, A.T.; Reddy, B.D.; Franz, T.; Spottiswoode, B.S. A validated patient-specific FSI model for vascular access in haemodialysis. *Biomech. Model. Mech.* **2018**, *17*, 479–497. [CrossRef] [PubMed]
28. Rabczuk, T.; Gracie, R.; Song, J.H.; Belytschko, T. Immersed particle method for fluid–structure interaction. *Int. J. Numer. Meth. Eng.* **2010**, *81*, 48–71. [CrossRef]
29. Zhong, D.H.; Li, M.C.; Liu, J. 3D integrated modeling approach to geo-engineering objects of hydraulic and hydroelectric projects. *Sci. China Ser. E Technol. Sci.* **2007**, *50*, 329–342. [CrossRef]
30. Baecher, G.B.; Lanney, N.A.; Einstein, H.H. Statistical description of rock properties and sampling. In Proceedings of the 18th US Symposium on Rock Mechanics (USRMS), Golden, Colorado, 22–24 June 1977; American Rock Mechanics Association: Golden, CO, USA, 1977. Available online: https://www.onepetro.org/conference-paper/ARMA-77-0400 (accessed on 29 January 2019).
31. Xu, C.S.; Dowd, P. A new computer code for discrete fracture network modelling. *Comput. Geosci.* **2010**, *36*, 292–301. [CrossRef]
32. Mauldon, M. Estimating mean fracture trace length and density from observations in convex windows. *Rock Mech. Rock Eng.* **1998**, *31*, 201–216. [CrossRef]
33. Huang, R.Q.; Xu, M.; Chen, J.P. *Meticulous Description of Complex Structure and Its Engineering Application*; Science Press: Beijing, China, 2004.
34. Kemeny, J.; Post, R. Estimating three-dimensional rock discontinuity orientation from digital images of fracture traces. *Comput. Geosci.* **2003**, *29*, 65–77. [CrossRef]
35. Miao, T.J.; Yu, B.; Duan, Y.G.; Fang, Q.T. A fractal analysis of permeability for fractured rocks. *Int. J. Heat Mass Transf.* **2015**, *81*, 75–80. [CrossRef]
36. Xu, S.S.; Nieto-Samaniego, A.F.; Alaniz-Álvarez, S.A.; Velasquillo-Martínez, L.G. Effect of sampling and linkage on fault length and length–displacement relationship. *Int. J. Earth Sci.* **2006**, *95*, 841–853. [CrossRef]
37. Zhu, J.T.; Cheng, Y.Y. Effective permeability of fractal fracture rocks: Significance of turbulent flow and fractal scaling. *Int. J. Heat Mass Transf.* **2018**, *116*, 549–556. [CrossRef]
38. Matthies, H.G.; Niekamp, R.; Steindorf, J. Algorithms for strong coupling procedures. *Comput. Methods Appl. Mech. Eng.* **2006**, *195*, 2028–2049. [CrossRef]
39. Chen, Y.G.; Price, W.G.; Temarel, P. An anti-diffusive volume of fluid method for interfacial fluid flows. *Int. J. Numer. Methods Fluids* **2012**, *68*, 341–359. [CrossRef]
40. Hirt, C.W.; Nichols, B.D. Volume of fuid (VOF) method for the dynamics of free boundaries. *J. Comput. Phys.* **1981**, *39*, 201–225. [CrossRef]
41. STAR-CCM + User Guide Version 10.02. CD-Adapco. 2015. Available online: https://mdx.plm.automation.siemens.com/star-ccm-plus (accessed on 29 January 2019).
42. Gopala, V.R.; Wachem, B.G.M.V. Volume of fluid methods for immiscible-fluid and free-surface flows. *Chem. Eng. J.* **2008**, *141*, 204–221. [CrossRef]

43. Giovannetti, L.M.; Banks, J.; Ledri, M.; Turnock, S.R.; Boyd, S.W. Toward the development of a hydrofoil tailored to passively reduce its lift response to fluid load. *Ocean Eng.* **2018**, *167*, 1–10. [CrossRef]
44. Rubio, J.E.; Schilling, P.J.; Chakravarty, U.K. Modal characterization and structural aerodynamic response of a crane fly forewing. *Acta Mech.* **2018**, *229*, 2307–2325. [CrossRef]
45. McVicar, J.; Lavroff, J.; Davis, M.R.; Thomas, G. Fluid–structure interaction simulation of slam-induced bending in large high-speed wave-piercing catamarans. *J. Fluids Struct.* **2018**, *82*, 35–58. [CrossRef]

![applied sciences logo] *applied sciences*

MDPI

Article

A Stochastic Bulk Damage Model Based on Mohr-Coulomb Failure Criterion for Dynamic Rock Fracture

Bahador Bahmani, Reza Abedi * and Philip L. Clarke

Department of Mechanical Aerospace and Biomedical Engineering, University of Tennessee Space Institute, Tullahoma, TN 37388, USA; bbahmani@vols.utk.edu (B.B.); pclarke1@vols.utk.edu (P.L.C.)
* Correspondence: rabedi@utk.edu

Received: 10 February 2019; Accepted: 20 February 2019; Published: 26 February 2019

Abstract: We present a stochastic bulk damage model for rock fracture. The decomposition of strain or stress tensor to its negative and positive parts is often used to drive damage and evaluate the effective stress tensor. However, they typically fail to correctly model rock fracture in compression. We propose a damage force model based on the Mohr-Coulomb failure criterion and an effective stress relation that remedy this problem. An evolution equation specifies the rate at which damage tends to its quasi-static limit. The relaxation time of the model introduces an intrinsic length scale for dynamic fracture and addresses the mesh sensitivity problem of earlier damage models. The ordinary differential form of the damage equation makes this remedy quite simple and enables capturing the loading rate sensitivity of strain-stress response. The asynchronous Spacetime Discontinuous Galerkin (aSDG) method is used for macroscopic simulations. To study the effect of rock inhomogeneity, the Karhunen-Loeve method is used to realize random fields for rock cohesion. It is shown that inhomogeneity greatly differentiates fracture patterns from those of a homogeneous rock, including the location of zones with maximum damage. Moreover, as the correlation length of the random field decreases, fracture patterns resemble angled-cracks observed in compressive rock fracture.

Keywords: bulk damage; brittle fracture; rock fracture; random fracture; Mohr-Coulomb; Discontinuous Galerkin

1. Introduction

Interfacial, *particle*, and *bulk* or *continuum* models form the majority of approaches used for failure analysis of quasi-brittle materials at continuum level. Interfacial models directly represent sharp fractures in the computational domain. Some examples are *Linear Elastic Fracture Mechanics* (LFEM), cohesive models [1,2], and interfacial damage models [3–7]. Since cracks are explicitly represented, interfacial methods are deemed accurate when crack propagation is the main mechanism of material failure. However, external criteria are needed for crack nucleation and propagation (direction and extension). Moreover, accurate representation of arbitrary crack directions can be cumbersome in computational settings. Mesh adaptive schemes [8–10], *eXtended Finite Element Methods* (XFEMs) [11–13], and *Generalized Finite Element Methods* (GFEMs) [14,15] address this problem to some extend. However, for highly dynamic fracture simulations and fragmentation studies, even these methods have challenges in accurate modeling of the fracture pattern. Particle methods such as *Peridynamics* [16–18] have been successfully used to model highly complex fracture patterns that are encountered in dynamic (rock) fracture. They model continua as a collection of interacting particles.

Bulk or *continuum* damage models approximate the effect of material microstructural defects and their evolution, e.g., microcrack nucleation, propagation, and coalescence, through the evolution of a damage parameter. Due to the implicit representation of microcracks and other defects, bulk damage models are more efficient than interfacial and especially particle methods. In addition, damage pattern is obtained as a part of the solution and no external criteria are needed for crack nucleation and propagation. Finally, since damage is a smooth field interpolated within finite elements, complex fracture patterns can be easily modeled by damage models, wherein the thickness of cracks is effectively regularized by the damage field.

Earlier bulk damage models, however, suffered from mesh sensitivity problem where the width of the localization and damaged region was proportional to element size; as a result, finer meshes resulted in a more brittle fracture response. This problem is related to the loss of ellipticity/hyperbolicity of the (initial) boundary value problem for the earlier formulations [19,20], and can be resolved by the introduction of an intrinsic length scale to the damage evolution formulation. In gradient-based models, this is achieved by adding higher order derivatives of the damage or strain fields to the damage evolution equation [21,22]. In nonlocal approaches, strain or damage field employed in a local damage formulation, is in turn computed over a neighborhood of finite size [23,24]. Finally, time-relaxed damage formulations possess an internal time parameter which through its interaction with elastic wave speeds introduce a finite length scale for the damage model in transient settings [25–28]. Related to these remedies is the *phase field* method which closely resembles a gradient-based damage model [29]. The sharper approximation of crack width is one of the main advantages of the phase field methods to gradient-based damage models [30].

We have presented a time-delay damage model for dynamic brittle fracture in [31]. The coupled elastodynamic-damage problem is solved by the *asynchronous spacetime Discontinuous Galerkin* (aSDG) method [32,33]. This damage model addresses the mesh sensitivity problem of the earlier damage models by the third approach discussed above, in that, damage evolution is governed by a time-delay model. In addition, the existence of a maximum damage evolution rate results in an increase in both the maximum attainable stress and toughness as the loading rate increases. This loading rate dependency of strength and toughness is experimentally verified; see for example [24,34]. Finally, the damage evolution law is an *Ordinary Differential Equation* in time. This greatly simplifies the damage model formulation and lends itself to the aSDG method; the aSDG method directly discretizes spacetime by elements that satisfy the causality constraint of the underlying hyperbolic problem being solved. The nonlocal damage models violate this causality constraint, whereas the majority of gradient-based damage models are not hyperbolic. In contrast, the time-delay damage model maintains the hyperbolicity of the elastodynamic problem. Besides, the ODE form of the governing equation greatly simplifies the application of initial and boundary conditions for the coupled problem.

The distribution of material defects at microstructure can have a great effect on macroscopic fracture response, particularly for quasi-brittle materials. Some examples are high variability in fracture pattern for samples with the same loading and geometry [35], high sensitivity of macroscopic strength and fracture toughness to microstructural variations [36], and the so-called size effect [37–39], i.e., the decrease of the mean and variations of fracture strength for larger samples. Weibull model [40,41] is one of the popular approaches for modeling the effect of defects in quasi-brittle fracture, particularly the size effect. We have used the Weibull model in the context of an interfacial damage model to capture statistical fracture response of rock, in hydraulic fracturing [42], fracture under dynamic compressive loading [43], and in fragmentation studies [44,45]. However, these models are computationally expensive due to the use of a sharp interfacial damage model.

In this manuscript, we propose a stochastic bulk damage model for rock fracture. There are two main differences to the damage model presented in [31]. First, in damage mechanics often only the spectral positive part of either strain or elastic stress tensor is used to drive damage accumulation. Moreover, upon full damage, only the negative part of the stress tensor is maintained in forming the effective stress. While these choices are appropriate for tensile-dominant fracture, they have some shortcomings for rock fracture under compressive loading. Specifically, using these models damage does not accumulate under compressive loading; even if it could, it would not have modeled the failure process as the effective stress remains the same as the elastic stress of the intact rock. Herein, we propose a new damage model based on the Mohr-Coulomb failure criterion and an effective stress that correctly represents rock failure in compression. Second, we employ a stochastic damage model wherein rock cohesion is treated as a random field. This aspect is important for the uniaxial compression examples considered, as due to the lack of macroscopic stress concentration points highly unrealistic fracture patterns will be obtained by using a homogeneous rock mass model. We note that the use of a bulk damage model makes the proposed approach significantly more efficient than the stochastic fracture problems [42–45] studies by the authors using an interfacial damage model.

The outline of the manuscript is as follows. The formulation of the stochastic damage model, its coupling to elastodynamic problem, and the aSDG method are discussed in Section 2. We use a dynamic uniaxial compressive example to demonstrate the effect of material inhomogeneity on fracture response in Section 3. Final conclusions are drawn in Section 4.

2. Formulation

The first three subsections are pertained to the formulation of damage model. In Section 2.1 the formulation of the damage force parameter based on the Mohr-Coulomb (MC) failure criterion and the damage evolution equation are provided. In Section 2.2 the coupling of elasticity and damage problems through the effective stress is described. Certain properties of the damage model are discussed in Section 2.3. A brief description of the aSDG method and the implementation of the damage model is provided in Section 2.4. Finally, the stochastic aspects of the damage model are explained in Section 2.5.

2.1. Bulk Damage Problem Description

2.1.1. Damage Driving Force

As will be discussed in Section 2.2, the damage parameter $D \in [0, 1]$ gradually reduces the elasticity stiffness in the process of material degradation. Damage evolution if generally driven by the strain field ϵ. For the remainder of the manuscript, we assume that the spatial dimension is two. The symmetric *elastic stress tensor* σ is defined as,

$$\sigma = \mathbf{C}\epsilon, \quad \text{where} \quad \sigma = \begin{bmatrix} \sigma_{xx} & \sigma_{xy} \\ \sigma_{yx} & \sigma_{yy} \end{bmatrix} \quad \text{and} \quad \epsilon = \begin{bmatrix} \epsilon_{xx} & \epsilon_{xy} \\ \epsilon_{yx} & \epsilon_{yy} \end{bmatrix} \tag{1}$$

are the expressions of stress and strain tensors in global coordinate system (x, y) and $\mathbf{C}$ is the elasticity tensor. Instead of ϵ, damage evolution can be expressed in terms of σ. This is more suitable for rock fracture given that many known *failure criteria* such as *Mohr-Coulomb* (MC) or Hoek-Brown [46] are expressed in terms of the stress tensor. Figure 1 shows the Mohr-Coulomb failure criterion in terms of normal σ and shear τ traction components on a fracture surface. We employ the tensile positive convention for σ. The failure criterion is determined by the *cohesion* $\underline{c}$ and *friction angle* $\phi = \tan^{-1}(k)$, where k is the friction coefficient. In the figure, the Mohr circle for a stress tensor A (red semi-circle) corresponding to principal stresses $\sigma_2 < \sigma_1$ is shown. Since only isotropic rocks are considered herein, $\underline{c}$ and ϕ are assumed to be constant with respect to the orientation of principal stresses (with respect to the global coordinate system axes). We define the *scalar stress* as,

$$c(\sigma, \phi) := \frac{R}{\cos \phi} + \sigma_{ave} \tan \phi \tag{2}$$

where as shown in Figure 1, c is the ordinate of the tangent line on the Mohr-circle with angle ϕ, and the radius R and average normal stress σ_{ave} are given by,

$$R := \frac{\sigma_1 - \sigma_2}{2} = \sqrt{\frac{(\sigma_{xx} - \sigma_{yy})^2}{4} + \sigma_{xy}^2}, \tag{3a}$$

$$\sigma_{ave} := \frac{\sigma_1 + \sigma_2}{2} = \frac{\sigma_{xx} + \sigma_{yy}}{2}, \tag{3b}$$

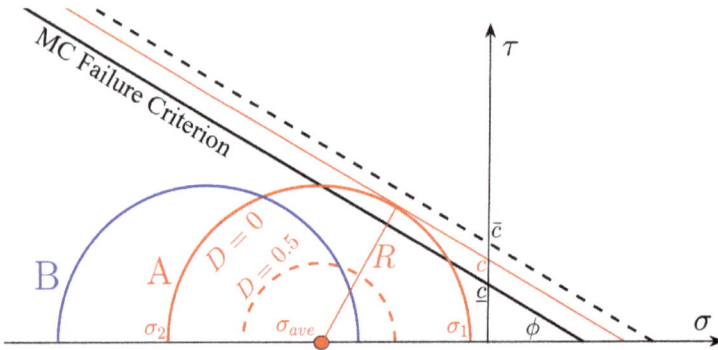

Figure 1. Mohr-Coulomb failure criterion and scalar stress c for a given stress state.

Figure 1 shows two stress states. For the stress state B, the entire Mohr circle is below the failure criterion curve, thus no degradation is expected. For the stress state A, the Mohr circle expands beyond the failure criterion curve; in a binary intact and failed classification, this stress state would be considered failed. These stages correspond to $c(\sigma, \phi) < \underline{c}$ and $c(\sigma, \phi) \geq \underline{c}$, respectively. Some specific strengths corresponding to the MC criterion $c(\sigma, \phi) = \underline{c}$ are shown in Figure 2 and are given by,

$$\underline{s}_{ht} = \frac{\underline{c}}{\tan \phi} \qquad \text{Hydrostatic tensile strength} \tag{4a}$$

$$\underline{s}_{at} = \frac{2\underline{c} \cos \phi}{1 + \sin \phi} \qquad \text{Uniaxial tensile strength} \tag{4b}$$

$$\underline{s}_s = \underline{c} \cos \phi \qquad \text{Shear strength} \tag{4c}$$

$$\underline{s}_{ac} = \frac{2\underline{c} \cos \phi}{1 - \sin \phi} \qquad \text{Uniaxial compressive strength} \tag{4d}$$

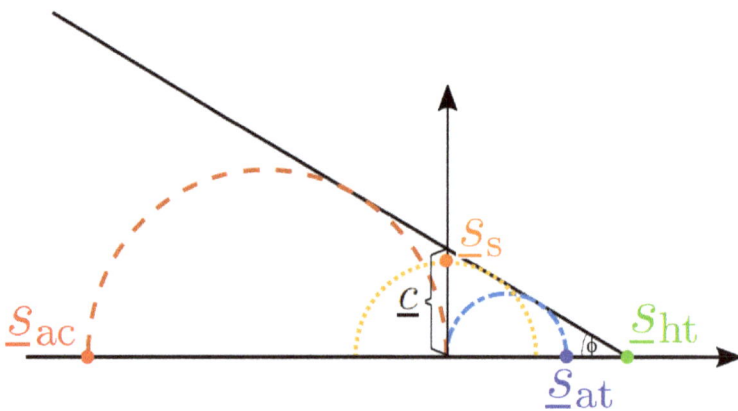

Figure 2. Relation of different fracture strengths, $\underline{s}_{ht}$, $\underline{s}_{at}$, $\underline{s}_s$, and $\underline{s}_{ac}$ to cohesion $\underline{c}$.

As will become clear later, the damage model, regularizes the process of failure. Otherwise, failure for a stress state occurs instantaneously once MC criterion $c(\sigma, \phi) = \underline{c}$ is satisfied; for example, when $\sigma_{xx} = \sigma_{yy} > 0$ reaches $\underline{s}_{ht}$ ($\sigma_{xy} = 0$). To facilitate this, the *damage force* is defined as,

$$
D_f(c, \underline{c}, \bar{c}) := \begin{cases} 0 & c \le \underline{c} \\ \frac{c - \underline{c}}{\bar{c} - \underline{c}} & \underline{c} < c < \bar{c} \\ 1 & \bar{c} \le c \end{cases} \tag{5}
$$

where $\bar{c}$ corresponds to the ordinate of the upper MC line shown in dashed line in Figure 1. The *brittleness* factor β defines a relation between the two MC lines through $\underline{c} = \beta \bar{c}$. In the absence of the damage model, complete failure occurs for any positive value of D_f as the Mohr circle expands over the failure criterion. However, in the context of the damage model, D_f corresponds to the quasi-static damage value for a given strain ϵ, which through (1) and (2) defines c. For example, for the strain (elastic stress) state A in Figure 1, $D_f = 0.5$.

2.1.2. Damage Evolution Law

The damage value can be taken to be equal to the damage force. However, this local definition of D has several shortcomings, as will be discussed below and in Section 2.3.2. We employ the time-delay model in [3,25] for damage evolution. The rate of damage evolution, $\dot{D}$, is given by,

$$
\dot{D} = \begin{cases} D_{\text{src}}(D, D_f) = \frac{1}{\tau_c}(1 - e^{-a\langle D_f - D \rangle}) & D < 1 \\ 0 & D = 1 \end{cases} \tag{6}
$$

where $D_{\text{src}}(D, D_f)$ is a general source term for the evolution equation $\dot{D} = D_{\text{src}}(D, D_f)$. This function can be calibrated from experimental strain-stress results, for example, for uniaxial tensile/compressive loading. The specific form of $D_{\text{src}}(D, D_f)$ is taken from [3,47] as it is claimed to accurately model materials' rate effect; cf. Section 2.3.2. In addition, τ_c is the *relaxation time*, a is the *brittleness exponent*, and $\langle . \rangle$ is the Macaulay positive operator.

Albeit its simplicity, this evolution model incorporates several essential characteristics of real materials. First, we observe that the damage evolution is governed by the difference of damage D and damage force D_f. The higher the difference, the higher the damage rate. Moreover, when $D = D_f$, damage evolution terminates. That is, D_f is the *target damage* value; if D is smaller than the target value, it evolves until it reaches D_f. Second, damage cannot instantaneously reach D_f given that $\dot{D}$ is bound by the maximum damage rate $1/\tau_c$. As will be discussed in Section 2.3.2, this results in the rate-sensitivity of strain-stress response. Third, the positive operator ensures that damage is a nondecreasing function in time (no material healing processes). Finally, Figure 3 shows the effect of a; for higher values of a, even small differences between D_f and D, quickly jumps up the damage rate close to its maximum value of $1/\tau_c$; implying a more brittle response.

2.2. Coupling of Damage and Elastodynamic Problems

The equation of motion, corresponding to strong satisfaction of the balance of linear momentum for elastodynamic problem, reads as,

$$
\nabla \cdot \sigma_{\text{eff}} + \rho \mathbf{b} = \dot{\mathbf{p}}, \tag{7}
$$

where σ_{eff}, $\mathbf{b}$, and $\mathbf{p}$ are the effective stress tensor, body force, and linear momentum density, respectively. The linear momentum density is defined as $\mathbf{p} = \rho \dot{\mathbf{u}}$, where ρ is the mass density. This equation is augmented by the compatibility equations between displacement, velocity, and strain, and initial/boundary conditions to form the elastodynamic initial boundary value problem.

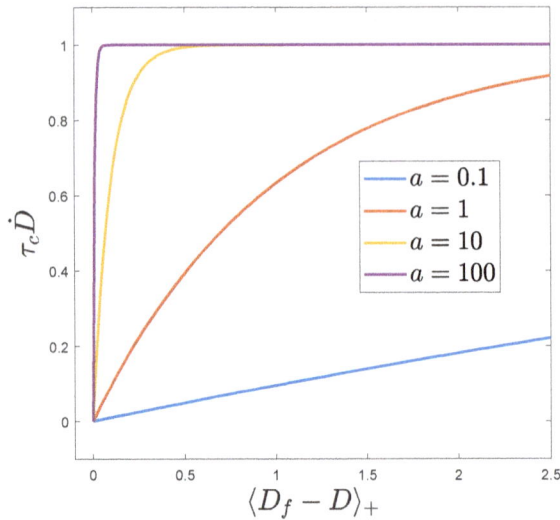

Figure 3. The effect of brittleness exponent a on the rate of damage evolution.

The coupling between damage and elastodynamic problems is through the effective stress tensor σ_{eff}. In the simplest form, the scalar damage parameter D linearly degrades the elasticity stiffness tensor, that is $\sigma_{\text{eff}} = (1 - D)\sigma = (1 - D)\mathbf{C}\epsilon$ [21]. However, in more advanced damage-elasticity constitutive equations, only certain parts of the elastic stress (or elastic strain) are degraded by D [48]. By inspecting Figures 1 and 2, it is observed that damage is induced by high tensile and shear stresses and no damage is induced by a hydrostatic compressive stress state ($\sigma_1 = \sigma_2 < 0$). Accordingly, we define a consistent damage-elasticity constitutive equation in which the entire elastic stress, except its hydrostatic compressive part, are degraded by D. That is,

$$\sigma_{\text{eff}} = (1 - D)\sigma_d + (1 - D)\langle\sigma_h\rangle + \langle\sigma_h\rangle_- \tag{8}$$

where σ_h and σ_d are hydrostatic and deviatoric parts of σ. The positive and negative ($\langle\sigma_h\rangle_- = \sigma_h - \langle\sigma_h\rangle$) parts of σ_h correspond to the hydrostatic tensile and compressive stresses of σ. For example, if $\sigma_2 \leq \sigma_1$ are the principal values of σ, σ_d, $\langle\sigma_h\rangle$, and $\langle\sigma_h\rangle_-$ have the principal values of $[(\sigma_2 - \sigma_1)/2, (\sigma_1 - \sigma_2)/2]$, $[\langle(\sigma_2 + \sigma_1)/2\rangle, \langle(\sigma_2 + \sigma_1)/2\rangle]$, and $[\langle(\sigma_2 + \sigma_1)/2\rangle_-, \langle(\sigma_2 + \sigma_1)/2\rangle_-]$, respectively, all with the same principal directions. Clearly, they correspond to the pure shear, tensile, and compressive parts of σ.

2.3. Properties of the Damage Model

We first discuss the properties of the damage force and effective stress models, concerning the mechanisms that drive damage and lead to the stress state at full damage. Next, we discuss how the damage evolution law captures material's stress rate effect and alleviates the mesh sensitivity problem of local damage models.

2.3.1. Damage Force and Effective Stress

Equations (5) and (8) determine under what strain (elastic stress) conditions damage initiates and how the effective stress evolves as D tends to unity. A common approach in continuum damage mechanics is to break the elastic stress tensor into its spectral positive and negative parts, and to express D_f and σ_{eff} as,

$$D_f(\sigma) = D_f(\sigma_+) \tag{9a}$$

$$\sigma_{\text{eff}} = (1-D)\sigma_+ + \sigma_- \tag{9b}$$

We note that alternative expressions exist where instead of σ, the spectral decomposition of strain is considered [21,49–51]; however, due to the use of σ in (5) and (8), the form (9) is preferred for the discussion in this section.

Rock fracture is often under compressive stress state. The shortcomings of (9) can be illustrated by referring to Figure 1. First, as can be seen a large difference between the principal stresses σ_1 and σ_2 corresponds to a large enough shear stress τ that can initiate damage evolution; see for example the stress state A. However, if (9a) is used, D_f (and damage) remain zero, since $\sigma_+ = 0$. Second, even if damage could evolve by an equation other than (9a), the stress would not degrade using (9b); that is, $\sigma_{\text{eff}} = \sigma_- = \sigma$ at $D = 1$. In contrast, stress state A induces a $D_f = 0.5$; cf. (5). Moreover, D_f is sensitive to the hydrostatic stress. For example, for the same maximum shear τ and higher compressive σ_{ave}, no damage occurs for the stress state B. Finally, through damage evolution, σ_{eff} tends to the hydrostatic compressive stress $\langle \sigma_h \rangle_-$ as $D \to 1$. This can be seen for stress state A and $D = 0.5$. In damage reaches unity, the effective stress state will correspond to the point σ_{ave} in the figure.

The two sets of equations for D_f and σ_{eff} predict a similar response for tensile dominant loading, i.e., when $\sigma_{ave} > 0$; while there are some differences in the details of damage evolution, in both cases $\sigma_{\text{eff}} \to 0$ as strain (proportionally) increases. There are, however, some differences in the failure damage state, $\sigma_{\text{eff}}(D = 1)$, for pure shear and compressive dominant mixed loading ($\sigma_2 < 0 < \sigma_1$ and $|\sigma_2| > \sigma_1$). In short, the proposed damage model based on the Mohr-Coulomb failure criterion is more appropriate for rock fracture, especially when compressive mode failure is concerned.

2.3.2. Damage Evolution: Rate Effects and Mesh Sensitivity

Figure 4 compares strain stress responses for three different model and loading scenarios. The loading considered can correspond to any of the strengths in (4). The nondimensional scalar elastic stress, strain, and effective stress are defined as $\sigma' = \sigma/\underline{\sigma}$, $\epsilon' = \sigma' = C\epsilon/\underline{\sigma}$, and $\sigma'_{\text{eff}} = \sigma_{\text{eff}}/\underline{\sigma}$, respectively, where σ, ϵ, and σ_{eff} are the scalar elastic stress, strain, and effective stress. (Note that the scalar elastic stress measure σ in this section is different from the normal stress component in the Mohr-Coulomb criterion; cf. Figure 1.) These scalar values, $\underline{\sigma}$, and stiffness C correspond to a particular loading condition; for example for uniaxial tensile loading $\sigma = \sigma_{xx}$, $\epsilon = \epsilon_{xx}$, $\underline{\sigma} = \underline{s}_{at}$, (cf. (4b)). The corresponding stiffness is $C = E$ and $E/(1-v^2)$, for plane stress and plane strain conditions, respectively, where E and v are the elastic modulus and Poisson ratio.

As loading (ϵ') increases, the scalar stress c increases until $c = \underline{c}$ in Figure 2 for the given loading condition. This corresponds to $\epsilon' = 1$. For the MC model, material is deemed to fail instantaneously, for $\sigma'_{\text{eff}} = \sigma' = 1$. This sudden failure is shown by the green circle in the figure. The damage model regularizes the MC failure criterion. For the quasi-static loading $\dot{D} \approx 0$, thus $D \approx D_f$ throughout the loading. Given the linear dependence of D_f on c in (5), σ_{eff} linearly decreases from unity to zero as ϵ' increases from unity to $\bar{c}/\underline{c} = 1/\beta$. As, $\beta \to 1$ the response of the damage model tends to that of the un-regularized MC model, clarifying why β is called the brittleness factor. Regardless of the rate of loading for ϵ', $\dot{D}$ remains bounded by $1/\tau_c$; cf. (6). This results in a delayed damage response where D falls far behind its quasi-static limit D_f for higher rates of loading for ϵ'. This, in turn, increases the maximum effective stress, $\max(\sigma'_{\text{eff}})$, failure strain, $\epsilon'(D = 1)$, and toughness, i.e., the area under the strain-stress curve. That is, the time-delay evolution law (6) can qualitatively model material's well-known stress rate effect. The dynamic solution in Figure 4 corresponds to a nondimensional strain rate of 3. For lower and higher nondimensional loading rates, the stress response gets closer to the quasi-static response and further expands, respectively.

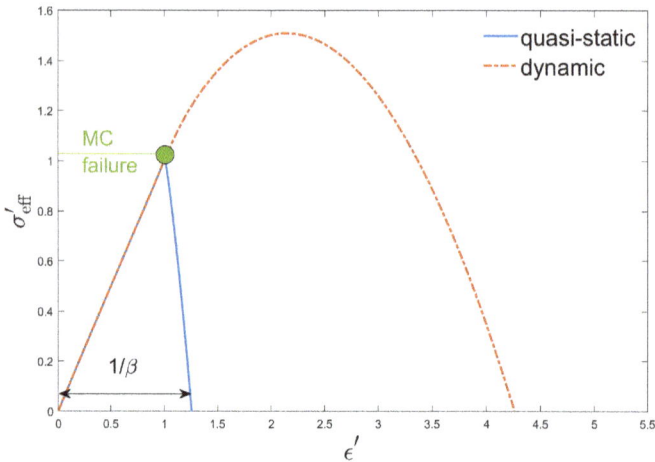

Figure 4. Sample quasi-static and dynamic strain versus stress responses.

If the quasi-static damage model $D = D_f$ where to be used, it would suffer the mesh-sensitivity problem of the early damage models. The introduction of an intrinsic length scale addresses this issue. The length scale l_d is either used in conjunction of added higher spatial order derivative terms in a local damage model [21,22] or by nonlocal integration of certain fields, e.g., strain, over neighborhoods of size l_d. However, both approaches are computationally expensive. The proposed damage model is much easier to implement, since it is simply an ODE in time. It also maintains the hyperbolicity of the elastodynamic problem which is critical for the solution of the coupled problem by the aSDG problem. Finally, the interaction of elastic wave speeds with the intrinsic time scale τ_c indirectly introduces a length scale l_d for the damage problem. While this length scale is not relevant for very low rate loading problems [52], at moderate to high loading rates it is expected to resolve the mesh sensitivity problem of local damage models.

2.4. aSDG Method

The *asynchronous Spacetime Discontinuous Galerkin* (aSDG) method, formulated for elastodynamic problem in [32], is use for dynamic fracture analysis. The *Tent Pitching* algorithm [53] is used to advance the solution in time by continuous erection of *patches* of elements whose exterior patch boundaries satisfy a special causality constraint. This results in a local and asynchronous solution process. In addition, since spacetime is directly discretized by finite elements, the order of accuracy can be arbitrarily high both in space and time directions. This is in contrast to conventional finite element plus time marching algorithms where increasing the order of accuracy in time is not straightforward.

In addition to the displacement field for the elastodynamic problem, the damage field D is discretized in spacetime. The finite elements solve the weak form of elastodynamic balance laws, cf. Section 2.2, and the damage evolution Equation (6), $\dot{D} - D_{\text{src}}(D, D_f) = 0$. Since the damage evolution is simply an ODE and maintains the hyperbolicity of the problem, the solution of the coupled elastodynamic-damage problem lends itself to the aSDG method. In addition, the satisfaction of balance laws per element for discontinuous Galerkin methods results in a very accurate discrete solution of the damage evolution equation. We refer the reader to [31] for more details on the aSDG implementation of the problem, including the specification of jump conditions, and initial/boundary conditions for the damage evolution equation.

2.5. Realization of Stochastic Damage Model Parameters

As discussed in Section 1, incorporating material inhomogeneity is quite important to capture realistic failure response of quasi-brittle materials. The inhomogeneity is both in elastic and fracture properties. If an isotropic material model is assumed at the mesoscale, often only the elastic modulus is deemed to be a random field as in [54]. However, in general the entire elasticity tensor should be considered as a tensorial random field. However, often due to the higher effect that fracture properties have on macroscopic failure response, only they are considered to be random and inhomogeneous.

For a general fracture model, strength, energy, and initial damage state are the main model parameters. For the MC model, the friction angle ϕ (or friction coefficient k) and cohesion $\underline{c}$ are the model parameters used to determine c and D_f from (2) and (5), respectively. Cohesion is the parameter that is associated with fracture strength. The relaxation time τ_c in (6) and brittleness factor β determine the area under the strain-stress curve for different loading rates in Figure 4. That is, they determine the fracture energy of the damage model. Finally, the initial condition for damage parameter, $D(\mathbf{x}, t = 0)$, corresponds to the initial state of material. In the present work, among strength, energy, and initial damage parameters, we consider inhomogeneity only in the strength property. This is in accord with a majority of similar studies in the literature such as [55–60].

Accordingly, the only random field in the present study is cohesion $\underline{c}$. For a macroscopically homogeneous material, the point-wise and two-point statistics of the random field are spatially uniform. For the point-wise statistics, the mean and standard deviation of the random field are the main parameters. For the two-point statistics the form of the correlation function and the correlation length, i.e., the length scale at which the field spatially varies are the main parameters. In [54], where the elastic modulus is considered to a random field, standard deviation and correlation length of the random field are considered as the main parameters that impact fracture response. The realization of random fields for fracture strength and the subsequent fracture analysis becomes more expensive as the correlation length tends to zero. In [54] it is shown that certain macroscopic fracture statistics converge as the correlation length tends to zero. That is, by maintaining sufficient level of material inhomogeneity through using a small enough correlation length, accurate representation of macroscopic fracture response can be obtained.

We treat cohesion as a stationary random field with certain standard deviation ς_c and correlation length l_c. The statistics of this random field can be systematically obtained by using *Statistical Volume Elements* (SVEs), as shown in [60,61]. However, for simplicity and better control on the effect of these parameters, we artificially manufacture random fields with certain ς_c and l_c. The distribution of $\underline{c}$ is assumed to follow a Lognormal(μ_c, ς_c) probability structure where μ_c and ς_c are the mean and standard deviation of the normal field. The corresponding mean and standard deviation of the log normal field for $\underline{c}$ are $M_c = \exp\left(\mu_c + \varsigma_c^2/2\right)$ and $\Sigma_c = \exp(\mu_c + \varsigma_c^2/2)\sqrt{\exp(\varsigma_c^2) - 1}$.

Once the underlying correlation function form and length, and point-wise *Probability Distribution Function* (PDF) are specified, there are a number of statistical methods to realize consistent random fields. We use the Karhunen-Loéve (KL) method [62,63] to realize a random field $\xi = \xi(\mathbf{x}, \omega)$ by an expansion of its covariance kernel; the field is described by the series,

$$\xi(\mathbf{x}, \omega) = \mu_\xi(\mathbf{x}) + \sum_{i=1}^{\infty} \sqrt{\lambda_i} b_i(\mathbf{x}) Y_i(\omega), \tag{10}$$

where the denumerable set of eigenvalues λ_i and eigenfunctions $b_i(\mathbf{x})$ are obtained as solutions of the Fredholm equation, i.e., the generalized eigenvalue problem (EVP), as detailed [64]. Since the eigenvalues monotonically decrease, the truncated series with an appropriate value of the upper limit n instead of ∞ in (10), can precisely represent the statics of the underlying random field. For practical use of the KL method, random variables Y_i should be statistically unrelated. This condition is automatically satisfied for Gaussian fields. Thus, we sample Gaussian random fields with the mean μ_c and standard deviation ς_c. To obtain the final random field for $\underline{c}$, we need to take the exponent of the realized

Gaussian random field. There are some technical challenges for using two distinct grids for the aSDG finite element solution in spacetime and the realized random field for $\underline{c}$ in a material grid. For more discussion on the use of KL method for fracture analysis and aSDG analysis of domains with random properties, we refer the reader to [65].

3. Numerical Results

We consider rock failure under dynamic compressive loading, and study the effect of mesh size, load amplitude, and material inhomogeneity on damage pattern. The geometry and loading description are shown in Figure 5, where a rectangular domain of width $w = 0.08$ mm and height $l = 2w = 0.16$ mm is subject to compressive loading $P(t)$ on top and bottom faces. The traction $P(t)$ ramps up from zero to the sustained value of P_{peak} in ramp time t_{ramp}. Zero tangential traction is applied on these faces to model a frictionless loading interface. A traction free boundary condition is applied on the vertical sides of the domain. We assume a 2D plain-strain condition with material properties reported in Table 1.

For this 2D problem, the spacetime mesh corresponds to a 2D × time grid of tetrahedron elements. The solution is advanced to the final time by an asynchronous patch-by-patch solution algorithm. The time increment of a pitched vertex is calculated based on the wave speed, spatial geometry, and sizes of elements around; cf. Section 2.4 and [32,53] for more details. We use third order polynomial basis functions for damage and displacement fields in space and time.

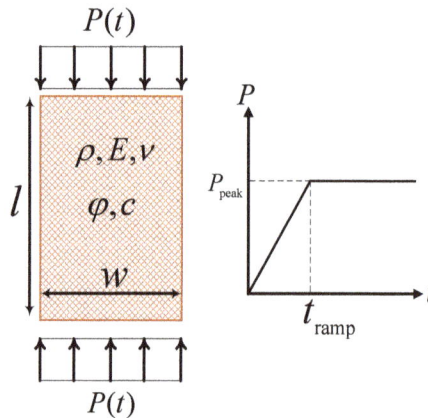

Figure 5. Problem description for a rectangle subject to a vertical compressive loading.

Table 1. Material properties.

Properties	Units	Values
E	GPa	65
ρ	kg/m^3	2650
τ_c	µs	30
t_{ramp}	µs	10
ν	-	0.27
c	MPa	4.7
ϕ	°	17
a	-	10

As shown in Figure 6, we use three different structured grids of 8 × 16, 16 × 32, and 32 × 64 squares, where each square is divided into two triangles. These are labeled as coarse, medium, and fine meshes, respectively. One of the numerical challenges in damage mechanics that affects the convergence of the *Newton-Raphson* method is the zero stiffness issue when damage is equal to unity. One way to avoid this problem is multiplying the damage value used in (8) by a positive reduction factor less than unity. Herein, we select a reduction factor of 93%.

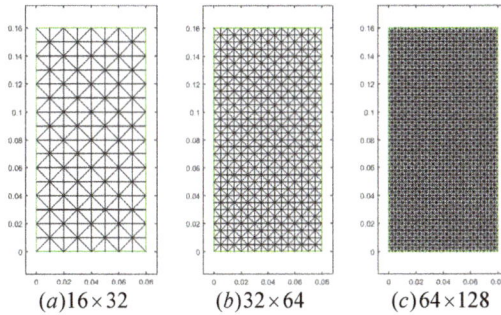

$(a)16 \times 32$ $(b)32 \times 64$ $(c)64 \times 128$

Figure 6. Initial meshes used for the simulations: (**a**) Coarse, (**b**) Medium, and (**c**) Fine.

3.1. Homogeneous Material

3.1.1. Mesh Sensitivity

The dependence of damage response on the resolution of the underlying discrete grid is a well-known problem for non-regularized continuum damage models. As described in Section 2.3.2, the proposed time-delay damage model introduces an inherent length scale proportional to the relaxation time and longitudinal elastic wave speed, i.e., $l_d \propto c_d \tau_c$. To show mesh-objectivity of the results, we compare the damage evolution for coarse and medium meshes in Figure 7. For this numerical example, material properties are homogeneous and listed in Table 1, and the loading magnitude is $P_{\text{peak}} = 13.5$ MPa.

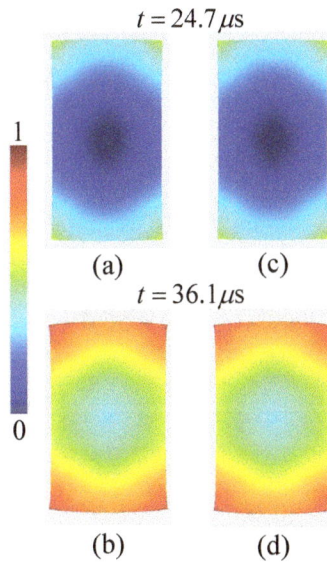

$t = 24.7\,\mu s$

(a)   (c)

$t = 36.1\,\mu s$

(b)   (d)

Figure 7. Damage responses at different times for two different meshes. Figures (**a**,**b**) correspond to the coarse mesh and figures (**c**,**d**) correspond to the medium mesh. The results are shown on the deformed mesh with a magnification factor of 300.

Figure 7 shows an excellent agreement between the solutions of the two meshes at early and evolved stages of damage evolution. We also refer the reader to [31] for a more detailed study of mesh objectivity for a tensile fracture problem where damage localization zone converges to a region of finite width. We reiterate that the time-delay formulation addresses the mesh-objectivity problem with much less computational difficulty than the non-local integration-based and gradient-based damage models. Moreover, it does not violate the hyperbolicity of the problem. This facilitates the use of the aSDG method and is consistent with the physical observation that damage propagates with a finite speed [66].

3.1.2. The Effect of Load Amplitude

In the previous example, the stress level was sufficiently high to initiate damage near the loading edges, from the early stages of the solution. The stress state in the middle of top and bottom faces is approximately similar to bi-axial compressive condition; material tends to expand in the horizontal direction because of the Poisson effect while the surrounding material prevents its deformation. However, the stress state around the corners is close to an unconfined uni-axial compressive condition because of the stress-free conditions at left and right boundaries. The higher differences between compressive stresses in the Mohr circle results in a higher value for c; cf. (2). Thus according to the MC failure criterion, the corner zones are more susceptible to an earlier time for damage initiation and higher damage values. This is verified by the higher damage values around the corners in Figure 7a,b. After the initiation of damage at corners, damage diffuses towards the middle of the domain.

To study the effect of load amplitude, we reduce the peak stress such that damage initiates in the middle of the domain. The vertical normal stress magnitude roughly doubles across the entire width when the stress waves collide in the middle of the domain. The load for this problem is chosen such that it is not large enough to initiate damage when the stress wave enters from the top and bottom edges, but is sufficient to cause damage in the middle of domain due to the doubling effect. We call this condition the low amplitude case, corresponding to $P_{peak} = 6$ MPa, and refer to the previous peak stress problem as the high amplitude case. As shown in Figure 8a, the initial damage occurs when the peak stress reaches the middle of the domain; i.e., at $t_{collision} \approx t_{ramp} + \frac{l}{2c_d} \approx 24$ μs which is well predicted by the numerical result. After the collision, the magnified reflected waves are sufficiently high to overcome the cohesion of rock. Thereafter, damage diffuses toward boundaries where the waves are propagating to; see Figure 8b–d. This failure mechanism is completely different from that of the high amplitude case where the damage initiates in a shear dominated regime at the corners. Therefore, load amplitude has a significant impact on damage pattern and failure mechanism. For a better comparison, we provide the damage response at various times for the high amplitude case in Figure 9.

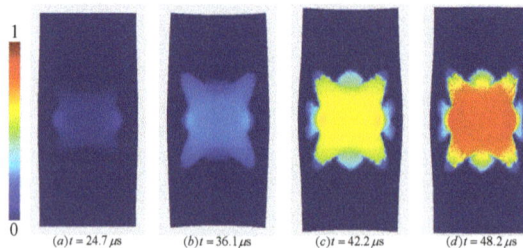

Figure 8. Damage evolution for the medium mesh and low amplitude load at times: (**a**) 24.7 μs, (**b**) 36.1 μs, (**c**) 42.2 μs, and (**d**) 48.2 μs. The results are shown on the deformed meshes with a magnification factor of 1000.

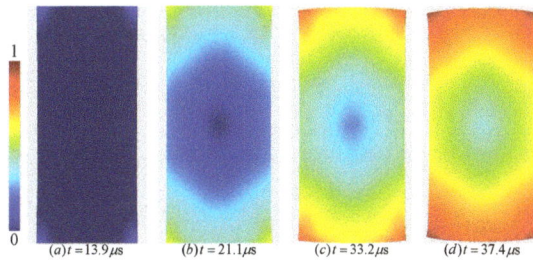

Figure 9. Damage evolution for the medium mesh and high amplitude load at times: (**a**) 13.9 µs, (**b**) 21.1 µs, (**c**) 33.2 µs, and (**d**) 37.4 µs. The results are shown on the deformed meshes with a magnification factor of 300.

3.2. Heterogeneous Material

As detailed in Section 2.5, for the analysis of inhomogeneous rock masses, we assume that cohesion is a random field. This analysis expands our preliminary comparison of the response of homogeneous and heterogeneous rock in [67]. We construct four random fields using the *KL* method with the mean cohesion value of $M_c = 4.7$ MPa, similar to the spatially uniform $\underline{c}$ used in the preceding examples for homogeneous rock. The standard deviation is set to $\Sigma_c = 2.35$ MPa. The correlation lengths of $l_c = 5$ mm, 10 mm, 20 mm, and 40 mm are used, where for each correlation length one random field realization is generated by the KL method. These random fields are shown in Figure 10. A smaller correlation length indicates faster variations in cohesion from one spatial point to another, so it corresponds to a more locally heterogeneous field. These random fields are constructed with the first 2000 terms of the KL series. For the following results, we use the fine mesh to have an adequate resolution for capturing the underlying inhomogeneity.

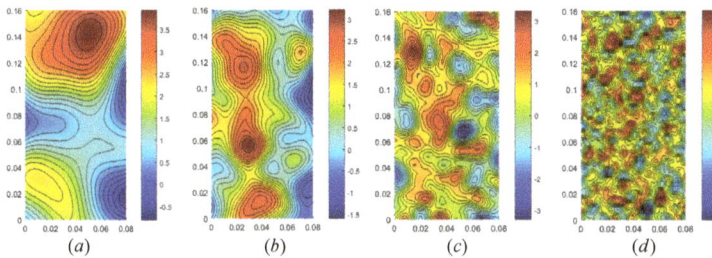

Figure 10. Random field realizations for cohesion with different correlation lengths, l_c, equal to: (**a**) 40 mm, (**b**) 20 mm, (**c**) 10 mm, and (**d**) 5 mm.

3.2.1. Low Amplitude Load

In this section, we study the effect of heterogeneity on damage response for the low amplitude condition. Figure 11 shows the damage response for $l_c = 40$ mm at various times. From the cohesion map in Figure 10a, we observe that $\underline{c}$ varies very slowly in space. It takes the highest values near the top boundary and the lowest ones at three spots close to the left and right boundaries; weak zones are colored by blue. In Figure 8a, the initial damage zone begins when the stress waves collide in the middle of the domain. The particular form of this realization for $\underline{c}$ actually favors damage accumulation in the center, given that a higher strength zone is near the top boundary in Figure 10a. As shown in Figure 11, damage initiates and accumulates both in this center location and in the three aforementioned weak sites close to the boundaries. Thus, the form of the failure pattern follows both the weak points in the material and locations with higher stress values in general. By comparison of Figures 8 and 11, we also observe that the earlier initiation of damage in weaker sites results in a response with more concentrated damage zones. Finally, the damage initiation time is almost the same as that for the homogeneous rock, and in both cases it is right after the collision of the waves at $t_{\text{collision}} \approx 24$.

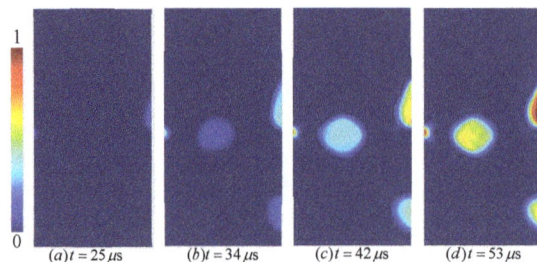

Figure 11. The evolution of damage field for the low amplitude load and cohesion realization with $l_c = 40$ mm at times: (**a**) 25 μs, (**b**) 34 μs, (**c**) 42 μs, and (**d**) 53 μs.

Figures 12–14 show the damage evolution for heterogeneous cohesion fields with correlation lengths equal to 20 mm, 10 mm, and 5 mm, respectively. According to Figures 12a–14a, the time for damage initiation decreases as the correlation length gets smaller, i.e., when the heterogeneity is increasing.

It is well accepted in the literature that one of the main reasons for localization and softening behavior in brittle materials is their heterogeneous structure at microscale [68–70]; the weaker points in material begin to fail earlier. This results in an increased stress concentration in the damaging zones and the shielding of the surrounding areas. That is, the inhomogeneity in material properties promotes inhomogeneity and localization in the stress field. Unlike ductile materials, there are not much energy dissipative reserves, for example from plasticity, to balance the stress field. Figures 11d–14d reveal a crucial impact of the correlation length on failure mechanism; this is a transition from diffusive damage propagation to a more localized response as the correlation length gets smaller. This agrees with the preceding discussion on the promotion of damage localization by material inhomogeneity. In fact, for the solutions with the lowest correlation length, even the mode and propagation of failure is significantly different than that of a homogenous material; in Figures 8d and 14d, the effect of the weakest point of the material is high to an extent that damage initiates and accumulates in a more distributed sense, as opposed to the damage accumulation in the central zone in Figure 8.

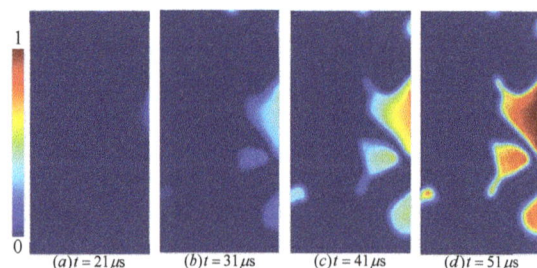

Figure 12. The evolution of damage field for the low amplitude load and cohesion realization with $l_c = 20$ mm at times: (**a**) 21 μs, (**b**) 31 μs, (**c**) 41 μs, and (**d**) 51 μs.

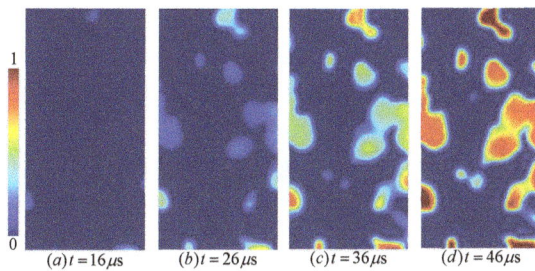

Figure 13. The evolution of damage field for the low amplitude load and cohesion realization with $l_c = 10$ mm at times: (**a**) 16 μs, (**b**) 26 μs, (**c**) 36 μs, and (**d**) 46 μs.

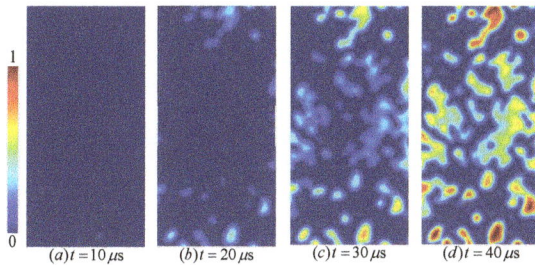

Figure 14. The evolution of damage field for the low amplitude load and cohesion realization with $l_c = 5$ mm at times: (**a**) 10 μs, (**b**) 20 μs, (**c**) 30 μs, and (**d**) 40 μs.

3.2.2. High Amplitude Load

Figures 15–18 show the evolution of the damage field for correlation lengths $l_c = 40$ mm to $l_c = 5$ mm. We observe a very good match between damage localization sites and the locations of material weak points in Figure 10b–d. Moreover, as we decrease the correlation length, the time of damage initiation decreases; cf. Figures 15a–18a.

From the final damage pattern in Figure 9d for the homogeneous domain, one observes that for high amplitude loading extensive damage is experienced almost everywhere, especially close to the top and bottom boundaries. There is little resemblance between this solution and those for high correlation random fields in Figures 15d and 16d. Similarly for the low amplitude load, high differences are observed between the solutions of homogeneous, Figure 8d, and inhomogeneous domains with high correlation lengths, Figures 11d and 12d. This is due to the fact that for such large correlation lengths, the large islands of low strength greatly impact the response.

In contrast, as the correlation length decreases, the overall material properties are almost the same in all areas, except the inhomogeneities that are observed at smaller length scales. Consequently, in comparison of damage patterns for the homogeneous rock in Figure 9d and rocks with small correlation length for $\underline{c}$ in Figures 17d and 18d, a very similar overall response is observed; in all cases, damage is widespread in the domain, with the top and bottom sides experiencing the highest damage. In contrast, there is no resemblance between the damage patterns of homogeneous domain in Figure 8d and those for low correlation length fields in Figures 13d and 14d. The reason is that for this low amplitude of load, damage can only accumulate in the center of the homogeneous domain, whereas for inhomogeneous domains damage can accumulate from weak points outside of this zone; this greatly affect the final damage pattern.

Appl. Sci. **2019**, *9*, 830

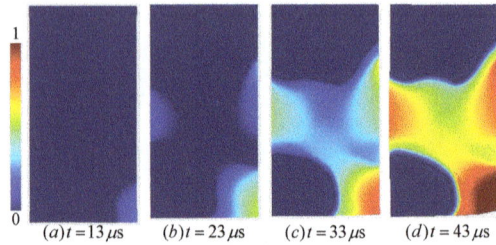

Figure 15. The evolution of damage field for the high amplitude load and cohesion realization with $l_c = 40$ mm at times: (**a**) 13 μs, (**b**) 23 μs, (**c**) 33 μs, and (**d**) 43 μs. The results are shown on the deformed mesh with a magnification factor of 100.

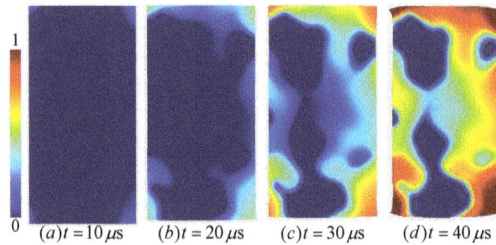

Figure 16. The evolution of damage field for the high amplitude load and cohesion realization with $l_c = 20$ mm at times: (**a**) 10 μs, (**b**) 20 μs, (**c**) 30 μs, and (**d**) 40 μs. The results are shown on the deformed mesh with a magnification factor of 100.

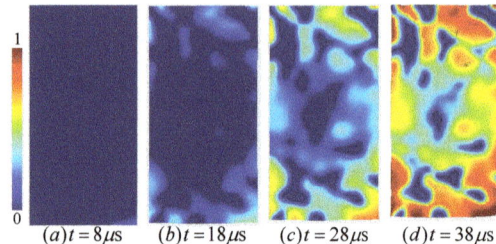

Figure 17. The evolution of damage field for the high amplitude load and cohesion realization with $l_c = 10$ mm at times: (**a**) 8 μs, (**b**) 18 μs, (**c**) 28 μs, and (**d**) 38 μs. The results are shown on the deformed mesh with a magnification factor of 100.

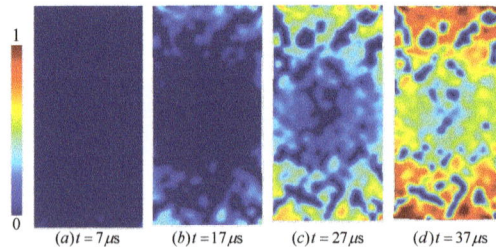

Figure 18. The evolution of damage field for the high amplitude load and cohesion realization with $l_c = 5$ mm at times: (**a**) 7 μs, (**b**) 17 μs, (**c**) 27 μs, and (**d**) 37 μs. The results are shown on the deformed mesh with a magnification factor of 100.

The statistical continuum damage model enhances the accuracy of conventional continuum damage models, and its solutions are more consistent with sharp interface fracture models. The reason are as follows. First, damage initiation zones from material weak points are more concentrated and better resemble crack nucleation events. Second, damaged zones tend to propagate in crack-like features with specific inclined directions rather than the diffuse response around the initiation points. For example, in Figure 18d, many localized zones resemble cracks at 45 degree and steeper relative to the vertical direction. This features qualitatively match other numerical and experimental observations [71–75]. Specifically, based on the MC failure criterion, cracks are formed at angles $\pm(45° + \phi/2)$ with respect to the compressive loading direction. This example demonstrates that a damage model based on uniform material properties not only misses crack-like damage localization features, but can also incorrectly predict the location of zones with the maximum overall damage accumulation (low load example).

3.2.3. Mesh Sensitivity

The mesh sensitivity of diffusive damage response for the sample with homogeneous properties was presented in Section 3.1.1. Here, we study the effect of mesh size for domains with heterogeneous cohesion that result in a localized damage response. Figure 19 compares damage responses for the domain with $l_c = 40$ mm at $t = 43$ μs. The results are presented for different load amplitudes and mesh sizes. The same results are presented in Figure 20 for the smallest correlation length $l_c = 5$ mm at $t = 36$ μs. While, there is a good agreement between the results obtained by medium and coarse meshes for both load conditions, the solutions for the largest correlation length in Figure 19 show a better agreement. This is due to the fact that the details of the solution are at the scale of the correlation length; thus, as smaller correlation lengths are used for material properties, finer finite elements should be used to accurately capture the details of the solution.

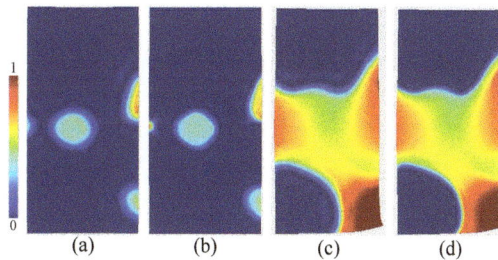

Figure 19. Damage responses at $t = 43$ μs for the domain with $l_c = 40$ mm with different meshes and load amplitudes: (**a**) low amplitude-medium mesh, (**b**) low amplitude-fine mesh, (**c**) high amplitude-medium mesh, and (**c**) high amplitude-fine mesh.

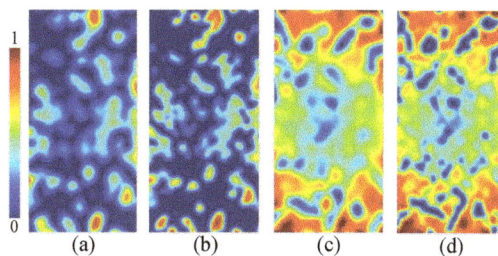

Figure 20. Damage responses at $t = 36$ μs for the domain with $l_c = 5$ mm with different meshes and load amplitudes: (**a**) low amplitude-medium mesh, (**b**) low amplitude-fine mesh, (**c**) high amplitude-medium mesh, and (**c**) high amplitude-fine mesh.

4. Conclusions

We presented a dynamic bulk damage model, based on the time-delay evolution law in [3]. The relaxation time τ_c indirectly introduces an intrinsic length scale for dynamic fracture problems. This resolves the mesh sensitivity problem of early local damage models. Moreover, by limiting the maximum damage rate, the model qualitatively captures stress rate effect, in that, both strength and toughness increase when the loading rate increases. The ODE form of the evolution model greatly simplifies the implementation of the damage model and maintains the hyperbolicity of the elastodynamic problem.

The coupled elastodynamic-damage problem was implemented by the aSDG method to solve a uniaxial compressive fracture problem for rock. The MC model is used to formulate a damage force model. In the process of damage accumulation, the effective stress tends from the initial elastic limit at $D = 0$ to its hydrostatic compressive value at $D = 1$. The MC model also captures rock strengthening effect as hydrostatic pressure increases. In contrast, damage models that are based on spectral positive and negative decomposition of strain (or stress) tensor, fail to model failure under compressive response.

To model the effect of material inhomogeneity, cohesion was assumed to be a random field. Two different macroscopic compressive load amplitudes were used for this study. For a homogeneous material, the higher load amplitude initiates damage as the compressive wave enters the domain, whereas for the lower load damage initiates only in the center of the domain where stress doubling effect occurs upon the intersection of compressive waves. Four lognormal fields with different correlation lengths l_c were generated for $\underline{c}$. It was shown that inhomogeneity could significantly alter the failure response of an otherwise homogeneous rock. For example, for the higher load amplitude, unlike the homogeneous case, damage initiates in the center of the domain. This is due to the particular form of the realized random field where a large zone of low $\underline{c}$ is sampled in the center of the domain. Moreover, for the lower load amplitude damage can initiate everywhere in the domain as the waves travel toward the center of the domain. This is due to the weaker sampled $\underline{c}$ at these locations, which does not require the stress wave doubling effect to initiate damage. Moreover, even the zones that eventually accumulate the highest damage can be significantly different between models with homogeneous and inhomogeneous properties, even as the correlation length tends to zero (low load amplitude example).

Another problem of using a homogeneous material model is the inability or difficulty of bulk damage models to capture sharp localization zones. In contrast, as lower correlation lengths were used for inhomogeneous domains, the fracture pattern became more realistic and resembled the results that are obtained by more accurate sharp interface models [43]. In particular, the MC model predicts fractures at $\pm(45 + \phi/2)$ degree angles with respect to the compressive load direction. For the lowest correlation lengths, localized damage zones with angles roughly in the range ±45 to $\pm(45 + \phi/2)$ are observed. These features are better resolved with the higher resolution finite element mesh, confirming that finer meshes are required for the solution of problems with more rapid variation of material properties.

There are several extensions to the present work. First, the form of effective stress (8) implies that friction coefficient is zero at complete damage ($D = 1$), whereas jointed (damaged) rock may still possess some residual friction coefficient. This will enhance the angle of localized regions in Figure 18d. Second, MC criterion is not appropriate for rock tensile fracture analysis and the damage force can be formulated by Hoek-Brown [76] and other more accurate models. Third, as shown in [77], rock anisotropy, for example induced by the existence of bedding planes, can affect fracture angle under compressive loading. Anisotropic failure criteria such as those in [78,79] can be used to formulate the damage force. Finally, mesh adaptive operations in spacetime [33] can drastically reduce the computational cost of the formulated aSDG method for this bulk damage model.

Author Contributions: B.B., R.A., and P.L.C. were responsible for the conceptualization, methodology, software development. P.L.C. was responsible for the random domain creation. B.B. was responsible for the formulation/implementation of the bulk damage model and dynamic fracture simulation/visualization. R.A. was responsible for aSDG software development writing review and editing, supervision, and project administration. The manuscript was written by R.A. and B.B.

Funding: The authors gratefully acknowledge partial support for this work via the U.S. National Science Foundation (NSF), CMMI—Mechanics of Materials and Structures (MoMS) program grant number 1538332 and NSF, CCF—Scalable Parallelism in the Extreme (SPX) program grant number 1725555.

Conflicts of Interest: The authors declare no conflict of interest.

Abbreviations

The following abbreviations are used in this manuscript:

MDPI	Multidisciplinary Digital Publishing Institute
DOAJ	Directory of open access journals
KL	Karhunen-Loève
aSDG	asynchronous Spacetime Discontinuous Galerkin Method

References

1. Dugdale, D.S. Yielding of steel sheets containing slits. *J. Mech. Phys. Solids* **1960**, *8*, 100–104. [CrossRef]
2. Barenblatt, G.I. The mathematical theory of equilibrium of cracks in brittle fracture. *Adv. Appl. Mech.* **1962**, *7*, 55–129.
3. Allix, O.; Feissel, P.; Thévenet, P. A delay damage mesomodel of laminates under dynamic loading: Basic aspects and identification issues. *Comput. Struct.* **2003**, *81*, 1177–1191. [CrossRef]
4. Alfano, G. On the influence of the shape of the interface law on the application of cohesive-zone models. *Compos. Sci. Technol.* **2006**, *66*, 723–730. [CrossRef]
5. Parrinello, F.; Failla, B.; Borino, G. Cohesive-frictional interface constitutive model. *Int. J. Solids Struct.* **2009**, *46*, 2680–2692. [CrossRef]
6. Nguyen, V.P. Discontinuous Galerkin/extrinsic cohesive zone modeling: Implementation caveats and applications in computational fracture mechanics. *Eng. Fract. Mech.* **2014**, *128*, 37–68. [CrossRef]
7. Abedi, R.; Haber, R.B. Spacetime simulation of dynamic fracture with crack closure and frictional sliding. *Adv. Model. Simul. Eng. Sci.* **2018**, *5*, 22. [CrossRef]
8. Spring, D.W.; Leon, S.E.; Paulino, G.H. Unstructured polygonal meshes with adaptive refinement for the numerical simulation of dynamic cohesive fracture. *Int. J. Fract.* **2014**, *189*, 33–57. [CrossRef]
9. Rangarajan, R.; Lew, A.J. Universal meshes: A method for triangulating planar curved domains immersed in nonconforming meshes. *Int. J. Numer. Methods Eng.* **2014**, *98*, 236–264. [CrossRef]
10. Abedi, R.; Omidi, O.; Enayatpour, S. A mesh adaptive method for dynamic well stimulation. *Comput. Geotech.* **2018**, *102*, 12–27. [CrossRef]
11. Belytschko, T.; Black, T. Elastic crack growth in finite elements with minimal remeshing. *Int. J. Numer. Methods Eng.* **1999**, *45*, 601–620. [CrossRef]
12. Moes, N.; Dolbow, J.; Belytschko, T. A finite element method for crack growth without remeshing. *Int. J. Numer. Methods Eng.* **1999**, *46*, 131–150. [CrossRef]
13. Khoei, A.; Vahab, M.; Hirmand, M. An enriched–FEM technique for numerical simulation of interacting discontinuities in naturally fractured porous media. *Comput. Methods Appl. Mech. Eng.* **2018**, *331*, 197–231. [CrossRef]
14. Duarte, C.A.; Babuška, I.; Oden, J.T. Generalized finite element methods for three-dimensional structural mechanics problems. *Comput. Struct.* **2000**, *77*, 215–232. [CrossRef]
15. Strouboulis, T.; Babuška, I.; Copps, K. The design and analysis of the Generalized Finite Element Method. *Comput. Methods Appl. Mech. Eng.* **2000**, *181*, 43–69. [CrossRef]
16. Silling, S.A.; Lehoucq, R. Peridynamic theory of solid mechanics. In *Advances in Applied Mechanics*; Elsevier: Amsterdam, The Netherlands, 2010; Volume 44, pp. 73–168.
17. Ha, Y.D.; Bobaru, F. Studies of dynamic crack propagation and crack branching with peridynamics. *Int. J. Fract.* **2010**, *162*, 229–244. [CrossRef]

18. Rabczuk, T.; Ren, H. A peridynamics formulation for quasi-static fracture and contact in rock. *Eng. Geol.* **2017**, *225*, 42–48. [CrossRef]

19. Lasry, D.; Belytschko, T. Localization limiters in transient problems. *Int. J. Solids Struct.* **1988**, *24*, 581–597. [CrossRef]

20. Loret, B.; Prevost, J.H. Dynamic strain localization in elasto-(visco-) plastic solids, Part 1. General formulation and one-dimensional examples. *Comput. Methods Appl. Mech. Eng.* **1990**, *83*, 247–273. [CrossRef]

21. Peerlings, R.; De Borst, R.; Brekelmans, W.; Geers, M. Gradient-enhanced damage modelling of concrete fracture. *Mech. Cohes.-Frict. Mater.* **1998**, *3*, 323–342. [CrossRef]

22. Comi, C. Computational modelling of gradient-enhanced damage in quasi-brittle materials. *Mech. Cohes.-Frict. Mater.* **1999**, *4*, 17–36. [CrossRef]

23. Peerlings, R.; Geers, M.; De Borst, R.; Brekelmans, W. A critical comparison of nonlocal and gradient-enhanced softening continua. *Int. J. Solids Struct.* **2001**, *38*, 7723–7746. [CrossRef]

24. Pereira, L.; Weerheijm, J.; Sluys, L. A new effective rate dependent damage model for dynamic tensile failure of concrete. *Eng. Fract. Mech.* **2017**, *176*, 281–299. [CrossRef]

25. Allix, O.; Deü, J.F. Delayed-damage modelling for fracture prediction of laminated composites under dynamic loading. *Eng. Trans.* **1997**, *45*, 29–46.

26. Lyakhovsky, V.; Hamiel, Y.; Ben-Zion, Y. A non-local visco-elastic damage model and dynamic fracturing. *J. Mech. Phys. Solids* **2011**, *59*, 1752–1776. [CrossRef]

27. Häussler-Combe, U.; Panteki, E. Modeling of concrete spallation with damaged viscoelasticity and retarded damage. *Int. J. Solids Struct.* **2016**, *90*, 153–166. [CrossRef]

28. Junker, P.; Schwarz, S.; Makowski, J.; Hackl, K. A relaxation-based approach to damage modeling. *Contin. Mech. Thermodyn.* **2017**, *29*, 291–310. [CrossRef]

29. de Borst, R.; Verhoosel, C.V. Gradient damage vs phase-field approaches for fracture: Similarities and differences. *Comput. Methods Appl. Mech. Eng.* **2016**, *312*, 78–94. [CrossRef]

30. Mandal, T.K.; Nguyen, V.P.; Heidarpour, A. Phase field and gradient enhanced damage models for quasi-brittle failure: A numerical comparative study. *Eng. Fract. Mech.* **2019**, *207*, 48–67. [CrossRef]

31. Bahmani, B.; Abedi, R. Asynchronous Spacetime Discontinuous Galerkin Formulation for a Hyperbolic Time-Delay Bulk Damage Model. *J. Eng. Mech.* **2019**, accepted.

32. Abedi, R.; Haber, R.B.; Petracovici, B. A spacetime discontinuous Galerkin method for elastodynamics with element-level balance of linear momentum. *Comput. Methods Appl. Mech. Eng.* **2006**, *195*, 3247–3273. [CrossRef]

33. Abedi, R.; Haber, R.B.; Thite, S.; Erickson, J. An *h*–adaptive spacetime–discontinuous Galerkin method for linearized elastodynamics. *Revue Européenne de Mécanique Numérique (Eur. J. Comput. Mech.)* **2006**, *15*, 619–642.

34. Bischoff, P.H.; Perry, S.H. Impact behavior of plain concrete loaded in uniaxial compression. *J. Eng. Mech.* **1995**, *121*, 685–693. [CrossRef]

35. Al-Ostaz, A.; Jasiuk, I. Crack initiation and propagation in materials with randomly distributed holes. *Eng. Fract. Mech.* **1997**, *58*, 395–420. [CrossRef]

36. Kozicki, J.; Tejchman, J. Effect of aggregate structure on fracture process in concrete using 2D lattice model. *Arch. Mech.* **2007**, *59*, 365–384.

37. Bazant, Z.; Novak, D. Probabilistic nonlocal theory for quasibrittle fracture initiation and size effect—I: Theory. *J. Eng. Mech.* **2000**, *126*, 166–174. [CrossRef]

38. Rinaldi, A.; Krajcinovic, D.; Mastilovic, S. Statistical damage mechanics and extreme value theory. *Int. J. Damage Mech. (USA)* **2007**, *16*, 57–76. [CrossRef]

39. Bazant, Z.P.; Le, J.L. *Probabilistic Mechanics of Quasibrittle Structures: Strength, Lifetime, and Size Effect*; Cambridge University Press: Cambridge, UK, 2017.

40. Weibull, W. A statistical theory of the strength of materials. *R. Swed. Inst. Eng. Res.* **1939**, *151*, 1–45.

41. Weibull, W. A statistical distribution function of wide applicability. *J. Appl. Mech.* **1951**, *18*, 293–297.

42. Abedi, R.; Omidi, O.; Clarke, P. Numerical simulation of rock dynamic fracturing and failure including microscale material randomness. In Proceedings of the 50th US Rock Mechanics/Geomechanics Symposium, Houston, TX, USA, 26–29 June 2016; ARMA 16-0531.

43. Abedi, R.; Haber, R.; Elbanna, A. Mixed-mode dynamic crack propagation in rocks with contact-separation mode transitions. In Proceedings of the 51th US Rock Mechanics/Geomechanics Symposium, San Francisco, CA, USA, 25–28 June 2017; ARMA 17-0679.

44. Abedi, R.; Haber, R.B.; Clarke, P.L. Effect of random defects on dynamic fracture in quasi-brittle materials. *Int. J. Fract.* **2017**, *208*, 241–268. [CrossRef]

45. Clarke, P.; Abedi, R.; Bahmani, B.; Acton, K.; Baxter, S. Effect of the spatial inhomogeneity of fracture strength on fracture pattern for quasi-brittle materials. In Proceedings of the ASME 2017 International Mechanical Engineering Congress & Exposition, IMECE 2017, Tampa, FL, USA, 3–9 November 2017; p. V009T12A045.

46. Hoek, E.; Brown, T. *Underground Excavations in Rock*; Geotechnics and Foundations; Taylor & Francis: Stroud, UK, 1980.

47. Suffis, A.; Lubrecht, T.A.; Combescure, A. Damage model with delay effect: Analytical and numerical studies of the evolution of the characteristic damage length. *Int. J. Solids Struct.* **2003**, *40*, 3463–3476. [CrossRef]

48. Murakami, S. *Continuum Damage Mechanics*; Springer: New York, NY, USA, 2012.

49. Mazars, J. Application de la Mécanique de l'Endommagement au Comportement non Linéaire et à la Rupture du Béton de Structure. Ph.D. Thesis, Universite Pierre et Marie Curie, Paris, France, 1984.

50. Jirásek, M.; Patzák, B. Consistent tangent stiffness for nonlocal damage models. *Comput. Struct.* **2002**, *80*, 1279–1293. [CrossRef]

51. Moreau, K.; Moës, N.; Picart, D.; Stainier, L. Explicit dynamics with a non-local damage model using the thick level set approach. *Int. J. Numer. Methods Eng.* **2015**, *102*, 808–838. [CrossRef]

52. Londono, J.G.; Berger-Vergiat, L.; Waisman, H. An equivalent stress-gradient regularization model for coupled damage-viscoelasticity. *Comput. Methods Appl. Mech. Eng.* **2017**, *322*, 137–166. [CrossRef]

53. Abedi, R.; Chung, S.H.; Erickson, J.; Fan, Y.; Garland, M.; Guoy, D.; Haber, R.; Sullivan, J.M.; Thite, S.; Zhou, Y. Spacetime meshing with adaptive refinement and coarsening. In Proceedings of the Twentieth Annual Symposium on Computational Geometry, SCG '04, Brooklyn, NY, USA, 8–11 June 2004; ACM: New York, NY, USA, 2004; pp. 300–309.

54. Dimas, L.; Giesa, T.; Buehler, M. Coupled continuum and discrete analysis of random heterogeneous materials: Elasticity and fracture. *J. Mech. Phys. Solids* **2014**, *63*, 481–490. [CrossRef]

55. Carmeliet, J.; Hens, H. Probabilistic nonlocal damage model for continua with random field properties. *J. Eng. Mech.* **1994**, *120*, 2013–2027. [CrossRef]

56. Zhou, F.; Molinari, J. Stochastic fracture of ceramics under dynamic tensile loading. *Int. J. Solids Struct.* **2004**, *41*, 6573–6596. [CrossRef]

57. Schicker, J.; Pfuff, M. Statistical Modeling of fracture in quasi-brittle materials. *Adv. Eng. Mater.* **2006**, *8*, 406–410. [CrossRef]

58. Levy, S.; Molinari, J. Dynamic fragmentation of ceramics, signature of defects and scaling of fragment sizes. *J. Mech. Phys. Solids* **2010**, *58*, 12–26. [CrossRef]

59. Daphalapurkar, N.; Ramesh, K.; Graham-Brady, L.; Molinari, J. Predicting variability in the dynamic failure strength of brittle materials considering pre-existing flaws. *J. Mech. Phys. Solids* **2011**, *59*, 297–319. [CrossRef]

60. Acton, K.A.; Baxter, S.C.; Bahmani, B.; Clarke, P.L.; Abedi, R. Voronoi tessellation based statistical volume element characterization for use in fracture modeling. *Comput. Methods Appl. Mech. Eng.* **2018**, *336*, 135–155. [CrossRef]

61. Bahmani, B.; Yang, M.; Nagarajan, A.; Clarke, P.L.; Soghrati, S.; Abedi, R. Automated homogenization-based fracture analysis: Effects of SVE size and boundary condition. *Comput. Methods Appl. Mech. Eng.* **2019**, *345*, 701–727. [CrossRef]

62. Karhunen, K.; Selin, I. *On Linear Methods in Probability Theory*; Rand Corporation: Santa Monica, CA, USA, 1960.

63. Loéve, M. *Probability Theory*; Springer: New York, NY, USA, 1977.

64. Ghanem, R.; Spanos, P. *Stochastic Finite Elements: A Spectral Approach*; Springer: New York, NY, USA, 1991.

65. Clarke, P.; Abedi, R. Fracture modeling of rocks based on random field generation and simulation of inhomogeneous domains. In Proceedings of the 51th US Rock Mechanics/Geomechanics Symposium, San Francisco, CA, USA, 25–28 June 2017; ARMA 17-0643.

66. Häussler-Combe, U.; Kühn, T. Modeling of strain rate effects for concrete with viscoelasticity and retarded damage. *Int. J. Impact Eng.* **2012**, *50*, 17–28. [CrossRef]

67. Bahmani, B.; Clarke, P.; Abedi, R. A bulk damage model for modeling dynamic fracture in rock. In Proceedings of the 52th US Rock Mechanics/Geomechanics Symposium, Seattle, Washington, USA, 17–20 June 2018; ARMA 18-826.

68. Bazant, Z.P.; Belytschko, T.B.; Chang, T.P. Continuum theory for strain-softening. *J. Eng. Mech.* **1984**, *110*, 1666–1692. [CrossRef]

69. Bažant, Z.P.; Belytschko, T.B. Wave propagation in a strain-softening bar: Exact solution. *J. Eng. Mech.* **1985**, *111*, 381–389. [CrossRef]

70. Xu, T.; Yang, S.; Chen, C.; Yang, T.; Zhang, P.; Liu, H. Numerical Investigation of Damage Evolution and Localized Fracturing of Brittle Rock in Compression. *J. Perform. Constr. Facil.* **2017**, *31*, 04017065. [CrossRef]

71. Tang, C.; Tham, L.; Lee, P.; Tsui, Y.; Liu, H. Numerical studies of the influence of microstructure on rock failure in uniaxial compression—Part II: Constraint, slenderness, and size effect. *Int. J. Rock Mech. Min. Sci.* **2000**, *37*, 571–583. [CrossRef]

72. Teng, J.; Zhu, W.; Tang, C. Mesomechanical model for concrete. Part II: Applications. *Mag. Concr. Res.* **2004**, *56*, 331–345. [CrossRef]

73. Li, G.; Tang, C.A. A statistical meso-damage mechanical method for modeling trans-scale progressive failure process of rock. *Int. J. Rock Mech. Min. Sci.* **2015**, *74*, 133–150. [CrossRef]

74. Dinç, Ö.; Scholtès, L. Discrete Analysis of Damage and Shear Banding in Argillaceous Rocks. *Rock Mech. Rock Eng.* **2018**, *51*, 1521–1538. [CrossRef]

75. Rangari, S.; Murali, K.; Deb, A. Effect of Meso-structure on Strength and Size Effect in Concrete under Compression. *Eng. Fract. Mech.* **2018**, *195*, 162–185. [CrossRef]

76. Hoek, E. Strength of jointed rock masses. *Geotechnique* **1983**, *33*, 187–223. [CrossRef]

77. Abedi, R.; Clarke, P.L. Modeling of rock inhomogeneity and anisotropy by explicit and implicit representation of microcracks. In Proceedings of the 52nd US Rock Mechanics/Geomechanics Symposium, Seattle, WA, USA, 17–20 June 2018; ARMA 18-151-0228-1094.

78. Pietruszczak, S.; Mroz, Z. On failure criteria for anisotropic cohesive-frictional materials. *Int. J. Numer. Anal. Methods Geomech.* **2001**, *25*, 509–524. [CrossRef]

79. Lee, Y.K.; Pietruszczak, S. Analytical representation of Mohr failure envelope approximating the generalized Hoek-Brown failure criterion. *Int. J. Rock Mech. Min. Sci.* **2017**, *100*, 90–99. [CrossRef]

applied
sciences

MDPI

Article

The Construction of Equivalent Particle Element Models for Conditioned Sandy Pebble

Panpan Cheng [1,2], Xiaoying Zhuang [1,2,*], Hehua Zhu [1,2] and Yuanhai Li [3]

[1] State Key Laboratory for Disaster Reduction in Civil Engineering, Shanghai 200092, China; chengpp1992@126.com (P.C.); zhuhehua@tongji.edu.cn (H.Z.)
[2] Key Laboratory of Geotechnical and Underground Engineering of the Ministry of Education, Tongji University, Shanghai 200092, China
[3] State Key Laboratory for Geomechanics and Deep Underground Engineering, China University of Mining and Technology, Xuzhou 221116, Jiangsu, China; lyh@cumt.edu.cn
* Correspondence: xiaoyingzhuang@tongji.edu.cn; Tel.: +86-21-65985140

Received: 11 February 2019; Accepted: 25 February 2019; Published: 18 March 2019

Abstract: When a shield tunneling machine based on earth pressure balance (EPB) bores through the sandy pebble stratum, the conditioned sandy pebble inside the soil cabin of shield machine is an aggregation of numerous granules with pebble grains as skeleton. It is essential to construct a reasonable particle element model of the conditioned sandy pebble before carrying out discrete element simulation of the soil cabin system. Sandy pebble belongs to a kind of frictional material, the friction behavior of which is highly sensitive to the angularity of the grains. In order to take the shape effect into account, two particle element models—single sphere with rolling resistance and cluster of particles—were attempted in this paper. The undetermined contact parameters in two models were calibrated by virtue of least squares support vector regression machine (LS-SVR). With the purpose of making both the flow behavior and mechanical properties of the modeled soil consistent with reality, the calibration targets the result of laboratory test of slump test and large-scale triaxial test as goals. The presented comparative analysis indicates that the two established particle models both can well describe the strength property and fluidity of the actual soil due to properly calibrated parameters. So, the rolling resistance and cluster models are two effective ways to incorporate the shape effect. Besides, because of the angularity of the nonspherical grains, there exists strong interlocking between clusters. So, in the cluster model, relatively smaller rolling friction coefficient and surface energy are required. It is also concluded that the single sphere model is more computationally efficient than the cluster model.

Keywords: EPB shield machine; conditioned sandy pebble; particle element model; parameters calibration

1. Introduction

Shield tunneling machines based on earth pressure balance (EPB) have been widely used in the construction of tunnels in urban areas. It is quite common that EPB type shield machines go through heterogeneous ground conditions, typically the sandy pebble stratum. In China, a total of 23 cities have constructed or to build metro tunnels in sandy pebble stratum, such as Chengdu, Beijing, Shenyang [1]. Japan also encountered sandy pebble in the construction of the Hiroshima Metro. Sandy pebble has attracted great interest among geotechnical researchers. Sandy pebble is a special type of geomaterial, a mixture of hard gravel and soft sand. The weakly weathered pebble grains, with high compressive strength and large grain size, are intermingled with soft and flowable sands. With pebble blocks as skeleton and weakly bonded interface, sandy pebble is a kind of loose and highly discrete noncontinuum material. Besides, it is sensitive to the external disturbance since the point-to-point contact between grains [2]. Obviously, sandy pebble soil cannot satisfy the requirements of EPB tunneling and it is essential to perform soil conditioning. Conditioned soil, which presents good fluidity and plasticity, can flow as well as form certain shape. So it is in a transient state between solid and fluid, and more like a solid-like material.

The soil cabin system of EPB shield machine is in charge of several important tasks, such as soil excavation and tunnel face stabilization, so its performance to a large degree determines the construction efficiency and safety. However, under the complex interaction between the soil and shield machine, the soil cabin system shows both time dependency and randomness. Therefore, it is challenging to investigate the dynamic behavior of the whole system. Classical computational methods for modeling such type of systems include the discrete element method (DEM) [1–3], meshless and particle methods [4–9], efficient remeshing techniques in the context of FEM [10–14], DH-PD [15,16] and specific multiscale methods [17–20].

In the DEM simulation, one of the difficult tasks is to construct the equivalent DEM model of the conditioned sandy pebble. Several types of particle are available in DEM simulation, typically single sphere, cluster of particles and bonded particles. Wherein, the single sphere is used for globular particles, while the other two for particles with irregular shape. In addition, the bonds within a cluster are impossible to break and the particles comprising the cluster remain at a fixed distance from each other, while the bonded particles are liable to separate once the suffered force exceeds its bonding strength. In reality, the grain shape of geomaterial is generally irregular. Especially for sandy pebble, which is a kind of frictional material, its friction behavior is highly sensitive to the angularity of the grains. Adopting single sphere, which is easy to roll without rolling resistance, the simulation may result in a deviation from the actuality. In contrast, the other two types can better characterize the microscopic shape of actual geomaterial. However, reconstructing the actual shape of all particles will make the complexity increase greatly; it is also impractical. Rahul et al. pointed out that nonspherical particles are more accurate than spherical particles. However, for nonspherical particles, when the sphericity is decreased to a certain value, the simulation results are similar. So there is no need to adopt highly precise particle shape. However, determining the parameters for discrete element model is another tricky task. Presently, there are mainly three approaches, namely analytical solution, experimental measurement and inverse analysis. Due to various simplifications of substantial material, the analytical solution proposed by Mindlin can only provide a rough estimation and qualitative analysis [21], which remains distant for engineering application. The relevant experiments include impact test [22,23], tribometer test [24,25], and drop particle test [26]. Besides, a set of device for calibrating contact parameters has been developed in Chalmers, Sweden [27]. However these experiments are usually applicable to large-size grains; these test devices are not widely available. The mostly employed method is inverse analysis method, which includes trial-and-error, fitting relation of macroscopic and mesoscopic parameters, and machine learning [28]. The trial-and-error is time-consuming due to repeated simulations and it is somewhat unreliable with strong subjective factors involved. Additionally, since highly nonlinear relationship exists between the macroscopic and mesoscopic parameters of the granular material, it is difficult to describe the relationship with explicit expressions. In the inverse analysis, the parameters are usually calibrated solely according to the macroscopic mechanical properties. Whereas, since the obtained parameters are not the unique solution, the strength-desired parameters can not necessarily meet the requirement of fluidity.

In this paper, the conditioned soil mixture of middle fine sand and pebble obtained from Beijing region is selected to construct its equivalent DEM model for further DEM simulation of soil cabin system within EDEM. The shape effects are considered by two approaches, namely defining rolling resistance and adopting cluster model. The machine learning method as adopted to determine the mesoscopic parameters in accordance to both the flow behavior and mechanical properties of the actual soil.

2. Properties of Conditioned Sandy Pebble

2.1. Conditioner Constituents

Based on the findings from slump tests in laboratory and tentative tunneling on site, the optimum injection ratio of foam (with expansion ratio of 15) was found to be approximately 45–60% subject to the condition that the addition ratio of bentonite slurry (with mass concentration of 12.5%) was 7%.

2.2. Grain Size Distribution

The granulometric curves of natural soil and conditioned soil are plotted in Figure 1. It can be found that the cumulative mass percentage corresponding to the grain size within the range between 0.32 mm and 5 mm is larger for the conditioned soil, which indicates the increase of fine-sized grains.

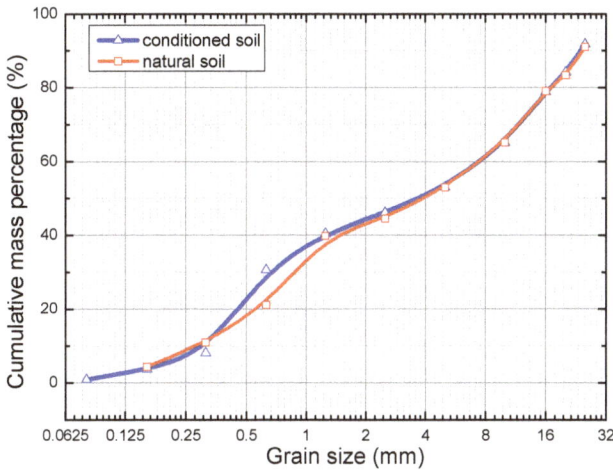

Figure 1. Granulometric curves of natural soil and conditioned soil.

2.3. Fluidity

During tunneling, the slump of soil in the discharging vehicle was tested at the interval of 3–5 rings of shield tunneling construction. Figure 2 shows the photo of slump test at the scene. It can be seen that the soil was transformed into a plastic paste with good fluidity and the fine particles wrap the larger grains completely due to enhanced viscosity. As shown in Figure 3, most of the measured slump magnitude ranges from 175 to 180 mm.

Figure 2. Slump test at the construction site.

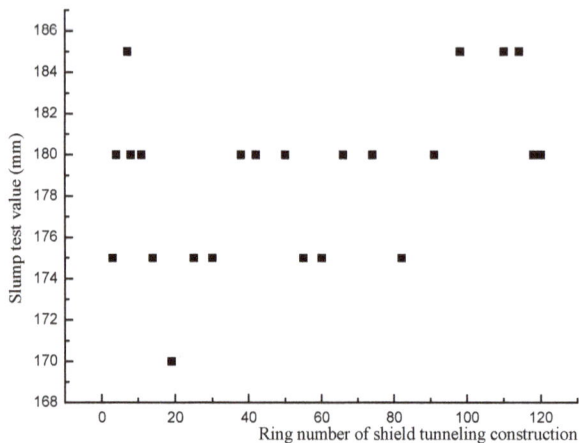

Figure 3. The slump value of soil from muck discharging vehicle.

2.4. Shear Strength

The triaxial shear tests under unconsolidated and undrained conditions for conditioned sandy pebble were conducted by using large triaxial shear apparatus (SJ-70, China Institute of Water Resources and Hydropower Research, Beijing, China), as shown in Figure 4. When making test samples, the prepared conditioned soil was compacted by means of vibration by five layers. The molded sample has a diameter of 300 mm, height of 700 mm, and wet density of 2 g/cm^3. During tests, the confining pressure was set to 0.4 MPa, 0.6 MPa, and 0.8 MPa, successively, and the shear rate was maintained at 1 mm/min. Figure 5 records the measured stress–strain curves. Since no peak can be seen in the stress–strain curve, the corresponding point with strain of 15% is treated as failure point. Subsequently, Morh's circles and their enveloping line are plotted in Figure 6, from which the total shear strength indexes are determined as cohesion $c_u = 26$kPa and internal friction angle $\varphi_u = 34.57°$.

Figure 4. Large-scale triaxial shear test.

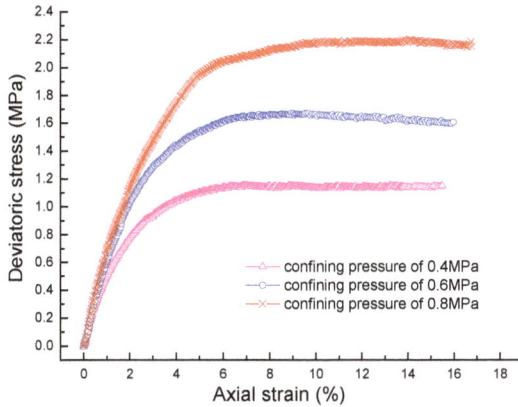

Figure 5. Stress–strain curves.

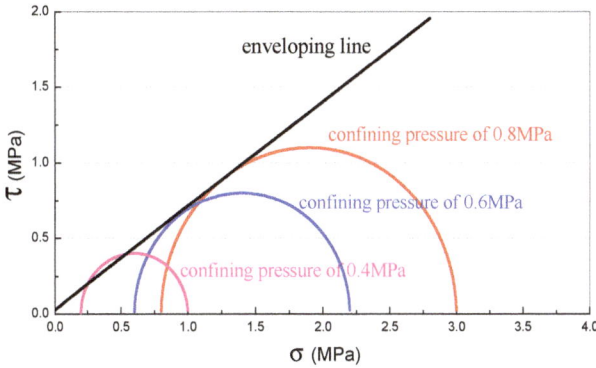

Figure 6. Morh's circles and enveloping line of conditioned sandy pebble.

3. Geometric Model of the Particle Element

The particle element model mainly involves geometric model and physical model. As for the geometric model, it includes three aspects: particle shape, particle size distribution, and particle spatial arrangement.

3.1. Particle Shape

To take into consideration the angularity effect induced by irregular shape, the alternative approaches include exerting rolling resistance on single sphere and adopting nonspherical particles. So this paper involves two types of shape—basic sphere and cluster of particles—furthermore, the simulation result and computational cost of the two established models can be comparatively analyzed. Figure 7 shows the adopted geometry template of cluster, which is randomly selected for the irregular pebble. And the template is fitted with four same spheres (see Figure 7). It can be seen that the constructed cluster is nearly an ellipsoid, and the ratio of its polar radius, short equatorial radius, and long equatorial radius is 1:1.19:1.69.

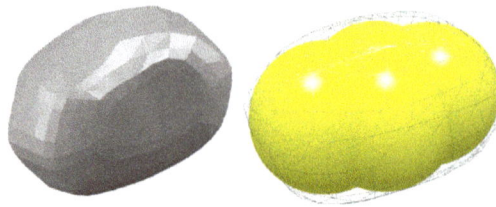

Figure 7. 3D geometry of realistic irregular pebble to the left and cluster model to the right.

3.2. Particle Size Distribution

Particle size distribution has certain effect on the macroscopic deformation and strength of bulk materials. Ideally, the particle size should be consistent with the substantial material. However, since the actual soil grains are generally too small, which will result in huge amount of particle elements and hence unacceptable computational cost, proper handling of the particle size is therefore necessary. It is generally accepted in the community that the simulation is aimed to investigate the macroscopic behavior of bulk materials instead of detailed interactions between particles. Thus, the particles of soil can be upscaled into surrogate particles which are much larger than the physical sizes of the grains in order to reduce the computational cost as long as a proper surrogate model can be found and validated.

From granulometric curve in Figure 1, it can be seen that the conditioned soil contains large amount of fine grains and are continuously distributed in numerous sizes. In order to reduce the particle number, the particle size is amplified here. Besides, since it is hardly possible to involve all particle sizes, only a few representative sizes are selected and herein the aperture diameters of the sieve meshes used in the sieving test are designated. Moreover, the large grains act as the main medium for force transmission in the soil mass, while most of the small grains only act as filler to fill the porous space. It can be concluded that omitting small particles has slight effect on the resultant mechanical response. So the original particle size is not only multiplied and discretized, but also truncated.

However, the determination of particle sizes and amplification factor should take into account computational cost, simulation accuracy, as well as opening size of shield machine. According to relevant experience, the number of particles should be controlled within 300,000. With single sphere model as example, Table 1 presents the particle number required for a soil box (24 m × 8 m × 2m) when adopting different size combinations and the corresponding minimum amplification factor to ensure an acceptable particle number. Additionally, the throughput capacity of the cutterhead and screw conveyor is also considerable. In view of the inner diameter of screw conveyor only 770 mm, the particle size cannot be too large: otherwise clogging and jamming accidents are easy to occur. On the premise of meeting the above two requirements and to model the material as realistically as possible, it is decided to magnify the particle 6 times and to select the size combination of 40 mm, 25 mm, and 20 mm. And, the particles of 240 mm, 150 mm, and 120 mm account for 10.82%, 9.02%, and 80.16% of the total mass, respectively.

Table 1. The required particle number when adopting different size combinations.

Combination of Particle Size (mm)	Particle Number ($\times 10^6$)	Minimum Amplification Factor	Particle Number after Amplification
40-25-20-16-10-5	2219	20	277,375
40-25-20-16-10	358.4	11	269,271
40-25-20-16	107.1	8	209,180
40-25-20	58.487	6	270,773
40-25	31.924	5	255,392

3.3. Particle Arrangement

In order to embody the significant anisotropy and spatial variability of soil mass, the particles in this simulation are arranged by random function, and they are oriented randomly when generated.

4. Physical Model between Particles

4.1. Contact Components

Compared with the natural sandy pebble, the cohesion among conditioned grains has been greatly enhanced under the action of foam and bentonite slurry. Thus cementitious interaction should be introduced in the contact model in addition to the standard contact action. In this study, the sphere and cluster employ the Hertz–Mindlin model (H-M) with the Johnson–Kendall–Roberts (JKR) cohesion model. The physical analogue characterized by several contact components for H-M with JKR cohesion model can be seen in Figure 8. Among them, the spring is used to simulate the elastic contact force between particles P_i and P_j The damper is a description of energy dissipation in the process of imperfect elastic collision. The frictional components include resistance devices to relative slide and rotation. The viscous components mainly provide resistance to tension, shear and torsion severally in the direction of normal, tangential and rotation. Those viscous components can simulate the viscous force exerted on pebble grains by foam and bentonite slurry. While the breaker implies the tension vanishes once the bonding between particles breaks.

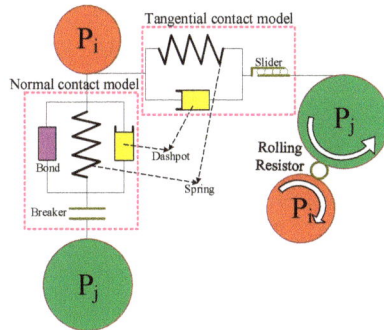

Figure 8. Physical contact models of conditioned soil.

4.2. Contact Constitutive Relation

A schematic diagram of contact vectors in the H-M model can be found in Figure 9. In the model, the calculation of normal elastic force $F_{nij}{}^e$ is based on Hertzian contact theory [29], while the tangential elastic force $F_{tij}{}^e$ was derived by Mindlin and Deresiewicz [21,30]. As for the energy dissipation in the contact process, Tsuji et al. [31] proposed a nonlinear normal damping force $F_{nij}{}^d$ and tangential damping force $F_{tij}{}^d$. Hence, the contact action exerted on particle i by particle j can be expressed as follows

$$\begin{cases} F_{nij}{}^e = \frac{4}{3}E^*\sqrt{R^*}u_n{}^{\frac{3}{2}} \\ F_{tij}{}^e = -k_t u_t \\ F_{nij}{}^d = -2\sqrt{\frac{5}{6}}\beta\sqrt{m^*k_n{}^*}v_n{}^{rel} \\ F_{tij}{}^d = -2\sqrt{\frac{5}{6}}\beta\sqrt{m^*k_t}v_t{}^{rel} \end{cases} \tag{1}$$

where, u_n and u_t denote the normal and tangential overlap of particle i and particle j, respectively; k_t is shear stiffness, $k_t = 8G^*\sqrt{R^*u_n}$; $k_n{}^*$ denotes equivalent normal stiffness, $k_n{}^* = 2E^*\sqrt{R^*u_n}$; E^*, R^*, m^*, and G^* denote Young's modulus, radius, mass, and shear modulus of particles in contact, respectively; β denotes damping coefficient, which is associated to the restitution coefficient e by

$\beta = \frac{\ln e}{\sqrt{\ln^2 e + \pi^2}}$; and $v_n{}^{\text{rel}}$ and $v_t{}^{\text{rel}}$ denote the normal and tangential components of relative speed of two particles, respectively.

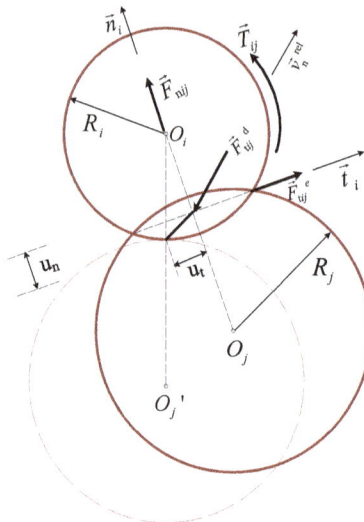

Figure 9. Schematic representation of Hertz–Mindlin contact model.

The tangential force obeys Coulomb law of friction [32], that is to say if $\left| \vec{F}_{tij}{}^e + \vec{F}_{tij}{}^d \right| >$ $F_\mu = \mu_s \left| \vec{F}_{nij} \right|$ (F_μ is friction force and μ_s denotes the coefficient of sliding friction), then slip occurs. While the rolling friction is accounted for by applying directional constant torque model [33], namely $T_\mu = -\mu_r \left| \vec{F}_{nij} \right| r_i \vec{\omega}_i$, where μ_r is the coefficient of rolling friction, r_i is the distance from the mass center of particle i to the contact point, ω_i denotes the unit vector of angular velocity.

To describe adhesive action, JKR theory enriched the Hertz–Mindlin model with normal cohesion force, and the normal elastic force $F_{nij}{}^e$ is revised to [34]

$$F_{nij}{}^{\text{JKR}} = -4\sqrt{\pi k E^*} \alpha^{\frac{3}{2}} + \frac{4E^*}{3R^*} \alpha^3 \tag{2}$$

where, k denotes surface energy per unit of contact area, measured in J/m^2; the relationship between coefficient α and normal overlap u_n can be expressed as $u_n = \frac{\alpha^2}{R^*} - \sqrt{\frac{4\pi k \alpha}{E^*}}$.

5. Calibration of Model Parameters

5.1. Undetermined Parameters

The macroscopic and mesoscopic parameters in DEM model can be subdivided into three categories. The intrinsic parameters, including shear modulus G, Poisson's ratio v, and density ρ, are inherent attribute of material itself and irrelevant to the external factors. The H-M contact parameters include restitution coefficient e, static friction coefficient μ_s, and rolling friction coefficient μ_r. The third category is about cohesion parameter and different model involves different cohesion parameters. JKR cohesion model only contains surface energy k. Table 2 lists all of the involved materials and contact parameters. Among them, some are either measured or obtained from literature and the others need to be calibrated. So there are altogether three undetermined parameters (e, μ_r, and k) in the H-M with JKR cohesion model.

Table 2. The input parameters for the simulation model.

Input Parameter	Unit	Value	Comment
Material Parameter			
Density ρ	kg/m^3	2700	Measured by overflow method
Poisson's ratio ν	n/a	0.15	Hu et al. (2013) [35]
Yong's modulus E	MPa	5×10^4	Hu et al. (2013) [35]
Shear modulus G	MPa	2.17×10^4	$G = \frac{E}{2(1+\nu)}$
H-M Contact parameter			
Restitution coefficient e	n/a	To be calibrated	
Static friction coefficient μ_s	n/a	0.6891	$\mu_s \approx \tan \varphi_u$
Rolling friction coefficient μ_r	n/a	To be calibrated	
Cohesion parameter (JKR model)			
Surface energy k	J/m^2	To be calibrated	

5.2. Method

With the purpose of making the flow behavior and mechanical properties of the modeled soil consistent with the reality, the calibration procedures are as follows. First of all, we established the least squares support vector machine (LS-SVM) with unknown parameters as input variables and the slump value as output variable. Then generate 10,000 parameter combinations at random and predict their corresponding slump value with the constructed LS-SVM. Subsequently, use the parameters that can turn out a desired slump value to simulate the large-scale triaxial shear test. Among all the triaxial tests, the one resulting in smallest deviation of stress–strain curve from measured curve in laboratory is regarded as calibration result.

5.3. Calibration Process

5.3.1. Range of Parameters

Before the establishment of LS-SVM, the correlations between the slump value and three undetermined parameters, as well as the value range of each parameter are analyzed by single-variable method. Figure 10 shows the established discrete element models for slump test. Similarly, the slump cone has been enlarged 6 times with top diameter of 600 mm, bottom diameter of 1200 mm, and height of 1800 mm. In the simulation, the lifting speed is 0.36 m/s and the whole lifting process is completed in 5 s.

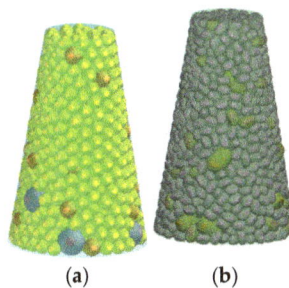

(a)   (b)

Figure 10. Slump test model of sphere (**a**) and cluster (**b**).

Figure 11 records the slump value against the surface energy k in the cases of single sphere (a) and cluster (b) when $e = 0.15$ and $\mu_r = 0.02, 0.05, 0.1, 0.15, 0.2$, successively. As is indicated in the figures, the slump value is in negative correlation to k. It can be explained by (1) low surface energy: the mutual cohesion is rather weak and the interparticle forces cannot resist the particle weight, therefore the aggregation is easy to scatter into single particles, which makes for a large slump value (Figure 12a,e). (2) When the surface energy improves, the internal contact force is strong enough to resist gravity, thus the aggregation can deform continuously and displays good fluidity and plasticity as the actual conditioned soil. Consequently, the slump value is reduced (Figure 12b,f). (3) If the surface energy is excessively high, the soil presents poor fluidity thus leading to a small slump (Figure 12c,g). In addition, it can be seen from Figure 11 that the slump value is also negatively correlated to μ_r. This is because the strong rolling friction contributes to a strong interlocking action and makes it harder to deform.

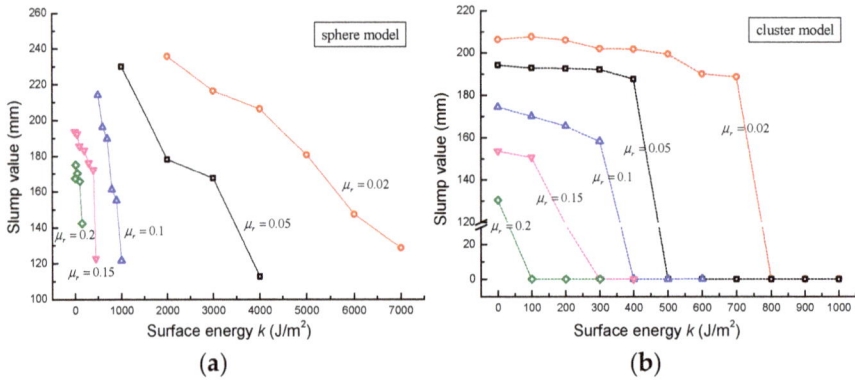

Figure 11. The variation of slump value versus surface energy: (**a**) sphere model and (**b**) cluster model.

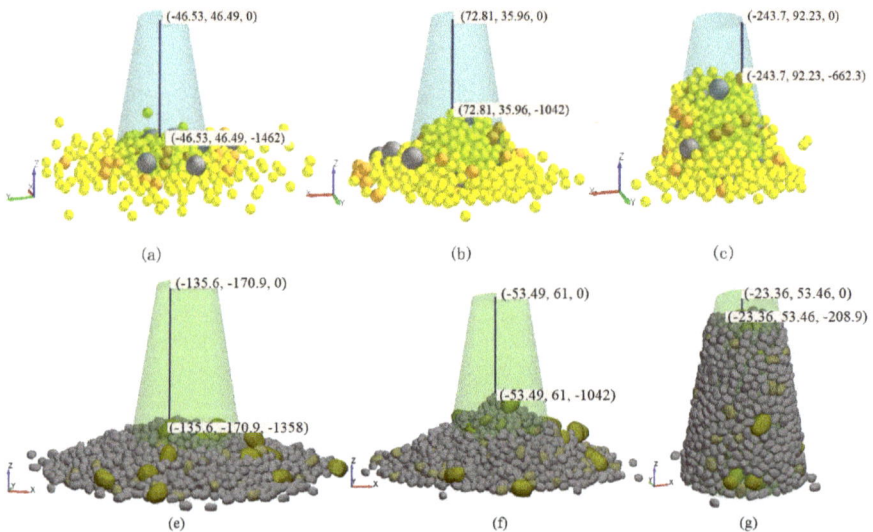

Figure 12. The simulation result of slump test: (**a–c**) sphere model and (**e–g**) cluster model.

Figure 13 shows the variation of slump value versus e in six different scenarios of μ_r and k (taking sphere model as example). It is obvious that no strict correlation exists between slump value and restitution coefficient e. On the whole, the slump value increases as e improves.

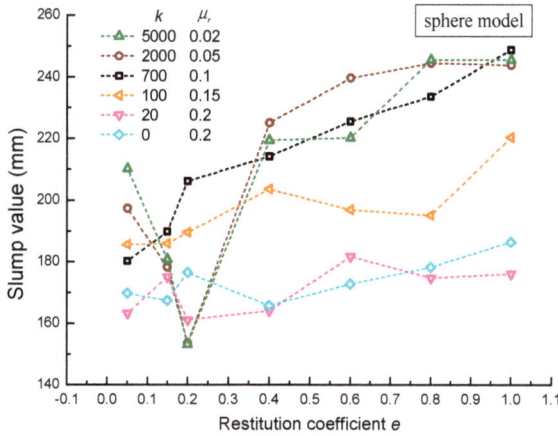

Figure 13. The slump value against restitution coefficient.

On the basis of literature research, the range of the restitution coefficient is set as $0 < e \leq 0.6$, while the scope of the other two parameters are determined according to their correlation with the slump value. Figure 14a,b shows the slump value against rolling friction coefficient when $k = 0$ respectively in the case of sphere and cluster. It can be concluded that once $\mu_r > 0.22$ for sphere or $\mu_r > 0.15$ for cluster, the slump value is less than 160mm regardless of the value of e. It means even without cohesive force between particles, the frictional interaction is strong enough to resist the flow of soil. In such a situation, the slump value can no more meet the requirements. Therefore, the rolling friction coefficient should be set within 0.22 and 0.15, respectively, for sphere and cluster. Similarly, as shown in Figure 15, when $\mu_r = 0$ and $k > 19,500$ for sphere ($k > 2660$ for cluster), the slump value cannot fall into the satisfactory range. So the reasonable scope of surface energy is $0 < k < 19,500$ for sphere and $0 < k < 2660$ for cluster.

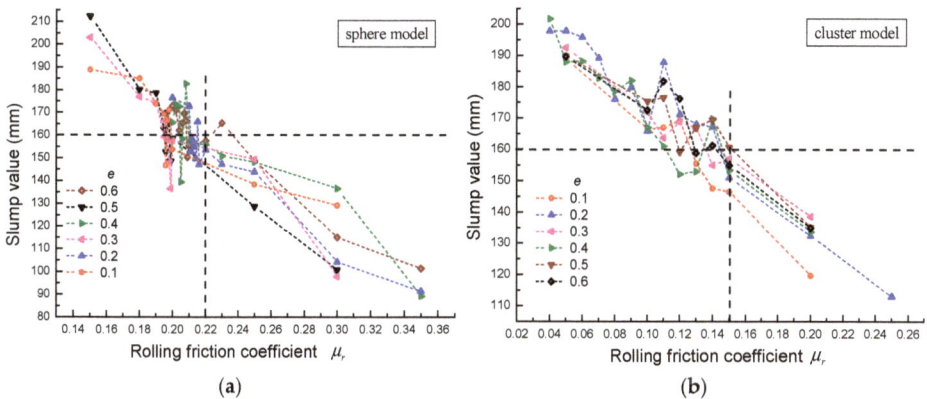

Figure 14. The slump value against rolling friction coefficient ($k = 0$): (**a**) sphere model and (**b**) cluster model.

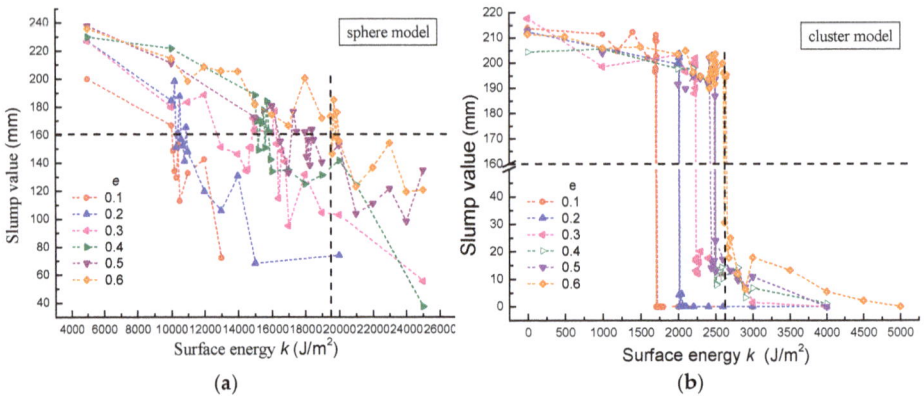

Figure 15. The slump value against surface energy ($\mu_r = 0$): (**a**) sphere model and (**b**) cluster model.

5.3.2. Parameter Sets with Satisfactory Fluidity

Then the LS-SVM that reflects the nonlinear relationship of slump value and three undetermined parameters can be established. As for training samples, some sets of unknown parameters are randomly generated within their respective range and the corresponding slumps are obtained through discrete element simulation. LS-SVR adopts a RBF kernel and its parameters are tuned by the GA-PSO collaborative algorithm. In the tuning process, the n-fold cross-validation error is chosen as a fitness function. For the sphere model, 440 training samples are collected altogether and n = 5. The parameters of LS-SVM are optimized as penalty parameter $\gamma = 281.87$ and kernel parameter $\sigma^2 = 0.824$. While in the cluster model, only 99 samples are collected and n is set as 9. And the optimization result is $\gamma = 7.3544$, $\sigma^2 = 0.1905$. For the two cases, the comparison between the fitted values and numerically calculated values can be seen in Figure 16a,b. The resultant root mean square errors (RMSE) are 12.53 and 9.68, respectively, indicating a satisfactory degree of accuracy.

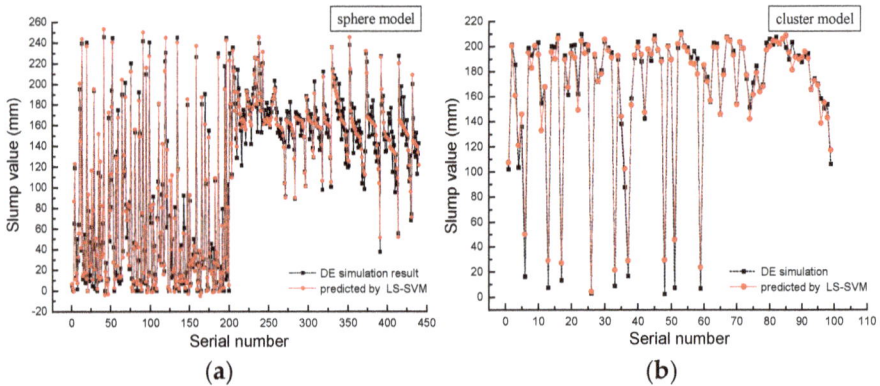

Figure 16. Comparison of fitted values and discrete element method (DEM) result: (**a**) sphere model and (**b**) cluster model.

Based on the established slump prediction model, the parameter sets that can result in a desired slump can be found. Firstly, randomly generate 10,000 parameter sets and predict their corresponding slump value by the established LS-SVM. Since the on-site measured slump ranges from 175 mm to 180 mm, and with a consideration of fitting error, the samples resulting in a slump between 170 mm and 185 mm are further checked by discrete element simulation. If the simulated slump can fall into the range of 175 to 180 mm, the corresponding parameter set is considered to be able to meet the requirement for fluidity. By regression analysis with LS-SVM, 86 (72) out of 10,000 sets of parameters get a satisfactory slump value in the case of sphere (cluster); Figure 17 shows comparison between fitted values and numerically calculated values of the 86 (72) sets, among which 22 (26) calculated values range from 175 mm to 180 mm. In addition, Figure 17 also indicates that the constructed LS-SVMs have satisfactory generalization ability and prediction accuracy.

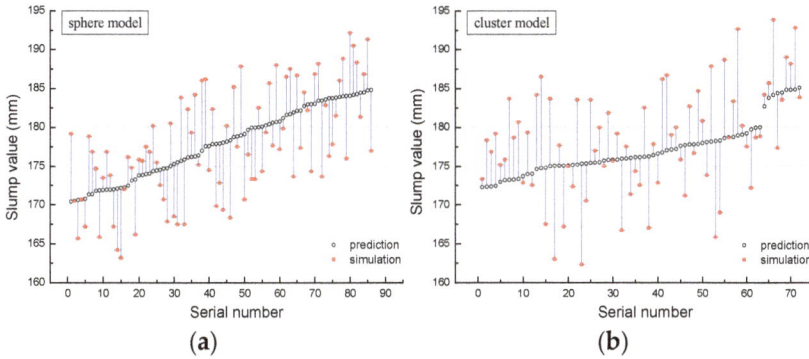

Figure 17. Comparison of fitted values and numerically calculated values: (**a**) sphere model; and (**b**) cluster model.

5.3.3. Strength Required Parameter Set

DE simulations of triaxial shear test are carried out successively with previously found parameter sets adopted, and the one with result closest to the experimental result is regarded as desired parameter set. Since the calculated static earth pressure at tunnel face ranges from 66.39 kPa to 92.98 kPa, and measured soil pressure at the bulkhead fluctuates between 30 kPa and 80 kPa, the soil inside shield chamber works in a low-pressure state. Therefore, the parameter sets are evaluated by the stress–strain curve measured under confining pressure of 0.4 MPa. Figure 18 presents the established discrete element models for the simulation of triaxial test. In the simulation, the confining pressure is realized by exerting body force on peripheral particles via developing custom plugin programmed in C++; the body force equals to the product of confining pressure and contact area between particle and boundary. According to Hu [7], the contact radius of particle and geometry can be approximately taken as particle radius.

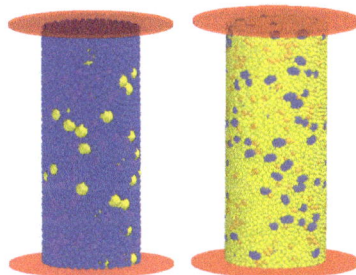

Figure 18. Discrete element models for triaxial test.

In the single sphere case, the 22 test specimens fall into three groups according to simulation result. The first group involves the 12th, 14th, 19th, 20th, 21st, and 22nd specimens. As for the six specimens, the measured stresses constantly remain zero except for the beginning of loading. This is because their surface energy is too low. With weak cohesive force between particles, the soil specimens have insufficient strength and self-stability. Therefore collapse occurs under the disturbance of axial load. The evolution of the 12th specimen along with loading can be seen in Figure 19.

Figure 19. The evolution of the 12th specimen as loading.

Whereas, the second group consists of the 2nd, 3rd, 4th, 6th, 9th, 11th, 15th, 16th, and 18th specimens. Compared with the first group, the nine soil specimens have developed some intensity because of the improved cohesive force between particles. Their stress–strain curves measured under the confining pressure of 0.4 MPa are shown in Figure 20. It can be seen that both the shape and value of the curves deviate greatly from the result of laboratory test (see Figure 4). Figure 21 records the evolution of the 2nd specimen during shear process. Obviously, the soil specimen expands continuously, leading to an ongoing reduce in interlocking induced shear resistance. Once the cohesive force and friction force are not enough to resist external force, local damage begins to occur on the specimen. Along with loading, the damage area gradually gets larger and only a part of soil specimen works in the end.

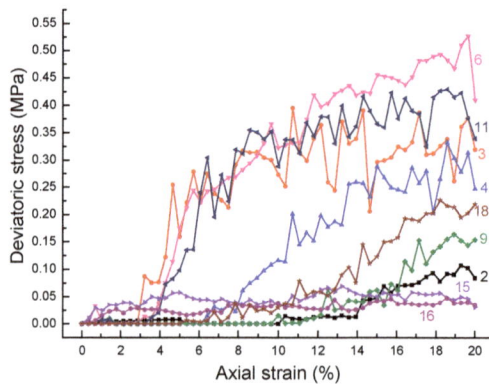

Figure 20. The stress–strain curve of specimens in the second group.

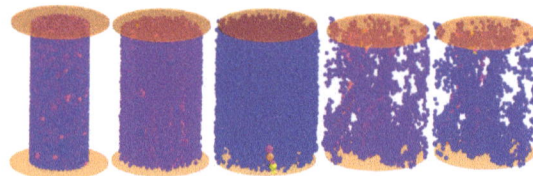

Figure 21. The evolution of the 2nd specimen as loading.

The third group, including the 1st, 5th, 7th, 8th, 10th, 13th, and 17th specimens, behaves similarly with the actual soil in laboratory. The stress–strain curves of the seven specimens (see Figure 22) generally experience three stages, namely initial stage, strain hardening stage and strain softening stage (or constant strain stage). At the beginning of the second stage, the deviatoric stress rises rapidly because the closely packed particles must overcome a strong interlocking resistance for shear dislocation, but the duration is short before shear dilation occurs. When the volume of specimen increases, the soil becomes loose, resulting in weakened interlocking effect between particles and thus a reduction in growth speed of stress. Then a peak, namely peak strength, can be seen in the curve with certain expanded specimen volume. Along with the increasing shear load, the curve turns into the third stage. In this stage, the specimen keeps expanding, but the shear strength decreases. Finally, the curve fluctuates at residual strength as a result of continuous stress relaxation. The residual strength is mainly shear resistance provided by sliding friction and rolling friction. Figure 23 shows the variation of the first specimen in the shear process. As the shear load increases, the specimen appears to undergo bulging deformation. It can be concluded from Figure 22; Figure 23 that discrete element simulation can embody the nonlinearity, plasticity, strain softening, and shear dilatancy of granular material. There are two possible reasons why the laboratory test cannot reflect the strain softening and shear dilatancy: (1) With improved gradation and vibration method adopted when making sample, the conditioned soil grains are closely arranged, therefore strong interlocking exists among grains. (2) Besides, the sample was wrapped with a rubber film when testing. Constrained by the rubber film, the soil cannot deform easily. Moreover, Figure 22 also shows that the 13th simulation has the most similar stress–strain curve to the laboratory test. Therefore, a decent set of parameters is found as $e = 0.3234$, $\mu_r = 0.0359$, and $k = 4523\text{J}/\text{m}^2$. In this way, the established discrete element model is almost consistent with the actual soil both with respect to strength property and fluidity.

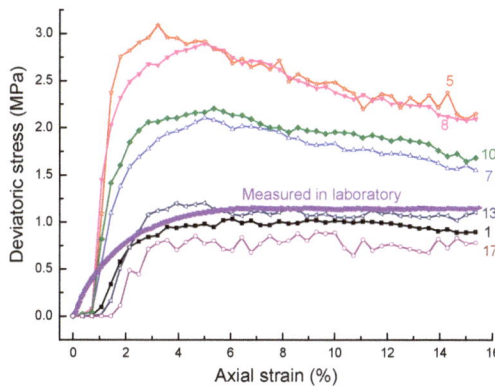

Figure 22. The stress–strain curve of specimens in the third group.

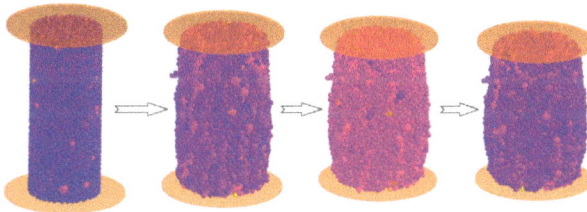

Figure 23. The evolution of the 1st specimen.

Similarly, the 26 cluster specimens fall into three groups. Figure 24 shows the stress–strain curves of the specimens in the third group, and Figure 25 presents the evolution of the 2nd specimen. So for the cluster model, the desired parameters are $e = 0.2569$, $\mu_r = 0.0249$, and $k = 1845J/m^2$.

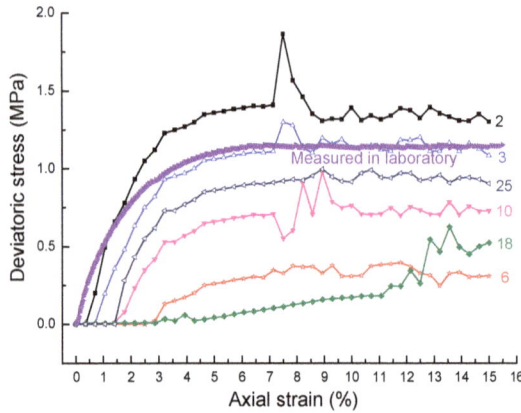

Figure 24. The stress–strain curve of cluster specimens.

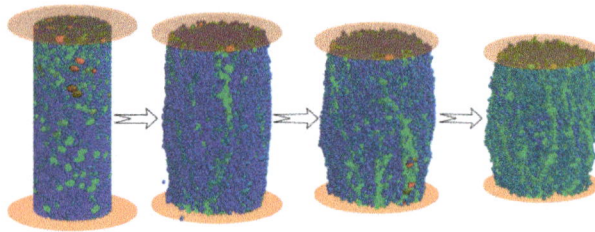

Figure 25. The evolution of the 2nd specimen.

From the above analysis it is notable that when cluster is adopted, lower values for rolling friction coefficient μ_r and surface energy k are required to obtain a good match between the experiment and DEM simulation. This is because there exists additional strength gained from interlocking action due to the irregular shape of the cluster. Besides, in the cluster case the CPU processing time for slump test and triaxial shear test are respectively 144 min and 1235 min, which are 4.97 times and 3.68 times higher than that in the sphere case.

6. Conclusions

This paper constructed two particle element models for conditioned sandy pebble. To take the shape effect into account, rolling resistance was exerted on the single sphere or the single sphere was replaced by the cluster of particles. Based on a laboratory test and DEM simulation of the slump test and large-scale triaxial test, the undetermined contact parameters in two models were calibrated by virtue of LS-SVR. Through comparative analysis, the following conclusions have been obtained.

(1) Because of the angularity of the nonspherical grains, there exists strong interlocking between clusters. So in the cluster model, relatively smaller rolling friction coefficient and surface energy are required.
(2) Since the model parameters are properly calibrated, the two established particle models both can well describe the strength property and fluidity of the actual soil. However, the cluster model takes ~4 times longer than the sphere model to execute the simulation.

It is worth noting that calibrating parameters by intelligent algorithm is still a time consuming task. There is an urgent need to conduct further research into multifunctional testing device for direct measurement of material parameters. An uncertainty analysis as done in [36–38] might be helpful.

Author Contributions: P.C. is responsible for the model setup and computational. X.Z. validated and checked the model. H.Z. is the PI of the project and supervised the work. Y.L. analysed the results and data.

Funding: The authors gratefully acknowledge the supports from the National Natural Science Foundation of China (Grant No. 41672360), Science and Technology Commission of Shanghai Municipality (Grant No. 17DZ1203800), and Shanghai Shentong Metro Group Co., Ltd. (Grant No. 17DZ1203804).

Conflicts of Interest: The authors declare no conflict of interest.

References

1. Zhang, P.; Jin, L.; Du, X.; Lu, D. Computational homogenization for mechanical properties of sand cobble stratum based on fractal theory. *Eng. Geol.* **2018**, *232*, 82–93. [CrossRef]
2. Yagiz, S. Brief notes on the influence of shape and percentage of gravel on the shear strength of sand and gravel mixtures. *Bull. Eng. Geol. Environ.* **2001**, *60*, 321–333. [CrossRef]
3. Maynar, M.J.M.; Rodríguez, L.E.M. Discrete numerical model for analysis of earth pressure balance tunnel excavation. *J. Geotech. Geoenviron. Eng.* **2005**, *131*, 1234–1242. [CrossRef]
4. Rabczuk, T.; Belytschko, T. A three dimensional large deformation meshfree method for arbitrary evolving cracks. *Comput. Methods Appl. Mech. Eng.* **2007**, *196*, 2777–2799. [CrossRef]
5. Rabczuk, T.; Belytschko, T. Cracking particles: a simplified meshfree method for arbitrary evolving cracks. *Int. J. Numer. Methods Eng.* **2004**, *61*, 2316–2343. [CrossRef]
6. Rabczuk, T.; Gracie, R.; Song, J.H.; Belytschko, T. Immersed particle method for fluid-structure interaction. *Int. J. Numer. Methods Eng.* **2010**, *81*, 48–71. [CrossRef]
7. Amiri, F.; Anitescu, C.; Arroyo, M.; Bordas, S.; Rabczuk, T. XLME interpolants, a seamless bridge between XFEM and enriched meshless methods. *Comput. Mech.* **2014**, *53*, 45–57. [CrossRef]
8. Rabczuk, T.; Bordas, S.; Zi, G. On three-dimensional modelling of crack growth using partition of unity methods. *Comput. Struct.* **2010**, *88*, 1391–1411. [CrossRef]
9. Rabczuk, T.; Zi, G.; Bordas, S.; Nguyen-Xuan, H. A simple and robust three-dimensional cracking-particle method without enrichment. *Comput. Methods Appl. Mech. Eng.* **2010**, *199*, 2437–2455. [CrossRef]
10. Areias, P.; Cesar de Sa, J.; Rabczuk, T.; Camanho, P.P.; Reinoso, J. Effective 2D and 3D crack propagation with local mesh refinement and the screened Poisson equation. *Eng. Fract. Mech.* **2018**, *189*, 339–360. [CrossRef]
11. Areias, P.; Rabczuk, T. Steiner-point free edge cutting of tetrahedral meshes with applications in fracture. *Finite Elem. Anal. Des.* **2017**, *132*, 27–41. [CrossRef]
12. Areias, P.; Rabczuk, T.; Msekh, M. Phase-field analysis of finite-strain plates and shells including element subdivision. *Comput. Methods Appl. Mech. Eng.* **2016**, *312*, 322–350. [CrossRef]
13. Areias, P.; Msekh, M.A.; Rabczuk, T. Damage and fracture algorithm using the screened Poisson equation and local remeshing. *Eng. Fract. Mech.* **2016**, *158*, 116–143. [CrossRef]
14. Areias, P.; Rabczuk, T.; Dias-da-Costa, D. Element-wise fracture algorithm based on rotation of edges. *Eng. Fract. Mech.* **2013**, *110*, 113–137. [CrossRef]
15. Ren, H.; Zhuang, X.; Rabczuk, T. Dual-horizon peridynamics: A stable solution to varying horizons. *Comput. Methods Appl. Mech. Eng.* **2017**, *318*, 762–782. [CrossRef]
16. Ren, H.; Zhuang, X.; Cai, Y.; Rabczuk, T. Dual-Horizon Peridynamics. *Int. J. Numer. Methods Eng.* **2016**, *108*, 1451–1476. [CrossRef]
17. Talebi, H.; Silani, M.; Rabczuk, T. Concurrent Multiscale Modelling of Three Dimensional Crack and Dislocation Propagation. *Adv. Eng. Softw.* **2015**, *80*, 82–92. [CrossRef]
18. Talebi, H.; Silani, M.; Bordas, S.; Kerfriden, P.; Rabczuk, T. A Computational Library for Multiscale Modelling of Material Failure. *Comput. Mech.* **2014**, *53*, 1047–1071. [CrossRef]
19. Budarapu, P.; Gracie, R.; Bordas, S.; Rabczuk, T. An adaptive multiscale method for quasi-static crack growth. *Comput. Mech.* **2014**, *53*, 1129–1148. [CrossRef]
20. Budarapu, P.; Gracie, R.; Shih-Wei, Y.; Zhuang, X.; Rabczuk, T. Efficient Coarse Graining in Multiscale Modeling of Fracture. *Theor. Appl. Fract. Mech.* **2014**, *69*, 126–143. [CrossRef]

21. Mindlin, R.D. Compliance of elastic bodies in contact. *J. Appl. Mech.* **1949**, *16*, 259–268.
22. Tavares, L.M.; King, R.P. Single-particle fracture under impact loading. *Int. J. Miner. Process.* **1998**, *54*, 1–28. [CrossRef]
23. Mishra, B.K.; Murty, C. On the determination of contact parameters for realistic DEM simulations of ball mills. *Powder Technol.* **2001**, *115*, 290–297. [CrossRef]
24. Jayasundara, C.T.; Yang, R.Y.; Yu, A.B.; Rubenstein, J. Effects of disc rotation speed and media loading on particle flow and grinding performance in a horizontal stirred mill. *Int. J. Miner. Process.* **2010**, *96*, 27–35. [CrossRef]
25. Rosenkranz, S.; Breitung-Faes, S.; Kwade, A. Experimental investigations and modelling of the ball motion in planetary ball mills. *Powder Technol.* **2011**, *212*, 224–230. [CrossRef]
26. Piechatzek, T.; Kwade, A. DEM based characterization of stirred media mills. In Proceedings of the European Symposium on Comminution, Espoo, Finland, 15–18 September 2009.
27. Quist, J.C.E. Device for calibrating of DEM contact model parameters. In Proceedings of the EDEM Conference, Edinburgh, UK, 5–7 April 2011.
28. Li, C.Q.; Liu, T.W.; Zhang, H.Y. Back-analysis on micromechanical parameters of soil mass using BP neural network. *J. Eng. Geol.* **2015**, *23*, 609–615. (In Chinese)
29. Hertz, H. On the contact of elastic solids. *J. Fur Die Reine Und Angew. Math.* **1882**, *92*, 156–171.
30. Mindlin, R.D.; Deresiewicz, H. Elastic spheres in contact under varying oblique forces. *ASME J. Appl. Mech.* **1953**, *20*, 327–344.
31. Tsuji, Y.; Tanaka, T.; Ishida, T. Lagrangian numerical simulation of plug flow of cohesionless particles in a horizontal pipe. *Powder Technol.* **1992**, *71*, 239–250. [CrossRef]
32. Cundall, P.A.; Strack, O.D.L. A discrete numerical method for granular assemblies. *Geotechnique* **1979**, *29*, 47–65. [CrossRef]
33. Sakaguchi, E.; Ozaki, E.; Igarashi, T. Plugging of the flow of granular materials during the discharge from a silo. *Int. J. Mod. Phys. B* **1993**, *7*, 1949–1963. [CrossRef]
34. Johnson, K.L.; Kendal, K.; Roberts, A.D. Surface energy and the contact of elastic solids. *Proc. R. Soc. A* **1971**, *324*, 301–313. [CrossRef]
35. Hu, M.; Xu, G.Y.; Hu, S.B. Study of equivalent elastic modulus of sand gravel soil with Eshelby tensor and Mori-Tanaka equivalent method. *Rock Soil Mech.* **2013**, *34*, 1437–1442. (In Chinese)
36. Vu-Bac, N.; Lahmer, T.; Zhuang, X.; Nguyen-Thoi, T.; Rabczuk, T. A software framework for probabilistic sensitivity analysis for computationally expensive models. *Adv. Eng. Softw.* **2016**, *100*, 19–31. [CrossRef]
37. Hamdia, K.; Zhuang, X.; Silani, M.; He, P.; Rabczuk, T. Stochastic analysis of the fracture toughness of polymeric nanoparticle composites using polynomial chaos expansions. *Int. J. Fract.* **2017**, *206*, 215–227. [CrossRef]
38. Hamdia, K.; Zhuang, X.; Alajlan, N.; Rabczuk, T. Sensitivity and uncertainty analyses for exoelectric nanostructures. *Comput. Methods Appl. Mech. Eng.* **2018**, *337*, 95–109. [CrossRef]

![applied sciences logo] *applied sciences*

MDPI

Article

Topological Photonic Media and the Possibility of Toroidal Electromagnetic Wavepackets

Masaru Onoda

Graduate School of Engineering Science, Akita University, 1-1 Tegatagakuen-machi, Akita 010-8502, Japan; onoda@gipc.akita-u.ac.jp

Received: 17 February 2019; Accepted: 3 April 2019; Published: 8 April 2019

Abstract: This study aims to present a theoretical investigation of a feasible electromagnetic wavepacket with toroidal-type dual vortices. The paper begins with a discussion on geometric phases and angular momenta of electromagnetic vortices in free space and periodic structures, and introduces topological photonic media with a review on topological phenomena of electron systems in solids, such as quantum Hall systems and topological insulators. Representative simulations demonstrate both the characteristics of electromagnetic vortices in a periodic structure and of exotic boundary modes of a topological photonic crystal, on a Y-shaped waveguide configuration. Those boundary modes stem from photonic helical surface modes, i.e., a photonic analog of electronic helical surface states of topological insulators. Then, we discuss the possibility of toroidal electromagnetic wavepackets via topological photonic media, based on the dynamics of an electronic wavepacket around the boundary of a topological insulator and a correspondence relation between electronic helical surface states and photonic helical surface modes. Finally, after introducing a simple algorithm for the construction of wavepacket solutions to Maxwell's equations with multiple types of vortices, we examine the stability of a toroidal electromagnetic wavepacket against reflection and refraction, and further discuss the transformation laws of its topological properties in the corresponding processes.

Keywords: geometric phase; photonic orbital angular momentum; topological insulator; topological photonic crystal

---

1. Introduction

Physical concepts proposed for a system are sometimes applicable to other systems that initially looks considerably different from the original system. Such concepts can be used to predict novel phenomena in the latter system and to explain a mechanism governing the phenomena. Such a mechanism could be conversely applied to the original system and predict similar phenomena in it. Finally, we come to realize their universality. Here we discuss about some interconnection among such concepts and mechanisms, e.g., band theory, geometric phase, Hall effect, topological phase and so on. "Energy bands" and "band gaps" were originally cultivated in the field of condensed matter theory which concerns electron systems in solids, e.g., natural crystals or artificial periodic structures. These concepts were applied to the old research theme [1] on electromagnetic waves in periodic structures composed of different kinds of dielectrics and magnetic materials, consequently establishing the concept of "photonic crystals" [2–4], which plays an important role in the realization and extension of "metamaterials" [5–8]. The concept of "geometric phase" was initially introduced in an electromagnetic system [9], and became clearly recognized in a quantum system with spin degrees-of-freedom (DOF) for electron systems in solids [10]. Interestingly, this clear-cut recognition was reapplied to an electromagnetic system, i.e., a photon system, and its validity

Appl. Sci. **2019**, *9*, 1468; doi:10.3390/app9071468

became clear in a photon system than in electron systems [11]. On the contrary, the vortex structure of an electromagnetic wave became widely recognized to closely relate with the orbital angular momentum of photons [12], a view currently being implemented in the optical and quantum information communication technology [13–15]. Moreover, electromagnetic vortices can appear in periodic structures, such as photonic crystals [16]. This suggests a new kind of internal orbital angular momenta of photon in such systems. These internal orbital angular momenta may be interpreted as quasi-spin DOF and potentially cause a variety of geometric phase effects. Specifically, an electromagnetic wavepacket composed from wave modes with such vortices can have orbital angular momentum perpendicular to its propagation direction. This relation between angular momentum and propagation direction for such a wavepacket is similar to that for an atmospheric tornado which shows unexpected exotic motions.

Herein, we theoretically investigate a possible electromagnetic wavepacket with toroidal-type dual vortices, i.e., having a ring vortex inside the wavepacket and a line vortex along its propagation direction. The line vortex resembles that of a Laguerre-Gaussian beam and suggests a finite orbital angular momentum of the wavepacket. This paper is organized as follows. In Section 2, we review the relation between geometric phases and angular momenta, followed by the discussion on electromagnetic vortices in periodic structures in Section 3, which further demonstrates the propagation characteristics of such electromagnetic vortices by conducting numerical simulations on Y-shaped waveguides. In Section 4, we introduce the topological photonic media, while reviewing topological phenomena of electron systems in solids, such as quantum Hall systems and topological insulators. Herein, a class of topological photonic media is interpreted as a photonic version of topological insulator and can be realized as an extension of photonic crystals accompanying the electromagnetic vortices. Moreover, we present another simulation of waveguide propagation via exotic boundary modes of such a medium. In Section 5, referring to the dynamics of an electronic wavepacket around the boundary between a topological insulator and conductor, we consider the possibility of toroidal electromagnetic wavepackets with an argument on a correspondence relation between electronic helical surface states of topological insulator and photonic helical surface modes of topological photonic media. In Section 6, we present an algorithm for constructing wavepacket solutions of Maxwell's equations with multiple types of vortices. Next, we numerically investigate the stability of the toroidal electromagnetic wavepacket in reflection and refraction at interfaces between homogeneous isotropic dielectrics, and reveal the transformation laws of topological charges of line and ring vortices.

In the next section and beyond, we adopt the natural system of units as $\hbar = 1$ ($\hbar$: Dirac constant or reduced Planck constant) and $c = 1$ (c: speed of light), unless those symbols are explicitly stated. We will not distinguish between the wavevector $\mathbf{k}$ of a plane wave and the momentum $\hbar\mathbf{k}$ of a quantum particle derived from second quantization of the wave. Likewise, the frequency ω of a harmonic wave and the energy $\hbar\omega$ of a corresponding quantum particle will not be distinguished. For convenience on later discussions, we introduce a spherical basis $\{\mathbf{e}_k, \mathbf{e}_\theta, \mathbf{e}_\phi\}$ ($\mathbf{e}_k = \mathbf{k}/k$) in wavevector space.

2. Geometric Phases and Angular Momenta of Electromagnetic Vortices

In this section, we first review the relation that exists between the spin angular momentum and geometric phase of an electromagnetic wavepacket, following those between polarization state and spin angular momentum and between polarization vector and geometric phase. Figure 1 describes the relation between polarization state and angular momentum [17] of a right circularly polarized wavepacket propagating in the z-direction (upward in the drawing). The arrows represent the deviation of the wavepacket energy flux density from the product of the energy density and the averaged velocity vector, whereas the hue represents the energy density (cold color < warm color). For simplicity, we considered the situation wherein the wavepacket spread is sufficiently large with respect to its central wavelength so that its deformation can be ignored. In Figure 1, we can find a clockwise vortex in the view facing the propagation direction (z-direction) of the wavepacket, indicating that the right circularly polarized wavepacket has a spin angular momentum in the propagation direction. By contrast, we could not find such structure of energy flux density in a similar plot of a linearly polarized wavepacket in the same scale as Figure 1.

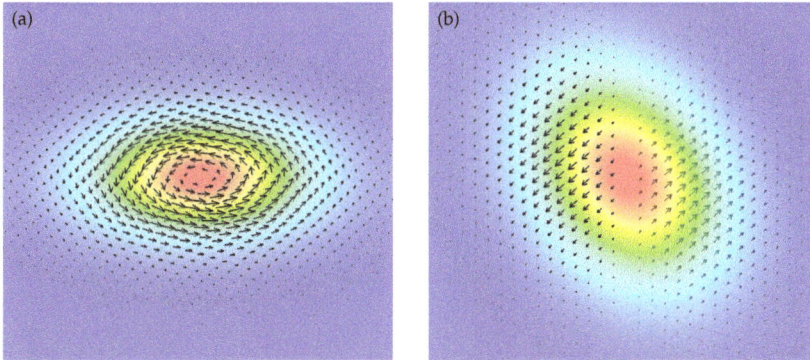

Figure 1. (a) *xy*- and (b) *yz*-cross-sections of a right circularly polarized Gaussian wavepacket.

Furthermore, once the well-known indications, i.e., the geometric phase appearing due to the change in polarization state [9] and the geometric phase due to orbital change of polarized beam [18], are accepted, the relation between spin and geometric phase looms into view. Herein, by specifically looking at the relation between polarization vector and geometric phase, we confirm the relation between the three concepts more explicitly. For that purpose, we introduce two quantities, the Berry connection and curvature, defined as

$$
[\mathbf{\Lambda}_k]_{\alpha\beta} = \begin{pmatrix} -ie_{k\alpha}^{\dagger} \cdot \frac{\partial}{\partial k_1} e_{k\beta} \\ -ie_{k\alpha}^{\dagger} \cdot \frac{\partial}{\partial k_2} e_{k\beta} \\ -ie_{k\alpha}^{\dagger} \cdot \frac{\partial}{\partial k_3} e_{k\beta} \end{pmatrix} = -ie_{k\alpha}^{\dagger} \cdot \nabla_k e_{k\beta}, \quad \mathbf{\Omega}_k = \nabla_k \times \mathbf{\Lambda}_k + i\mathbf{\Lambda}_k \times \mathbf{\Lambda}_k, \tag{1}
$$

where $\{e_{k\alpha}\}$ is an orthonormal basis of polarization vectors normal to the wavevector k ($e_{k\alpha}^{\dagger} \cdot e_{k\beta} = \delta_{\alpha\beta}$, and $e_{k\alpha}^{\dagger} \cdot e_k = 0$); the symbol α is an indicator of polarization state; and $e_{k\alpha}^{\dagger}$ is the complex conjugate transpose (the Hermitian conjugate) of $e_{k\alpha}$. Herein, we also introduce the Jones vector corresponding to this orthonormal basis as $|z_k\rangle$. With $\langle z_k|$ as the Hermitian conjugate of $|z_k\rangle$, they satisfy $\langle z_k|z_k\rangle = \sum_{\alpha} |z_{k\alpha}|^2 = 1$. Although representations of $\mathbf{\Lambda}_k$ and $\mathbf{\Omega}_k$ depend on how the basis is selected, $\langle z_k|\mathbf{\Omega}_k|z_k\rangle$ is uniquely determined once a state is given. On the basis of right and left circular polarizations, the Berry curvature is represented as $k/k^3\sigma_3$, where σ_3 is the third component of Pauli matrices $\sigma = (\sigma_1, \sigma_2, \sigma_3)^{\top}$ and is diagonal in the standard representation. On the contrary, $\langle z_k|\mathbf{\Omega}_k|z_k\rangle$ can be expressed as $(|z_{kR}|^2 - |z_{kL}|^2)/k^2 e_k$ using the corresponding representation $[z_{kR}, z_{kL}]^{\top}$ of $|z_k\rangle$. Since the expected value of the spin angular momentum per photon $\langle s_k\rangle$ is evaluated as $\langle s_k\rangle = (|z_{kR}|^2 - |z_{kL}|^2)k/k$, a close relation, $\langle s_k\rangle = k^2\langle z_k|\mathbf{\Omega}_k|z_k\rangle$, exists between the two quantities.

We can extend the above discussion to a case accompanied by orbital angular momentum. A close relation between the internal orbital angular momentum and Berry curvature is derived in the same way as above, while its discussion gets a little bit complex. Herein, we consider a beam of a central wavevector k_c and introduce the extension

$$e_{k\alpha} \to \exp\left(il_\alpha \varphi_{k;k_c}\right) e_{k\alpha}, \quad \varphi_{k;k_c} = \arctan \frac{e_{\phi_c} \cdot k}{e_{\theta_c} \cdot k}, \tag{2}$$

where l_α corresponds to the vorticity of the α-polarized component and is an integer number, i.e., $l_\alpha \in \mathbb{Z}$. For simplicity, we consider only a class of beams that are a superposition of a given polarization component with the vorticity l and its orthogonal component with l'. In other words, we restrict ourselves to a finite subspace of infinite whole space of states. On the basis of right and left circular polarizations, the Berry connection and curvature of this subspace are represented as,

$$\Lambda_k = -\frac{\cos\theta}{k\sin\theta}(l_{\text{OAM}} + \sigma_3)e_\phi, \quad \Omega_k = \frac{1}{k^2}(l_{\text{OAM}} + \sigma_3)e_k, \tag{3}$$

where l_{OAM} is a 2×2 Hermitian matrix with a pair of integer eigenvalues (l, l'). On the other hand, the expected value of the total angular momentum per photon $\langle j_k \rangle$ is evaluated as $\langle j_k \rangle = (z_k|l_{\text{OAM}} + \sigma_3|z_k)e_k$. We can find a close relation $\langle j_k \rangle = k^2(z_k|\Omega_k|z_k)$ again.

Next, we consider the meaning of the form of the Berry curvature, $k/k^3\sigma_3$, on the basis of circular polarizations which we shall call helicity basis hereafter. As we shall see, this form reflects the photon characteristics of a spin-1 gauge symmetric massless boson. To this end, we introduced a degenerate two-band model of a spin-1/2 fermion system with conical dispersions, similar to relativistic electrons. The Hamiltonian of this model and its projection to the subspace of definite wavevector k are given by

$$H = -iv\nabla_r \cdot \alpha + \Delta\beta \to H_k = vk \cdot \alpha + \Delta\beta, \tag{4}$$

where α and β are Dirac matrices and v and Δ are parameters of effective phase velocity and band gap, respectively. The band gap of this model plays a similar role as the mass gaps in relativistic theories; hence, we shall refer to the limit $\Delta \to 0$ as massless limit. The eigenvalue problem of H_k can be easily solved. The eigenvalues of the upper and lower bands are obtained as $\pm\sqrt{v^2k^2 + \Delta^2}$; only the upper bands will be considered hereafter. The degenerate eigenstates of the upper band $|u_{k\lambda}\rangle$ in terms of the spherical coordinate of wavevector space (k, θ, ϕ) are expressed as

$$|u_{k\pm}\rangle = \frac{1}{\sqrt{2E_k}} \begin{pmatrix} \sqrt{E_k + \Delta}\chi_{k\pm} \\ \pm\sqrt{E_k - \Delta}\chi_{k\pm} \end{pmatrix}, \quad \chi_{k+} = \begin{pmatrix} e^{-i\frac{\phi}{2}}\cos\frac{\theta}{2} \\ e^{i\frac{\phi}{2}}\sin\frac{\theta}{2} \end{pmatrix}, \quad \chi_{k-} = -i\sigma_2\overline{\chi}_{k+}, \tag{5}$$

where $E_k = \sqrt{v^2k^2 + \Delta^2}$ and the index λ corresponds to the degrees of helicity, i.e., spin component in the direction of k. Note that the above expressions are well-defined only in regions excluding the points $\theta = 0$ and π. However, we shall not step into regularization at these points but only mention that expressions valid at $\theta = 0$ or π are obtained by some gauge transformations of the above expressions. One can easily confirm that the above expressions satisfy the time-independent Schrödinger equation in k-subspace,

$$E_k|u_{k\lambda}\rangle = \mathcal{H}_k|u_{k\lambda}\rangle. \tag{6}$$

At this point, we would like to mention that the electromagnetic polarization vectors $e_{k\lambda}$ can be interpreted as solutions of a similar Schrödinger-type matrix equation, which is a transcription of Maxwell's equations. The correspondence relation is confirmed by the replacements, $E_k = k/(\sqrt{\epsilon}\sqrt{\mu})$, and

$$\mathcal{H}_k = \begin{pmatrix} 0 & ie^{-\frac{1}{2}}k \cdot S\mu^{-\frac{1}{2}} \\ -i\mu^{-\frac{1}{2}}k \cdot S\epsilon^{-\frac{1}{2}} & 0 \end{pmatrix}, \quad [S_i]_{jk} = -i\epsilon_{ijk}, \quad |u_{k\lambda}\rangle = \frac{1}{\sqrt{2}}\begin{pmatrix} e_{k\lambda} \\ e_k \times e_{k\lambda} \end{pmatrix}, \tag{7}$$

where ϵ, μ, and ϵ_{ijk} are the relative permittivity, relative permeability, and the Levi-Civita symbol, respectively. For simplicity, we assumed a homogeneous and isotropic background medium here. Therefore, by means of the state vectors $|u_{k\lambda}\rangle$, we can define the Berry connection in a form common in electronic and electromagnetic systems,

$$[\mathbf{\Lambda}_k]_{\alpha\beta} = -i\langle u_{k\alpha}|\boldsymbol{\nabla}_k|u_{k\beta}\rangle. \tag{8}$$

Now, let us get back to the degenerate two-band model. On the helicity basis, the Berry connection and curvature of the upper band are expressed by

$$\mathbf{\Lambda}_k = \tfrac{1}{2k}\left\{\tfrac{\Delta}{E_k}\left(-\sigma_2 e_\theta + \sigma_1 e_\phi\right) - \tfrac{\cos\theta}{\sin\theta}\sigma_3 e_\phi\right\}, \quad \mathbf{\Omega}_k = \tfrac{v^2}{2E_k^2}\left\{\tfrac{\Delta}{E_k}\left(\sigma_1 e_\theta + \sigma_2 e_\phi\right) + \sigma_3 e_k\right\}. \tag{9}$$

In the massless limit $\Delta \to 0$, this coincides with the Berry curvature of photons except for the overall coefficient due to a different spin magnitude. In the non-relativistic limit, i.e., an increasing $|\Delta|$ for a fixed k, the Berry curvature decreases in the form $1/\Delta^2$, similar to the scale of the spin-orbit interaction. Next, let us consider the relation between the spin angular momentum and the geometric phase in this spin-1/2 massive fermionic system. The spin operator s and its projection to the upper band s_k are given as follows:

$$s = \begin{pmatrix} \sigma & 0 \\ 0 & \sigma \end{pmatrix} \to [s_k]_{\alpha\beta} = \langle u_{k\alpha}|s|u_{k\beta}\rangle, \quad s_k = \frac{1}{2}\left\{\frac{\Delta}{E_k}\left(\sigma_1 e_\theta + \sigma_2 e_\phi\right) + \sigma_3 e_k\right\}. \tag{10}$$

In the massless limit $\Delta \to 0$, s_k matches its photonic version except for the coefficient 1/2 that comes from the spin-1/2 nature of the present fermionic system. We can again find the simple relation between the Berry curvature and the spin angular momentum, $s_k = (E_k^2/v^2)\mathbf{\Omega}_k$.

As for the electromagnetic waves in periodic structures, we can develop the same discussion by replacing the polarization vectors with eigenstate vectors expressed using Bloch wave functions, as in the example of a spin-1/2 fermion system discussed above [19,20]. We do not intend to have an unnecessary abstract debate using "Berry connection" and "Berry curvature". Thus, whereas they were introduced in a way that makes the theories under consideration abstract, we can use a common principle that is independent of the details of electron or photon systems for better understanding. Although the definition of angular momentum in a periodic system is accompanied by ambiguity, the Berry curvature can be uniquely defined apart from the freedom-of-choice of basis. Our knowledge of phenomena or effects in a given system are easily applicable to the realization of analogous phenomena or effects in other systems. From this viewpoint, information on the Berry curvature of each band helps us organize the relation between photonic bands in wavevector space and vortices in real space, and serves as guide for controlling vortices.

3. Electromagnetic Vortices in Periodic Systems

The spin DOF of photon is no longer well-defined in the presence of periodic structure, while we shall need an alternative concept of photonic spin to discuss the possibility of toroidal electromagnetic wavepackets in Section 5, based on an electron dynamics around the interface of a topological insulator where the concept of electronic spin still works well. (The photonic version of topological insulator will be introduced in Section 4.) One possibility of the alternative is the internal rotational motion of electromagnetic Bloch modes with vortices. In this section, we step further into electromagnetic vortices in periodic structures such as photonic crystals. After looking back briefly on the relation between the Berry curvature and real-space vortex structure in periodic systems, we present a numerical simulation on propagation modes around a standing vortex mode in a photonic crystal, indicating the effect of the internal rotation in the propagation process.

Useful functionalities of a photonic crystal, such as light confinement and waveguide, are realized by adjusting energy (frequency) dispersion relations and forbidden bands by the periodic structure and symmetry design of the system. The required design procedures stem from the band theory common to wave phenomena in periodic structures. Moreover, based on the studies of geometric Hall effect in electron systems [21,22], we can also use such design to control the Berry curvature of each band [23,24]. For instance, by employing a two-dimensional (2D) periodic system in the xy direction, we can consider a situation where two bands are immediately prior to touching each other at a point k_0 in a wavevector space. Given an approximate description, $f_k \pm \sqrt{v^2|k - k_0|^2 + \Delta^2}$ (f_k: a function of wavevector k) for the local energy dispersion of each, the z-component of the Berry curvature of each band is estimated as $\Omega_{k\pm}^z = \pm v^2\Delta/\{2(v^2|k - k_0|^2 + \Delta^2)\}^{\frac{3}{2}}$. In other words, we can control the Berry curvature by adjusting the level repulsion Δ. We confirmed this mechanism in a more strict theory exactly treating the periodic structure, as well as the relation among the Berry curvature, the angular momentum [19,20] and the real-space vortex structure [16,25]. The electromagnetic vortex can be also controlled through the adjustment of Δ. Figure 2 shows an example of a set of such periodic structure, band diagram, and electromagnetic vortex. As for the 2D photonic crystals, we considered only photonic modes of two distinct polarizations, i.e., transverse-electric (TE) and transverse-magnetic (TM), which propagate strictly parallel to the plane with a 2D periodicity. Herein, we adopted this definition: TE modes have magnetic fields normal to the plane and electric fields in the plane; conversely, TM modes have electric fields normal to the plane and magnetic fields in the plane.

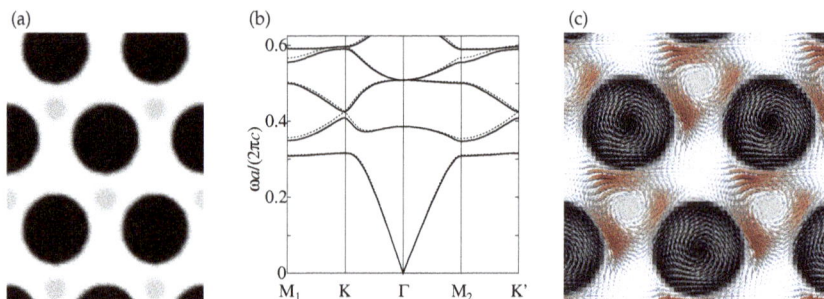

Figure 2. (**a**) A sample inversion asymmetric 2D photonic crystal for relative permittivity of 1, 3, and 12 in white, gray, and black regions, respectively. (**b**) Band diagram of TE modes for (**a**). Dotted lines show the case where the relative permittivity of gray rods is set to one. The vertical axis represents the dimensionless frequency $\omega a/(2\pi c)$ (*a*: lattice constant; *c*: speed of light). (**c**) A sample optical tornado: energy flux density of a state at a K-point of the TE 2nd band.

The electromagnetic vortex shown in Figure 2c corresponds to a standing wave mode of a zero group velocity. On the other hand, modes around it have finite group velocities in addition to the vortex structure, and can propagate through the crystal with rotational motion. Next, we describe the propagation characteristics of such modes in a Y-shaped waveguide in Figure 3a composed of the crystal in Figure 2a and a block layer with a sufficiently large band gap covering the relevant frequency range. Since it was not easy to excite a specific electromagnetic vortex mode in a real-time-and-space simulation, we adopted an excitation using a linear source with a line width of a few percent around the central frequency of the targeted vortex modes. Moreover, the linear source was set at the left end of the left branch of the waveguide and the vortex modes were excited by electric field oscillations along the source. Figure 3b shows the z-component of the magnetic field H_z and Figure 3c displays the transmission spectra measured at the ends of the upper right and lower right branches. As the vertical axis is in arbitrary unit, we also plotted the spectrum (blue and red broken lines) of the case where the Y-shaped region is replaced by vacuum, for comparison. Two broken lines overlapped each other, and only the blue broken line is visible. The transmission spectra of the target system in Figure 3c is extremely asymmetric, whereas we find only weak asymmetry in the real-space image of Figure 3b. Figure 3b also shows that this system contains accidental edge modes localized around interfaces aside from bulk vortices; therefore, the asymmetry could not be attributed solely to the bulk vortex modes. An additional simulation (not shown here) confirmed that the edge modes in this system are strongly reflected at the bents of the Y-shaped waveguide; therefore, we conclude that the bulk vortex modes contribute primarily to the asymmetric propagation.

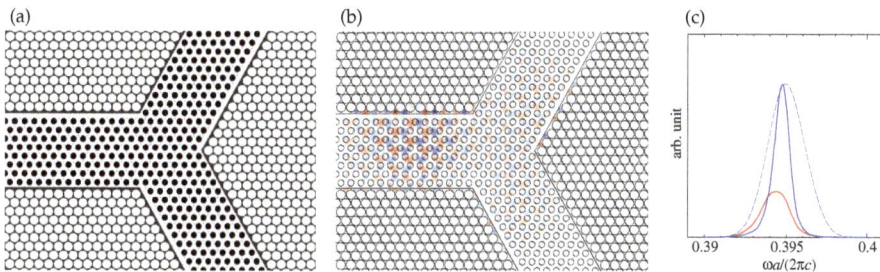

Figure 3. (a) Y-shaped waveguide composed of the photonic crystal in Figure 2. The relative permittivity of the gray region of block layer around the waveguide is set to be 9. (b) z-component of magnetic field H_z. (c) Spectra of the transmissions to the upper right (blue line) and to the lower right (red line) branches. The broken lines represent the transmission spectra for a vacuum Y-shaped region.

4. Topological Photonic Media

Based on the pioneering studies on quantum Hall effect in 2D electron systems [26–28], a topological invariant, known as Chern number or index, is assigned to an isolated band via integration of its Berry curvature over the entire first Brillouin zone. Depending on total Chern numbers of bulk bands below a bulk gap, exotic states localize at the edge of a finite system and form edge bands traversing over the bulk gap [29,30]. Furthermore, each state works a one-way waveguide; therefore, is called a chiral edge state. For a Chern number of an isolated band to be nonzero, time-reversal symmetry breaking is necessary. This symmetry may be broken not only by an externally applied magnetic field but also by a spontaneously induced magnetic order [28]. When the quantum Hall effect is induced by the latter mechanism, it is sometimes called the spontaneous quantum Hall effect, as distinguished from the original one. To realize a similar situation in photon systems, the above mechanism requires isolated bands to form with band gaps stemming from time-reversal symmetry breaking. For instance, a magnetic body of a complex permittivity tensor with imaginary off-diagonal components breaks time-reversal symmetry for the photon system. At least the necessary conditions are satisfied by designing a 2D periodic structure made of such a material to form isolated photonic bands. This analogy was the basis for a photonic version of the quantum Hall system theoretically proposed [31–33] and experimentally confirmed [34,35]. Presently, a clear-cut demonstration has also been made on nonreciprocal lasing from chiral edge modes surrounding a network of topological cavities in arbitrary geometry [36].

In reverse, a one-way propagation of a chiral edge mode inevitably breaks the time-reversal symmetry. (Time-reversal symmetry breaking is a necessary condition for the presence of a single chiral edge state; conversely, the presence of a single chiral edge state is a sufficient condition for the symmetry breaking.)

Fortunately, however, chiral edge states and their time-reversal partners can simultaneously exist in a single system preserving time-reversal symmetry in its entirety [37]. As an extension of quantum Hall system to the case with time-reversal symmetry, an insulator with a topologically-protected pair of edge states has been proposed [38]. Such insulator and edge states are respectively called topological insulator and helical edge states [39]. The propagation direction of a helical edge state is selectively governed by its spin polarization. In a naive picture, topological insulator is understood as a superposition of a spontaneous quantum Hall system spin-polarized in a specific direction and its time-reversal partner spin-polarized in the opposite direction which is necessary to maintain time-reversal symmetry. Based on this situation, it was initially called quantum spin Hall system. More precisely, the parity of the numbers of Kramers pairs of helical edge states is critical [38] and corresponds to the topological invariant called $\mathbb{Z}_2$ index. As the index can be calculated by means of the bulk states of a system with periodic boundary conditions, topological insulators can be distinguished from non-topological insulators even with bulk information alone. Furthermore, topological crystalline insulators were proposed by introducing combinations of crystalline symmetries and time-reversal symmetry [40], extending to their photonic version [41–43]. A wider range of topological materials, including superconductors, have been systematically classified based on symmetry and dimensionality [44,45].

Typical physical conditions under which topological insulators emerge are as follows: (1) two pairs of bands with different parities opposite to each pair, which are energetically close to each other, hybridize with each other through a strong spin-orbit interaction and; (2) the resultant level repulsion forms an enough sized bulk gap [46]. By contrast, for photon systems in periodic structures, what kind of DOF should be regarded as spin DOF remains unclear. Nevertheless, if the difference between certain degenerate modes is approximately regarded as pseudo-spin DOF and the coupling between electric and magnetic fields introduced by an artificial chiral medium are regarded as effective spin-orbit interaction of photon, then a similar mechanism could be applied to photon systems in periodic structures. Photonic versions of topological crystalline insulator using metamaterials as artificial chiral media have been proposed [41,42]. After such proposals, it has been pointed out that those topological photonic media could be realized even by photonic crystals composed of only ordinary dielectrics [43], whereas such systems needed to introduce some complication to the unit cell of the crystal. Therefore, we should regard the chiral medium as an example of effective spin-orbit interaction implementation, and not an item of necessity. Figure 4a,b show examples closely related to the all-dielectric topological photonic crystals in Ref. [43]. The structure of Figure 4a is an inversion asymmetric deformation of a topological photonic crystal; hence, we shall call it a quasi-topological photonic crystal for convenience. Compared to the case of Figure 2a, the degree of the symmetry breaking is so weak that it is not easy to distinguish two kinds of rods colored by dark gray and black. The bulk bands are almost unchanged from the symmetric case as depicted in Figure 4c. However, as we shall see below, this symmetry breaking clearly resolves the degeneracy of edge modes and opens a recognizable gap in edge bands. The crystal comes into the topological phase when the inversion symmetry is restored by setting the same values of relative permittivity, i.e., 10, in the dark gray and black regions. Contrastingly, the crystal of Figure 4b is in non-topological phase. Figure 4c is the band diagram of the TM modes of the quasi-topological photonic crystal in Figure 4a. We can find sufficiently-sized band gap at approximately 0.5 in the unit $\omega a/(2\pi c)$. Figure 5a displays the unit cell of the superlattice composed of the crystals of Figure 4a,b. Figure 5b is a closeup of the projected band diagram of TM modes in the superlattice. The edge modes are emphasized in red. The red-dotted lines are the edge modes when the inversion symmetry of the middle part is restored. In this superlattice, the structure around the edge part also breaks the inversion symmetry of the whole system, as evidenced by a small gap in the edge band modes even when the middle part is in the topological phase. The explicit breaking of the inversion symmetry in the middle part increases the size of this gap as well as resolves the degeneracy of the edge modes. Figure 5c gives the energy flux density of an edge mode belonging to the lowest branch. We can see that the mode is well confined around the boundary and accompanies some eddies.

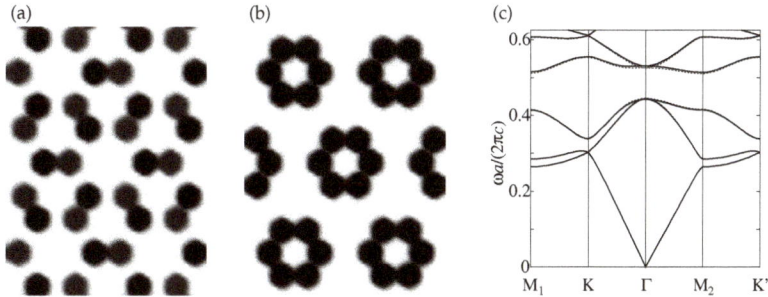

Figure 4. (**a**) A sample inversion asymmetric 2D quasi-topological photonic crystal for relative permittivity of 1, 9, and 11 in white, dark gray, and black regions, respectively. When the inversion symmetry is restored, the crystal can be in the topological phase. (**b**) A sample 2D photonic crystal in non-topological phase for relative permittivity of 1 and 10 in white and black regions, respectively. (**c**) Band diagram of TM modes of the photonic crystal in (**a**); dotted lines show the case where the relative permittivities of the gray and black rods are set to 10. The vertical axis represents the dimensionless frequency $\omega a/(2\pi c)$. (*a*: lattice constant and *c*: speed of light).

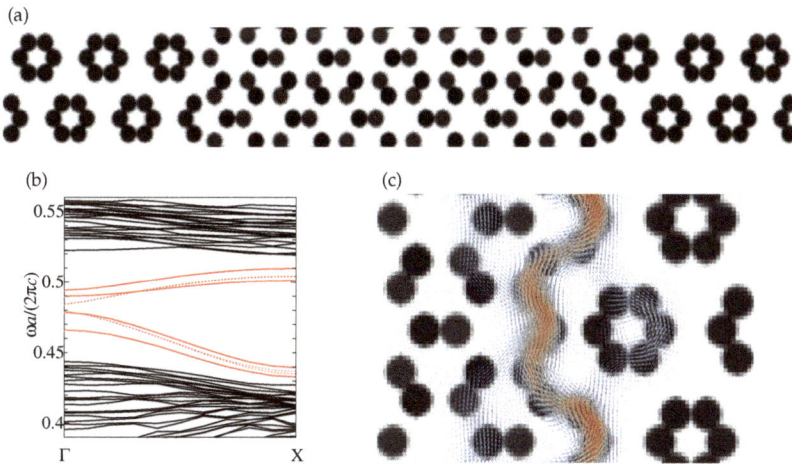

Figure 5. (**a**) Unit cell of a 2D superlattice composed of quasi-topological and non-topological photonic crystals in Figure 4. (**b**) Projected band diagram of TM modes of (**a**). Edge modes are emphasized in red. The red-dotted lines show the edge modes where the relative permittivities of the dark gray and black rods in the middle part are set to 10. The vertical axis represents dimensionless frequency $\omega a/(2\pi c)$. (*a*: lattice constant and, *c*: speed of light) (**c**) Energy flux density of an edge mode ($\omega a/(2\pi c) = 0.455$) belonging to the lowest branch (the lowest red curve) in (**b**).

The propagation characteristics of the edge modes were demonstrated in a Y-shaped waveguide in Figure 6a, where the Y-shaped part is composed of the crystal in Figure 4a, whereas the block layer is composed of the crystal in Figure 4b with a sufficiently large band gap to cover the relevant frequency range. The linear source was set at the left end of the left branch of the waveguide, and the vortex modes were excited by electric field oscillations along the rods, or specifically, in the vertical direction to the page. Figure 6b shows the z-component of electric field E_z. Here, we can see an extremely anisotropic propagation through helical edge channels. Figure 6c shows the transmission spectra (blue and red solid lines) measured at the ends of the upper and lower right branches along with the spectra (blue and red broken lines) of the case where the Y-shaped region is replaced by vacuum, for comparison. The blue and red broken lines should overlap for the idealistic simulation treating each rod as a material with an exactly sharp boundary. However, smearing each boundary was introduced in the real simulation and the order of introducing the parts influenced the actually simulated structure. For the present case, the simulated structure of the block layer weakly broke the space inversion symmetry; hence, a small discrepancy appeared between the blue and red broken lines. By contrast, a large difference appeared between the blue and red solid lines: the transmission characteristics of the target system were quite asymmetric.

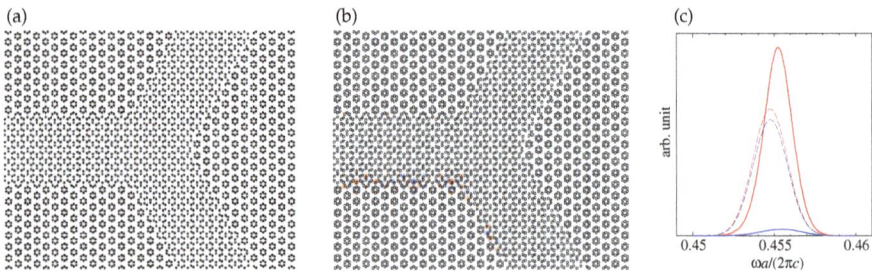

Figure 6. (**a**) Y-shaped waveguide composed of quasi-topological and non-topological photonic crystals in Figure 4. (**b**) z-component of electric field E_z. (**c**) Spectra of the transmissions to the upper right (blue line) and lower right (red line) branches. The broken lines represent the transmission spectra for a vacuum Y-shaped region.

5. Electronic State with Twisted Spin-polarization

Focusing on the highly-resolved spin selectivity of helical edge states of topological insulators, we proposed a spin filter using one of the edge states as a conduction channel and a spin control method using hybridization between the edge states and conduction electrons in References [47,48]. For example, we simulated the reflection of an electronic wavepacket at a boundary between the 2D conductor (left side) and topological insulator (right side), as shown in Figure 7, using an effective tight-binding lattice model. For the convenience of numerical treatment, the model was constructed on a simple square lattice, $r = n_1 a_1 + n_2 a_2$, where n_1 and n_2 are integers and a_1 and a_2 are primitive lattice vectors of the square lattice. As our focus was on a single-particle state, in principle the first quantization formalism is sufficient. Nevertheless, we introduced the second quantized formalism as a convenient representation method, which enables us to represent operators in compact forms. The Hamiltonian of the conductor part was modeled following a simple tight-binding model,

$$H_{2DC} = \sum_r \left[\left\{ -t_0 \left(c_{r+a_1}^\dagger c_r + c_{r+a_2}^\dagger c_r \right) + \text{H.c.} \right\} + 4t_0 c_r^\dagger c_r \right], \qquad (11)$$

where $c_r^\dagger$ and c_r are the creation and annihilation operators at a lattice site, and H.c. is Hermitian conjugate. Here, c_r is a spinor operator consisting of up and down spin components, i.e., $c_r = (c_{r\uparrow}, c_{r\downarrow})^\top$. We introduced the last term to adjust the bottom of the conduction band to the origin of energy $E_{k=0} = 0$. Under periodic boundary conditions, the energy dispersion of this model is derived as $E_k = 2t_0[2 - \cos(k \cdot$

$a_1) - \cos(k \cdot a_2)]$, and which mimics a conventional k-square dispersion around the origin of k-space, i.e., $E_k \cong k^2/(2m^*)$ $(m^* = 2t_0 a^2)$. In modeling the topological insulator, we introduced a spin-dependent π-flux per square plaquette by the nearest-neighbor hopping of the magnitude $|t_n|$ and classified all the lattice points alternatively to the sub-lattices, A $(n_1 + n_2 \in \text{even})$ and B $(n_1 + n_2 \in \text{odd})$. (The unit cell is doubled, and the primitive vectors of each sub-lattice are given by $a_1 + a_2$ and $-a_1 + a_2$.) Next, we introduced the next-nearest-neighbor hopping of the magnitude $|t_{nn}|$ with alternating signs depending on the sub-lattices and a staggered potential of magnitude $|v_s|$. The Hamiltonian of the topological insulator part is represented by

$$
\begin{aligned}
H_{2\text{DTI}} &= \sum_r (-1)^r t_n \left[\left(c^{\dagger}_{r+a_1} e^{i\frac{\pi}{4}(-1)^r \sigma_3} c_r - c^{\dagger}_{r+a_2} e^{-i\frac{\pi}{4}(-1)^r \sigma_3} c_r \right) + \text{H.c.} \right] \\
&+ \sum_r (-1)^r t_{nn} \left[\left(c^{\dagger}_{r+a_1+a_2} c_r + c^{\dagger}_{r-a_1+a_2} c_r \right) + \text{H.c.} \right] + \sum_r (-1)^r v_s c^{\dagger}_r c_r,
\end{aligned}
\tag{12}
$$

where $(-1)^r = (-1)^{n_1+n_2} = \pm 1$ for A and B sub-lattices, respectively. The Hamiltonian $H_{2\text{DTI}}$ is time-reversal invariant as a whole as well as $H_{2\text{DC}}$, because each of the spin sectors is a time-reversal partner of the other. In the parameter range $4|t_{nn}| > |v_s|$, each of the spin sectors comes in a quantum Hall phase. The spin-resolved quantized Hall conductances have a common absolute value and different signs. They cancel each other so as to preserve time-reversal symmetry. A homogeneously spin-polarized incident wavepacket is illustrated in Figure 7a. The packet is incident from the conductor (left) perpendicularly to the boundary (red vertical line); its spin polarization is uniformly pointing in the incident direction. On the other hand, the spin-polarization state of the reflected wavepacket is depicted in Figure 7b. The spin density of the top white area faces in this side of the page, whereas that of the bottom black area faces the back. Polarization in the vicinity of the wavepacket center has the same state as the pre-incidence. (See Figure 7c for details about the correspondence between spin density and color space.) These results suggest that an electronic state with twisted spin-polarization can be generated from a homogeneously spin-polarized state via topological interface between a conductor and a topological insulator, where helical edge states run along the boundary.

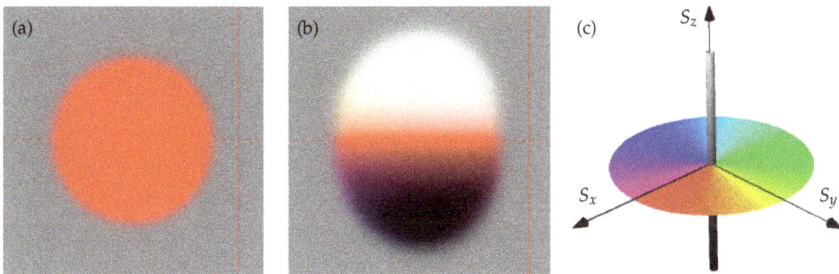

Figure 7. (a) An incident wavepacket with homogeneous polarization along the x-direction and (b) a reflected wavepacket with twisted spin texture. (c) Relative correspondence between spin density S and hue, lightness, and saturation (HLS) color space. Approximately, HLS correspond to the azimuth angle, polar angle, and magnitude of spin density, respectively.

The concept of topological insulator extends to the three-dimensional (3D) system, where helical surface/interface states traversing bulk band gaps emerge [49,50]. Figure 8a shows a conceptual diagram of idealistic helical surface states. The green balls depict the electrons, whose propagation direction and spin angular momentum are indicated by each set of green and black arrows, respectively. For example, the 2D topological insulator model in Equation (12) can also be extended to 3D versions with some generalizations. A sequence of 3D models is constructed on a simple cubic lattice $r = \sum_\mu n_\mu a_\mu$ $(n_\mu \in \mathbb{Z})$ with an orthogonal set of unit lattice vectors a_μ $(\mu = 1,2,3)$. Every site is classified into either A or B sub-lattice as $r \in A(B)$ when $\sum_\mu n_\mu$ = even(odd), in which a sign symbol $(-1)^r$ can be introduced as $(-1)^r = (-1)^{\sum_\mu n_\mu}$. Moreover, each sub-lattice forms a face-centered cubic lattice. The sequence of 3D models is characterized by three types of parameters t_μ, $t_{\mu\nu}(= t_{\nu\mu})$, and v_s, along with SU(2) matrices $\{U_\mu\}$ $(\mu, \nu = 1,2,3)$ representing the spin-precession processes in the μ-directional nearest-neighbor hoppings. The Hamiltonian of the sequence is given by

$$
\begin{aligned}
\hat{H}_{\text{3DTI}} &= \sum_r \sum_\mu (-1)^r t_\mu \left[\hat{c}_{r+a_\mu}^\dagger U_\mu^{(-1)^r} \hat{c}_r + \text{H.c.} \right] \\
&+ \sum_r \sum_{\mu<\nu} (-1)^r t_{\mu\nu} \left[\left(\hat{c}_{r+a_\mu+a_\nu}^\dagger \hat{c}_r + \hat{c}_{r-a_\mu+a_\nu}^\dagger \hat{c}_r \right) + \text{H.c.} \right] + \sum_r (-1)^r v_s \hat{c}_r^\dagger \hat{c}_r.
\end{aligned}
\tag{13}
$$

This sequence is advantageous in that the edge states of a member with open boundary condition can be analytically investigated in some parameter regions, provided that $\{U_\mu\}$ satisfies the conditions $U_\mu^\dagger U_\nu + U_\nu^\dagger U_\mu = 2\delta_{\mu\nu}\sigma_0$ and $\sigma_2 \overline{U}_\mu \sigma_2 = U_\mu$. Here, $U_\mu^\dagger$ and $\overline{U}_\mu$ are Hermitian and complex conjugates of U_μ, respectively. The symbol σ_0 stands for the 2×2 unit matrix in the spin space. Unfortunately, there remain unresolved issues in plausible modeling of the interface between a member of this sequence and conductor. The analysis also accompanies technical complications and will be given elsewhere. Besides, 3D versions of topological photonic crystals have also been proposed [51,52]. Although the relation between a photon's pseudo-spin and actual angular momentum in a periodic structure remains ambiguous currently, examples of energy flux densities of photonic chiral edge modes in References [25,33] and photonic helical edge modes in Figure 5 suggest that the former corresponds to a vortex structure stemming from the latter. A schematic of the ideal photonic helical surface modes is shown in Figure 8b. Here, each set of yellow and black arrows represent the propagation direction and local angular momentum density of a photonic helical surface mode, respectively, whereas the yellow circles containing arrows represent the local vortex structures of the modes.

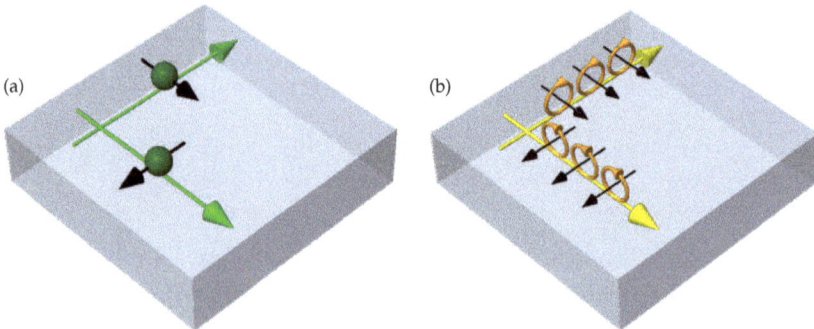

Figure 8. Conceptual diagrams of (**a**) electronic helical surface states and (**b**) photonic helical surface modes.

Let us look back to the electronic wavepacket with a twisted spin structure in Figure 7b and consider its 3D extension. Suppose a case exists where a wavepacket homogeneously polarized in the propagation direction is perpendicularly incident on the surface of an idealistic 3D topological insulator depicted in Figure 8a. Since in this case we can find the rotational symmetry around the incident axis, the reflected wavepacket is expected to accompany a toroidally-twisted spin texture derived by rotating

Figure 7b around the incident axis. Therefore, the question that arises is how do we extend the above discussion of electoronic wavepacket to its photonic version. Section 4 argues that various types of topological photonic media can be proposed based on the same idea in electron systems. As speculated in Figure 8b, the electronic spin would be replaced by a photonic vortex structure. It is reasonable to replace the incident electronic wavepacket homogeneously spin-polarized to the propagation direction by an electromagnetic wavepacket with photonic orbital angular momentum in the propagation direction, i.e., by that as depicted in Figure 9. Similarly, a simple thinking on the reflected wavepacket would suggest that a toroidally-twisted spin structure can be replaced by a toroidal vortex structure, as depicted in Figure 10. (The hue and the arrows in Figures 9 and 10 represent the energy density and the deviation of energy flux density, respectively, as in Figure 1.) This speculation appears to be extremely naive, because in general, the vortex structure of a photonic helical surface mode is complicated, as displayed in Figure 5c. Nevertheless, as long as the focus is on the topological information of wavepackets, e.g., a set of topological charges of multiple-vortex structure, some realistic vortices are very likely to belong to the same topological class as in Figure 5c, as was the case for Laguerre-Gaussian beams. Therefore, studying the possibility and stability of a photonic/electromagnetic wavepacket with such toroidal vortex structure is worthwhile, not only from an academic point-of-view but also from an application perspective.

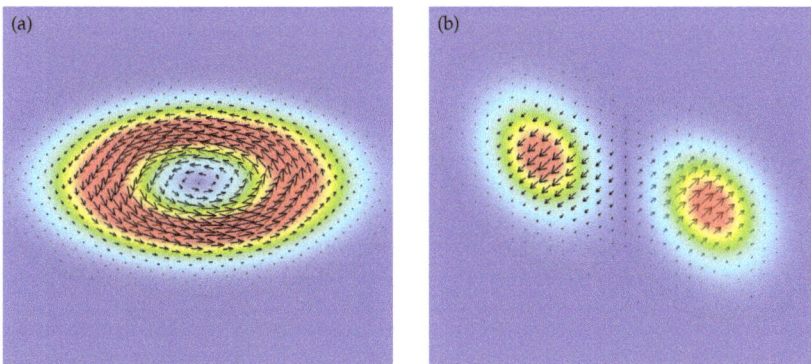

Figure 9. (**a**) xy- and (**b**) yz-cross-sections of a linearly polarized Laguerre-Gaussian wavepacket with orbital angular momentum directed to the positive z-axis.

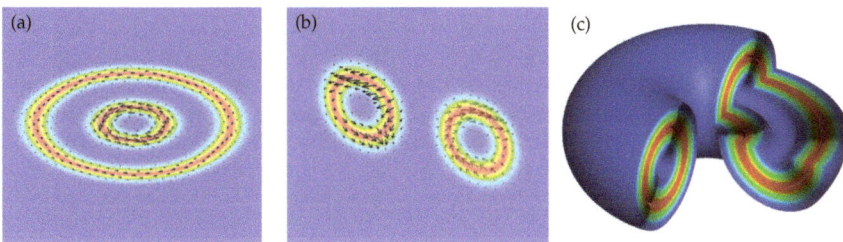

Figure 10. (**a**) xy- and (**b**) yz-cross-sections, and (**c**) isosurface of the energy density of a linearly polarized toroidal wavepacket with orbital angular momentum directed to the positive z-axis.

6. Propagation Characteristics of Toroidal Electromagnetic Wavepacket

The electromagnetic vortices shown in Figure 9 can be implemented into a quantum digit for information communication [13] and mode division multiplexing in telecommunication technology [14,15].

These applications use the fact that there are multiple quasi-orthogonal modes in a narrow frequency band. Hence, it is a meaningful task to devise various extensions of such electromagnetic vortices. This section brings up the toroidal vortex structure in Figure 10 as one of those extensions. Moreover, it aims to answer questions such as whether the electromagnetic wavepacket with a toroidal vortex can exist as a solution to Maxwell's equations and how stable it is when it can be present.

To answer the initial question, we shall present a procedure to construct wavepacket solutions with generic vortex structures. In the process, we assume that we already have the information of Fourier components $e_{k\alpha}$ of plane wave solutions of mode α with eigenfrequency $\omega_{k\alpha}$, and that the set $\{e_{k\alpha}\}$ $(\alpha = 1, 2, \cdots)$ constitutes a perfect orthonormal system, at least at a practical approximation level. The construction procedure consists of four steps as follows:

1. Construct normalized scalar wavepackets $\{f_\alpha(r)\}$ with trial vortex structures for mode α.
2. Calculate the Fourier transform $\{\tilde{f}_{k\alpha}\}$ of $\{f_\alpha(r)\}$.
3. Construct the solution of electromagnetic field $\tilde{E}(k, t)$ in k-space by

$$\tilde{E}(k, t) = \sum_\alpha \tilde{f}_{k\alpha} z_{k\alpha} e_{k\alpha} \exp(i\omega_{k\alpha} t), \tag{14}$$

where the set of parameters $\{z_{k\alpha}\}$ reduces to Jones vector in simple cases.

4. Calculate the inverse Fourier transform of $\tilde{E}(k, t)$, and take its real part as the solution $E(r, t)$.

Generally, the above procedure can contain numerical calculations and is inevitably accompanied by approximation due to discretization of both real and wavevector spaces. Nonetheless, in principle the obtained solution converges to an exact solution in the continuous limit. Electromagnetic wavepackets with any vortex structure can actually exist, whereas the stability remains uncertain. In other words, this procedure is applicable as long as the quasi-complete set of $\{e_{k\alpha}\}$ is obtained by either an analytical or numerical method. For instance, a typical case for the former may include the reflection and refraction of linearly polarized plane waves at a flat interface between two different media of homogeneous isotropic permittivity ϵ and permeability μ. Analytical expressions of $\{e_{k\alpha}\}$ are given by Fresnel's equations, where the index α stands for either P- or S-polarization, while the index k can represent the wavevector of an incident plane wave. As for the latter, we can consider an extension to periodic systems by replacing momentum k by crystal momentum and making mode index α include a band index, along with a degenerate mode index, as in $\alpha \rightarrow n\lambda$ (n: band index and λ: degenerate mode index). Finally, note that the center of a wavepacket can be easily shifted by r_0 through the replacement $\{f_\alpha(r)\} \rightarrow \{f_\alpha(r - r_0)\}$.

To simplify the discussion, we shall omit the mode dependence of trial functions introduced above, and pick up only linearly-polarized wavepackets here. Figure 10 provides a sample electromagnetic wavepacket constructed from the procedure above, and which propagates at a positive z-direction. The wavepacket has an energy density distributed in a hollow torus-shape, and the wavepacket has a ring-shaped vortex along the internal hollow part of the torus, in addition to a line-shaped vortex associated with the orbital angular momentum directed to positive z-direction, which penetrates through the central hole of the torus. Figure 10a represents the xy-cross-section of the wavepacket, where we can find the eddy structure of energy flux density corresponding to the orbital angular momentum. On the other hand, Figure 10b represents the yz-cross-section of the wavepacket, where we can find another vortex structure whose core corresponds to the hollow part inside the torus. Let us take a closer look at a trial scalar wavepacket with toroidal-type vortex structure, $f_{\text{TWP}}(r)$. This function contains seven types of parameters, namely m_{line}: vorticity of line vortex; m_{ring}: vorticity of ring vortex; k_c: central wavevector; ℓ_R: radius of the central ring inside the hollow region; ℓ_r: radius of torus-type tube; ℓ_Δ: thickness of surface layer of hollow torus and; ℓ_v: size of vortex core, where we set the core sizes of two kinds of vortices to be the same. By introducing a right-handed basis set $\{e_1, e_2, e_3\}$ with the condition $e_3 = k_c/|k_c|$, we can give an example such as

$$f_{\text{TWP}}(r) = \mathcal{N}\left(\frac{z_{\text{line}}}{|z_{\text{line}}|}\right)^{m_{\text{line}}}\left(\frac{z_{\text{ring}}}{|z_{\text{ring}}|}\right)^{m_{\text{ring}}}\tanh\left(\frac{|z_{\text{line}}|}{\ell_v}\right)\tanh\left(\frac{|z_{\text{ring}}|}{\ell_v}\right)$$

$$\times \exp\left\{-\frac{1}{2\ell_\Delta^2}\left(|z_{\text{ring}}| - \ell_r\right)^2 + ik_c \cdot r\right\}, \tag{15}$$

$$z_{\text{line}} = (e_1 + ie_2)\cdot r, \quad z_{\text{ring}} = (|z_{\text{line}}| - \ell_R) + ie_3 \cdot r, \tag{16}$$

where $\mathcal{N}$ is a normalization factor. Figure 10 corresponds to the case where $m_{\text{line}} = 1$, $m_{\text{ring}} = -1$, $\lambda = 2\pi/|k_c| = 0.445\ell_0$, $\ell_R = 6\ell_0$, $\ell_r = 3\ell_0$, $\ell_\Delta = \ell_0$, and $\ell_v = 0.25\ell_0$ where ℓ_0 is a unit of length scale. Herein, we shall consider only this set of parameters for toroidal wavepackets, as our focus is limited on the topological properties of toroidal wavepackets, and does not extend to details of their shape deformations. For better understanding by way of comparison, we present trial scalar functions $f_{\text{GWP}}(r)$ and $f_{\text{LGWP}}(r)$ for Gaussian and Laguerre-Gaussian wavepackets, respectively, defined by

$$f_{\text{GWP}}(r) = \mathcal{N}\exp\left(-\frac{r^2}{2\ell_R^2} + ik_c \cdot r\right), \tag{17}$$

$$f_{\text{LGWP}}(r) = \mathcal{N}\left(\frac{z_{\text{line}}}{|z_{\text{line}}|}\right)^{m_{\text{line}}}\tanh\left(\frac{|z_{\text{line}}|}{\ell_v}\right)\exp\left(-\frac{|z_{\text{ring}}|^2}{2\ell_r^2} + ik_c \cdot r\right), \tag{18}$$

The Gaussian wavepacket in Figure 1 and the Laguerre-Gaussian wavepacket in Figure 9 correspond to the cases with $\ell_R = 6\ell_0$ and with $m_{\text{line}} = 1$, $\ell_R = 6\ell_0$, $\ell_r = 2\ell_0$, $\ell_v = 0.25\ell_0$, respectively. In both cases, the wavelength is set at $\lambda = 2\pi/|k_c| = 0.445\ell_0$.

For the second question, we shall consider the stability of the toroidal wavepacket against reflection and refraction on flat interfaces between different kinds of homogeneous isotropic media. The dielectric constants on the lower and upper sides are represented by the symbols ϵ_1 and ϵ_2, respectively. Figures 11–13 show the time lapses for cases with $\epsilon_2/\epsilon_1 = 0.40, 0.75$, and 2.50, respectively. The incident angle is set at $45°$. The time is measured in units of $\ell_0\sqrt{\epsilon_1\mu_0}$. The dimensionless time τ of each frame is $\tau = -16, -8, 0, +8, +16$ from left to right. Figure 14 shows the incident-angle dependence for $\epsilon_2/\epsilon_1 = 2.50$. The incident angle θ of each frame is $\theta = 0°, 15°, 30°, 45°, 60°$ from left to right, and the dimensionless time τ is $\tau = +16$ in every frame. In all cases, the magnetic permeability is set at $\mu = \mu_0$ everywhere, and only the xz-cross-sections are depicted. (x- and z-axes correspond to horizontal and vertical directions, respectively.) We adopt quasi-P-type configuration for the polarization state of every incident wavepacket. (The mean magnetic field of every incident wavepacket is parallel to the interface and normal to the quasi-incident-plane.) From the result in Figure 11, we can presume that the toroidal vortex is stable against reflection. On the other hand, two cases for refraction emerge. First, as the refractive index at the transmission side (Figure 12) decreases, the wavepacket shape stretches in a similar manner to its central wavelength, leading to an unstable ring vortex. Second, as the refractive index at the transmission side (Figure 13) increases, the wavepacket compresses in a similar manner to its central wavelength, resulting in a stable ring vortex at least up to $\theta = 60°$ (Figure 14).

Finally, we would like to mention the transformation laws of the wavepacket topological properties. The vorticity m_{line} of the linear vortex corresponding to the orbital angular momentum changes as $m_{\text{line}} \rightarrow -m_{\text{line}}$ in reflection, while it does not in refraction, suggested by our analogy with the Laguerre-Gaussian beam with an orbital angular momentum. By contrast, the vorticity m_{ring} of the ring vortex does not change in both reflection and refraction. In general, recognizing a wavepacket as a particle-like object may give odd results at first glance. However, the wavepacket is actually a wave phenomenon, and it transforms to get turned inside out in reflection. Since both the rotational flow stemming from the ring vortex and the propagation direction change, the vorticity m_{ring} defined based on the propagation direction remains unchanged. We would like to conclude this section with a note. As for the cases of partial reflection in Figures 12 and 13, it is not easy to identify reflected wavepackets in the present color contrast due to their weak intensities. Vague reflected wavepackets should appear after extremely increasing the contrast of these figures. On the other hand, in the incident-angle dependence of Figure 14, it becomes possible to recognize reflected wave packets as the incident angle gets away from Brewster's angle for $\epsilon_2/\epsilon_1 = 2.50$ ($\sim$57.7°).

Figure 11. Reflection of a toroidal-vortex at the interface of $\epsilon_2/\epsilon_1 = 0.40$ with the incident angle of 45°.

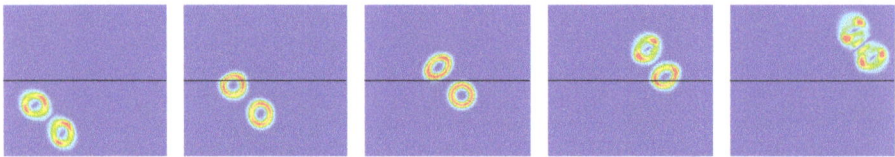

Figure 12. Refraction of a toroidal-vortex at the interface of $\epsilon_2/\epsilon_1 = 0.75$ with the incident angle of 45°.

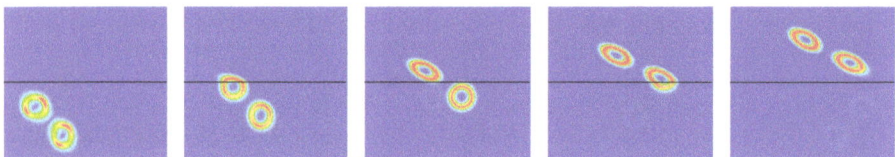

Figure 13. Refraction of a toroidal-vortex at the interface of $\epsilon_2/\epsilon_1 = 2.50$ with the incident angle of 45°.

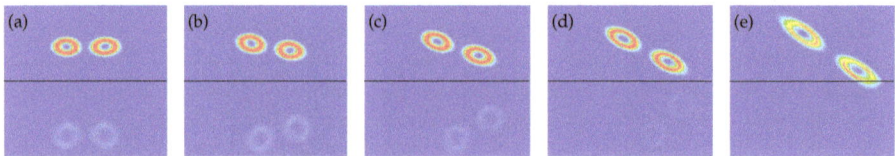

Figure 14. Incident-angle dependence of refracted and reflected toroidal-vortices for $\epsilon_2/\epsilon_1 = 2.50$. The incident angle is (**a**) 0°, (**b**) 15°, (**c**) 30°, (**d**) 45° and (**e**) 60° from left to right.

7. Discussion

Inspired by electromagnetic vortices in free space and periodic structures, and by exotic boundary modes of topological photonic media, we theoretically investigated the topological characteristics and feasibility of a toroidal electromagnetic wavepacket. Our proposal was also based on the numerical analysis of an electronic wavepacket with toroidally-twisted spin structure, generated by a reflection at the interface between an electronic topological insulator and a conductor. We recognized a class of topological photonic media as the photonic version of the electronic topological insulator and further interpreted it as an extension of a class of photonic crystals where various types of electromagnetic vortex modes emerge with their time-reversal partners. Furthermore, we referred to the fact that modern information transmission technology via electromagnetic waves started to pay attention to photonic orbital angular momentum in a unique way. For instance, optically-based communication technologies have demonstrated the use of photonic orbital angular momentum in the realization of quantum digits for single-photon communication and in the development of a new scheme for multiplexing signals in telecommunications. A key concept common in both examples is the presence of multiple nearly-orthogonal modes within a narrow range of frequencies. From this point of view, we stressed the meaningful benefits of investigating the extensions of this concept, and proposed the toroidal electromagnetic wavepacket as a fusional application with the exotic surface modes of topological photonic media. The electromagnetic wavepacket with toroidal-type dual vortices is an extension of the Laguerre-Gaussian wavepacket whose line vortex corresponds to the photonic orbital angular momentum. Herein, we presented the procedure to construct the solutions of Maxwell's equations with multiple types of vortices. Afterward, we numerically examined the stability of the toroidal electromagnetic wavepacket against reflection and refraction at flat interfaces between the homogeneous isotropic media. Finally, we derived the transformation laws of topological charges of line and ring vortices in these processes.

Funding: This work was supported by JSPS KAKENHI Grant Number JP16K05467.

Acknowledgments: Band calculations and time-domain simulations for photon systems in periodic structures were performed with free and open-source software packages, MPB [53] and MEEP [54], respectively.

Conflicts of Interest: The author declares no conflict of interest. The funders had no role in the design of the study; in the collection, analyses, or interpretation of data; in the writing of the manuscript, or in the decision to publish the results.

References

1. Rayleigh, L. XXVI. On the remarkable phenomenon of crystalline reflexion described by Prof. Stokes. *Lond. Edinb. Dublin Philos. Mag. J. Sci.* **1888**, *26*, 256–265. [CrossRef]
2. Ohtaka, K. Energy band of photons and low-energy photon diffraction. *Phys. Rev. B* **1979**, *19*, 5057–5067. [CrossRef]
3. Yablonovitch, E. Inhibited spontaneous emission in solid-state physics and electronics. *Phys. Rev. Lett.* **1987**, *58*, 2059–2062. [CrossRef] [PubMed]
4. John, S. Strong localization of photons in certain disordered dielectric superlattices. *Phys. Rev. Lett.* **1987**, *58*, 2486–2489. [CrossRef]
5. Veselago, V.G. The electrodynamics of substances with simultaneously negative values of ϵ and μ. *Sov. Phys. Uspekhi* **1968**, *10*, 509–514. [CrossRef]
6. Smith, D.R.; Padilla, W.J.; Vier, D.C.; Nemat-Nasser, S.C.; Schultz, S. Composite medium with simultaneously negative permeability and permittivity. *Phys. Rev. Lett.* **2000**, *84*, 4184–4187. [CrossRef]
7. Smith, D.R.; Kroll, N. Negative refractive index in left-handed materials. *Phys. Rev. Lett.* **2000**, *85*, 2933–2936. [CrossRef]
8. Pendry, J.B. Negative refraction makes a perfect lens. *Phys. Rev. Lett.* **2000**, *85*, 3966–3969. [CrossRef] [PubMed]
9. Pancharatnam, S. Generalized theory of interference and its applications. *Proc. Indian Acad. Sci. Sect. A* **1956**, *44*, 398–417. [CrossRef]
10. Berry, M.V. Quantal phase factors accompanying adiabatic changes. *Proc. R. Soc. A Math. Phys. Eng. Sci.* **1984**, *392*, 45–57. [CrossRef]
11. Tomita, A.; Chiao, R.Y. Observation of Berry's topological phase by use of an optical fiber. *Phys. Rev. Lett.* **1986**, *57*, 937–940. [CrossRef]

12. Allen, L.; Beijersbergen, M.W.; Spreeuw, R.J.C.; Woerdman, J.P. Orbital angular momentum of light and the transformation of Laguerre-Gaussian laser modes. *Phys. Rev. A* **1992**, *45*, 8185–8189. [CrossRef]

13. Mair, A.; Vaziri, A.; Weihs, G.; Zeilinger, A. Entanglement of the orbital angular momentum states of photons. *Nature* **2001**, *412*, 313–316. [CrossRef]

14. Wang, J.; Yang, J.Y.; Fazal, I.M.; Ahmed, N.; Yan, Y.; Huang, H.; Ren, Y.; Yue, Y.; Dolinar, S.; Tur, M.; et al. Terabit free-space data transmission employing orbital angular momentum multiplexing. *Nat. Photonics* **2012**, *6*, 488–496. [CrossRef]

15. Bozinovic, N.; Yue, Y.; Ren, Y.; Tur, M.; Kristensen, P.; Huang, H.; Willner, A.E.; Ramachandran, S. Terabit-scale orbital angular momentum mode division multiplexing in fibers. *Science* **2013**, *340*, 1545–1548. [CrossRef] [PubMed]

16. Onoda, M.; Ochiai, T. Designing spinning Bloch states in 2D photonic crystals for stirring nanoparticles. *Phys. Rev. Lett.* **2009**, *103*, 033903. [CrossRef]

17. Poynting, J.H. The wave motion of a revolving shaft, and a suggestion as to the angular momentum in a beam of circularly polarised light. *Proc. R. Soc. A Math. Phys. Eng. Sci.* **1909**, *82*, 560–567. [CrossRef]

18. Berry, M.V. The adiabatic phase and Pancharatnam's phase for polarized light. *J. Mod. Opt.* **1987**, *34*, 1401–1407. [CrossRef]

19. Onoda, M.; Murakami, S.; Nagaosa, N. Hall effect of light. *Phys. Rev. Lett.* **2004**, *93*, 083901. [CrossRef]

20. Onoda, M.; Murakami, S.; Nagaosa, N. Geometrical aspects in optical wave-packet dynamics. *Phys. Rev. E* **2006**, *74*, 066610. [CrossRef] [PubMed]

21. Karplus, R.; Luttinger, J.M. Hall effect in ferromagnetics. *Phys. Rev.* **1954**, *95*, 1154–1160. [CrossRef]

22. Adams, E.; Blount, E. Energy bands in the presence of an external force field—II. *J. Phys. Chem. Solids* **1959**, *10*, 286–303. [CrossRef]

23. Onoda, M.; Nagaosa, N. Topological nature of anomalous Hall effect in ferromagnets. *J. Phys. Soc. Jpn.* **2002**, *71*, 19–22. [CrossRef]

24. Jungwirth, T.; Niu, Q.; MacDonald, A.H. Anomalous Hall effect in ferromagnetic semiconductors. *Phys. Rev. Lett.* **2002**, *88*, 207208. [CrossRef]

25. Hayashi, K.; Takemura, T.; Onoda, M. Numerical analysis of spin-filter effect of optical tornados and chiral edge states in two-dimensional photonic crystal waveguides. In Proceedings of the SICE Annual Conference (SICE), Akita, Japan, 20–23 August 2012; pp. 1050–1055.

26. Thouless, D.J.; Kohmoto, M.; Nightingale, M.P.; den Nijs, M. Quantized Hall conductance in a two-dimensional periodic potential. *Phys. Rev. Lett.* **1982**, *49*, 405–408. [CrossRef]

27. Kohmoto, M. Topological invariant and the quantization of the Hall conductance. *Ann. Phys. (N. Y.)* **1985**, *160*, 343–354. [CrossRef]

28. Haldane, F.D.M. Model for a quantum Hall effect without Landau levels: Condensed-matter realization of the "parity anomaly". *Phys. Rev. Lett.* **1988**, *61*, 2015–2018. [CrossRef] [PubMed]

29. Laughlin, R.B. Quantized Hall conductivity in two dimensions. *Phys. Rev. B* **1981**, *23*, 5632–5633. [CrossRef]

30. Halperin, B.I. Quantized Hall conductance, current-carrying edge states, and the existence of extended states in a two-dimensional disordered potential. *Phys. Rev. B* **1982**, *25*, 2185–2190. [CrossRef]

31. Haldane, F.D.M.; Raghu, S. Possible realization of directional optical waveguides in photonic crystals with broken time-reversal symmetry. *Phys. Rev. Lett.* **2008**, *100*, 013904. [CrossRef]

32. Wang, Z.; Chong, Y.D.; Joannopoulos, J.D.; Soljačić, M. Reflection-free one-way edge modes in a gyromagnetic photonic crystal. *Phys. Rev. Lett.* **2008**, *100*, 013905. [CrossRef]

33. Ochiai, T.; Onoda, M. Photonic analog of graphene model and its extension: Dirac cone, symmetry, and edge states. *Phys. Rev. B* **2009**, *80*, 155103. [CrossRef]

34. Wang, Z.; Chong, Y.; Joannopoulos, J.D.; Soljačić, M. Observation of unidirectional backscattering-immune topological electromagnetic states. *Nature* **2009**, *461*, 772–775. [CrossRef]

35. Poo, Y.; Wu, R.x.; Lin, Z.; Yang, Y.; Chan, C.T. Experimental realization of self-guiding unidirectional electromagnetic edge states. *Phys. Rev. Lett.* **2011**, *106*, 093903. [CrossRef]

36. Bahari, B.; Ndao, A.; Vallini, F.; El Amili, A.; Fainman, Y.; Kanté, B. Nonreciprocal lasing in topological cavities of arbitrary geometries. *Science* **2017**, *358*, 636–640. [CrossRef]

37. Onoda, M.; Nagaosa, N. Spin current and accumulation generated by the spin Hall insulator. *Phys. Rev. Lett.* **2005**, *95*, 106601. [CrossRef]

38. Kane, C.L.; Mele, E.J. Z_2 topological order and the quantum spin Hall effect. *Phys. Rev. Lett.* **2005**, *95*, 146802. [CrossRef]

39. Moore, J.E. The birth of topological insulators. *Nature* **2010**, *464*, 194–198. [CrossRef]

40. Fu, L. Topological crystalline insulators. *Phys. Rev. Lett.* **2011**, *106*, 106802. [CrossRef]

41. Yannopapas, V. Gapless surface states in a lattice of coupled cavities: A photonic analog of topological crystalline insulators. *Phys. Rev. B* **2011**, *84*, 195126. [CrossRef]

42. Khanikaev, A.B.; Hossein Mousavi, S.; Tse, W.K.K.; Kargarian, M.; MacDonald, A.H.; Shvets, G.; Mousavi, S.H.; Tse, W.K.K.; Kargarian, M.; MacDonald, A.H.; et al. Photonic topological insulators. *Nat. Mater.* **2013**, *12*, 233–239. [CrossRef]

43. Wu, L.H.; Hu, X. Scheme for achieving a topological photonic crystal by using dielectric material. *Phys. Rev. Lett.* **2015**, *114*, 223901. [CrossRef]

44. Schnyder, A.P.; Ryu, S.; Furusaki, A.; Ludwig, A.W.W. Classification of topological insulators and superconductors in three spatial dimensions. *Phys. Rev. B* **2008**, *78*, 195125. [CrossRef]

45. Ryu, S.; Schnyder, A.P.; Furusaki, A.; Ludwig, A.W.W. Topological insulators and superconductors: Tenfold way and dimensional hierarchy. *New J. Phys.* **2010**, *12*, 065010. [CrossRef]

46. Bernevig, B.A.; Hughes, T.L.; Zhang, S.C. Quantum spin Hall effect and topological phase transition in HgTe quantum wells. *Science* **2006**, *314*, 1757–1761. [CrossRef]

47. Ohmura, Y.; Shimokawa, T.; Hosaka, S.; Onoda, M. Spin-dynamics of a quantum wave-packet via helical edge states and spin-filter effect. In Proceedings of the SICE Annual Conference (SICE), Akita, Japan, 20–23 August 2012; pp. 1044–1049.

48. Onoda, M. Implementation of a unitary algorithm in the analysis of quantum dynamics at the interface between a topological insulator and a conductor. In Proceedings of the SICE Annual Conference (SICE), Akita, Japan, 20–23 August 2012; pp. 376–381.

49. Fu, L.; Kane, C.L.; Mele, E.J. Topological insulators in three dimensions. *Phys. Rev. Lett.* **2007**, *98*, 106803. [CrossRef]

50. Fu, L.; Kane, C.L. Topological insulators with inversion symmetry. *Phys. Rev. B* **2007**, *76*, 045302. [CrossRef]

51. Lu, L.; Fang, C.; Fu, L.; Johnson, S.G.; Joannopoulos, J.D.; Soljačić, M. Symmetry-protected topological photonic crystal in three dimensions. *Nat. Phys.* **2016**, *12*, 337–340. [CrossRef]

52. Slobozhanyuk, A.; Mousavi, S.H.; Ni, X.; Smirnova, D.; Kivshar, Y.S.; Khanikaev, A.B. Three-dimensional all-dielectric photonic topological insulator. *Nat. Photonics* **2017**, *11*, 130–136. [CrossRef]

53. Johnson, S.; Joannopoulos, J. Block-iterative frequency-domain methods for Maxwell's equations in a planewave basis. *Opt. Express* **2001**, *8*, 173–190. [CrossRef]

54. Oskooi, A.F.; Roundy, D.; Ibanescu, M.; Bermel, P.; Joannopoulos, J.; Johnson, S.G. Meep: A flexible free-software package for electromagnetic simulations by the FDTD method. *Comput. Phys. Commun.* **2010**, *181*, 687–702. [CrossRef]

applied
sciences

MDPI

Article

Prediction of Shape Change for Fatigue Crack in a Round Bar Using Three-Parameter Growth Circles

Yali Yang [1,2], Seokjae Chu [1,*] and Hao Chen [2]

[1]   School of Mechanical Engineering, University of Ulsan, Ulsan 680-749, Korea; carolyn71@163.com
[2]   School of Mechanical and Automotive Engineering, Shanghai University of Engineering Science,
    Shanghai 201620, China; pschenhao@163.com
*   Correspondence: sjchu@ulsan.ac.kr; Tel.: +82-52-259-2141

Received: 26 March 2019; Accepted: 22 April 2019; Published: 27 April 2019

Featured Application: Life prediction for engineering materials.

Abstract: The conventional method for predicting the shape change of a surface crack in a round bar simply utilizes the Paris-Erdogan law with the least squares method using a certain shape assumption with excessive constraints. In this paper, a three-parameter model for a round bar subjected to tension is developed with fewer shape assumption restraints by employing a fatigue crack growth circles method. The equivalent stress intensity factor ΔK_e based on both stress intensity factors along the current and new crack front is used to reduce the total number of increments. The results show that the proposed method has a good convergence speed and accurate prediction of crack shapes. The present method is validated by comparing the solution with other simulation solutions and experimental data.

Keywords: fatigue crack growth; surface crack; crack shape change; three-parameter model

1. Introduction

The propagation analysis of a surface crack is a critical capability for structural integrity prediction of cylindrical metallic components (bolts, screws, shafts, etc.) Part-through flaws appear on the free surface of a smooth round bar and the front of a growing crack can be considered as a so-called 'almond' shape by extensive experimental works [1–6].

Attempts to predict fatigue growth of a surface crack in a round bar have been reported. Some investigators have employed a circular arc to describe the crack front [5,7,8], then the hypothesis that an actual part-through crack can be replaced by an equivalent elliptical arc edge flaw has been widely applied. A. Carpinteri [9–13], as one of the most representative researchers on this topic, conducted extensive studies related to this configuration. However, regardless of whether they used a circular arc or elliptical arc, most researchers employed a certain shape with a fixed center, which reduced the fatigue calculations to one-or two-dimensional problems. Few efforts have been made utilizing a three-parameter model. Although A. Carpinteri [14] mentioned the three-parameter model previously, the fatigue crack propagation was simply examined by applying the Paris-Erdogan law with the least square method as in almost all previous studies [15–17]. In addition to experimental backtracking technique [18,19] and normalized area-compliance method [20], there is no further research regarding the method of surface crack prediction.

The objective of this paper was to predict the shape change of a fatigue crack in a round bar subjected to tension by employing fatigue crack growth circles, based on a three-parameter model using finite element analysis. In this paper, a fewer shape restraints model with part-elliptical cracks whose center was allowed to move along the vertical axis was built, which could be more precise for expressing the actual crack shape front. The nominal aspect ratio of an ellipse, which is more meaningful, is proposed for the three-parameter model. Meanwhile, the fatigue crack growth circles, which are on a tangent to both current and new crack fronts, were developed to predict the crack path. The equivalent stress intensity factor ΔK_e based on both stress intensity factors along the current and new crack fronts was proposed to reduce the number of modeling computations with only a few iterations. The validity of the present method will be shown by comparing its results with a simulation solution and experimental results.

2. Numerical Propagation Process

2.1. Three-Parameter Model

A surface crack in a smooth round bar with diameter D_0 subjected to fatigue tension are taken into consideration. A part-elliptical surface flaw is defined by three parameters: (1) major axis of an ellipse a, (2) minor axis of an ellipse b, and (3) center of ellipse O_y (Figure 1).

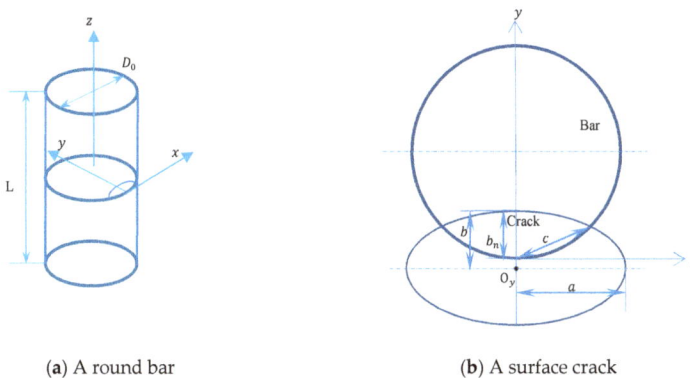

(a) A round bar   (b) A surface crack

Figure 1. Definition of the geometrical parameters for a surface crack in a round bar.

2.2. Fatigue Crack Propagation

The propagation of a surface crack in a round bar under cyclic tension is predicted by employing fatigue crack growth circles [21] (Figure 2). If the crack presents an ellipse shape up to the i-th loading step, the initial ellipse whose center is located on the surface of the specimen can be defined with given a_i and b_i, as represented by the following equation

$$\frac{x^2}{a_i^2} + \frac{y^2}{b_i^2} = 1.$$ (1)

Points O, A, B, C and D in Figure 2 with coordinates (x_{ji}, y_{ji}) are deployed equidistantly along the current crack front, where the subscript j refers to the points O, A, B, C and D.

The growth of a new crack front lying on an ellipse with semi-axes a_{i+1}, b_{i+1}, and O_{i+1} after one cyclic loading step to a new configuration can be described by the following equation

$$\frac{x^2}{(a_{i+1})^2} + \frac{\left(y - O_{y,i+1}\right)^2}{(b_{i+1})^2} = 1.$$ (2)

The assumed crack growth circles, which pass points O, A, B, C and D, respectively, are tangent to both current and new crack fronts. The new crack front points O', A', B', C' and D' with coordinates $(x_{j,(i+1)}, y_{j,(i+1)})$ are the points of tangency between crack growth circles and the new crack front. Meanwhile the centers of crack growth circles can be determined as $(x_{j,c}, y_{j,c})$.

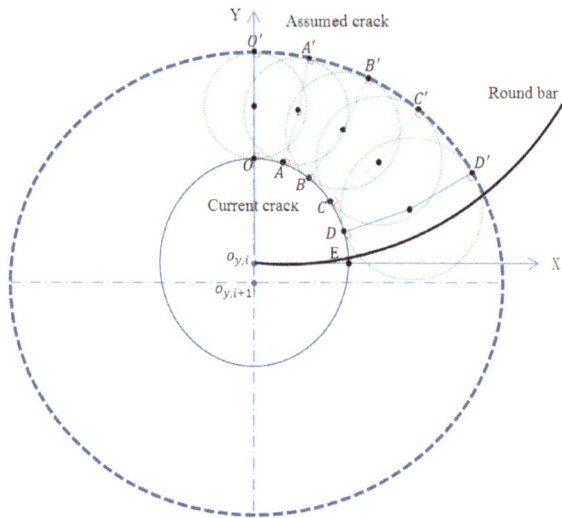

Figure 2. Determination of a new crack front by fatigue crack growth circles.

The crack growth increment among these points can be determined by applying the Paris-Erdogan law

$$\frac{da}{dN} = C(\Delta K)^m,$$ (3)

where da/dN is crack growth rate, ΔK is the stress intensity factor range, and C and m are material constants.

After each computed crack configuration, an increment of crack growth at the interior point O' is given. The crack growth length of other points A', B', C', and D' can be determined

$$\Delta l_j = (y_{O,i+1} - y_{O,i})\frac{\left(\Delta K_{ej}\right)^m}{\left(\Delta K_{eO}\right)^m}.$$ (4)

Here, ΔK_e stands for the equivalent stress intensity factor related to the stress intensity factors of both current and new crack fronts.

The stress intensity factor K is assumed to be a liner function of crack growth increment. An arbitrary number of crack growth steps can be assumed. Using

$$da = C(\Delta K)^m dN,$$ (5)

the crack growth length is increased to $a + da$ repeatedly in each step to the last step by adjusting material constant C. The equivalent stress intensity factor ΔK_e with stepping coefficient μ can be obtained appropriately through the crack growth plot of da/dN vs. N.

$$\Delta K_{ej} = \mu\left(K_{ij}\right)^m + \left(\frac{1}{2}\left(K_{i,j} + K_{(i+1),j}\right)\right)^m + (1-\mu)\left(K_{(i+1),j}\right)^m \quad 0 < \mu < 1$$ (6)

At the beginning of iteration, sometimes a relatively large value of μ can be used to avoid diverging.

The distance from the center of crack growth circles to points O', A', B', C' and D' along the new crack front are calculated using the geometrical relationship

$$\Delta d_j = \sqrt{\left(x_{j,(i+1)} - x_{j,c}\right)^2 + \left(y_{j,(i+1)} - x_{j,c}\right)^2}.$$ (7)

An error equation can be derived as

$$\text{Error} = \sum \left| \Delta d_j - \Delta l_j / 2 \right|. \tag{8}$$

The values of a_{i+1} and O_{i+1}, which minimize the error equation, are based on iterative methods and repeat all of the above steps based on the obtained crack front. The parameters of the ellipse for each new crack front can be determined until the results converge.

2.3. Numerical Simulation

The typical model of a round bar with diameter D_0 and length L that contains a surface crack in its median cross section has been used in many experimental tests and numerical simulations. F.P. Yang [19] presented the experimental results of fatigue crack growth for a straight-fronted edge crack in an elastic bar under axial loading with a diameter of 12 mm, a length of 90 mm, and carbon steel S45 as the material. Table 1 lists material parameters for steel S45. A. Carpinteri [11,12] calculated the surface cracks in round bars with 50 mm diameters through finite-element analysis. Since the propagation of crack shape is defined by the crack configuration for a given loading type [15], in the present paper, the models are established for different values of these initial parameters to compare the fatigue crack propagation with the experimental and simulation results from F.P. Yang [19] and A. Carpinteri [11,12].

Table 1. Material parameters for steel S45.

Monotonic Tensile Yield Strength σ_0	Nominal Ultimate Tensile Strength σ_m	True Ultimate Tensile Strength σ_f	Young's Modulus E	Poisson's Ratio v	Crack Growth Parameter m
635.07 MPa	775.65 MPa	2101.65 MPa	2.06×10^5 MPa	0.33	3

The finite element analysis software Abaqus^{TM} (France) is used to simulate the scenario. Since the bar geometry and applied loads present two planes of symmetry, 3D finite element analysis was performed by modeling a quarter of the round bar, as shown in Figure 3. About 350,000–380,000 quadratic hexahedral elements have been employed in each model. The 1/4-node displacement method and fine meshing with a 0.02 mm mesh size has been used around the crack front to model the stress field singularity and improve the accuracy of the contour integral calculation.

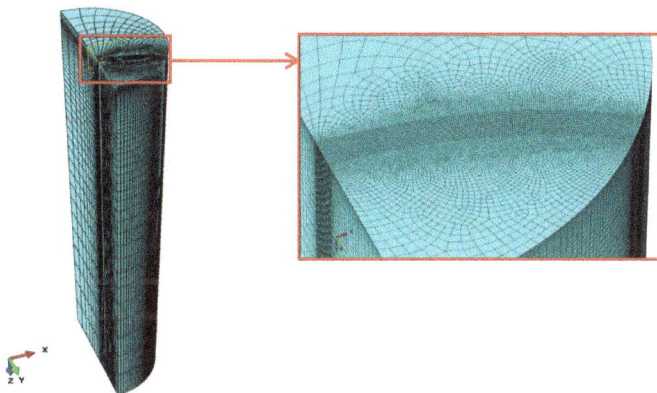

Figure 3. The finite element models of a surface-cracked round bar.

For each crack configuration defined by parameters b/D_0 and b/a, the stress-intensity factor $K_j (j = O, A, B, C, D)$ along the crack front is obtained through the above described finite-element analysis. The fatigue growth for the initial defects with $b/D_0 = 0.05$, 0.08, 0.1 and $b/a = 0$, 1 is considered in this paper.

3. Results and Discussion

3.1. Evolution of the Crack Shape

Figure 4 illustrates the fatigue shape evolution by the crack growth circles in a round bar subjected to tension. The seven crack front profiles displayed are deduced from roughly 30 crack growth circles in less than 20 iterations. The outermost crack growth circle rolls along the internal profile of the round bar in an approximate manner. When the point of tangency between crack growth circle with crack front approaches very closely to the surface of the bar, such as crack front 6 in Figure 4, the outermost crack growth circle will disappear in the next propagation. The rate of crack propagation can be observed intuitionally by the size of crack growth circles. As shown in Figure 5, the optimum simulation result for the center of an ellipse is not fixed on the surface of the bar, but is reciprocating along the y-axis. Therefore, the actual crack shape can be accurately expressed by the three-parameter model.

In the simulation process, notice that several different ellipses with the same chord length can be replaced to describe one actual crack front, since only part of an ellipse is used, once the center is not fixed (Figure 6). A large variation of ellipse actual aspect ratio is obtained with undifferentiated iteration error, as shown in Figure 7. Hence, the actual aspect ratio of the ellipse semi-axis is meaningless for the three-parameter model to describe the crack front.

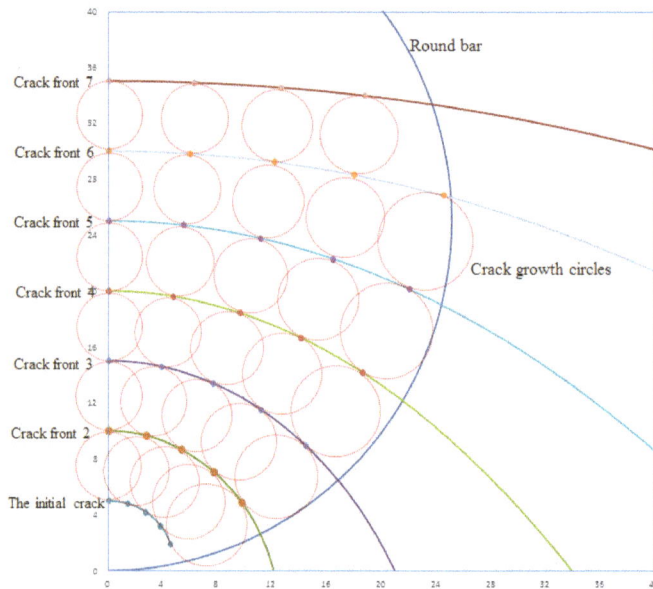

Figure 4. Successive determination of crack fronts by the crack growth circles with initial crack $b_0/a_0 = 1, b_0/D_0 = 0.1, m = 2$.

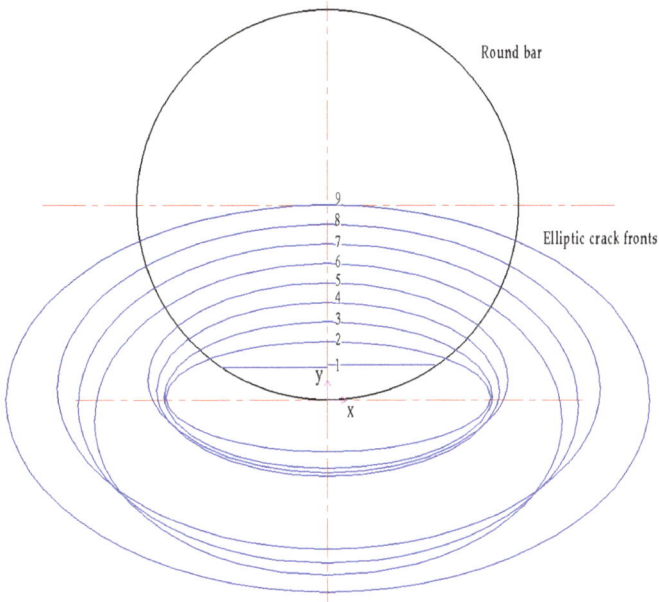

Figure 5. Ellipses used to determine crack fronts when initial crack $b_0/a_0 = 0, b_0/D_0 = 0.08, m = 2$.

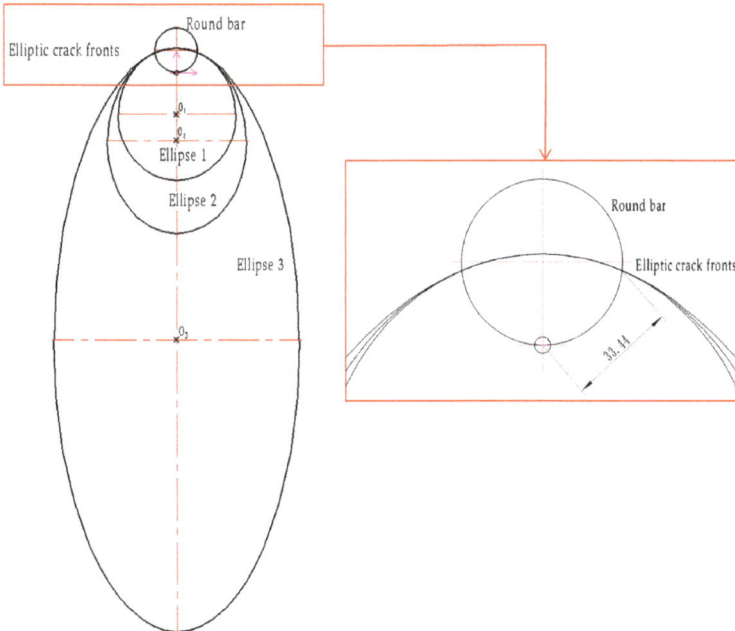

Figure 6. Crack front as a part of an ellipse.

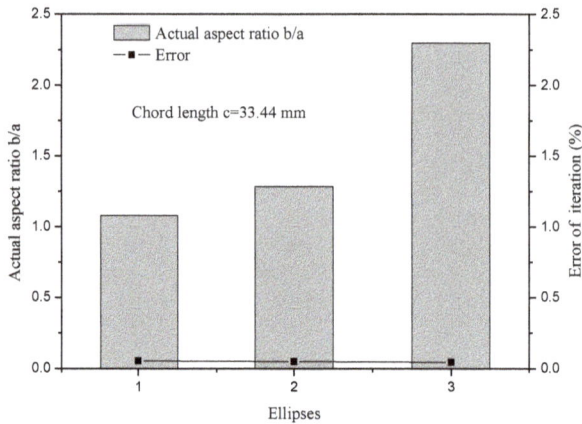

Figure 7. Change of actual aspect ratio with the same chord length *c*.

Figure 8 illustrates the fatigue shape evolution for five cases. The aspect ratio of initial ellipses $b_0/a_0 = 0$, 1, and relative crack depth $b_n/D_0 = 0.05$, 0.08, and 0.1, while the material constants in the Paris-Erdogan law are assumed to be $m = 2$, 3, and 4. The trends of crack propagation are adequately demonstrated.

As mentioned previously, the nominal aspect ratio of an ellipse, which is the ratio of the maximum crack depth to the chord length *c*, b_n/c can be considered here. It is noteworthy that, as shown in Figure 9, both initial crack dimensions and Paris law exponent m have an effect on the evolution of different parameters. The crack propagation trends are consistent with the same initial crack aspect ratio when using the same material, although the beginning propagation is affected by the crack depth provisionally. Meanwhile a difference of transition can be noticed between the crack propagation with different Paris law exponent *m* values. In Figure 9, it can be found that the nominal aspect ratio change is very sensitive to the initial crack geometry during early growth, and then the nominal aspect ratios for all cases are converged and become constant around $b_n/D_0 \approx 0.4$. It shows the flaws tend to follow preferential propagation paths that flatten gradually when the crack depth become larger.

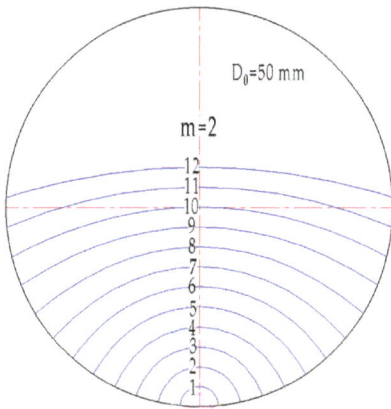

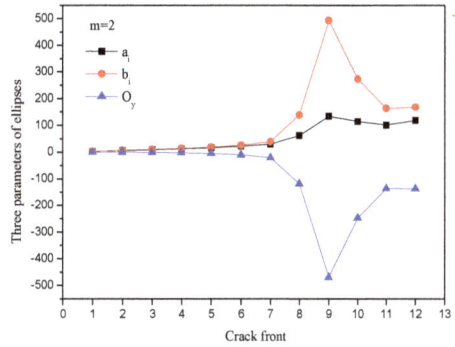

(**a**) $b_0/a_0 = 1, b_0/D_0 = 0.05$

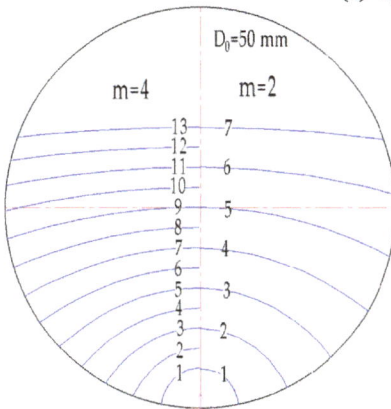

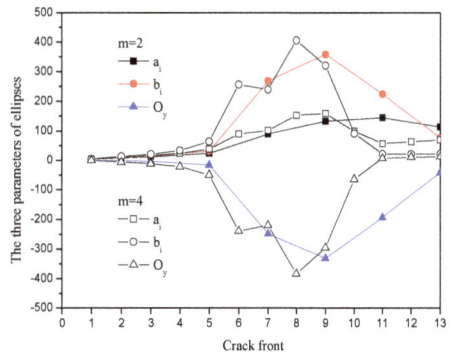

(**b**) $b_0/a_0 = 1, \quad b_0/D_0 = 0.1$

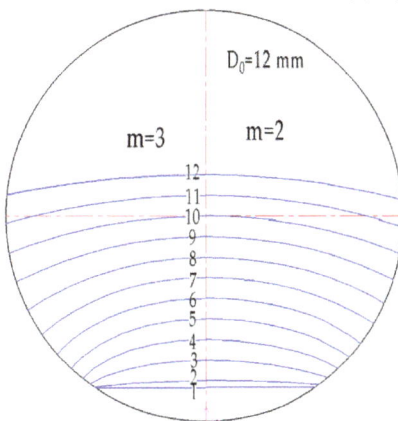

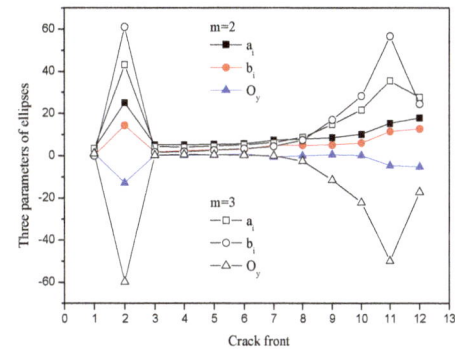

(**c**) $b_0/a_0 = 0, b_0/D_0 = 0.08$

Figure 8. Shape change of different initial crack for different fatigue crack growth exponent m values.

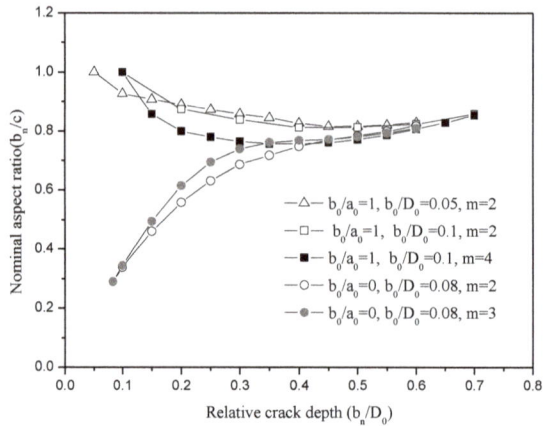

Figure 9. Nominal aspect ratio vs. relative crack depth.

The fatigue crack developments b_n/D_0 with c/D_0 under cyclic for different initial parameters are shown in Figure 10. It can be seen that the crack propagation paths differ with different initial flaws, but will converge asymptotically. Furthermore, in the process of expansion, the crack growth rate for center and outermost points are variable, which is deduced from the gradient of two type lines with initial flaws $b_0/a_0 = 0$, 1. This can be seen more precisely in Figure 11. For the case of an initial crack $b_0/a_0 = 1$ shown in Figure 11a, the ratio of crack growth (db/dc) is always less than 1 for most propagation processes, which means the crack growth rate for the central point is always slower than the outermost point until the relative crack depth $b_n/D_0 \approx 0.6$. However, the change in growth ratio will slow down from the beginning to the stage of $b_n/D_0 \approx 0.6$ for all the cases with initial flaws $b_0/a_0 = 1$, and then increase distinctly. For the case of an initial crack with $b_0/a_0 = 0$, as shown in Figure 11b, the crack growth along the vertical central line is always greater than the growth adjacent to the horizontal surface until the relative crack depth satisfies $b_n/D_0 \approx 0.4$, since the gradient line exceeds 1. Furthermore, the rate decreases sharply at the beginning propagation, especially for $m = 3$. Larger values of Paris law exponent m convey more drastic changes. It can be deduced that in the early propagation stage, the exponent m in the Paris law have a distinct effect on the evolution of the crack. The change of crack growth rate for central point is bigger for large value of *m*. It is considered to be related to plasticity which suppress the crack propagation on the outermost surface.

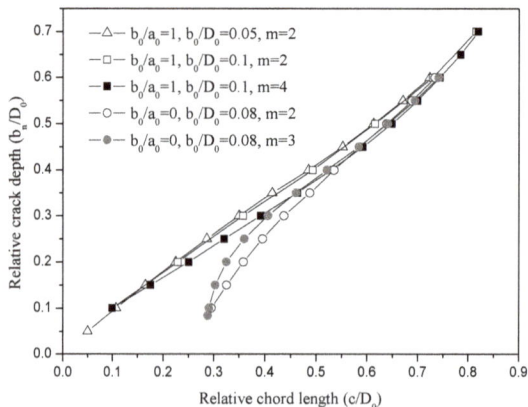

Figure 10. Relative crack depth vs. relative chord length with different initial parameters.

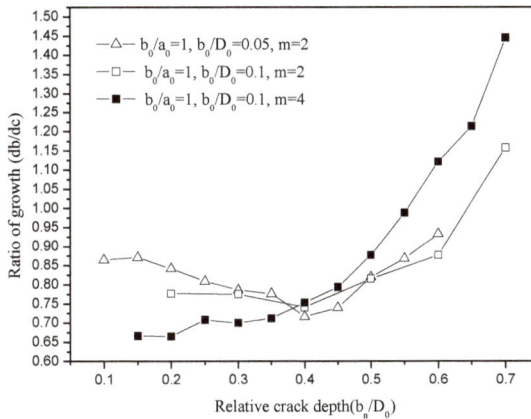

(a) $b_0/a_0 = 1$

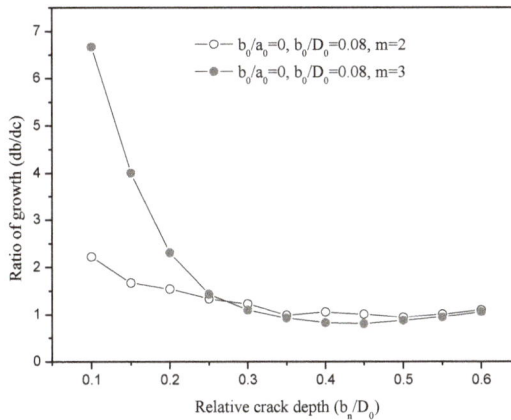

(b) $b_0/a_0 = 0$

Figure 11. Ratio of crack growth along the vertical central line and toward the horizontal surface.

3.2. Comparison with Other Numerical Solutions and Experimental Results

In Figure 12, the fatigue propagation of the initial crack $b_0/a_0 = 1$, $b_0/D_0 = 0.05$ and 1 is compared with numerical solutions by A. Carpinteri [11,12]. The curves in the present results are similar for all cases. However, some discrepancy between the present result and Carpinteri can be seen, especially for the initial crack $b_0/a_0 = 1$, $b_0/D_0 = 0.05$. It should be pointed out that in the above comparison, the deviation is mainly due to the difference in the crack growth method adopted and the idealized crack front geometry. A two-parameter elliptical-arc shape with fixed center is assumed only by employing the Paris-Erdogan law ordinarily by Carpinteri [11,12]. The two-parameter shape assumption method mentioned above can simplify the fatigue calculations, but it is also clear that better predications should be obtained if the shape restraint can be reduced, such as those generated by the present method. Moreover, the crack growth circles, which are tangent to the new crack front as well as to the current crack front, can accurately represent the real path of the fatigue crack and thus yield more accurate results. In addition, the better mesh refinement demonstrated in this paper also leads to improved prediction accuracy.

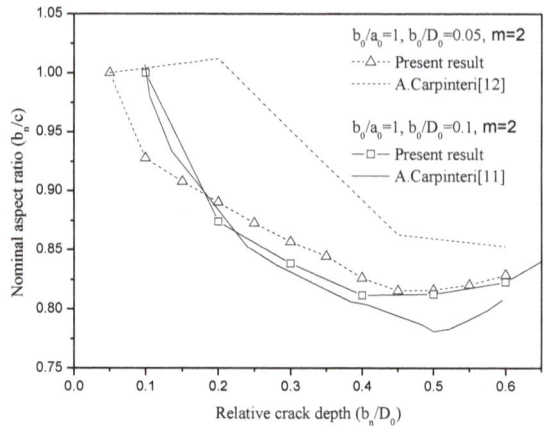

Figure 12. Crack propagation patterns compared with numerical solutions.

Figures 13 and 14 compare the crack propagation result with the experimental data deduced from F.P. Yang [19]. It is shown that the present results agree well with the experimental data. The experimental result deviates abnormally around the relative crack depth of $b_n/D_0 = 0.4$ in Figure 14. The maximum discrepancies are approximately 12%. The deviation of the two solutions are acceptable since as the fracture begins to happen in the experimental method approach, the relative crack depth $b_n/D_0 = 0.4$. It is confirmed that the present method could provide more accurate results.

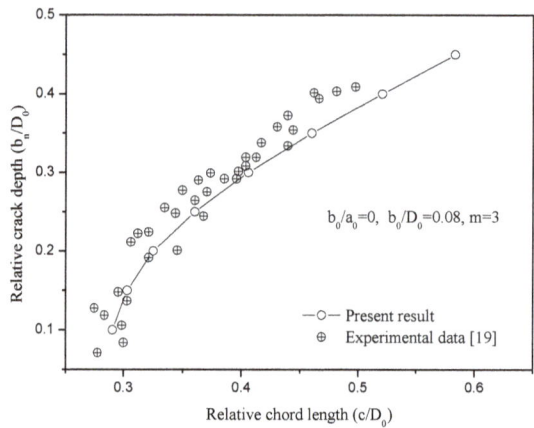

Figure 13. Relationship of crack propagation with depth and chord length compared with experimental data.

Appl. Sci. **2019**, *9*, 1751

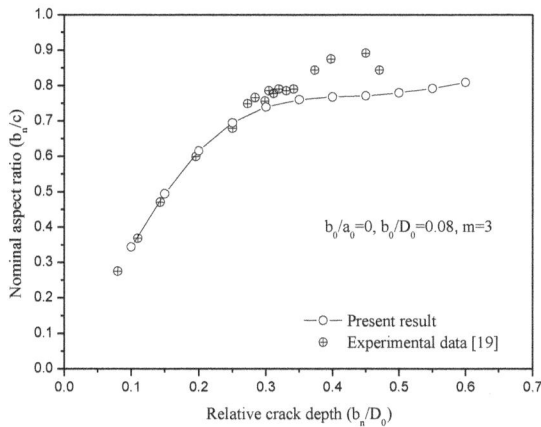

Figure 14. Crack propagation patterns compared with experimental data.

4. Conclusions

The fatigue propagation of a surface crack in a round bar subjected to tension loads has been investigated by using crack growth circles. The present results demonstrate that the experimental method had good convergence speed and accurate prediction of crack shape patterns. The following conclusions can be drawn:

- The crack growth circles method is developed for the surface cracks of a round bar, and the circles are tangent to both current and new crack fronts. In this way, good simulation accuracy can be achieved with fewer iterations.
- A three-parameter model with fewer shape restraints whose center is allowed to move along the vertical axis is built, and the shape change of a fatigue crack is predicted more precisely. The nominal aspect ratio of an ellipse, which is the ratio of the maximum crack depth to the chord length c, b_n/c, is considered, instead of the actual aspect ratio of an ellipse semi-axis.
- A relatively large crack growth increment can be used by adopting the equivalent stress intensity factor ΔK_e based on the stress intensity factors along the current and new crack fronts.
- The crack propagation process is described accurately based on the ratio of vertical growth toward the horizontal surface. It can be seen that the crack propagation paths differ with different initial flaws, but will converge asymptotically. The ratio of crack growth is always less than 1 for the case of initial crack $b_0/a_0 = 1$, and the crack growth along the vertical central line is always greater than the growth toward the horizontal surface. For the case of an initial crack $b_0/a_0 = 0$, a greater Paris law exponent m value generates more drastic change.
- The present solutions are compared with other numerical solutions and experimental data. Comparison shows that the present solutions agree well with the experimental data and are better than other numerical solutions.

Author Contributions: Conceptualization, S.C. and Y.Y.; methodology, S.C. and Y.Y.; software, Y.Y.; validation, S.C. and Y.Y.; formal analysis, S.C. and Y.Y.; investigation, H.C.; resources, H.C.; data curation, Y.Y.; writing—original draft preparation, S.C. and Y.Y.; writing—review and editing, S.C. and Y.Y.; visualization, S.C. and Y.Y.; supervision, S.C.; project administration, S.C.; funding acquisition, S.C. and H.C.

Funding: This research was funded by the Natural Science Foundation of Shanghai (18ZR1416500) and Development Fund for Shanghai Talents.

Conflicts of Interest: The Authors declare no conflicts of interest.

Nomenclature

D_0	Diameter of round bar	ΔK	Stress intensity factor range
a	Major axis of an ellipse	ΔK_e	Equivalent stress intensity factor
b	Minor axis of an ellipse	μ	Stepping coefficient
O_y	Center of ellipse	Δl_j	Crack growth length in Equation (4)
c	Chord length of an ellipse	Δd_j	Distance in Equation (7)
a_i, b_i	Semi-axes of ellipse for i-th loading step	b/a	Actual aspect ratio
a_{i+1}, b_{i+1}	Semi-axes of ellipse for $i + 1$-th loading step	b_n/c	Nominal aspect ratio
$O_{y,i+1}$	Center of ellipse for $i + 1$-th loading step	b_n/D_0	Relative crack depth
x_{ji}, y_{ji}	Coordinate for points O, A, B, C and D	c/D_0	Relative chord length
da/dN	Crack growth rate	db/dc	Ratio of growth
C, m	Constants of the Paris–Erdogan law		

References

1. Athanassiadis, A.; Boissenot, J.M.; Brevet, P.; Francois, D.; Raharinaivo, A. Linear elastic fracture mechanics computations of cracked cylindrical tensioned bodies. *Int. J. Fract.* **1981**, *17*, 553–566. [CrossRef]
2. Nezu, K.; Machida, S.; Nakamura, H. Stress intensity factor of surface cracks and fatigue crack propagation behavior in a cylindrical bar. In Proceedings of the 25th Japan Congress on Material Research, Metallic Metals, Kyoto, Japan, 25–28 March 1982; pp. 87–92.
3. Mackay, T.L.; Alperin, B.J. Stress intensity factors for fatigue cracking in high-strength bolts. *Eng. Fract. Mech.* **1985**, *21*, 391–397. [CrossRef]
4. Lorentzen, T.; Kjaer, N.E.; Henriksen, T.K. The application of fracture mechanics to surface cracks in shafts. *Eng. Fract. Mech.* **1986**, *23*, 1005–1014. [CrossRef]
5. Forman, R.G.; Shivakumar, V. Growth behavior of surface cracks in the circumferential plane of solid and hollow cylinders. In *Fracture Mechanics: Seventeen Volume*; American Society of Testing and Materials: Conshohocken, PA, USA, 1986; pp. 59–74.
6. Caspers, M.; Mattheck, C.; Munz, D. Propagation of surface cracks in Notched and Unnotched Rods. In *Surface-Crack Growth: Models, Experiments, and Structures*; American Society of Testing and Materials: Conshohocken, PA, USA, 1990; pp. 365–389.
7. Caspers, M.; Mattheck, C. Weighted averaged stress intensity factors of circular-fronted cracks in cylindrical bars. *Fatigue Eng. Mater. Struct.* **1987**, *9*, 329–341. [CrossRef]
8. Ael din, S.S.; Lovegrove, J.M. Stress intensity factors for fatigue cracking of round bars. *Int. J. Fatigue* **1981**, *3*, 117–123.
9. Carpinteri, A. Elliptical-arc surface cracks in round bars. *Fatigue Eng. Mater. Struct.* **1992**, *15*, 1141–1153. [CrossRef]
10. Carpinteri, A.; Brighenti, R. Fatigue propagation of surface flaws in round bars: A three-parameter theoretical model. *Fatigue Eng. Mater. Struct.* **1996**, *19*, 1471–1480. [CrossRef]
11. Carpinteri, A.; Brighenti, R.; Vantadori, S. Surface cracks in notched round bars under cyclic tension and bending. *Int. J. Fatigue* **2006**, *28*, 251–260. [CrossRef]
12. Carpinteri, A. Shape change of surface cracks in round bars under cyclic axial loading. *Int. J. Fatigue* **1993**, *15*, 21–26. [CrossRef]
13. Carpinteri, A.; Ronchei, C.; Vantadori, S. Stress intensity factors and fatigue growth of surface cracks in notched shell and round bars: Two decades of research work. *Fatigue Fract. Eng. Mater. Struct.* **2013**, *36*, 1–13. [CrossRef]
14. Carpinteri, A.; Vantadori, S. Surface crack in round bars under cyclic tension or bending. *Key Eng. Mater.* **2008**, *378–379*, 341–354. [CrossRef]
15. Couroneau, N.; Royer, J. Simplified model for the fatigue growth analysis of surface cracks in round bars under mode I. *Int. J. Fatigue* **1998**, *20*, 711–718. [CrossRef]
16. Carpinteri, A.; Vantadori, S. Sickle-shaped surface crack in a notched round bar under cyclic tension and bending. *Fatigue Fract. Eng. Mater. Struct.* **2009**, *32*, 223–232. [CrossRef]
17. Ayhan, A.O. Simulation of three-dimensional fatigue crack propagation using enriched finite elements. *Comput. Struct.* **2011**, *89*, 801–812. [CrossRef]

Appl. Sci. **2019**, *9*, 1751

18. Shin, C.S.; Cai, C.Q. Experimental and finite element analyses on stress intensity factors of an elliptical surface crack in a circular shaft under tension and bending. *Int. J. Fatigue* **2004**, *129*, 239–264. [CrossRef]
19. Yang, F.P.; Kuang, Z.B.; Shlyannikov, V.N. Fatigue crack growth for straight-fronted edge crack in a round bar. *Int. J. Fatigue* **2006**, *28*, 431–437. [CrossRef]
20. Cai, C.Q.; Shin, C.S. A normalized area-compliance method for monitoring surface crack development in a cylindrical rod. *Int. J. Fatigue* **2005**, *27*, 801–809. [CrossRef]
21. Liu, C.; Chu, S. Prediction of shape change of corner crack by fatigue crack growth circles. *Int. J. Fatigue* **2015**, *75*, 80–88. [CrossRef]

![applied sciences logo] *applied sciences*

MDPI

Article

Discrete and Phase Field Methods for Linear Elastic Fracture Mechanics: A Comparative Study and State-of-the-Art Review

Adrian Egger [1], Udit Pillai [2], Konstantinos Agathos [1], Emmanouil Kakouris [2], Eleni Chatzi [1], Ian A. Aschroft [3] and Savvas P. Triantafyllou [2,*]

[1] Institute of Structural Engineering, D-BAUG, 8093 ETH Zurich, Switzerland; egger@ibk.baug.ethz.ch (A.E.); agathos@ibk.baug.ethz.ch (K.A.); chatzi@ibk.baug.ethz.ch (E.C.)
[2] Centre for Structural Engineering and Informatics, Faculty of Engineering, The University of Nottingham, NG7 2RD Nottingham, UK; ezzup@exmail.nottingham.ac.uk (U.P.); evxek3@exmail.nottingham.ac.uk (E.K.)
[3] Centre for Additive Manufacturing, Faculty of Engineering, The University of Nottingham, NG7 2RD Nottingham, UK; Ian.Aschroft@nottingham.ac.uk
* Correspondence: Savvas.Triantafyllou@nottingham.ac.uk; Tel.: +44-115-951-4108

Received: 31 March 2019; Accepted: 23 May 2019; Published: 14 June 2019

Abstract: Three alternative approaches, namely the extended/generalized finite element method (XFEM/GFEM), the scaled boundary finite element method (SBFEM) and phase field methods, are surveyed and compared in the context of linear elastic fracture mechanics (LEFM). The purpose of the study is to provide a critical literature review, emphasizing on the mathematical, conceptual and implementation particularities that lead to the specific advantages and disadvantages of each method, as well as to offer numerical examples that help illustrate these features.

Keywords: LEFM; XFEM/GFEM; SBFEM; Phase field

---

1. Introduction

The need for lighter, stronger, and resilient structures across multiple engineering domains, e.g., the aerospace, automotive, and construction industries, necessitates a robust, economical and high-fidelity simulation of failure processes [1–3]. Failure in structural components subjected to static and/or dynamic loading is commonly associated with complex phenomena, i.e., crack nucleation, propagation, branching, merging and arrest [4,5]. These phenomena emerge from micro-material discontinuities, which under the action of external stimuli accumulate to cracks and evolve across several length scales eventually leading to structural failure. From a computational standpoint, these physics of damage evolution have proven challenging to resolve.

Over the past 20 years, the eXtended Finite Element method (XFEM), the Scaled Boundary Finite Element method (SBFEM) and most recently the Phase Field method (PFM), have emerged as distinct methodologies with the common objective of resolving fracture propagation. In this work, we provide a comparative platform for these methodologies pertinent to both the mathematical treatment of damage evolution and the corresponding algorithmic implications within the framework of Linear Elastic Fracture Mechanics (LEFM).

LEFM methods describe damage initiation and propagation within the remit of brittle and quasi-brittle material response. LEFM has been traditionally treated within two distinct methodological frameworks, i.e., computational fracture mechanics (see, e.g., [6]) and continuum damage mechanics (see, e.g., [7]). In the former, damage is explicitly defined as a discrete topological discontinuity. In the latter, damage is effectively homogenised over a representative volume. Diffuse crack approaches effectively lie in the boundary of the two aforementioned methods. The need to predict damage related phenomena precisely, accurately, and economically within the context of LEFM has spurred research into an extensive suite of alternative methodologies.

The finite element method (FEM), a representative of the discrete methods class, has reached a mature development status, effectively becoming the industry standard in numerical methods. Yet, challenges remain when characterizing singularities or propagation due to discrete cracks. This is a direct consequence of the following, select FEM shortcomings. The first four challenges primarily originate from the discretization method itself, while the remaining two pertain to difficulties associated with integration of LEFM into the discretization process:

1. A conforming mesh topology is required to represent the associated crack.
2. The typical polynomial-based interpolation functions cannot reproduce the singular stress field.
3. Tracking crack paths and incorporating branching and merging behaviour is algorithmically challenging.
4. Mesh dependant projection errors arise within the context of nonlinear and dynamic analyses.
5. Nucleation, branching and merging of cracks cannot be treated in a uniform and theoretically sound manner.
6. Calculation of the stress intensity factors (SIFs) requires additional post-processing methods.

Several techniques have been developed to tackle the aforementioned issues. First, sophisticated remeshing algorithms [8–10] and tools [11,12] have been introduced to model the singular stress field. The utilization of special element types or the introduction of a fine mesh around crack tips contribute to tackling this challenge. Second, specially developed quarter-point elements [13], which are placed around the crack tip, to accurately capture the crack tip singularity. Third, diverse techniques have been proposed to determine the fracture parameters, such as the SIFs. This includes path-independent integrals [14–17], the virtual crack closure technique [18–20], the hybrid-element approach [21,22], and the Irwin's crack closure integral [23]. The computational toll for such analyses is significant, with the majority of the effort stemming from the remeshing algorithm and the need for a fine mesh in the vicinity of the crack tip. Due to these limitations several novel numerical methods treating discrete cracks, such as meshless methods (MM), material point methods (MPM), boundary element methods (BEM), the extended/generalized finite element method (XFEM/GFEM), and the scaled boundary finite element method (SBFEM) have been applied, all distancing themselves from FEM in the way they define their support.

MM [24–26] were conceived with the aim of eliminating difficulties associated with the reliance on a mesh. Hence, the interpolation in MMs is solely based on a set of distributed nodes, thus eliminating FEM issues commonly associated with mesh distortion and remeshing. Crack path extensions are effortlessly accounted for by introducing additional nodes. However, certain drawbacks remain. The MM shape functions require higher order integration and the treatment of essential boundary conditions is intricate, since the shape functions do not necessarily satisfy the Kronecker delta property. Generally, the computational toll of MMs results higher to that of the FEM [27].

MPM [28] is an extension to Particle-In-Cell methods [29], which efficiently treat history-dependent variables. In MPM, the continuum is represented by a set of material points that are moving within a non-deformable (Eulerian) computational grid where contrary to MM, solution of the governing equations is performed. Treatment of discrete cracks is accounted for by the introduction of multiple velocity fields [30] or more recently phase fields [31–33]. MPM has been found to offer significant computational advantages when compared to purely meshless methods since it does not require time-consuming neighbour searching.

The BEM [34] solves initial value problems described as boundary integral equations hence reducing dimensionality by one. This significantly reduces the complexity of mesh generation, since only the boundary and the crack front need be discretised. Furthermore, compared to the FEM, BEM can often achieve greater accuracy, due to the nature of integrals used in the problem description. However, this is simultaneously the source of disadvantages. This formulation results in fully populated, dense matrices necessitating tailored numerical methods [35,36] to efficiently solve the resulting discrete equations. The introduction of isogeometric analysis (IGA) [37,38] suggests profound implications on practical engineering design. The key concept entails employing the Non-Uniform Rational B-spline (NURBS) not only for the geometric representation, but also for the discretization employed in the subsequent analysis. NURBS substitute standard FEM shape functions with the solution obtained on their support. A hybrid isogeometric boundary element method has been proposed [39–41] coupling many of the advantages of its parent methods. The direct adoption of the geometry representation as given by CAD software, greatly facilitates the integration of design and analysis, since no volume parametrization is required for crack propagation. Additionally, when applied to fracture [42,43], the delivered higher-continuity can increase the accuracy of the stress field around the crack tip.

An effective means of tackling the issues of mesh dependency and treatment of singularities, is provided by the extended/generalized [44,45] finite element method (XFEM/GFEM), whose use is wide spread both in academia and industry. The most characteristic trait of this method is the use of partition of unity (PU) enrichment [46–48], to incorporate known features of the solution in the finite element approximation space through appropriate enrichment functions. For fracture mechanics problems, discontinuous and singular enrichment functions are employed locally, i.e., in the vicinity of the crack, to allow the representation of discrete cracks independently of the underlying mesh. This in turn significantly decreases or even removes the remeshing burden, while also increasing the accuracy with which asymptotic fields are represented. Alternatively, the scaled boundary finite element method (SBFEM) [49] naturally incorporates the singular stress field, providing an elegant extraction of the generalized stress intensity factors (gSIFs) in post-processing at negligible additional computational cost [50]. This is a consequence of SBFEM retaining an analytical solution in the radial direction, while only requiring discretization along the tangential boundary in the standard FEM sense. However, double nodes are introduced to accommodate strong discontinuities. This is partially mitigated due to the polytope nature of SBFEM, which only imposes the condition of star-convexity on elements. Exploiting balanced quadtrees as hierarchical meshes in conjunction with polygon clipping the majority of meshing effort is circumvented [51]. XFEM and SBFEM receive in-depth treatment in Sections 3 and 4.

Alternative discrete fracture methods based on cohesive theories have been utilized to overcome stress singularities in LEFM also accounting for the nonlinear separation phenomena [6]. Barenblatt [52] originally introduced the cohesive zone method (CZM) to model fracture in brittle materials. Later, Dugdale [53] extended the CZM to simulate the plastic fracture process zone around the crack tips. In cohesive fracture theory, the material is not considered perfectly brittle as in Griffith's theory. Rather, there is a small zone in front of the crack that can exhibit some ductility. The fracture energy is gradually released at the crack tip based on the crack opening and equals the critical fracture energy at full crack opening. If the cohesive zone is sufficiently small, the ductility zone becomes unimportant and the theory of LEFM can be applied. The CZM has been employed within a FEM, see, e.g., [54,55] and a BEM, see, e.g., [56] setting, also in conjunction with a partition of unity approach [47]. Furthermore, CZM has been introduced within a particle based approach as in the case of SPH [57], reproducing kernel particles [58], and the Element-Free Galerkin method [59].

A popular partition of unity approach and a reasonable extension of the CZM is the Cohesive Segments Method (CSM) [60]. The CSM introduces arbitrary cohesive segments within the finite elements that act as discontinuities in the displacement field hence alleviating the CZM requirement for the definition of cohesive elements at the finite element interface. In CSM, the cracks are modelled as a set of overlapping cohesive segments with their support nodes being enriched with jump and tip enrichment functions similar to the XFEM. A combination of overlapping crack cohesive segments results in a continuous crack. Remmers et al. [60] originally applied the CSM in quasi-static brittle fracture problems mainly focused on mode I separation problems and further extended the method for the simulation of dynamic crack propagation problems [61]. Using the CSM as point of departure, various PUM with cohesive theories have been successfully introduced with a meshless discritization approach, see, e.g., [62,63].

Rather than attempting to model the actual, discrete crack topology, either as a strong discontinuity in the displacement field (e.g., XFEM) or an explicitly defined boundary (e.g., SBFEM), diffuse approximations of cracks incorporate the effects associated with the crack formation, e.g., the stress release or the stiffness degradation into the constitutive model [64]. Such approaches initiated with the pioneering work of Rashid [65], who defined a cracking criterion for pre-stressed concrete pressure vessels on the basis of loss of material stiffness in the direction normal to a crack as this evolves. In the past 10 years, several methodologies pertinent to diffuse crack models emerged, such as gradient enhanced damage methods [66,67], Thick Level Set methods [68], and Phase field methods [69]. In the taxonomy of damage theories, diffuse crack approximations fall within the family of Continuum Damage Mechanics, where however particular treatment of strain localisation is implicitly performed. de Borst and Verhoosel [70], see, also Mandal et al. [71] highlighted the similarities between gradient enhanced damage methods and phase field methods. An insightful discussion on the similarities and differences between thick level sets and phase fields is provided in [72].

Phase field methods (PFM) for brittle fracture arose from the pioneering work of Francfort and Marigo [73], who treated elastic fracture as an energy minimization problem within a robust variational setting. Bourdin et al. [69] used the Mumford-Shah potential [74] to provide a regularization of this variational formulation. In this, brittle fracture is numerically treated as a coupled, i.e., displacement and phase field problem; the latter accounts for the crack interface geometry. To this point, finite element-based phase field formulations have been introduced to treat brittle/fatigue [75–78], ductile [79,80], and hydraulic fracture [81–85]. Very recently, the phase field method has been introduced within an MPM [32] and a Virtual Element framework [75].

This paper delivers a critical comparison among numerical methods relying on discretisation, namely XFEM/GFEM and SBFEM, and the PFM, which belongs in the class of diffuse methods. The latter has as of late garnered much attention, not only limited to the field of LEFM. Specifically, we compare the potential of these methods in accurately and efficiently predicting crack propagation, paths and arrest. Additionally, we remark on the overall computational effort involved in the analysis and the inherent capabilities/limitations of each method within the LEFM context.

The manuscript is organized as follows. In Section 2 the LEFM problem statement is introduced. Subsequently, methods relying on discretisation are discussed, with the XFEM/GFEM variants over-viewed in Section 3, and the SBFEM treated in Section 4. Section 5 offers an overview of phase field methods, as a salient representative of the diffuse methods class. The workings of the methods are illustrated by means of four numerical examples, described in Section 6, while Section 7 provides a methodological comparison and concluding remarks.

2. LEFM Problem Statement

To formulate the linear elastic fracture mechanics (LEFM) problem, we consider the two dimensional cracked domain Ω shown in Figure 1. The boundary Γ consists of the parts Γ_0, where free surface boundary conditions apply, Γ_u, where displacements $\bar{u}$ are prescribed and Γ_t where the surface tractions $\bar{t}$ are applied as Neumann conditions. The domain includes a crack under the assumption of free surface conditions Γ_c. As depicted in Figure 1, the domain boundary is decomposed as $\Gamma = \Gamma_0 \cup \Gamma_u \cup \Gamma_t \cup \Gamma_c$. Then, the elasticity equations shown in Equation (1) with their corresponding boundary conditions hold:

$$\nabla \cdot \sigma + b = 0 \quad \text{in} \quad \Omega \tag{1a}$$

$$u = \bar{u} \quad \text{on} \quad \Gamma_u \tag{1b}$$

$$\sigma \cdot \mathbf{n} = \bar{t} \quad \text{on} \quad \Gamma_t \tag{1c}$$

$$\sigma \cdot \mathbf{n} = 0 \quad \text{on} \quad \Gamma_0 \tag{1d}$$

where σ is the Cauchy stress tensor, $\mathbf{n}$ is the unit outward normal to the boundary, $\mathbf{b}$ is the applied body force per unit volume, $\mathbf{u}$ is the displacement field and ∇ is the gradient operator.

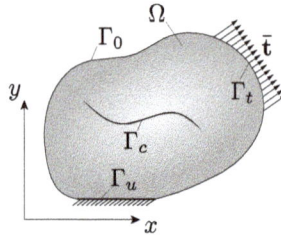

Figure 1. Cracked Body and boundary conditions.

If small deformations are assumed, then the strain field ε can be described as as the symmetric gradient of the displacement field:

$$\varepsilon = \nabla_s u \tag{2}$$

Furthermore, if linear elastic material behavior is assumed, stresses can be obtained from strains through Hooke's law:

$$\sigma = D : \varepsilon \tag{3}$$

where D is the elasticity tensor, which in case of two dimensional problems assumes the following form:

$$D = \frac{E}{1-v^2} \begin{bmatrix} 1 & v & 0 \\ v & 1 & 0 \\ 0 & 0 & \frac{1-v}{2} \end{bmatrix}, \quad \text{for plane stress} \tag{4a}$$

$$D = \frac{E}{(1+v)(1-2v)} \begin{bmatrix} 1-v & v & 0 \\ v & 1-v & 0 \\ 0 & 0 & \frac{1-2v}{2} \end{bmatrix}, \quad \text{for plane strain} \tag{4b}$$

with E and v denoting Young's modulus and Poisson's ratio respectively.

A decisive quantity in classic fracture mechanics [86] is the energy release rate, defined as:

$$\mathcal{G} = -\frac{\partial \Pi}{\partial a} \tag{5}$$

where Π is the total potential energy and a is the crack length (or area for three dimensional problems). Then, based on the energy release rate criterion, crack propagation will occur when:

$$\mathcal{G} \geq \mathcal{G}_c \tag{6}$$

where $\mathcal{G}_c$ is the critical energy release rate or fracture toughness, which can be considered as a material parameter.

For a pure mode I problem, the mode I stress intensity factor (SIF) is related to the energy release rate as follows:

$$\mathcal{G} = \frac{K_I^2}{E'} \tag{7}$$

where K_I is the stress intensity factor and E' is the effective elastic modulus:

$$E' = \begin{cases} E & \text{for plane stress} \\ \dfrac{E}{1-v^2} & \text{for plane strain} \end{cases} \tag{8}$$

Based on this, the critical stress intensity factor is defined as:

$$K_c = \sqrt{E'\mathcal{G}_c} \tag{9}$$

The corresponding relation for mixed mode planar problems is:

$$\mathcal{G} = \frac{1}{E'}\left(K_I^2 + K_{II}^2\right) \tag{10}$$

where K_{II} is the mode II SIF. The square root of the quantity $\left(K_I^2 + K_{II}^2\right)$ can be considered as an equivalent SIF:

$$K_{eq} = \sqrt{K_I^2 + K_{II}^2} \tag{11}$$

Then, the energy release rate criterion of Equation (6) can be written in terms of the SIFs as:

$$K_{eq} \geq K_c \tag{12}$$

Several criteria have been proposed to determine the direction of crack propagation, such as the maximum circumferential stress or maximum hoop-stress criterion [87], the maximum energy release rate criterion [88] and the minimum strain energy density criterion [89]. To what concerns the XFEM/SBFEM analyses carried out in this work, we adopt the latter criterion, which results in the following expression for the angle of crack propagation:

$$\theta_c = 2\tan^{-1}\left[\frac{-2K_I/K_{II}}{1 + \sqrt{1 + 8(K_I/K_{II})^2}}\right] \tag{13}$$

3. The Extended/Generalized Finite Element Methods (XFEM/GFEM)

As further mentioned in Section 1, one of the main difficulties associated with the modeling of fracture by means of conventional finite element methods lies in the fact that a new mesh is needed at each propagation step. This, apart from increasing the computational cost, significantly limits the degree of automation that can be achieved in such simulations. The introduction of the partition of unity method (PUM) [46–48] has provided the background for the subsequent development of a suite of methods, including the extended [44] and generalized [45] finite element methods that have managed to overcome, to a large extent, this difficulty. In the following subsections, we provide a brief overview of these methods with the focus shed onto the methodological and implementational aspects relating to crack propagation problems. For a more detailed exposition of the methods and their applications we refer the reader to the several review papers available in existing literature, as for instance References [90–92] and more recently [93], as well as the references therein.

3.1. Partition of Unity Enrichment

Partition of unity enrichment, in general, allows the incorporation of known features of the solution in the numerical approximation in the form of enrichment functions. If finite elements are used as the basis for the numerical approximation, then partition of unity enrichment can be realized as follows:

$$\mathbf{u}(\mathbf{x}) = \underbrace{\sum_{\forall I} N_I(\mathbf{x})\,\mathbf{u}_I}_{\text{FE approximation}} + \underbrace{\sum_{\forall I} N_I^*(\mathbf{x})\,\Phi(\mathbf{x})\,\mathbf{b}_I}_{\text{enriched part}} \tag{14}$$

where $N_I(\mathbf{x})$ are the FE interpolation functions, $\mathbf{u}_I$ are FE degrees of freedom (dofs), $N_I^*(\mathbf{x})$ is a basis of functions that form a partition of unity, $\Phi(\mathbf{x})$ is the enrichment function and $\mathbf{b}_I$ are the enriched degrees of freedom.

Most commonly, finite element shape functions are employed to form the partition of unity basis:

$$N_I^*(\mathbf{x}) \equiv N_I(\mathbf{x}) \tag{15}$$

Alternative PU bases have can also be found in the literature Zhang et al. [94], Griebel and Schweitzer [95], Hong and Lee [96], aiming mostly at improving specific aspects of the method, such as conditioning of the resulting system matrices.

3.2. XFEM/GFEM Enrichment Functions for LEFM

In the original partition of unity finite element method (PU-FEM) [48], enrichment functions were used as a means of improving the overall accuracy of the approximation, thus enrichment was applied globally, i.e., for all nodes of the FE mesh. For problems involving localized phenomena, such as fracture, enrichment functions are only needed locally, thus, in the XFEM [44,97] only a subset of the nodes is enriched to increase the efficiency of the method. This type of enrichment was subsequently also adopted in the GFEM rendering the two methods almost identical. In fact, in more recent publications [92] almost no distinction is made between the two methods.

For LEFM problems, two types of enrichment functions, i.e., specializations of function $\Phi(\mathbf{x})$, are most commonly used to represent the discontinuities and singularities introduced in the solution by the crack. In the following, these enrichment functions are presented along with possible alternatives from the literature. Furthermore, some common problems, associated with their use, are identified and possible remedies discussed.

3.2.1. Jump Enrichment

The first type of enrichment functions consists of modified Heaviside step functions, usually referred to as jump enrichment functions, which allow representing the displacement jump along the crack surface:

$$H(\mathbf{x}) = \begin{cases} 1 & \text{above the crack} \\ -1 & \text{below the crack} \end{cases} \tag{16}$$

These functions were introduced in the work of Moës et al. [44], and constitute perhaps the most distinctive feature of XFEM. Enrichment with these functions is realized locally, only for nodes whose nodal support is completely split in two by the crack.

Other types of discontinuous enrichment include the alternative formulation of Hansbo and Hansbo [98] and higher order discontinuous enrichment functions found both in the XFEM [99–101] and GFEM [102] literature. In the context of fracture mechanics, special discontinuous functions have also been proposed to handle branched and intersecting cracks [103].

3.2.2. Tip Enrichment

The second type of enrichment functions is a set of asymptotic functions, also referred to as tip enrichment functions, that allow representing the discontinuity at the crack tip or front:

$$F_j(r, \theta) = \left\{ \sqrt{r}\sin\frac{\theta}{2}, \sqrt{r}\cos\frac{\theta}{2}, \sqrt{r}\sin\frac{\theta}{2}\sin\theta, \sqrt{r}\cos\frac{\theta}{2}\sin\theta \right\} \tag{17}$$

where r, θ are spatial coordinates of a polar system with its origin at the crack tip/front. These functions were introduced by Belytschko and Black [97] and form a basis that can exactly represent the analytical solution of the Westergaard problem.

Initially [44], the use of asymptotic enrichment was limited to elements containing the crack tip/front, however in the works of Stazi et al. [104] and Laborde et al. [105] it was shown that this would lead to suboptimal convergence rates. In order to obtain the same rate of convergence as for smooth problems, the use of asymptotic enrichment in a domain of fixed size around the crack front is necessary [105,106]. This alternative enrichment scheme was termed "geometrical enrichment" while the initial scheme is referred to as "topological enrichment". Usually, the domain where asymptotic enrichment is used is defined as the set of points whose distance from the crack tip/front is smaller that a predefined length r_e, called the enrichment radius.

An alternative to the enrichment functions of Equation (17) consists of using the displacement expression of the Westergaard solution directly as an enrichment function. This approach was introduced by Duarte et al. [107] and subsequently adopted in several works in the XFEM [108–110] and GFEM [111,112] literature. This kind of enrichment results in different enrichment functions in each spatial dimension, thus in some works [113,114] it was termed "vector enrichment" as opposed to "scalar enrichment" where the same enrichment functions are used in all spatial dimensions. As a disadvantage of this approach it is mentioned that it could complicate the implementation, especially in existing codes. On the other hand it leads to a decreased number of degrees of freedom compared to scalar enrichment and it can allow the direct estimation of stress intensity factors. Typically, to increase the accuracy of this estimation, higher order terms of the asymptotic expansion are also used as enrichment.

3.2.3. Kronecker Delta Property

From Equation (14) it can be easily deduced that for enriched nodes, the FE degrees of freedom will no longer correspond to displacements at the nodes. To restore this desirable property, enrichment functions can be modified such that they vanish at nodal points. A simple way to accomplish that is through enrichment function "shifting" [115], which consists of subtracting from the enrichment functions, their values at the nodal points:

$$\Phi_I(\mathbf{x}) = \Phi(\mathbf{x}) - \Phi(\mathbf{x}_I) \tag{18}$$

where $\Phi_I(\mathbf{x})$ is the modified enrichment function and $\mathbf{x}_I$ are the spatial coordinates of nodal point I.

From the above, it becomes clear that shifting results in a different enrichment function for each node. Furthermore, when applied to the jump enrichment functions of Equation (16), it causes the functions to vanish for elements that do not contain the crack, thus simplifying the implementation.

The Kronecker delta property can also be preserved by employing the stable GFEM [111,112,116], a technique where the FE interpolant of the enrichment functions is subtracted from the enrichment functions themselves. The main advantage of this technique however, lies in the fact that it can considerably improve the conditioning of the resulting stiffness matrices.

3.2.4. Blending

As already mentioned, enrichment in the XFEM and GFEM is mostly performed locally to increase efficiency. This leads to situations where only some of the nodes of an element are enriched with a specific enrichment function and the remaining nodes are either not enriched at all or enriched with a different enrichment function. In these elements, the shape functions pre-multiplying the enrichment functions no longer form a partition of unity leading to increased errors, also called "blending" errors. For the enrichment functions used in LEFM, these errors only result in some loss of accuracy, leaving the convergence rates unaffected. For other types of enrichment functions however, the convergence rate can also be affected [117].

Due to the above reasons, the "blending" problem has been extensively studied and several solutions have been proposed involving a variety of techniques such as assumed/enhanced strain formulations [117–119], directly matching displacements between the enriched and non enriched part of the approximation [105,120,121] and the use of weight functions [122–124] to smoothly blend different parts of the approximation. The later approach, also known as the corrected XFEM, is likely the most successful due to its relative simplicity and effectiveness.

3.2.5. Ill-Conditioning

An additional problem related to enrichment is the linear dependence between the enriched and standard part of the approximation. As far as jump enrichment is concerned, linear dependence may arise if the crack either intersects, or lies very close to a node. Then, the enriched shape function of this node is identical or very close to its standard FE shape function leading to linear dependence problems. A commonly used technique to avoid this problem is "snapping", which consists of not enriching nodes if they are very close to the crack surface [103]. Other approaches involve pre-conditioning [125,126] and stabilization in the element [127] or global equilibrium equations [128] level.

With respect to tip enrichment, ill-conditioning can arise when geometrical enrichment is used due to the fact that away from the singularity the tip enrichment functions tend to become linearly dependent both with respect to the FE part of the approximation and each other [101,129]. To overcome this issue several alternatives have been proposed such as altering the partition of unity basis used to pre-multiply the tip enrichment functions [105,121,130], preconditioners [106,125], stabilization [127] and enrichment function orthogonalization [101,129]. Moreover, vector enrichment functions have been shown to lead to improved conditioning [114], and if further combined to the stable GFEM [111,112] they can lead to optimal growth rates of the scaled condition number.

3.3. Displacement Approximation

Using the enrichment functions of the previous subsection, the XFEM/GFEM displacement approximation can be obtained:

$$\mathbf{u}\left(\mathbf{x}\right)=\underbrace{\sum_{I\in\mathcal{N}}N_{I}\left(\mathbf{x}\right)\mathbf{u}_{I}}_{\text{FE approximation}}+\underbrace{\sum_{J\in\mathcal{N}^{j}}N_{J}\left(\mathbf{x}\right)H\left(\mathbf{x}\right)\mathbf{b}_{J}}_{\text{jump enriched part}}+\underbrace{\sum_{T\in\mathcal{N}^{t}}\sum_{j}N_{T}\left(\mathbf{x}\right)F_{j}\left(\mathbf{x}\right)\mathbf{c}_{Tj}}_{\text{tip enriched part}} \tag{19}$$

where $\mathbf{b}_{J}$, $\mathbf{c}_{Tj}$ are enriched degrees of freedom.

The nodal sets of Equation (19) are defined as follows:

$\mathcal{N}$ is the set of all nodes in the FE mesh.
$\mathcal{N}^{j}$ is the set of jump enriched nodes. This nodal set includes all nodes whose support is split in two by the crack.
$\mathcal{N}^{t}$ is the set of tip enriched nodes. This nodal set includes all nodes whose support includes the crack front.

The method resulting from the above approximation does not involve any modifications, for instance dealing with blending or conditioning issues, and is thus often referred to as the "standard XFEM".

3.4. Weak Form and Discretised Equilibrium Equations

For LEFM problems, the standard weak formulation for linear elasticity is typically used:

Find $u \in \mathcal{U}$ such that $\forall v \in \mathcal{V}^{0}$

$$\int_{\Omega}\sigma(u):\varepsilon(v)\,d\Omega=\int_{\Omega}b\cdot v\,d\Omega+\int_{\Gamma_{t}}\bar{t}\cdot v\,d\Gamma \tag{20}$$

where :

$$\mathcal{U}=\left\{u|u\in\left(H^{1}\left(\Omega\right)\right)^{3},u=\bar{u}\text{ on }\Gamma_{u}\right\} \tag{21}$$

and

$$\mathcal{V}^{0}=\left\{v|v\in\left(H^{1}\left(\Omega\right)\right)^{3},v=0\text{ on }\Gamma_{u}\right\} \tag{22}$$

Functions of $H^1(\Omega)$ are implicitly discontinuous along the crack surface.

By introducing the constitutive relationship of Equation (3), the problem can be written as: Find $u \in \mathcal{U}$ such that $\forall v \in \mathcal{V}^0$:

$$\int_\Omega \varepsilon(u) : D : \varepsilon(v) \, d\Omega = \int_\Omega b \cdot v \, d\Omega + \int_{\Gamma_t} \bar{t} \cdot v \, d\Gamma \tag{23}$$

The above equation can be discretised using the approximation of Equation (19) to produce the discretised equilibrium equations.

3.5. Crack Representation

To allow the evaluation of the enrichment functions as well as the definition of the nodal sets involved in the enriched approximation, some kind of geometrical representation of the crack is necessary. In early XFEM works, as well as some GFEM publications, crack surfaces were explicitly represented as a series of linear segments (2D) or triangles (3D) [44,131,132]. However, the combination of this kind of representation to the XFEM can render the implementation quite involved by requiring for instance the computation of intersections of the crack with elements of the FE mesh.

3.5.1. The Level Set Method

An approach that is much better suited for combination to the XFEM, is the implicit representation of cracks using the level set method [133,134]. Due to this fact, the method has been extensively used in the XFEM framework in 2D [135] and 3D [136–138] applications.

To implicitly represent closed surfaces, such as cracks, two level set functions are needed:

- The normal level set ϕ, defined as the signed distance from the crack surface.
- The tangent level set ψ, defined as the signed distance from a surface that is normal to the crack surface and intersects the crack surface at the crack tip/front.

The crack surface is then defined as the set of points for which the normal level set is equal to zero and the tangent level set assumes negative values.

Typically, these level set functions are only computed at nodal points and interpolated for the rest of the domain using the FE shape functions:

$$\phi = \phi(\mathbf{x}) = \sum_{\forall I} N_I(\mathbf{x}) \phi_I, \qquad \psi = \psi(\mathbf{x}) = \sum_{\forall I} N_I(\mathbf{x}) \psi_I \tag{24}$$

where ϕ_I, ψ_I are the nodal values of the level set functions.

From the above expressions, spatial derivatives of the level set functions can be conveniently obtained through the spatial derivatives of the FE shape functions. Also evaluation of the enrichment functions can be significantly simplified. More specifically, jump enrichment functions can be directly computed as functions of the first level set, while the polar coordinates of Equation (17), needed for the tip enrichment functions, can be computed as:

$$r = \sqrt{\phi^2 + \psi^2}, \qquad \theta = \arctan\left(\frac{\phi}{\psi}\right) \tag{25}$$

For the general case of evolving surfaces, level sets are usually updated based on some velocity field by integrating the Hamilton-Jacobi equation. The case of propagating cracks however, requires several additional steps due to the nature of the problem and the fact that cracks are closed surfaces. Firstly, the velocity field, needed to update the crack, is only known at the crack tip/front, thus an additional step is required to extend the field to the whole domain. Then, the crack surface that has already formed should remain unaffected by the level set update, thus the velocity field should be appropriately modified. Finally, an orthogonalization step is necessary to ensure that the two level sets are normal after the update. To simplify the above procedure, several approaches were proposed in the work of Duflot [139] that allowed the update of level set descriptions for cracks without requiring the integration of evolution equations. In Elguedj et al. [140] a similar approach was proposed and applied to dynamic 3D crack propagation. It should be noted that both of these simplified methods rely on some geometric operations and are in fact very similar to methods from the category of the following paragraph.

3.5.2. Hybrid Implicit/Explicit Methods

As an alternative, aiming at combining advantages of both explicit and implicit representations, Fries and Baydoun [141] proposed a method where level set functions were directly computed from explicit crack representations using linear segments (2D) or triangles (3D). Similarly, in the vector level set method [142–144] linear segments (2D) or quadrilaterals (3D) are used to update the level set description of the crack and are subsequently discarded. Another instance of a method combining elements from both types of representations is the method of Sadeghirad et al. [145], where an explicit representation is constructed in order to correct the level set representation by removing disconnected parts of the crack.

3.6. Numerical Integration

Another challenge, associated with the use of discontinuous and singular enrichment functions, lies in the numerical integration of the weak form of Equation (23). Since the functions to be integrated are not smooth, standard Gauss quadrature cannot be used and more sophisticated tools need to be employed.

For the discontinuous jump enrichment functions, the most common approach would be element partitioning where elements are divided into integration sub-cells based on the crack geometry [44,132]. Extensions of this technique have also been proposed for higher order elements [100,146,147]. Alternatively, other works completely avoid the use of element partitioning by employing either equivalent polynomials [148,149], or the Schwarz–Christoffel conformal mapping [150].

As far as asymptotic enrichment functions are concerned, the most widely used solution would involve element partitioning combined with some transformation aiming at removing the singularity. Several such transformations have been proposed, e.g., the almost polar mapping of Laborde et al. [105], the parabolic transformation of Béchet et al. [106], and the Duffy transformation by Mousavi and Sukumar [151]. Element partitioning is used to divide the element containing the crack tip in triangles with one node lying on the singularity, thus also accounting for the discontinuity present in this element. Subsequently, the transformation is used to map quadrilateral elements to the constructed triangles leading to an accumulation of Gauss points around the crack tip and additionally removing the singularity. A promising solution, also including the above steps, is the algorithm introduced in Chevaugeon et al. [114], where a mapping is used for all asymptotically enriched elements, rather than just the ones containing the crack tip, and an adaptive strategy is devised to determine the number of Gauss points required for each element. Similar element partitioning algorithms [152] and mappings [153] have also been introduced for the three dimensional case.

3.7. Crack Propagation

The methods presented so far in this section mainly deal with discretising cracked domains using fixed meshes. For propagating cracks, principles of classic linear elastic fracture mechanics, as presented in Section 2, can be applied. Within this framework, stress intensity factors (SIFs) are the main tool used to both indicate the occurrence and determine the direction of crack propagation under certain loading conditions.

3.7.1. Stress Intensity Factors

One of the most widely used techniques for the extraction of SIFs in extended, generalized or standard finite element simulations, involves the use of the interaction integral. This can be derived by initially converting the J integral in a domain form and subsequently evaluating it for a stress state resulting from the superposition of an auxiliary stress state and the computed numerical solution. Then the interaction term of the integral, for two dimensional problems, assumes the form:

$$I = -\int_V q_{,j} \left(\sigma_{kl} \varepsilon_{kl}^{\mathrm{aux}} \delta_{1j} - \sigma_{kj}^{\mathrm{aux}} u_{k,1} - \sigma_{kj} u_{k,1}^{\mathrm{aux}} \right) dV \tag{26}$$

where $\varepsilon^{\mathrm{aux}}$, σ^{aux} and u^{aux} are the auxiliary stress, strain and displacement fields respectively which can be defined as in Moës et al. [44] and q is a virtual velocity field. Typically, q is chosen to assume a value of one for nodes within a disc of radius r_d around the crack tip and a value of zero for the remaining nodes.

In the interior of the elements, the values of q are interpolated using the FE basis functions. As a result, the expression of Equation (26) needs to be evaluated only in a "ring" or layer of elements around the crack tip. The components of the tensors of Equation (26), refer to a basis aligned with the crack, which for implicit crack representations can be conveniently defined using the level sets [136,137]. By considering the relation between the J integral and the SIFs it is straightforward to show that with an appropriate selection of the SIF values of the auxiliary state, the SIFs can be directly obtained from the interaction integral.

It should be noted that in the derivation of Equation (26) it has been assumed that the crack is straight. Of course, the expression can also be used for curved cracks, perhaps with some loss of accuracy, provided that the curvature of the crack is not very pronounced within the interaction integral domain. Alternatively, a more complicated formulation can be used [154], leading to more accurate results.

For three dimensional problems a more complicated expression for the interaction integral needs to be used as, for instance, in Gosz and Moran [15]. Furthermore, different alternatives exist for the definition of the virtual velocity field and the domain of integration [121,132,155] as well as the basis on which the tensor components refer to [154,155].

Alternative methods of SIF extraction, employed in the XFEM/GFEM context, include direct extraction based on the enriched degree of freedom values [108–110,114], Irwin's integral [23,156–159], and extraction through crack opening displacements [160]. The former technique relies on the fact that when vector enrichment is used, the physical meaning of the enriched degrees of freedom corresponding to the tip enriched nodes is by definition equivalent to the SIFs.

In several works [108–110], the technique is combined to degree of freedom gathering and the use of higher order terms of the Williams expansion to increase the accuracy of the extracted SIFs. Similarly, extraction using Irwin's integral also requires higher order enrichment. A relative advantage of both of these methods is their low computational cost and the fact that they do not require the use of auxiliary fields as in the interaction integral method. Extraction through crack opening displacements [160] does not require the use of higher order enrichment functions and is computationally inexpensive, it does however employ auxiliary fields. Finally, it should be mentioned that even though some of the above methods might be advantageous for certain problems, the interaction integral method is in general more accurate and has in general a wider field of applicability since domain integral formulations are available also for problems outside the LEFM domain.

3.7.2. Determination of the Crack Propagation Increment

While the direction of crack propagation can be obtained using the SIFs through one of the available criteria, as mentioned in Section 2, the length of the propagation increment is typically predefined and constant during the simulation. Nevertheless, this length is probably the parameter with the more pronounced effect on the crack paths obtained and should be set as small as possible. On the other hand, the length of this increment Δa is subject to the following constraint [161,162]:

$$\Delta a > r_d > 1.5h \tag{27}$$

where h is the mesh size. This constraint is necessary to ensure that the crack will be indeed straight within the domain of integration, whose radius in turn needs to be larger than $1.5h$ in order to include a ring of elements around the crack tip. Thus, the length of the crack increment is essentially determined by the mesh size. Nonetheless, if an alternative interaction integral formulation or extraction method is used, as discussed in the previous section, the constraint could be removed or at least relaxed allowing reducing the length of the crack increments without refining the mesh.

For the case of multiple cracks [163], a stability analysis is usually conducted to determine active cracks at each step, while in the three dimensional case, Paris's law is a common choice [137] for determining the propagation increment for different points along the crack front.

3.8. Applications in Fracture Mechanics and Extensions

As a result of the extensive research conducted in almost two decades, the method has reached a level of maturity that allows its application in a wide range of problems of both academic and industrial interest. Some representative applications would include damage tolerance assessment of aerospace structures [164] and hydraulic fracturing [165]. Significant research effort has also been devoted in implementing the method both in a procedural [161,166] and object oriented framework [167,168]. Thus, implementations of the method can be found in several open source libraries and commercial software packages such as Ansys and Abaqus.

The range of possible applications includes problems far more challenging than two-dimensional linear elastic crack propagation. For instance, the method can be extended to three dimensions in a straightforward way [132,136,137], while the treatment of problems involving multiple cracks [163,166,169,170] is also possible. The extension to dynamic crack propagation can be challenging, it is however possible and has been studied in several works, for instance references [171–173]. The method's flexibility also allows for application to problems involving different types of material models, for instance orthotropic [174], or in the nonlinear domain hyperelastic [175] and elastic-plastic [176]. Finally, other models for fracture, such as the cohesive zone model [115,177], can also be incorporated with relative ease.

4. The Scaled Boundary Finite Element Method (SBFEM)

4.1. An Abridged Literature Review of Advancements in SBFEM Fracture Modeling

The SBFEM belongs to the class of semi-analytical methods and is therefore related to the thin layer method [178], the Trefftz method [179], the BEM [34], Spectral elements [180] and the semi-analytical finite elements [181]. SBFEM's key feature lie in introduction of a scaling center, which has been pioneered in the context of different domains, such as the solution of electric field problems [182]. Dasgupta et al. [183] refined and tailored the approach, which they termed the "cloning algorithm", to solid mechanics of unbounded media. Wolf and Song [184] subsequently adopted a similar formulation, which they termed the "consistent infinitesimal finite-element cell method". They later developed a standardized derivation relying on use of a weighted residual method [185,186], and first coined the term "SBFEM". Later work by Deeks and Wolf [187] enabled broader adoption of the SBFEM method by introducing a virtual work based formulation.

Although much of the early research focused on the treatment of unbounded domains, it was soon discovered that SBFEM is more effective at modelling bounded domains [186], particularly in the context of LEFM. This is apparent, since the fracture parameters, e.g., SIFs, T-stress as well as the coefficients of higher order terms, can be directly extracted from the singular components of the stress field [188,189]. The method is able to robustly transition between power and power-logarithmic singularities [189]. It has thus been applied for computing the order of singularity and SIFs in multi-material plates under both static and dynamic loading [190], for predicting the crack propagation direction at bi-material notches [191], and for determining the free-edge stresses about holes in laminated composites [192].

Yang et al. [193] first modelled crack propagation via use of SBFEM and a few large sized subdomains, whose initial meshes were manually specified. This approach was extended to model nonlinear cohesive fracture in concrete [194–198], dynamic fracture [199,200] and crack propagation in reinforced concrete [201]. Reaching the limits of the laborious meshing approach, fully automated modelling of crack propagation was achieved by repurposing newly proposed meshers [202] for polygonal elements [203]. Currently, the most widely adopted meshing procedure combines the use of a quadtree decomposition with polygon clipping, to accurately represent curved geometries [51] with coarser meshes. The advantage of adopting balanced quadtree meshes as a basis lies in the limited amount of possible element realizations, whose pre-computation greatly enhances computational efficiency. Having resolved most mesh related issues, SBFEM was most recently extended to treat functionally graded materials [204,205] and non-local damage [206,207].

An interesting development pertains to fusion of scaled boundary principles with IGA (SBIGA), which is shown to provide lower error in displacement and energy norm per degree of freedom. The method ensures exact treatment of curved boundaries [208,209], delivers additional refinement possibilities and the ability to adjust continuity as required. However, the computational costs increased as compared against the standard SBFEM due to the integration procedure associated with IGA [210] partially due to the NURBS basis forming a larger support for the calculation of element related quantities [211]. When contrasted to established methods (e.g., FEM, IGA), this draw-back is negated as only the boundary need be discretised.

4.2. Principles of the Scaled Boundary Finite Element Method

The characteristic trait of SBFEM, setting it apart from other numerical methods, and enabling an elegant computation of the gSIFs, is the introduction of a scaling center. Each polygonal SBFEM element, referred to as a subdomain, may only contain one scaling center, from which the whole boundary must be visible. By consequence, a new reference system is introduced with a radial coordinate ζ and a local tangential coordinate η (Figure 2a). These resembles a polar reference frame, and are termed the *scaled boundary coordinates*.

The theoretical basis of the SBFEM is summarized in [186] and more recently and extensively in [212]. In this work, only the fundamental features are discussed. A thorough and more extensive treatment of the latest SBFEM-advancements in the context of LEFM can be found in the recent review paper by Song et al. [213].

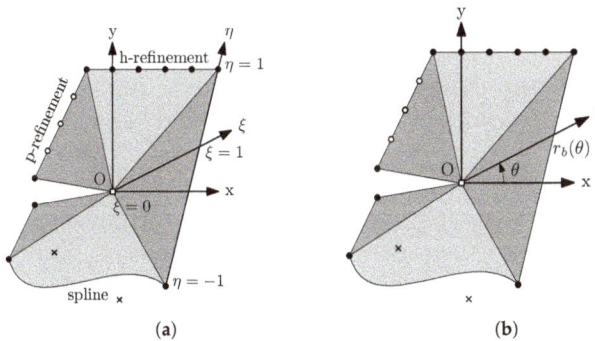

Figure 2. Domain discretisation, scaling center O and introduction of scaled boundary coordinates. (a) Polygon domain with scaled boundary coordinates; (b) Transformation of singular stress field around crack tip to polar coordinates.

Considering 2D bounded domains, the radial coordinate ranges from $0 < \xi < 1$, initiating at the scaling center and ending on the boundary. This component is kept analytical throughout the analysis, thereby reducing the dimensionality of the problem by one. Hence, only the boundary of the subdomain requires discretization, in the finite element sense, into independent line elements. For each line element, a separate local coordinate η is introduced with $-1 < \eta < 1$. The mapping between Cartesian (x, y) and scaled boundary coordinates $(x(\xi, \eta), y(\xi, \eta))$ is achieved by scaling points (x_b, y_b) on the boundary: For a given set of nodal coordinates x_b, y_b and conventional finite element shape functions $N(\eta)$ the below mapping results in:

$$x(\xi, \eta) = \xi x_b(\eta) = \xi N(\eta) x_b \tag{28}$$
$$y(\xi, \eta) = \xi y_b(\eta) = \xi N(\eta) y_b \tag{29}$$

which employs ξ as a scalar.

Similarly the displacements contain an analytical and an interpolatory component:

$$u(\xi, \eta) = N^u(\eta) u(\xi) = (N_1(\eta) I, ..., N_n(\eta) I) u(\xi) \tag{30}$$

The subscript n, denotes the degrees of freedom (DOF) present in the line element. In 2D, I is a 2×2 identity matrix and $u(\xi)$ are nodal displacement functions along a line connecting the scaling center and the boundary. Consequently, the displacements on the boundary are synonymous with $u = u(\xi = 1)$. The expression of the stress follows as [214]:

$$\sigma(\xi, \eta) = D(B^1(\eta) u(\xi)_{,\xi} + B^2(\eta) u(\xi) / \xi) \tag{31}$$

D represents the constitutive matrix. $B^1(\eta)$ and $B^2(\eta)$ together describe the strain-displacement relation [212]. Once the governing differential equation is rewritten in scaled boundary coordinates, standard techniques such the Galerkin's weighted residual method [186], the principle of virtual work [187] or the Hamiltonian principle [215] may be applied in the circumferential direction η giving rise to the two governing equations of SBFEM, i.e., Equations (32) and (33).

$$E^0 \xi^2 u(\xi)_{,\xi\xi} + (E^0 - E^1 + E^{1^T}) \xi u(\xi)_{,\xi} - E^2 u(\xi) = 0 \tag{32}$$
$$P = E^0 \xi u_{,\xi} + E^{1^T} u \text{ or in modal form } q = E^0 \xi u(\xi)_{,\xi} + E^{1^T} u(\xi) \tag{33}$$

Equation (32), termed the scaled boundary finite element equation in displacement describes the behavior within the domain. Equation (33) expresses the behavior on the boundary, where P comprises the vector of nodal forces.

The coefficient matrices E^0, E^1, E^2 are conceptually analogous to a subdomain stiffness matrix in the FEM: They are calculated for each element individually and then assembled. A general solution to the scaled boundary finite element equation in displacements, i.e., the homogeneous set of Euler-Cauchy differential equations in ξ, is sought in its simplest form as a power series:

$$u(\xi) = c_1 \xi^{-\lambda_1} \phi_1 + ... + c_n \xi^{-\lambda_n} \phi_n = \phi \xi^{-\lambda} c \tag{34}$$

The calculation of the eigenvalues, λ_i, and eigen-vectors, ϕ_i, by means of eigen-decomposition can result in numerical errors, when near parallel eigen-vector pairs are present [216]. To alleviate this, the block-diagonal Schur decomposition may be adopted [217]. The displacement solution is obtained as a superposition (Figure 3) of the modes, with associated scaling values, and constrained by integration constants c_i, as obtained from the imposed boundary conditions.

$$u(\xi) = \Psi^{(u)} \xi^{-S} c = \sum_{i=1}^{n} \Psi_i^{(u)} \xi^{-S_i} c_i \tag{35}$$

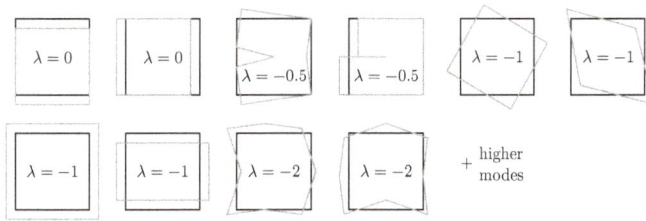

Figure 3. Graphical representation of modes. In black the original domain with linear elements and in gray the modes with corresponding values.

The transformation matrix $\mathbf{\Psi}$ and block diagonal real Schur form S are derived from recasting the system of first order differential equations (Equations (32) and (33)) as:

$$\xi \begin{Bmatrix} u(\xi) \\ q(\xi) \end{Bmatrix}_{,\xi} = -Z \begin{Bmatrix} u(\xi) \\ q(\xi) \end{Bmatrix} \tag{36}$$

with the Hamiltonian coefficient matrix Z defined by

$$Z = \begin{bmatrix} E^{0-1}E^{1^T} & -E^{0-1} \\ -E^2 + E^1 E^{0-1} E^{1^T} & -E^1 E^{0-1} \end{bmatrix} \tag{37}$$

so that Equation (36) is decoupled by the block-diagonal Schur decomposition.

$$Z\mathbf{\Psi} = \mathbf{\Psi} S \tag{38}$$

The columns of the transformation matrix contain the modes, whereas the diagonal blocks of the real Schur form contain the corresponding eigenvalues. However, Equation (36) results in doubling the amount of DOFs present in the solution, which can be shown to contain a bounded response ($0 < \xi < 1$ and negative eigenvalues) and an unbounded response ($1 < \xi < \infty$ and positive eigenvalues). S and $\mathbf{\Psi}$ are sorted in ascending order and partitioned accordingly.

$$S = diag(S_{neg}, S_{pos}) \tag{39}$$

$$\mathbf{\Psi} = \begin{bmatrix} \mathbf{\Psi}_{neg}^{(u)} & \mathbf{\Psi}_{pos}^{(u)} \\ \mathbf{\Psi}_{neg}^{(q)} & \mathbf{\Psi}_{pos}^{(q)} \end{bmatrix} \tag{40}$$

By expressing the nodal forces on the boundary with enforced integration constants (Equation (35) in Equation (33)), an expression for the stiffness matrix of the subdomain is obtained and a displacement solution is calculated analogous to FEM:

$$K_{bounded} = \mathbf{\Psi}_{pos}^{(q)} \mathbf{\Psi}_{neg}^{(u)-1} \tag{41}$$

Finally, the stresses are obtained by substituting Equation (35) into Equation (31):

$$\sigma(\xi, \eta) = \mathbf{\Psi}_{\sigma}(\eta) \xi^{-S_{neg}-I} c \tag{42}$$

where $[\mathbf{\Psi}_{\sigma i}(\eta)]$ is the stress mode of the corresponding displacement mode $[\mathbf{\Psi}_i^{(u)}]$.

$$\mathbf{\Psi}_{\sigma}(\eta) = D(-B^1(\eta)\mathbf{\Psi}_{neg}^{(u)} S_{neg} + B^2(\eta)\mathbf{\Psi}_{neg}^{(u)}) \tag{43}$$

The calculation of the stress on the domain boundary ($\xi = 1$) does not require the evaluation of the matrix exponential ξ^{-S-I}. This is beneficial when sufficient discretisation of the domain is achieved, e.g., via use of quadtree meshes. However, in case the complete domain is represented by a single large-sized SBFEM cell, the evaluation of displacements and stresses at internal points can become computationally intensive.

4.3. Calculation of SIFs

Since the general solution to the SBFEM equation is extracted as a power series, the singular modes are readily identified: By inspection of S_i any $-1 < real(\lambda) < 0$ will result in a singularity at $\xi = 0$. Placement of the scaling center at a crack tip may be exploited to calculate the generalized SIFs (Figure 2a). By including a double node at the crack mouth, two additional modes, the singular modes, arise (Figure 3), whose eigen-vectors resemble the mode I and mode II fracture cases. The singular stress field is extracted from the general solution (Equation (42)), where the superscript $^{(s)}$ denotes the singular quantities:

$$\sigma^{(s)}(\xi, \eta) = \Psi_\sigma^{(s)}(\eta)\xi^{-S^{(s)}-I}c^{(s)} \tag{44}$$

For consistency with other numerical methods and experimental reporting, a characteristic length L is introduced and a transformation to polar coordinates is sought (Figure 2b):

$$\xi = \frac{r}{r_b(\theta)} = \frac{L}{r_b(\theta)} \times \frac{r}{L} \tag{45}$$

The singular stress field is equivalently expressed in polar coordinates as:

$$\sigma^{(s)}(r, \theta) = \Psi_L^{(s)}(\theta)(\frac{r}{L})^{-S^{(s)}-I}c^{(s)} \tag{46}$$

implying the corresponding stress modes $\Psi_L^{(s)}(\theta)$ given by:

$$\Psi_L^{(s)}(\theta) = \Psi_\sigma^{(s)}(\eta(\theta))(\frac{L}{r_b(\theta)})^{-S^{(s)}-I} \tag{47}$$

For the case of 2D elastostatics, two singular stress modes exist. Hence, $S^{(s)}$ and $\Psi_L^{(s)}(\theta)$ reduce to matrices of size (2×2), while both $c^{(s)}$ and $\sigma^{(s)}(r, \theta)$ form vectors of size (2×1). More specifically, only the components of $\sigma^{(s)}(r, \theta) = (\sigma_\theta^{(s)}(r, \theta), \tau_{r\theta}^{(s)}(r, \theta))^T$ are retained, which correspond to mode I and II cracks, for which the formal definition of the gSIFs at angle θ is given as [50]:

$$\begin{Bmatrix} \sigma_\theta^{(s)}(r, \theta) \\ \tau_{r\theta}^{(s)}(r, \theta) \end{Bmatrix} = \frac{1}{\sqrt{2\pi L}}(\frac{r}{L})^{-\tilde{S}^{(s)}(\theta)} \begin{Bmatrix} K_I(\theta) \\ K_{II}(\theta) \end{Bmatrix} \tag{48}$$

The *matrix of orders of singularity* $\tilde{S}^{(s)}(\theta)$ is introduced such that:

$$\tilde{S}^{(s)}(\theta) = \Psi_L^{(s)}(\theta)(S^{(s)} + I)\Psi_L^{(s)}(\theta)^{-1} \tag{49}$$

$$\begin{Bmatrix} \sigma_\theta^{(s)}(r, \theta) \\ \tau_{r\theta}^{(s)}(r, \theta) \end{Bmatrix} = (\frac{r}{L})^{-\tilde{S}^{(s)}(\theta)}\Psi_L^{(s)}(\theta)c^{(s)} \tag{50}$$

Comparing Equation (48) with Equation (50) permits the evaluation of the gSIFs as:

$$\begin{Bmatrix} K_I(\theta) \\ K_{II}(\theta) \end{Bmatrix} = \sqrt{2\pi L}\Psi_L^{(s)}(\theta)c^{(s)} \tag{51}$$

The use of the matrix order of singularity automatically accounts for special cases in material interfaces. This is achieved by its off-diagonal terms [50]. Consequently, the SBFEM does not pose any a priori assumption on the type of singularity, which greatly facilitates the simulation of crack propagation through heterogeneous media.

Enhancing SIFs

Since the SIFs are directly evaluated using singular stress modes, standard recovery techniques may be applied, in order to improve on the solution during post-processing. Two pertinent methods are the Superconvergent patch recovery (SPR) theory [218–220] and curve fitting by splines [221]. In the former, an improved estimation of the singular stresses is obtained by smoothing the singular stress modes by means of SPR theory. The main benefit originates in the availability of error estimators [217] and the theoretical underpinning of the method. The latter is highly pragmatic and empirically offers comparable accuracy at reduced computational cost. Differing from the SPR method, the singular stresses computed at the Gauss points are fitted using a spline.

4.4. Balanced Hybrid-Polygon Quadtrees

Early efforts in SBFEM were limited due to the lack of specialized meshers. With the advent of polygon and virtual finite element methods [203,222,223], this was partially remedied, allowing for treatment of more involved and practical numerical examples. Specifically the use of the quadtree decomposition [51] has established itself as the predominant mesh choice [205,224,225], since it elegantly complements SBFEM's polygon underpinnings. By restricting the differences in cell sizes between neighbors to a ratio of 2:1, i.e., by enforcing balanced quadtrees (Figure 4b), it suffices to precompute only 16 realizations of SBFEM subdomains, while issues commonly associated with hanging nodes are alleviated.

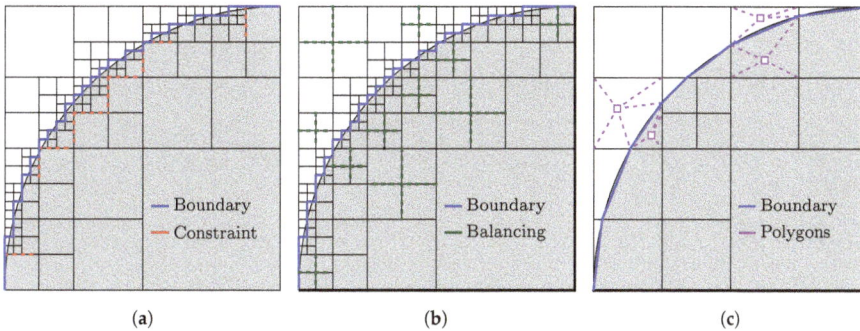

Figure 4. Polygon clipping operating on a balanced quadtree decomposition enables accurate geometry representation with coarser meshes. (**a**) Conventional FEM-based quadtree decomposition; (**b**) Balanced quadtree decomposition; (**c**) Balanced hybrid-polygon quadtree decomposition.

Features which do not align with the square grid of the quadtree decomposition require special treatment. In the standard FEM, this is achieved by mans of refinement near boundaries, until the lower threshold to a user-specified block size is reached. Generally, this results in step-like boundaries (Figure 4a) and excessively fine meshes. This is mitigated in SBFEM by employing polygon clipping. Consequently, the mesh consists of (a) standard square cells and (b) clipped polygon cells (Figure 4c). So-called hybrid quadtree meshes combine both types of cells, with the benefit of improved approximation of the geometry, at coarser discretisation levels. Standard FEM quadtree decompositions are nonetheless also adopted in SBFEM analyses, mostly in the context of automated image-based stress analysis [226], where the input data (pixel information) is inherently jagged by nature.

In conclusion, balanced quadtree meshes are economical to construct, automatically provide a certain degree of adaptivity around changing domain features and permit efficient analysis using the SBFEM by exploiting precomputation.

Crack Propagation

Cracks are introduced into the hybrid balanced quadtree mesh by polygon clipping [51]. Traversed blocks are split into two parts, by introducing a double node. Blocks containing a crack tip are augmented with an additional node, where the crack enters, and the scaling center is placed to coincide with the crack tip (Figure 5a). Discretisation of the crack tip segment is not required, since its solution is included in the radial and therefore analytic portion of the SBFEM solution. Specifically, discretisation of the crack tip segment is not permitted, due to the Jacobian of the respective element becoming zero.

In the case of crack propagation, the SIFs have to be calculated with sufficient accuracy. Since a simply cracked block does not permit sufficient resolution of the singular stress field or its radial distribution, a region surrounding the crack tip is homogenized (Figure 5b,c). The crack is then propagated by imposing a suitable criterion, e.g., Equation (13), with which the new crack tip is then projected (Figure 5d).

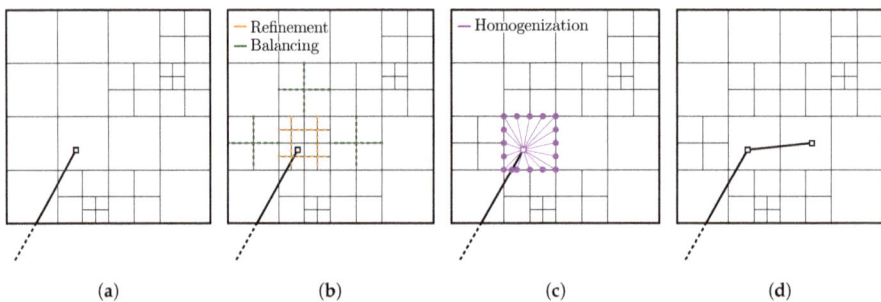

(a)   (b)   (c)   (d)

Figure 5. Main steps in SBFEM crack propagation scheme. (**a**) Crack entering existing balanced quadtree region.; (**b**) Refinement and balancing around crack tip; (**c**) Unifying cells into SBFEM macro element around crack tip; (**d**) New crack tip projected by gSIFs and Δa.

5. Phase Field Methods

5.1. Overview

PFM emerged as an alternative to discrete fracture aiming to address some of the challenges of computational fracture mechanics, e.g., automatic crack initiation, robust resolution of branching and merging and also the treatment of curved crack paths. The PFM diffusive crack interface is represented by a scalar variable, i.e., the phase field. The latter evolves according to a set of governing equations arising from a robust variational structure. As a result, the method does not require numerical tracking of the evolving discrete crack topologies and complex problems as in the case of 3D crack paths (see, e.g., [227–230]) and dynamic fragmentation are naturally resolved [231].

PFM emerged from the pioneering work of Francfort and Marigo [73] who proposed a variational theory of fracture based on energy minimization principles. Bourdin et al. [69] provided a regularised formulation by introducing a length scale parameter that rendered the approach more suitable for numerical approximations. The variational formulation was further modified and extended to multi-dimensional mixed-mode dynamic brittle fractures [228,232,233] also targeting the response of high performance composites [234–236]. The PFM for brittle fracture has been implemented in the commercial software Abaqus [237] via a User Element subroutine by Msekh et al. [64], which was later extended by Liu et al. [227]. Li et al. [238], see also [239], combined the variational phase field model of brittle fracture with an extended Cahn-Hilliard model [240,241], and formulated a fourth-order phase field model suitable resolving crack propagation in anisotropic materials. Rate-dependent PFM models for modelling fracture in visco-elastic solids [242] have also been established.

The phase field representation of fracture has been extended to the ductile regime [79,80,243,244] also within the context finite strains. The PFM has found application in the simulation of fractures in plates and shells [245–247], which involve a 3-D degradation of induced stresses whereas the element kinematics and damage are defined at the mid-surface. Attempts to experimentally validate the method have also been provided (see, e.g., [79]).

Verhoosel and de Borst [248] attempted to model cohesive fractures in composite materials using PFM by casting the cohesive zone approach in an energetic framework and introducing an auxiliary field in addition to the displacement and phase field which represents the jump in displacement across the cracked domain. The motivation to use an auxiliary field is to define the crack opening in cohesive fracture as a properly defined kinematic quantity, rather than an internal discontinuity as in the case of brittle fracture. Vignollet et al. [249] further extended the phase field-based cohesive fracture formulation for the case of propagating cracks. This approach succeeds in achieving convergence with lesser number of elements and in contrast to brittle fracture, confines the length scale parameter only to topological approximations hence rendering it uninfluential for the mechanical behaviour of the structure. Nguyen et al. [42] proposed a new phase field formulation which could model the interaction between interfacial damage and bulk brittle damage for complex topologies arising from voxel-based models of microtomography images. The formulation used a level-set method to describe the diffused jump in displacement field and used the phase field variable, instead of an additional internal variable as in [248], to model crack opening and reclosure during cohesive fractures.

There have been several recent efforts emphasizing the requirement of a generalized cohesive description of fracture using the phase field method [250,251], see, also Lorentz [252]. More specifically, Wu and Nguyen [251] proposed a unified phase field theory, namely the PF-CZM, for brittle and quasi-brittle fractures which converges to a cohesive zone model within the limits of a vanishing length-scale parameter. More importantly, the authors provided a method for the precise fitting of linear, exponential, and hyperbolic softening laws. PF-CZM was compared to the XFEM in [253] and further extended to the case of dynamic fracture in [254]. Furthermore, Geelen et al. [250] extended the work introduced in [255] to a dynamic cohesive fracture model incorporating phase field formulations.

The fundamental features of the phase field method are discussed in the following section.

5.2. PFM Variational Formulation

Griffith [86] postulated that the total potential energy Π of an elastic body undergoing elastic fracture comprises the contributions of the elastic strain energy Π_e and the fracture energy Π_f

$$\Pi\left(\mathbf{u}, \Gamma\right) = \Pi_e + \Pi_f + W_{ext} = \int_\Omega \psi_e d\Omega + \int_\Gamma \mathcal{G}_c d\Gamma + W_{ext} \tag{52}$$

where ψ_e is the elastic energy density and $\mathcal{G}_c$ is the critical fracture energy density, and W_{ext} is the work done by the external forces. The elastic energy density for the case of an isotropic medium is defined as

$$\psi_e\left(\varepsilon\right) = \frac{1}{2}\lambda[\text{Tr}\left(\varepsilon\right)]^2 + \mu\left[\text{Tr}\left(\varepsilon^2\right)\right] \tag{53}$$

where λ and μ are the Lamé constants.

Phase field modelling of fracture approximates the fracture surface integral expression introduced in Equation (52) with a volume integral defined over the entire deformable domain Ω according to Equation (54) below.

$$\int_\Gamma \mathcal{G}_c d\Gamma \approx \int_\Omega \mathcal{G}_c F_\Gamma\left(c, \nabla c\right) d\Omega \tag{54}$$

where $c = c\left(\mathbf{x}\right) \in [0, 1] \; \forall \mathbf{x} \in \Omega$ is the scalar phase field representing crack.

Using Equation (54), the expression of the potential energy of the elastic deformable body introduced in Equation (52) can be modified into the following form

$$\Pi \approx \int_{\Omega} \psi_e d\Omega + \overbrace{\int_{\Omega} \mathcal{G}_c F_\Gamma d\Omega}^{\text{Fracture Energy Approximation}} - \left(\int_{\Omega} u_i b_i d\Omega + \int_{\partial \Omega_{\bar{t}}} u_i \bar{t}_i d\Omega_{\bar{t}} \right) \tag{55}$$

The functional F_Γ assumes the following generic form

$$F_\Gamma = \frac{1}{c_w} \left(\frac{1}{2l_0} \omega(c) + 2l_0 |\nabla c|^2 \right), \tag{56}$$

where $l_0 \in \mathbb{R}$ is a length scale parameter and $\omega(c)$ and c_w are the generic crack geometric function and associated constant; these assume different expressions based on the type of fracture surface energy approximation used.

With the introduction of the crack surface density function in Equation (56), the discrete description of a sharp crack Γ_c in Figure 1 is transformed onto a diffused crack description as shown in Figure 6 via the regularized crack functional $\Gamma_{l_0}(c)$ which is scaled by the length-scale parameter l_0 (57).

$$\Gamma_{l_0}(c) = \int_{\Omega} F_\Gamma(c, \nabla c) \, d\Omega \tag{57}$$

The length scale parameter l_0 is the regularisation length over which damage diffuses as shown in Figure 6. In the conventional phase field formulation, originally presented in Bourdin et al. [69], the peak force reached before the onset of fracture depends on the value of length-scale parameter l_0. Higher values of the length-scale parameter lead to lower peak forces and vice versa. In recent formulations, see, e.g., Wu and Nguyen [251] and Geelen et al. [250] this is alleviated, hence providing a significant advantage in enhancing the critical-stress predicting capabilities of the phase field method. In Miehe et al. [256], generalized crack-driving forces with a failure criteria based on the maximum principal stress were introduced which also succeeded in predicting critical fracture loads unaffected by the length-scale parameter. However in notched structures, a crack nucleation principle based purely on the maximum principal stress criteria suffers from the curse of stress singularity at the notch-tip as also highlighted in [257].

Providing different expressions for $\omega(c)$ and c_w results in variants of the phase field approximation; key variants, i.e., the second and fourth order quadratic approximations and the second order linear phase field approximation are discussed in Sections 5.2.1, 5.2.2, and 5.2.3, respectively. A schematic of the variation of the phase field c in the direction normal to the crack surface for all phase field variants considered as compared to the discrete fracture case is provided in Figure 7. In all cases, the phase field value $c = 1$ corresponds to an un-cracked region, whereas $c = 0$ corresponds to a cracked region.

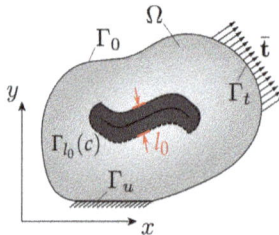

Figure 6. Description of diffused crack scaled by the length-scale parameter l_0 and boundary conditions.

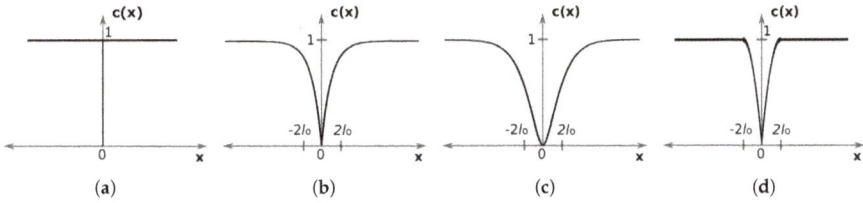

Figure 7. 1-D spatial variation of phase-field $c(x)$ for (**a**) Discrete crack (**b**) Diffused crack with second-order quadratic approximation (**c**) Diffused crack with fourth-order quadratic approximation, and (**d**) Diffused crack with second-order linear approximation.

Remark 1. *From a geometric standpoint, the length scale parameter regularises the width of the crack as shown in Figure 6 in accordance with [69], see, also Borden et al. [229]. It is of interest to note that the length scale considered in Miehe et al. [232] (see, also, [228,258]) is double the size of the one adopted in [69,229]. Of course, both implementations are equivalent; one however should be careful to appropriately adapt the length scale parameter when comparing between the two. In this work, we comply with the former definitions.*

5.2.1. Second-Order Quadratic Approximation

For the second-order quadratic approximation, the 1-D spatial variation of phase-field variable $c(x)$ can be expressed as (Figure 7b):

$$c(x) = 1 - e^{-|x|/2l_0} \tag{58}$$

It is straight-forward to show that the width of diffusion zone decreases with decreasing the value of length-scale parameter l_0, which can also be seen in Figure 8.

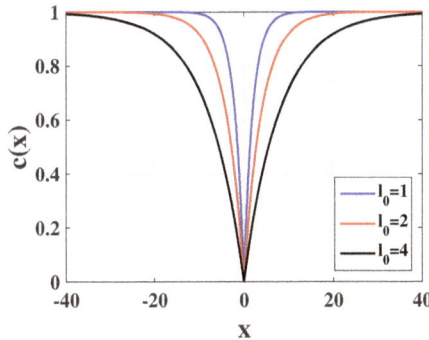

Figure 8. Second-order quadratic approximation: Effect on length-scale parameter l_0 on the width of diffusion.

The specific second order functional proposed in Bourdin et al. [69] can be retrieved by modifying the general form of Equations (54)–(56) and considering the following definitions in Equation (59)

$$
\begin{aligned}
c_w &= 2 \\
w(c) &= (c-1)^{2^{\cdot}}
\end{aligned}
\tag{59}
$$

Hence, the crack surface energy approximation assumes the following form

$$F_\Gamma = \left[\frac{(c-1)^2}{4l_0} + l_0|\nabla c|^2\right]$$

$$\int_\Gamma \mathcal{G}_c d\Gamma \approx \int_\Omega \mathcal{G}_c \left[\frac{(c-1)^2}{4l_0} + l_0|\nabla c|^2\right] d\Omega \tag{60}$$

5.2.2. Fourth-Order Quadratic Approximation

A fourth-order quadratic approximation is established considering the definition introduced in [259], i.e.,

$$\int_\Gamma \mathcal{G}_c d\Gamma \approx \int_\Omega \mathcal{G}_c \left[\frac{(c-1)^2}{4l_0} + \frac{l_0}{2}|\nabla c|^2 + \frac{l_0^3}{4}(\Delta c)^2\right] d\Omega \tag{61}$$

The expression for $c(x)$ for the fourth-order quadratic approximation can be given as (also shown in Figure 7c):

$$c(x) = 1 - e^{-|x|/l_0}\left(1 + \frac{|x|}{l_0}\right) \tag{62}$$

The effect of the length-scale parameter on the diffusion width is illustrated in Figure 9. The higher-order term introduced in Equation (61) leads to greater regularity of the phase-field solution, and improves its convergence rate and accuracy. However due to increased continuity requirements of the solution, the basis functions used for numerical interpolation must be at least (C^1) continuous, for e.g., hierarchically refined B-splines used within an isogeometric analysis framework [259]. It should also be noted that the use of 4th-order model leads to a more accurate approximation of stresses, which in turn facilitates higher rates of crack growth. More applications of higher-order phase-field models can be found in [259–261].

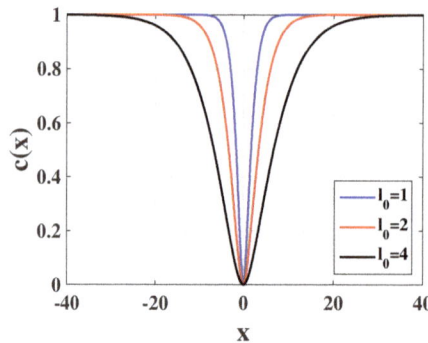

Figure 9. Fourth-order quadratic approximation: Effect of the length-scale parameter l_0 on the width of diffusion.

5.2.3. Linear Approximation

In the quadratic approximations shown in Sections 5.2.1 and 5.2.2, the phase field variable and therefore the degradation function evolve as soon as the structure is loaded. This is clearly not the case in purely elastic brittle materials that demonstrate a linear elastic behavior until a crack initiates.

Pham et al. [262] addressed this issue by employing a linear approximation of the surface energy integral to achieve a diffused localization band and a purely elastic global response until the onset of damage. The 1-D expression for $c(x)$ in this case can be given as in Equation (63), which is also illustrated in Figure 7d (See also Figure 10).

$$c(x) = 1 - \left(\frac{|x|}{2l_0} - 1\right)^2 \tag{63}$$

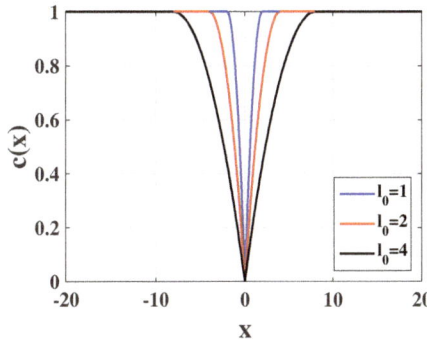

Figure 10. Second-order linear approximation: Effect of the length-scale parameter l_0 on the width of diffusion.

More recently, Geelen et al. [250] provided an analogous linear approximation based on the following expressions for c_w and $\omega(c)$

$$\begin{aligned} c_w &= \frac{16}{3} \\ \omega(c) &= 4(1-c) \end{aligned} \tag{64}$$

which result in the following definition of the crack functional

$$F_\Gamma = \frac{3}{8l_0}\left[1 - c + l_0^2|\nabla c|^2\right]. \tag{65}$$

In view of Equation (66), the approximation of the surface energy integral in Equation (54) assumes the following form

$$\int_\Gamma \mathcal{G}_c d\Gamma \approx \int_\Omega \frac{3\mathcal{G}_c}{8l_0}\left[1 - c + l_0^2|\nabla c|^2\right] d\Omega \tag{66}$$

The linear approximation in Equation (66) differs from the corresponding formulation in [250] in the sense that a fully cracked-state in the current study is represented by $c = 0$ in the current study, as opposed to $c = 1$ in [250]. In addition, the total diffusion width in the current model (Equation (66) and Figure 7d) is twice the diffusion width in [250] to maintain consistency with other models.

It is of interest to note that the quadratic form (Equations (60) and (61)) implicitly guarantees the boundedness of the phase field variable c within the limits $[0, 1]$. However, the solution obtained by Equation (65) is not intrinsically bounded within this interval, and additional constraints must be imposed to ensure boundedness.

This is achieved by employing a Penalty (see, e.g., [263]) or a Lagrange multiplier method (see, e.g., [250]). In both methods, a staggered iterative scheme is required for the solution of the resulting constrained system of governing phase-field equation. To guarantee both the boundedness and irreversibility of the phase field variable, Gerasimov and De Lorenzis [263] proposed a method to choose the value of an optimal or lower bound of the penalty parameter beyond which adequate constraint enforcement can be ensured.

5.3. Material Degradation

The expression of the potential energy introduced in Equation (52) implies that in a given conservative system, any increase in the fracture energy due to a unit increase in the fracture surface has to be compensated by a corresponding decrease in the elastic strain energy. Hence, the expression of the elastic energy must be coupled to the evolution of the phase field c as the latter dictates the value of the fracture energy. In physical terms, the phase field has to account for the gradual degradation of material stiffness as cracks propagate through the medium.

Mathematically, this has been expressed through the definition of a degradation function, $g(c)$, which is then used to reduce the value material elastic energy density giving rise to the so-called isotropic phase field methods. Driven from the fact that such an approach led to unrealistic and in cases erroneous results, e.g., cracks initiating and propagating due to pure compression later attempts postulated material degradation on the basis of an energy split, i.e.,

$$\psi_e = g(c)\,\psi_e^+ + \psi_e^- \tag{67}$$

where ψ_e^+ and ψ_e^- are the elastic strain energy densities whose expressions are specific to the type of energy split adopted, see, e.g., Miehe et al. [228] for an energy decomposition based on the spectral decomposition of the strain tensor and Amor et al. [264] for a volumetric/deviatoric decomposition giving rise to the so-called anisotropic degradation models. It is of interest to note that although anisotropic models mitigated the unrealistic crack patterns derived from the isotropic ones for most typical stress states, the problem is not yet fully resolved. The volumetric split defined in [264] may still result in degradation under a pure compressive stress state. The spectral decomposition model defined in [228] leads to a strongly non-linear stress-strain relation that has been shown to be computationally taxing (see e.g. [265] for a detailed comparison of these two models).

The expression of the degradation function $g(c)$ is not unique see, e.g., [80,250,255,266–271]. A widely used definition for the degradation function that is compatible with the first and second order quadratic approximations provided in Equations (60) and (61), respectively is

$$g(c) = \left[(1-k)\,c^2 + k\right] \tag{68}$$

where k in Equation (68) is a model parameter utilized in several applications, see, e.g., [264,272] as a way to avoid ill-posedness. Geelen et al. [250] introduced a quasi-quadratic definition of $g(c)$ to be employed in conjuction with the linear approximation defined in Equation (65) that is defined as

$$g(c) = \frac{c^2}{c^2 + m(1-c)[1 + p(1-c)]} \quad \text{with } p \geq 1 \ \text{ and } l_0 < \frac{3E\mathcal{G}_c}{4(p+2)\sigma_c^2} \tag{69}$$

where $m = (3\mathcal{G}_c)/(8l_0\psi_c) = g'(c_0)$ is the initial slope of the degradation function $g(c)$ and p provides the initial slope and shape parameters for the softening curve assuming $c_0 = 1$ as the initial phase-field. Here, $\psi_c = (\sigma_c^2)/(2E)$ is the critical fracture energy per unit volume of the material, in which σ_c and E represent the critical tensile strength and Young's modulus of the material respectively. This definition, however, comes with an additional upper bound restriction on the value of length-scale parameter l_0 which is necessary to achieve optimal convergence. The upper bound on the regularization length is related to the characteristic length of the fracture process zone $l_{FPZ} = (E\mathcal{G}_c)/(\sigma_c^2)$, see [250,273] for details.

Substituting Equation (67) in Equation (55), the expression for the brittle fracture potential energy assumes the following form

$$\Pi \approx \int_\Omega g(c)\psi_e^+ d\Omega + \int_\Omega \psi_e^- d\Omega + \int_\Omega \mathcal{G}_c F_\Gamma d\Omega - \left(\int_\Omega u_i b_i d\Omega + \int_{\partial\Omega_{\bar{t}}} u_i \bar{t}_i d\Omega_{\bar{t}} \right) \tag{70}$$

where definitions of ψ_e^+ and ψ_e^- are specific to the energy split adopted and $g(c)$, F_Γ may be chosen based on the Table 1.

Table 1. Definition variants for the degradation function $g(c)$ and the functional F_Γ.

$g(c)$		F_Γ		Reference
$[(1-k)c^2 + k]$	Equation (68)	$\frac{(c-1)^2}{4l_0} + l_0|\nabla c|^2$	Equation (60)	Borden et al. [229]
$\frac{c^2}{c^2 + m(1-c)[1+p(1-c)]}$	Equation (69)	$\frac{3}{8l_0}\left[1 - c + l_0^2|\nabla c|^2\right]$	Equation (65)	Geelen et al. [250]

5.4. PFM Strong Form

The Euler-Lagrange equations of the displacement $\mathbf{u}(\mathbf{x},t)$ and phase field $c(\mathbf{x},t)$ coupled formulation of the Lagrangian functional are employed to derive the strong form of the quasi-static brittle-fracture phase field formulation. The latter assumes the following general form:

$$\nabla\sigma + \mathbf{b} = 0, \qquad \text{on } \Omega \qquad (71)$$
$$\mathcal{G}_c\delta_c(F_\Gamma) = -g'(c)\tilde{D}, \qquad \text{on } \Omega \qquad (72)$$

where $\delta_c(F_\Gamma)$ denotes the derivative of surface energy approximation function F_Γ with respect to the phase field variable c, and $\tilde{D}$ is the energetic crack-driving force which depends on the phase field formulation used. A detailed description on the different crack-driving forces that can be employed in conjuction with Equation (72) is provided in Miehe et al. [256].

The coupled field Equations (71) and (72) are subject to the boundary conditions introduced in Equation (1) supplemented by

$$\frac{\partial c}{\partial x_i}n_i = 0 \qquad \text{on} \qquad \Gamma_c^t. \qquad (73)$$

where $n_i, i=1\ldots r$ is the outward-pointing normal vector to the crack boundary. The Cauchy stress tensor $\sigma \in R^{r\times r}$ is defined as

$$\sigma_{ij,e} = \frac{\partial\psi_e}{\partial\varepsilon_{ij}} \qquad (74)$$

Hence, substituting Equation (67) into Equation (74) gives rise to the degraded Cauchy stress tensor

$$\sigma = \sigma_{ij} = g(c)\frac{\partial\psi_e^+}{\partial\varepsilon_{ij}} + \frac{\partial\psi_e^-}{\partial\varepsilon_{ij}} = g(c)\sigma^+ + \sigma^- \qquad (75)$$

where $g(c)$ takes one of the forms shown in Equations (68) and (69) depending upon the formulation.

5.5. Derivation of the Phase Field Evolution Equation in from the General Form

The phase field evolution equation employed in Borden et al. [229] can be obtained from the general expression of the strong form (Equations (71) and (72)), considering the expressions for F_Γ and $g(c)$ from Equation (60) and (68), i.e.,

$$F_\Gamma = \left[\frac{(c-1)^2}{4l_0} + l_0|\nabla c|^2\right] \quad ; \quad \delta_c(F_\Gamma) = \left[\frac{(c-1)}{2l_0} - 2l_0\Delta c\right]$$
$$g(c) = (1-k)c^2 + k \quad ; \quad g'(c) = 2(1-k)c \qquad (76)$$

In the original formulations of Miehe et al. [232], which is later also adopted in [229], the crack driving force $\tilde{D}$ was the positive part of the elastic strain energy density, i.e.,

$$\tilde{D} = \psi_e^+ \qquad (77)$$

where ψ_e^+ is the tensile part of strain energy density taken from [228].

Substituting Equation (77) in Equation (72) and considering also Equation (76) the following evolution equation is derived, i.e.,

$$\left(\frac{4l_0\,(1-k)\,\psi_e^+}{\mathcal{G}_c}+1\right)c-4l_0^2\Delta c=1,\text{ on }\Omega \tag{78}$$

which is a linear differential equation with respect to c. It is of interest to note that the Laplacian of the phase field in Equation (78) is scaled by the squared value of the length scale parameter hence it rapidly vanishes for small values of l_0 compared to the c.

5.6. Derivation of the Cohesive Phase Field Evolution Equation in from the General Form

The phase field evolution equation presented in Geelen et al. [250] can be obtained from the general expression of the coupled strong form considering the following expressions for F_Γ, $\delta_c(F_\Gamma)$, and $g(c)$

$$F_\Gamma = \frac{3}{8l_0}\left[1-c+l_0^2|\nabla c|^2\right]\quad;\quad \delta_c(F_\Gamma)=\frac{3}{8l_0}\left[-1-2l_0^2\Delta c\right] \tag{79}$$

$$g(c)=\frac{c^2}{c^2+m(1-c)[1+p(1-c)]}\quad\text{with }p\geq1\text{ and }l_0<\frac{3E\mathcal{G}_c}{4(p+2)\sigma_c^2}$$

Substituting Equation (79) into Equation (72) and performing the necessary algebraic manipulations results in the following expression Geelen et al. [250].

$$\frac{3\mathcal{G}_c}{8l_0}\left[2l_0^2\Delta c+1\right]-g'(c)\tilde{D}=0,\text{ on }\Omega \tag{80}$$

where

$$\tilde{D}=\max(\psi_c,\psi_e^+) \tag{81}$$

and $\psi_c=\sigma_c^2/2E$. Specific to this formulation, an additional augmented Lagrange constraint is incorporated to ensure the smooth monotonic evolution of the phase field variable c, such that $\dot{c}\leq0$. In view of this, Equation (80) transforms into the following expression:

$$\frac{3\mathcal{G}_c}{8l_0}\left[2l_0^2\frac{\partial^2 c}{\partial x_i^2}+1\right]-g'(c)\tilde{D}+\langle\lambda+\gamma(c-c^{n-1})\rangle_+=0,\text{ on }\Omega \tag{82}$$

where $\lambda\in L^2(\Omega)$ are Lagrange multipliers and $\gamma\in R_{>0}$ is the penalty kernel. c^{n-1} is the value of phase field at preceding $(n-1)^{th}$ time-increment.

5.7. Irreversibility Conditions

The expression of the potential energy defined in Equation (70) implies that regardless of the value of the degradation function, the fracture energy would need to further increase in the case of unloading to compensate for the corresponding elastic energy decrease. This is also derived on the basis of Equations (78), i.e., the strong form of the coupled system. In particular, the second of Equations (78) would result in an increasing value of the phase field for decreasing values of the elastic energy potential in the case of unloading. This would correspond to a reduction in the crack length, thus negating the irreversibility condition

$$\Gamma^{(t+\Delta t)}\supseteq\Gamma^{(t)} \tag{83}$$

Among the various irreversibility constraints proposed within the phase field literature, the history variable approach given by Miehe et al. [232] is most widely applied. Based on the theoretical arguments provided in [232], irreversibility is enforced by introducing a so-called history variable such that the following Kuhn-Tucker conditions hold

$$\psi_e^+ - \mathcal{H} \leq 0 \quad \dot{\mathcal{H}} \geq 0 \quad \dot{\mathcal{H}}\left(\psi_e^+ - \mathcal{H}\right) = 0 \tag{84}$$

where $\mathcal{H}$ is a history field.

Some other recent works have also proposed penalty and augmented Lagrange methods for imposing the irreversibility constraints on the phase field equations, see e.g., [250,263], so that the monotonicity of the phase field variable constantly holds. It is to be noted that these methods provide a more natural way of imposing the constraints, and do not disrupt the original variational nature of the phase field equations. Equation (82) employs such an augmented Lagrange constraint to ensure the monotonic evolution of phase field variable.

5.8. Effective Critical Energy Release-Rate

In the original variational formulation proposed by Bourdin et al. [69], it was shown that the fracture energy is slightly overestimated during simulations and the amount of this amplification depends upon the size of elements in the overall finite-element discretization. This amplification effect must be compensated by defining an effective critical energy release rate G_c^{eff} for the purpose of phase-field simulation (see also [274]).

$$G_c^{eff} = \frac{G_c^{actual}}{1 + (h/4l_0)} \tag{85}$$

where G_c^{actual} and G_c^{eff} are the actual and effective critical energy release rates respectively. It must be emphasized that using the amplified value of material fracture energy G_c^{actual} leads to overestimation of critical fracture loads in comparison to discrete fracture methods, and hence for all practical purposes G_c^{eff} must be used while solving the phase-field evolution equation. This would also be highlighted in detail in the numerical examples section.

5.9. Galerkin Approximation

The strong form of the coupled governing Equations (78) and (82) are set in a discrete form following standard Galerkin approximation. In this setting, the trial solution spaces are defined as

$$S_u = \left\{ \mathbf{u} \in \left(H^1\left(\Omega\right)\right)^d \middle| \mathbf{u} = \bar{\mathbf{u}} \text{ on } \partial\Omega_b \right\} \tag{86}$$

and

$$S_c = \left\{ c \in H^1\left(\Omega\right) \right\} \tag{87}$$

for the displacement field and the phase field respectively. Corresponding weighting functions spaces are further defined as

$$W_u = \left\{ \mathbf{w}^u \in \left(H^1\left(\Omega\right)\right)^d \middle| \mathbf{w}^u = \bar{\mathbf{w}}^u \text{ on } \partial\Omega_b \right\} \tag{88}$$

and

$$W_c = \left\{ w^c \in H^1\left(\Omega\right) \right\} \tag{89}$$

Multiplying Equation (71) with the weighting functions (88) and performing the necessary integration by parts leads to the standard weak form of the equilibrium equation

$$\int_\Omega \sigma \cdot \nabla \mathbf{w}^u d\Omega - \int_\Omega \mathbf{b} \cdot \mathbf{w}^u d\Omega - \int_{\partial \Omega_{\bar{t}}} \bar{t} \cdot \mathbf{w}^u d\partial \Omega_{\bar{t}} = 0 \tag{90}$$

Multiplying Equation (78) with the weighting functions (89) and performing the necessary algebraic manipulation gives rise to the phase field weak form employed in [229]

$$\int_\Omega \left(\left[\frac{4l_0 (1-k)\, \mathcal{H}}{\mathcal{G}_c} + 1 \right] c, w^c \right) d\Omega + \int_\Omega \left(4l_0^2 \nabla c, \nabla w^c \right) d\Omega - \int_\Omega (1, w^c)\, d\Omega = 0 \tag{91}$$

Similarly, the cohesive phase field weak form derived from Equation (82) assumes the following form

$$\int_\Omega (g'(c)\tilde{D}, w^c)\, d\Omega + \int_\Omega \frac{3\mathcal{G}_c}{8l_0} \left[-(1, w^c) + \left(2l_0^2 \nabla c, \nabla w^c \right) \right] d\Omega$$
$$+ \int_\Omega (\langle \lambda + \gamma(c - c^{n-1}) \rangle_+, w^c)\, d\Omega = 0 \tag{92}$$

The weak forms introduced in Equation (91) or (92) can be further discretised employing either mesh-based, i.e., the FEM, mesh-less methods, see, e.g., [275] or MPM [32]. The resulting discrete equations are then solved in an incremental fashion. Due to the nonlinear nature of $g(c)$, the resulting discrete problem is a nonlinear one, even for the case of elastic fracture, hence necessitating the use of iterative solvers.

6. Numerical Examples

In this section, four numerical examples are presented, allowing for a comparison in terms of the modeling capabilities of the investigated methods. The first two examples consider a square plate, first under tension, then under shear loading, with both setups having been studied extensively in existing literature. Although analytical solutions for these two setups do not exists, the geometry can be modelled by one SBFEM subdomain and therefore a high-fidelity reference solution can be constructed for the peak load and displacements following the first crack increment. For the last two examples, the notched plate with hole and L-shaped panel, respectively, there exist experimentally obtained crack paths to compare against. Furthermore, the test setups closely mimic crack propagation scenarios under real world conditions. For the former numerical example, modelling the complete crack path by discrete crack methods is particularly challenging, since they do not provide the capability to nucleate cracks. The later numerical example presents a similar issue, however, modelling by discrete crack methods is achieved by placing the crack tip at the re-entrant corner, effectively circumventing the nucleation issue manually. To this end, we first outline the implementation details adopted for each numerical method, then proceed to the numerical examples.

6.1. Implemented Variants

For the numerical examples presented in this section, the standard XFEM with shifted enrichment functions is employed. The enrichment radius assumes a value equal to $r_e = 3.5$ h, with h denoting the element size, while the radius used for the interaction integral is $r_d = 1.5$ h. Element partitioning and almost polar integration are employed for the integration of jump and tip enriched elements respectively. Finally, levels sets are updated using the $\phi \psi r \theta$ method from the work of Duflot [139].

The specific realization of SBFEM employed in the presented examples is based on balanced hybrid-polygon quadtrees, unless otherwise explicitly stated, and thus discretises the boundary with linear line elements. The Gauss-Lobatto integration scheme is employed, to offset computational effort for the numerical examples where hp-refinement is introduced (Section 6.2). Decoupling of the linear system of ordinary differential equations (Equation (36)) is performed by block diagonal Schur decomposition. The gSIFs are estimated by means of the spline fitting approach. For the case of the tension test, the domain is approximated via use of a single subdomain with hp-refinement on the boundary to produce gSIFs of highest possible accuracy. Results obtained by this variant are termed SBFEM hi-fi, acknowledging the high fidelity solutions they produce [276].

For the PF-FEM case, 4-noded quadrilateral plane strain/stress elements with bilinear basis functions and based on a full integration technique have been adopted. A displacement-controlled nonlinear static analysis scheme is utilized with constant displacement increments. Displacement is monitored and controlled at any single node on the loading edge, to which all other nodes on the edge are kinematically coupled in the direction of loading. Unless explicitly stated, the solution is implemented within a stagger phase-field solution algorithm with a single prediction step ($N_{staggs} = 1$) and a displacement norm convergence tolerance $tol_u = 10^{-5}$. In all the numerical experiments conducted in this work, the mesh size is consistently smaller than the length scale, i.e., $h \leq l_0$ to accurately resolve the crack path.

Remark 2. *In case the phase field functional definition and associated length scale parameter initially adopted by [232] is employed (see, also Remark 1), then the corresponding mesh size inequality becomes $h \leq l_0/2$.*

6.2. Numerical Example 1: Single Edge-Notched Tension Test

This example considers mode-I fracture behavior of a square panel, with geometric description of the domain, boundary conditions and material parameters as defined in Figure 11. A state of plane strain is assumed, the specimen thickness is $t = 1$ mm. The Young's modulus, Poisson's ratio, length scale, fracture energy density and crack propagation length are chosen as $E = 210$ kN/mm^2, $v = 0.30$, $l_0 = 0.0075$ mm, $\mathcal{G}_c = 0.0027$ kN/mm, $\sigma_c = 2.5$ kN/mm^2 and $\Delta a = 0.02$ mm, where applicable. The bottom edge of the specimen is clamped in both x and y directions, such that $u_x = 0; u_y = 0$. The loads and boundary conditions of the top edge by discrete and PFM are enforced differently, yet with equivalent outcome; for XFEM and SBFEM a prescribed displacement of $u = u_y \geq 0$ is imposed on the top edge, while for PFM, a quasi-static displacement control analysis procedure is implemented considering a concentrated load applied at point C and kinematic coupling of the vertical displacement DOF along the top edge, such that $u = u_y \geq 0$ is obtained. The analysis procedures for each approach as described in Section 6.1 apply. Two different solution procedures based on standard and cohesive phase field approaches, as described in Sections 5.5 and 5.6 respectively, are studied within this example. The resulting load deflection paths for all methods are shown in Figure 12; the standard SBFEM and XFEM implementations match the deflections and peak load, while the phase field method with G_c^{eff} approximates only the peak load closely.

The nucleation and propagation of the crack at successive time-increments is shown in Figure 13. The nucleation of the crack automatically occurs at the notch-tip, and then this propagates linearly in the direction perpendicular to the applied load. It is known that the value of the length scale parameter l_0 not only controls the width of the phase field diffusion zone, but also affects the peak fracture force values. This is illustrated in Figures 14 and 15, where a decreasing the value of l_0 leads to sharper crack topologies and higher peak fracture forces, thus showcasing a more brittle-like fracture behaviour. It can be inferred from Figure 15 that if l_0 is chosen sufficiently small, i.e., in the limit $l_0 \to 0$, the force-displacement curves converge towards the discrete solution, i.e., Griffith's description of brittle fracture; a property well-known as Γ-convergence of regularized phase field fractures. However, an important point to note is that a formal proof of Γ-convergence of anisotropic strain-energy splits (detailed in [232,264]) towards Griffith's theory is not available yet, as also stated in [277].

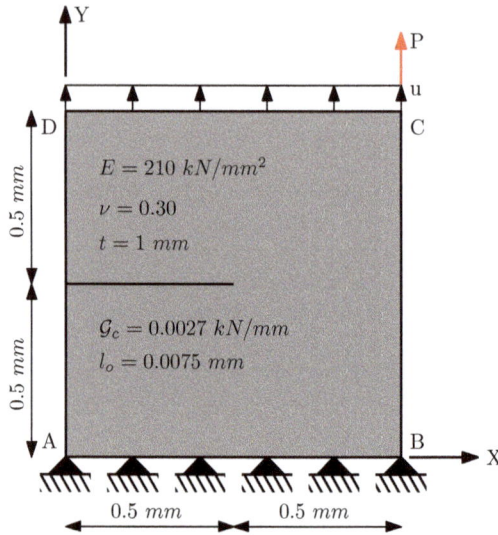

Figure 11. Tension test geometry, material parameters, loading and boundary conditions.

It is evident from Figure 12 that both discrete crack methods, i.e., XFEM/SBFEM, predict similar fracture characteristics, whereas the critical fracture force obtained from phase-field method is slightly overestimated when the actual value of $G_c^{actual} = 0.0027$ kN/mm is used. Considering $h_{PFM} = 0.005$ mm and $l_0 = 0.0075$ mm which have been used for the current analysis, an effective fracture energy $G_c^{eff} = 0.00231$ kN/mm can be calculated based on Equation (85). The critical fracture load thus obtained using G_c^{eff} shows very good agreement with those predicted by discrete methods XFEM/SBFEM. The difference in the elastic stiffness of the material between XFEM/SBFEM and PF-FEM cases is due the fact that in conventional PF-FEM formulations, as in [69], the phase-field variable evolution and consequently stress degradation start as soon as the material is loaded and hence, prevents recovery of a pure linear elastic limit. The crack paths, however, coalign as expected, although for the PF-FEM the resulting displacements are over-estimated as the fracture must initiate at the same critical load for a given value of G_c. An alternate approach, which is highly effective in determining accurate gSIFs [276], may be applied when the domain is star convex with regards to the crack tip, and is introduced here as a high fidelity reference solution (SBFEM hi-fi). Although by hp-refinement on the boundary, the gSIFs are accurately determined utilizing only a few DOFs, and thus minimal computational resources, this approach is only applicable to crack propagation in a select few cases, such as in this symmetric tension test, where the crack path remains straight. The SIFs obtained by discrete crack methods coincide to the fourth significant figure.

For comparison purposes, the tension test is also performed using the cohesive phase field method shown in Equation (92). The fracture response in this case depends on the shape parameter p, which controls the shape of cohesive stress-crack opening curve. Increasing the value of p enables faster degradation of stresses as soon as the critical stress limit is reached, however, too large p may lead to poor convergence. Figure 16 shows the dependence of load-displacement responses and critical loads on the choice of shape parameter p. The length-scale parameter l_o for each case is chosen based on its upper bound value in Equation (69). A cohesive phase-field model is highly useful when the size of fracture process zone (FPZ) is large enough, and the Griffith's description of purely brittle fracture becomes inadequate [250]. In such cases, the numerical phase-field model can be calibrated with the specific material responses by making an optimal choice for the parameter p.

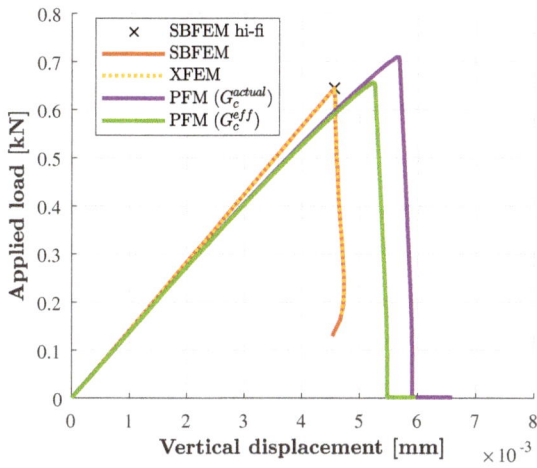

Figure 12. Tension test load-deflection curves.

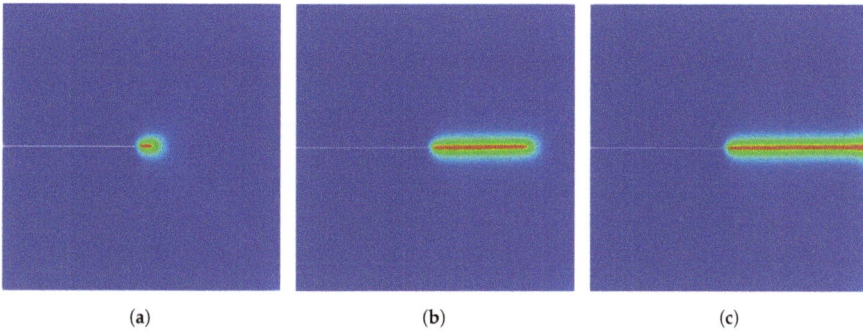

Figure 13. Tension test phase field evolution for (**a**) $u = 0.0057$ mm (**b**) $u = 0.00585$ mm (**c**) $u = 0.00595$ mm, with displacement increment $\Delta u = 1 \times 10^{-6}$ mm.

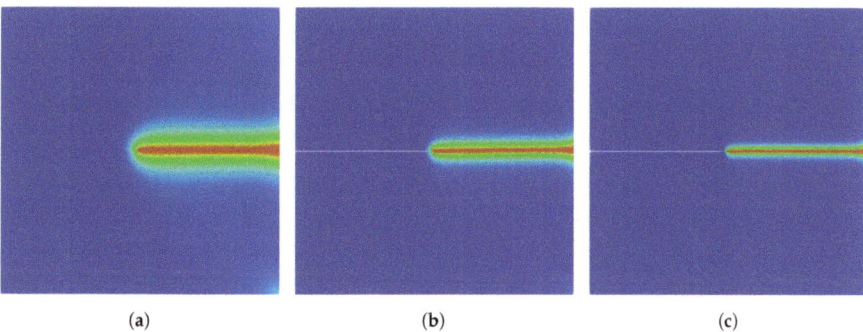

Figure 14. Tension test comparison of phase field diffusion widths employing (**a**) $l_0 = 0.015$ mm (**b**) $l_0 = 0.0075$ mm (**c**) $l_0 = 0.00375$ mm.

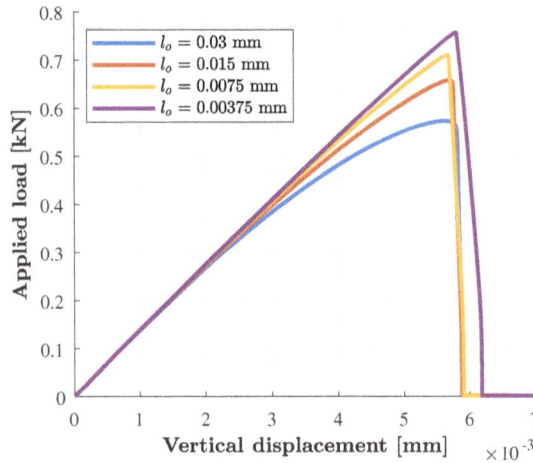

Figure 15. Tension test effect of length-scale parameter variation on load displacement curves.

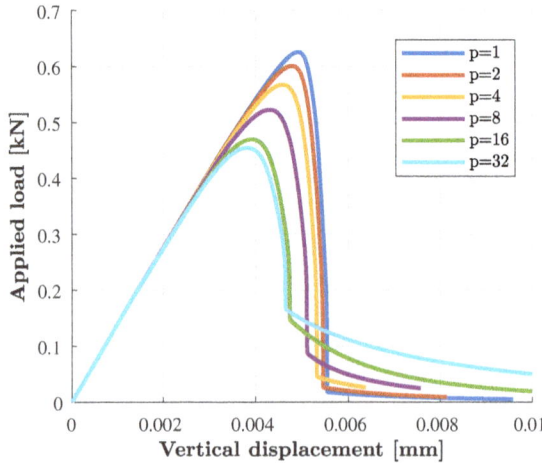

Figure 16. Tension test with cohesive phase field formulation studying effect of shape parameter p on the peak fracture loads

6.3. Numerical Example 2: Single Edge-Notched Shear Test

In the present example, the mode-II fracture behavior of a square panel is examined, with geometric description of the domain and boundary conditions as shown in Figure 17. This is a standard benchmark test to evaluate damage characteristics under shear loads, and has been analyzed extensively in the literature, see for e.g., [229,232]. The specimen thickness is $t = 1$ mm and a state of plane-strain is assumed. The material parameters are chosen as $E = 210$ kN/mm^2, $v = 0.30$, $l_0 = 0.0075$ mm, $\mathcal{G}_c = 0.0027$ kN/mm with crack propagation increment $\Delta a = 0.02$ mm, in accordance with [232]. For the phase-field analysis, the mesh is refined with $h_{PFM} = 0.005$ mm in the regions where the crack is expected to propagate. Zero y-displacement boundary conditions are enforced ($u_y = 0$, Figure 17) on all outer edges of the plate. Furthermore, the bottom edge of the specimen is retrained in the horizontal direction ($u_x = 0$). For the discrete crack methods, a horizontal displacement $u = u_x \geq 0$ is imposed on the top edge of the specimen, while the PFM applies a concentrated load P at point C, kinematically couples the horizontal DOF on the top edge and solves enforcing quasi static displacement control. The second order quadratic phase field formulation described in Section 5.2.1 is employed in this example. The analysis procedures for each approach are as described in Section 6.1. The corresponding load-deflection paths are shown in Figure 18.

The shear test results in a biaxial stress state developed at the notch-tip which leads to an inclined crack propagation at an angle 45° to the horizontal.

The crack paths are closely aligned (Figure 19), however, the origin of the discontinuity differs slightly between the PFM and the discrete crack methods, resulting in a slight differentiation of the crack paths upon crack propagation. Such behaviour is a consequence of the discrete crack methods mandating the crack propagate starting from the proceeding crack tip, whereas the PFM permits the evolution along the notch. Various stages of the phase field evolution are shown in Figure 20.

Further discrepancy is also observed in the significantly differentiated behaviour of the associated load-deflection curve. The higher peak load obtained by discrete crack methods and the snap back behaviour is not mirrored in the PFM result. The difference in snap back behaviour between the SBFEM and XFEM is attributed to the adaptivity of the SBFEM mesh about the crack tip, while the XFEM relies on the initial mesh topology. After this oscillatory step, the respective load deflection curves coincide closely.

Contrary to the discrete methods where the equilibrium path is derived from sequential linear solutions, PFM relies on incremental iterative solvers; hence the snap back response would not be captured with a displacement control nonlinear analysis procedure; rather, a generalized, e.g., arc-length, analysis is required. Eventhough the PFM results shown in Figure 19 are identical to the results provided in the literature (see, e.g., [232,265]), the 8% difference in the peak load compared to discrete methods highlights the importance of the length scale parameter on the solution. The effect of the length scale l_0 on the crack topology and the peak fracture loads is shown in Figures 21 and 22, respectively. It can be noted that the shear crack paths and load-displacement curves show a similar trend as already seen in Section 6.2, wherein decreasing l_0 leads to sharper and more brittle cracks with higher peak fracture forces which converge to the discrete fracture solution.

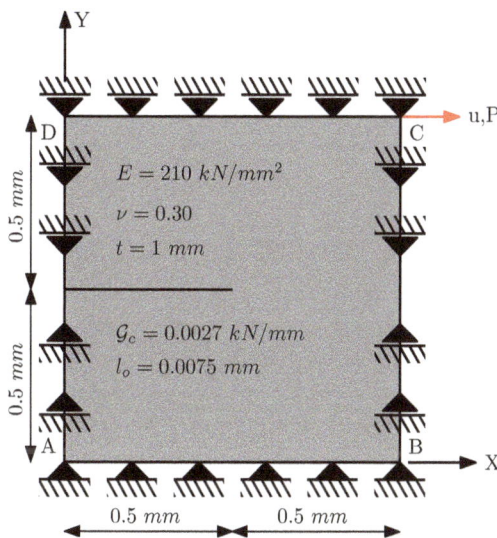

Figure 17. Shear test geometry, material parameters, loading and boundary conditions.

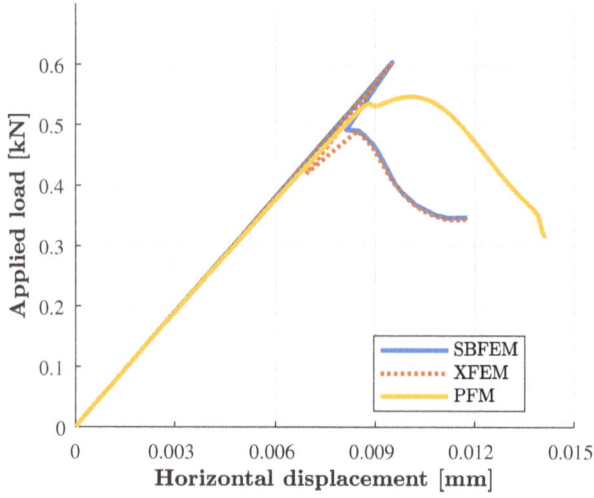

Figure 18. Load-deflection curves of the shear test.

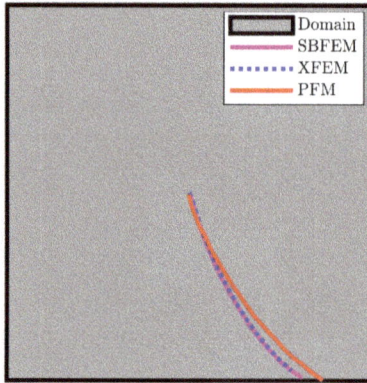

Figure 19. Shear test crack-paths obtained from SBFEM, XFEM and PFM-based crack propagation analysis.

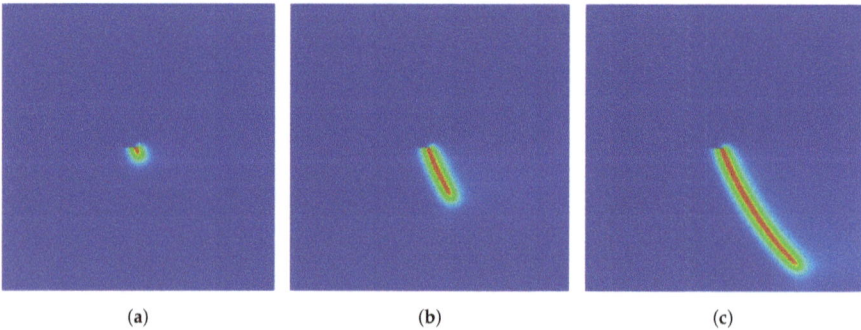

(a) (b) (c)

Figure 20. Shear test phase field evolution at (**a**) $u = 0.009$ mm (**b**) $u = 0.011$ mm (**c**) $u = 0.013$ mm, with displacement increment $\Delta u = 1e^{-6}$ mm.

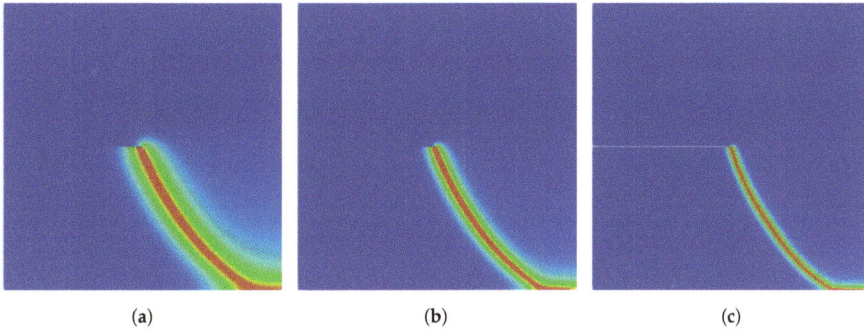

(a)   (b)   (c)

Figure 21. Shear test comparison of phase field diffusion widths with respect to decreasing l_o, where (a) $l_o = 0.015$ mm (b) $l_o = 0.0075$ mm (c) $l_o = 0.00375$ mm.

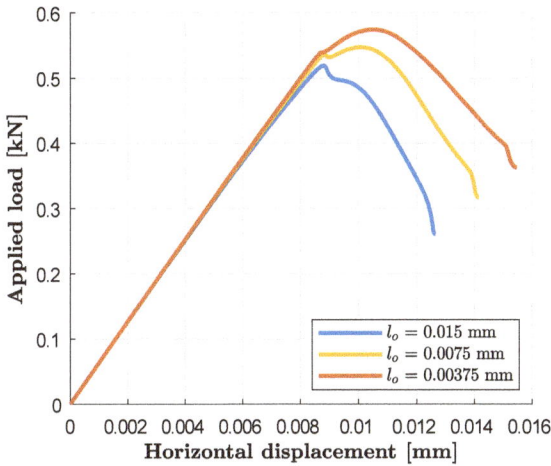

Figure 22. Shear test effect of length-scale parameter l_o variation on load displacement curves

6.4. Numerical Example 3: Notched Plate with Hole (NPwH)

A notched plate containing a hole is considered with geometric description of the domain, boundary conditions and material parameters as defined in Figure 23. In [32,243], a similar example has been analyzed previously. The specimen thickness is $t = 15$ mm and a state of plane-stress is treated. The Young's modulus, Poisson's ratio, length scale, fracture energy density and crack propagation length are chosen as $E = 5.98$ kN/mm^2, $v = 0.221$, $l_0 = 0.35$ mm, $\mathcal{G}_c = 0.00228$ kN/mm and $\Delta a = 2$ mm, where applicable. For the PFM, the mesh-size is kept at a value of $h_{PFM} \approx 0.34$ mm in the crack propagation region. A zero displacement boundary condition ($u_x = 0; u_y = 0$) is enforced on the bottom pin, whereas a vertical displacement $u = u_y \geq 0$ is imposed on the top pin. The numerically predicted crack path is compared with the experimental results presented in [265].

Comparing PFM to discrete crack methods, the obtained peak load is similar (Figure 24), however, the crack paths differ significantly (Figures 25 and 26). Since the discrete methods do not possess an intrinsic method to nucleate cracks, once the crack tip has propagated into the hole, the algorithm terminates. This is apparent, since both XFEM and SBFEM report a final vertical displacement of approximately 0.33 mm. Due to this inherent limitation, expert judgment is required to interpret crack propagation results stemming from discrete crack methods as their termination is indistinguishable from crack arrest, when inspecting conventional results. The phase field methods circumvent these issues resulting in a highly flexible and generalized method, at the cost of significantly increased computational effort.

For the PF-FEM case, the effect of the number of staggered phase field iterations on the accuracy of the predicted peak fracture loads is examined. Four different cases with constant displacement increments $\Delta u = 10^{-2}$, $\Delta u = 5 \times 10^{-3}$ mm, $\Delta u = 10^{-3}$ mm and $\Delta u = 5 \times 10^{-4}$ mm are considered and the corresponding load-deflection paths are shown in Figure 27a. In all cases, the phase-field solution is predicted using a single staggered iteration step $N_{staggs} = 1$ and a tolerance of $tol_u = 10^{-5}$ is maintained. It can be seen that solution accuracy improves when the size of displacement increments Δu is sufficiently small, and convergence is achieved for $\Delta u = 1 \times 10^{-3}$ mm. Further reduction of Δu marginally affects the results at the cost of increased number of calculations, with $\Delta u = 5 \times 10^{-4}$ mm and $\Delta u = 1 \times 10^{-3}$ mm yielding almost similar load-displacement curves.

In Figure 27b the converged solution of Figure 27 is compared against the solution with $\Delta u = 5 \times 10^{-3}$ mm when (i) only a single staggered iteration is performed and (ii) staggered iterations are performed until the phase field solution converges. It is evident that the peak fracture loads obtained in converged staggered iteration case is lower as compared to the $N_{staggs} = 1$ case, and are actually closer to the converged solution shown in Figure 27a. The evolution of phase field at successive monitored displacements is shown in Figure 28; results are obtained using single staggered iteration $N_{staggs} = 1$ and a constant displacement increment $\Delta u = 10^{-3}$ mm. The crack paths obtained from phase field calculations (Figure 29) show good agreement with the experimental fracture results presented from [265].

Furthermore, the analysis has been conducted using two different anisotropic strain energy splits widely used within the phase field literature (Figure 30):

1. Spectral decomposition of strains proposed in [232]
2. Volumetric Deviatoric strain split proposed in [264]

The crack path predicted via the spectral strain decomposition [232] appears closer to the experimentally observed crack than the volumetric-deviatoric strain split [264] (Figure 30a,b). These minor differences are also reflected to the equilibrium paths shown in Figure 30c. Since, contrary to the spectral strain decomposition split, the volumetric-deviatoric split only partially prohibits degradation due to purely compressive stresses, a higher amount of material is overall degraded in the latter case; hence the peak force is indeed expected to be lower. However, the spectral strain decomposition leads to a highly nonlinear formulation and therefore increased computational costs—see also [265] for a hybrid procedure to alleviate these. This highlights the significance of choosing the appropriate split and hence the level of expert judgment required when employing PFM for LEFM.

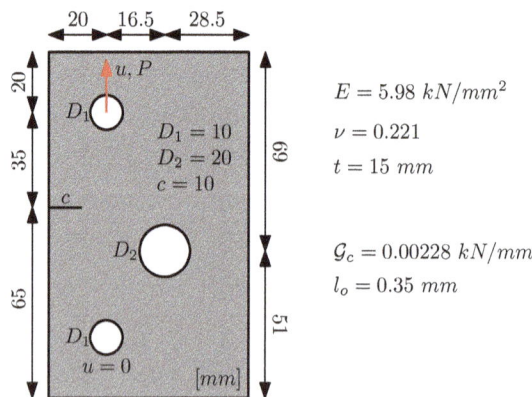

Figure 23. NPwH geometry, material parameters, loading and boundary conditions.

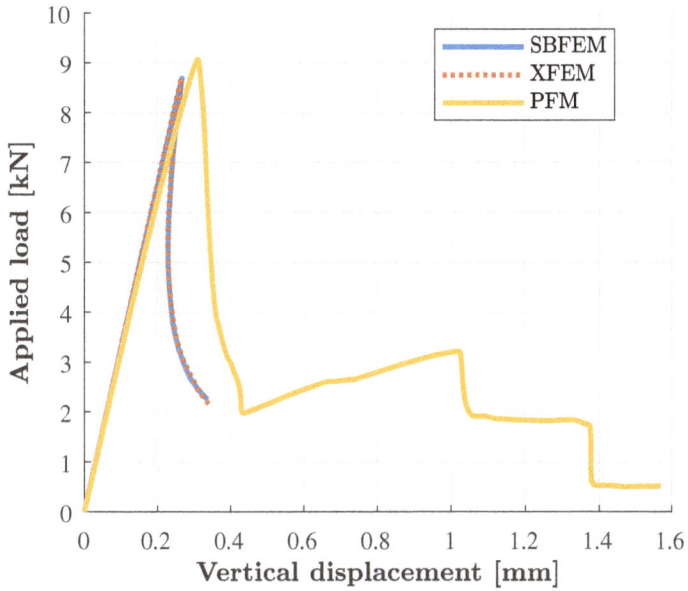

Figure 24. NPwH load-deflection curves.

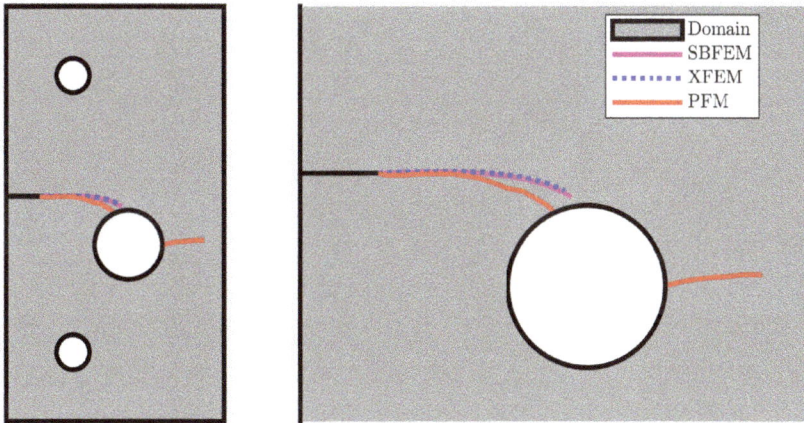

Figure 25. NPwH crack-paths obtained from SBFEM, XFEM and PFM-based crack propagation analysis.

Figure 26. NPwH meshes for SBFEM (**top**) and XFEM (**bottom**), with focus on crack path region. The last crack propagation step prior to the cracks reaching the hole is depicted.

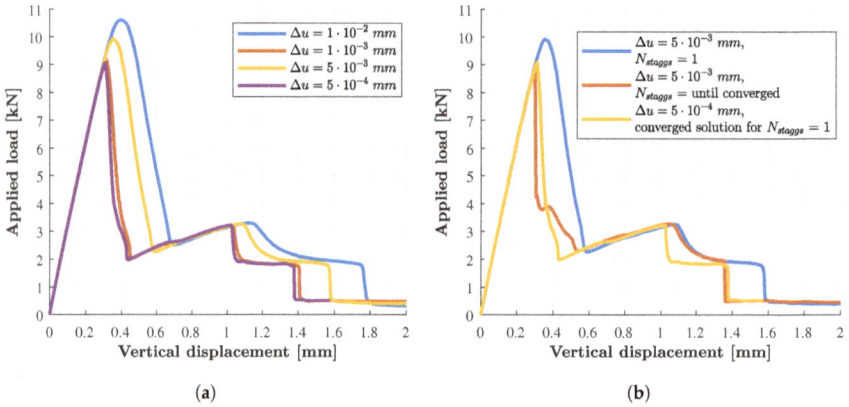

(**a**) (**b**)

Figure 27. NPwH PFM force displacement response illustrating the dependence of peak fracture force on (**a**) Δu for Nstaggs = 1 (**b**) the number of staggered iterations.

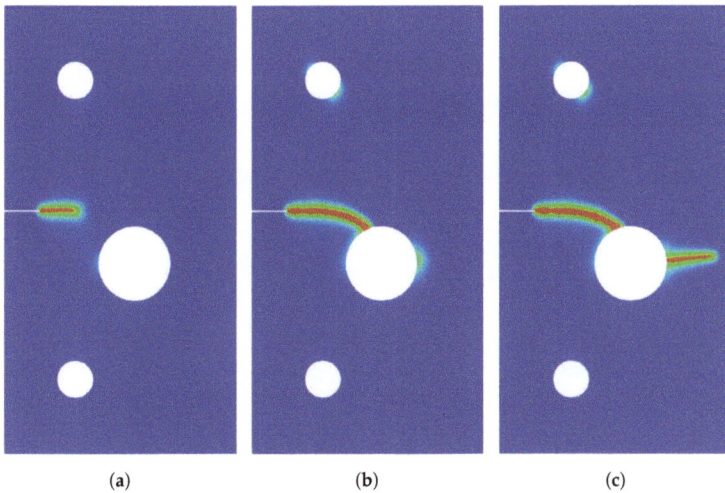

Figure 28. NPwH phase field at monitored displacement (**a**) $u = 0.35$ mm (**b**) $u = 0.96$ mm (**c**) $u = 1.20$ mm, with a displacement increment $\Delta u = 1e^{-3}$ mm and a single stagger iteration.

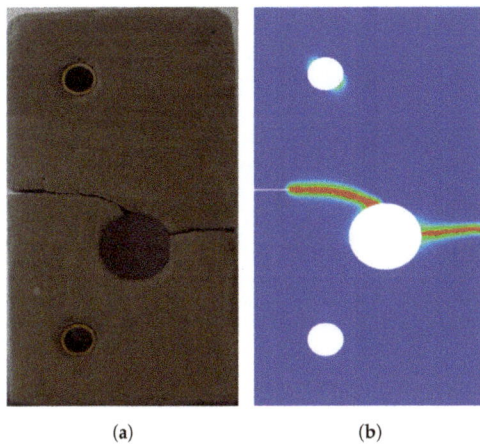

Figure 29. NPwH comparison of crack topologies depicting experiments from (**a**) [243] vs. (**b**) phase field simulations.

6.5. Numerical Example 4: L-Shaped Panel (LSP) Test with Crack at Re-Entrant Corner

Figure 31b depicts the geometric description of the domain, boundary conditions and material parameters for an L-shaped panel. A state of plane stress is considered with specimen thickness $t = 100$ mm. The Young's modulus, Poisson's ratio, length scale, fracture energy density and crack propagation length are chosen as $E = 5.98$ kN/mm^2, $v = 0.2$, $l_0 = 2.5$ mm, $\mathcal{G}_c = 0.0089$ kN/mm, $h_{PFM} \approx 1.4$ mm and $\Delta a = 10$ mm, where applicable. A zero displacement boundary condition ($u_x = 0; u_y = 0$) is enforced on the bottom side, while a cyclic imposed displacement envelope is considered at a distance $d_l = 30$ mm from the rightmost edge of the panel with a constant displacement increment $\Delta u = 10^{-3}$ mm and the load history as shown in Figure 31a. The analysis procedures described in Section 6.1 for each method apply. Through this application, we simulate the experimental program undertaken in [278] which has also been investigated in previous publications pertinent to computational fracture mechanics [265]. Since the discrete crack methods do not intrinsically posses the capability to avoid crack over-closure and interpenetration, without introducing contact, the numerical simulations employing XFEM and SBFEM follow a modified loading path (Figure 31 left) starting from time step 1000.

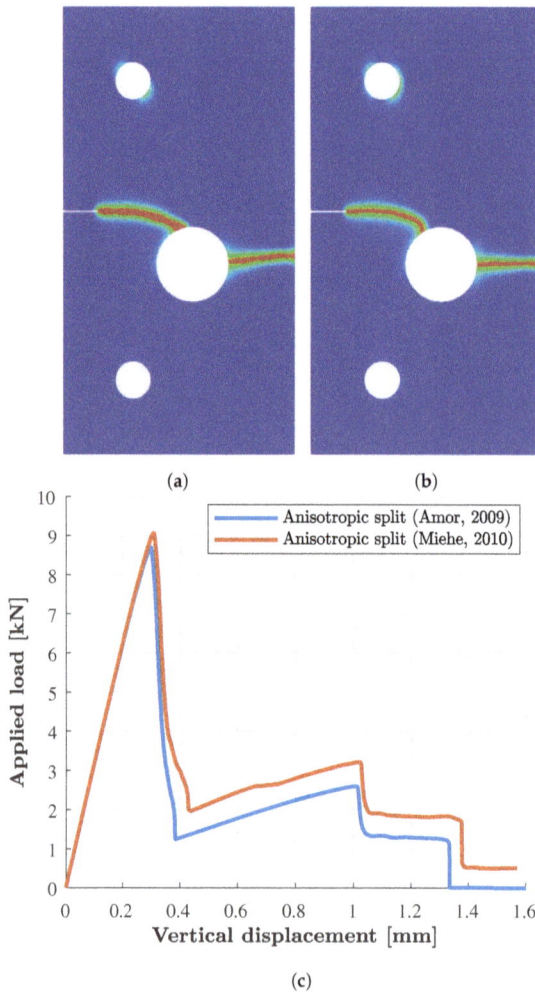

(a)   (b)

(c)

Figure 30. NPwH comparison between anisotropic phase field models with strain energy splits proposed in [232,264]. (**a**) Crack path from analysis implementing the anisotropic split proposed in [232]; (**b**) Crack path from analysis implementing the anisotropic split proposed [264]; (**c**) Force-displacement response comparison between the anisotropic phase field models.

The load-displacement curve and the peak fracture force (Figure 32) are in accordance with existing literature [265] and a good agreement is observed between all methods. The crack paths obtained from all methods remain within the envelope of the experimental results (Figures 33 and 34). Furthermore, the crack path obtained in Figure 35 coincides with the experimentally observed crack in [278]. For the case of SBFEM, the crack tip does not coincide with the re-entrant corner, since the implementation requires the crack tip to reside within the domain and not on the boundary. Hence, the crack tip was perturbed by a small value and thus the peak load is slightly overestimated.

Furthermore, a comparison is drawn between the load-displacement curves obtained using the spectral strain decomposition [232] and the constrained hybrid phase field model proposed in [265].

The resulting load-deflection curves are shown in Figure 36. It is of interest to note that the anisotropic spectral split [232] naturally avoids crack face overlapping during crack closure when cyclic loads are considered. On the other hand, the hybrid phase-field model in [265] requires an additional constraint to prohibit interpenetration of crack faces during compression phase.

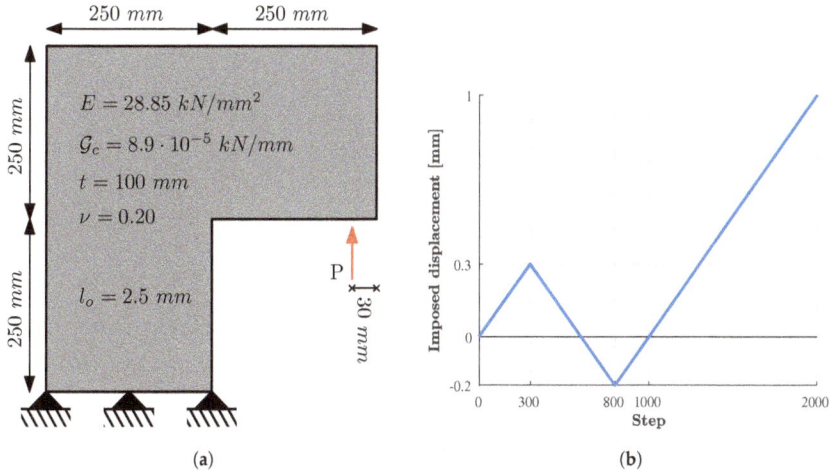

(a) (b)

Figure 31. (a) LSP geometry, material parameters, loading and boundary conditions (b) Cyclic envelope of monitored displacement.

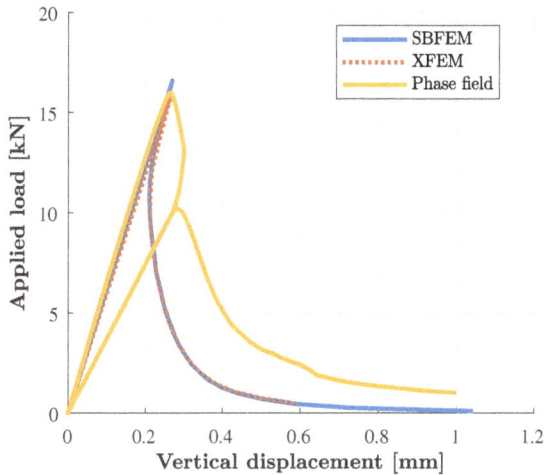

Figure 32. LSP load-deflection curves.

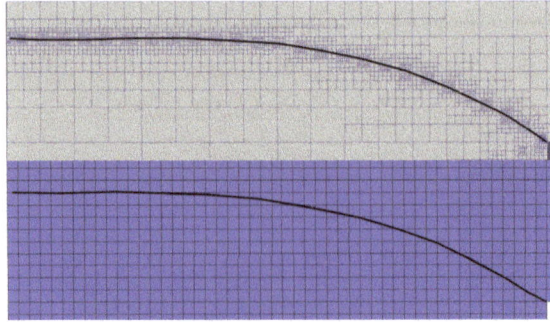

Figure 33. LSP meshes for SBFEM (**top**) and XFEM (**bottom**), with focus on crack path region.

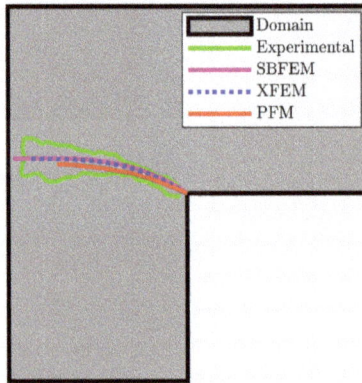

Figure 34. LSP crack paths for SBFEM, XFEM and PFM.

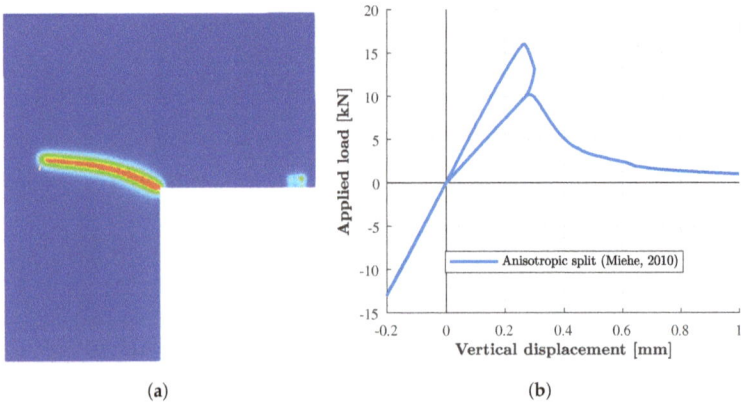

(**a**)   (**b**)

Figure 35. (**a**) LSP phase field (**b**) load-deflection response under the cyclic loading defined in Figure 31**b**.

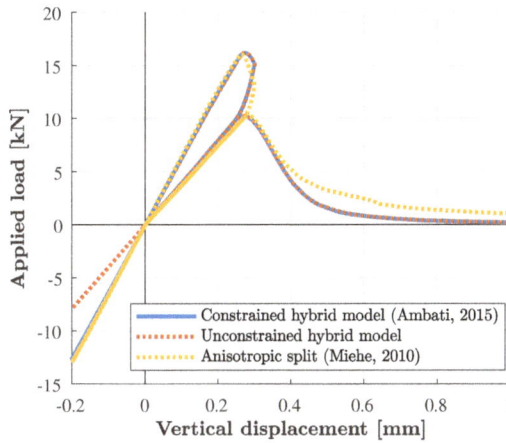

Figure 36. LSP comparison of load-displacement curves implementing the anisotropic spectral split vs. hybrid phase field models.

6.6. Numerical Example 5: Plate with Two Holes and Edge Cracks (PwHC)

The case of the plate shown in Figure 37 is considered here. This numerical example is studied, since it poses challenges for both diffuse and discrete methods as discussed in [277]. The boundary conditions and material parameters are also shown in Figure 37 according to [9].

A state of plane strain is considered. The Young's modulus, Poisson's ratio, length scale, fracture energy density and crack propagation length are $E = 210$ kN/mm^2, $v = 0.3$, $l_0 = 0.1$ mm, $\mathcal{G}_c = 1.0$ N/mm, $h_{PFM} \approx 0.06$ mm, and $\Delta a \leq 1$ mm, where applicable. Furthermore, for the phase-field analysis, a volumetric-deviatoric strain decomposition (similar to Amor et al. [264]) is employed. The bottom edge of the plate is clamped, while on the top edge a prescribed displacement is applied in the vertical direction and displacements in the horizontal direction are prohibited ($u_x = 0; u_y > 0$). The specimen thickness is $t = 1$ mm.

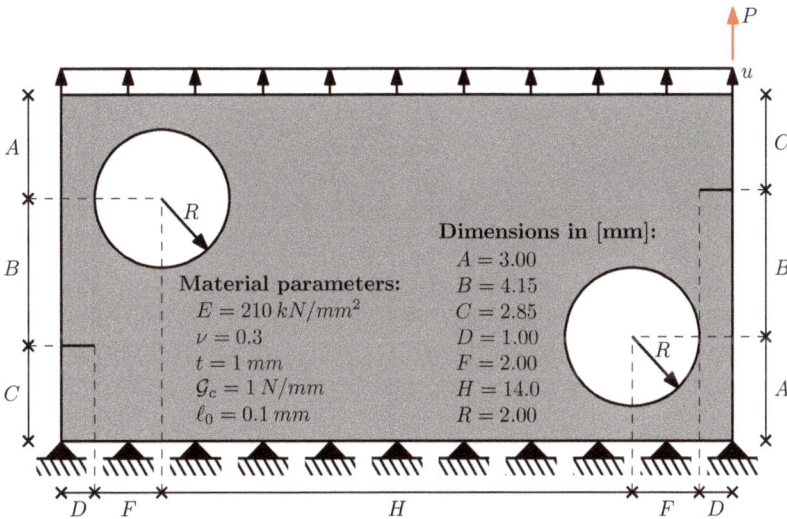

Figure 37. PwHC geometry, material parameters, loading and boundary conditions.

In the presence of multiple cracks inside a domain, methods employing discrete crack representations typically implement a stability analysis [163] to ascertain the propagating cracks at each step. However, in this specific case, this involved procedure can be circumvented, due to the symmetric test setup. Nevertheless, the naive approach of simply running the analysis will result in an undesirable outcome, since slight numerical imbalances can result in asymmetric and erroneous results. To counteract these effects, symmetric meshes are employed in the XFEM analysis, while the SBFEM analysis enforces symmetric gSIFs about the diagonal. An average of the gSIFs is calculated to determine the crack propagation angle.

Solving this example using the phase-field method produces interesting characteristics with respect to the crack initiation location and crack-paths. It is observed that when there is no restriction imposed on the crack from initiating near the holes, the phase-field initiates simultaneously and symmetrically at the top and bottom hole edges and then propagates almost horizontally as if no notches were present in the structure (Figure 38). However, when the crack evolution is restricted near the hole boundary, e.g., by imposing a very high G_c in the surrounding region, the crack initiates at both notch tips and propagates towards the hole edges simultaneously (Figure 39a). Further loading leads to evolution of multiple cracks initiating at the edges of holes which ultimately merge in the centre of the structure (Figure 39d). This observation is similar to what has been previously reported in [277]. However in the absence of experimental results for this problem, it is currently difficult to deduce which method predicts a realistic crack pattern. Hence, we refrain from reporting the typical load-deflection curves and focus only on the crack paths.

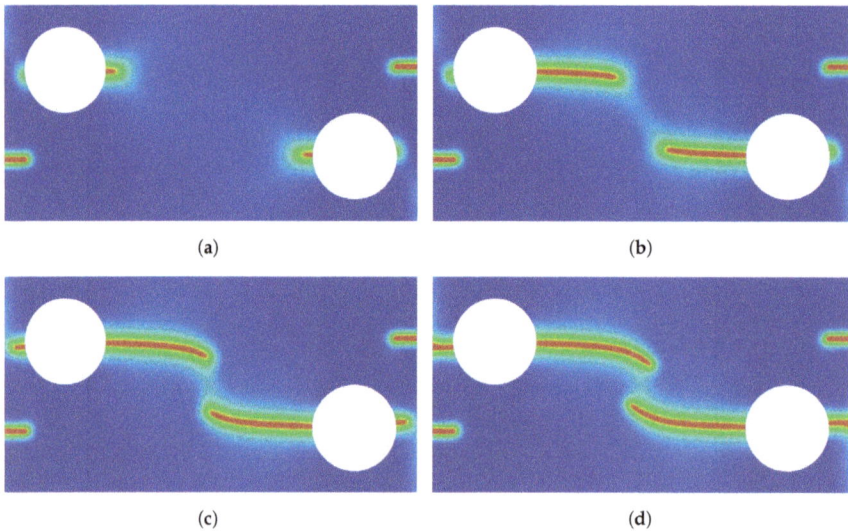

Figure 38. PFM crack path without restricting nucleation at the holes. (**a**) Cracks initiating at the holes; (**b**) Growth of cracks originating from the holes; (**c**) Additional cracks nucleate at the holes; (**d**) Nucleated cracks reach the domain boundary.

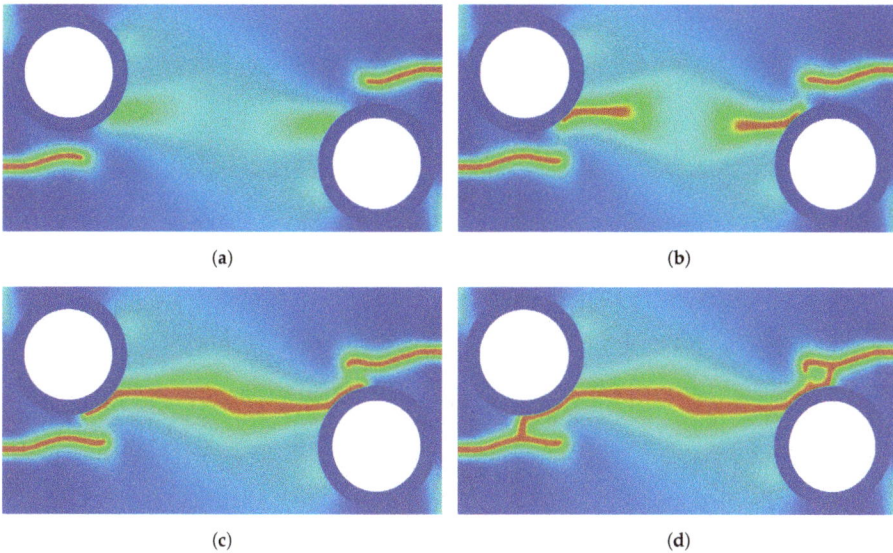

(a)   (b)

(c)   (d)

Figure 39. PFM crack path when restricting the nucleation at the holes. (**a**) Crack growth at the notches; (**b**) Crack nucleation and growth at the holes; (**c**) Joining of nucleated cracks at the holes; (**d**) Merging of notch and hole cracks.

Since the crack paths derived from XFEM/SBFEM have been shown to coincide very well when employing similar discretization levels and crack propagation increments, modified mesh discretizations and crack propagation increments are sampled (Figure 40). The crack paths for all three variants align very well for the initial portion, while separating slightly as they approach the holes due to the crack propagation increment and mesh density variations.

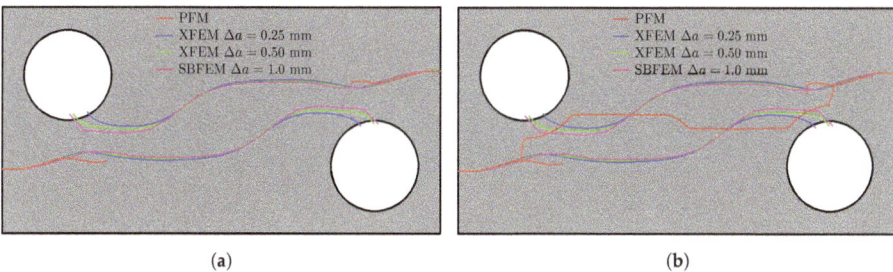

(a)   (b)

Figure 40. Crack path overlay for three variants: XFEM employing a fine mesh with $\Delta a = 0.25$ mm (pink), a coarse mesh with $\Delta a = 0.50$ mm (green) and SBFEM employing an adaptive mesh with $\Delta a = 1.00$ mm (orange). (**a**) Prior to crack nucleation at the holes; (**b**) After crack merging.

7. Discussion and Conclusions

This section initiates by detailing the steps involved in a crack propagation analysis, attempted by each of the described methods. Emphasis is placed on identifying sources of computational effort, while illustrative flowcharts are provided for each method. This visual representation of the methods then serves as a basis for the discussion on the merits and drawbacks of each individual method within the context of LEFM.

7.1. Crack Propagation by XFEM/GFEM

A conceptual representation of the steps involved in a typical crack propagation analysis with the XFEM/GFEM is provided in Figure 41. As should be obvious based on Section 3, enriched finite element methods are essentially discretisation schemes and, as such, require coupling to appropriate criteria in order to model crack propagation. In the present case these are provided by the LEFM framework. The flowchart of Figure 41 involves elastic solution steps followed by the evaluation of a crack propagation criterion. This is common for most LEFM schemes relying on discretizstion, such as for instance FEM or SBFEM. The coupling to further schemes for crack propagation, such as the cohesive zone model, is also possible, in which case the steps of Figure 41 would have to be modified.

The enriched finite element schemes contained within the XFEM/GFEM family of numerical methods permit the treatment of discontinuities and singularities independently of the mesh, while preserving the convergence rates of the underlying FE method. Hence, conventional meshers are employed, yet enriched node and element sets need to be specified and their contributions to the equilibrium equations need to be assembled. This, apart from introducing additional DOFs associated with the enrichment functions (Equation (19)) and potential conditioning problems, requires the use of more involved numerical integration schemes leading to an increased computational toll. Nevertheless, these operations are only performed on a small part of the domain, thus minimizing this additional cost. As mentioned in Section 3, several techniques are available that allow performing the required tasks in a robust and automated manner.

For the calculation of the SIFs, elements within the interaction integral domain are identified and their contributions are assembled. A suitable crack propagation criterion is applied in order to evaluate the propagation direction, and together with a user-specified crack propagation increment Δa determine the new crack tip location. Since implicit crack representation has become an almost integral part of enriched finite element methods, the next step would involve the update of this representation. This task might introduce additional challenges, however, significant work has been carried out in this direction, with several methods available for tackling this issue in a simplified manner.

```
                  ┌─ input: mesh, boundary conditions
        pre-      │  initial crack
     processing   ├─→ for each node
                  │   └─  initialize level sets
                  └─→ determine propagation increment
```

```
                  ┌─ for each crack propagation step
                  │  ┌─ for each element
                  │  │  ┌─ if cut by crack
                  │  │  │     flag element as cut
                  │  │  └─    flag nodes as jump enriched
                  │  │  └─→ if contains tip or nodes within r_e
                  │  │        flag nodes as tip enriched
                  │  │     └─ flag element as tip enriched
                  │  ├─→ for each element
                  │  │  ┌─ if enriched
                  │  │  │     obtain integration points
                  │  │  │     compute enriched stiffness matrix
                  │  │  │     compute enriched load vector
                  │  │  └─    assemble
    crack         │  │  └─→ else
  propagation     │  │        compute standard stiffness matrix
                  │  │        compute standard load vector
                  │  │     └─ assemble
                  │  ├─→ solve for displacements
                  │  ├─→ for each element
                  │  │  ┌─ if in interaction integral domain
                  │  │  │     compute interaction integral contribution
                  │  │  └─ assemble
                  │  ├─→ compute SIFs
                  │  ├─→ evaluate criterion for propagation direction
                  │  ├─→ advance crack tip in the computed direction
                  │  └─→ for each node
                  │      └─  update level sets
                  └─→ postprocessing
```

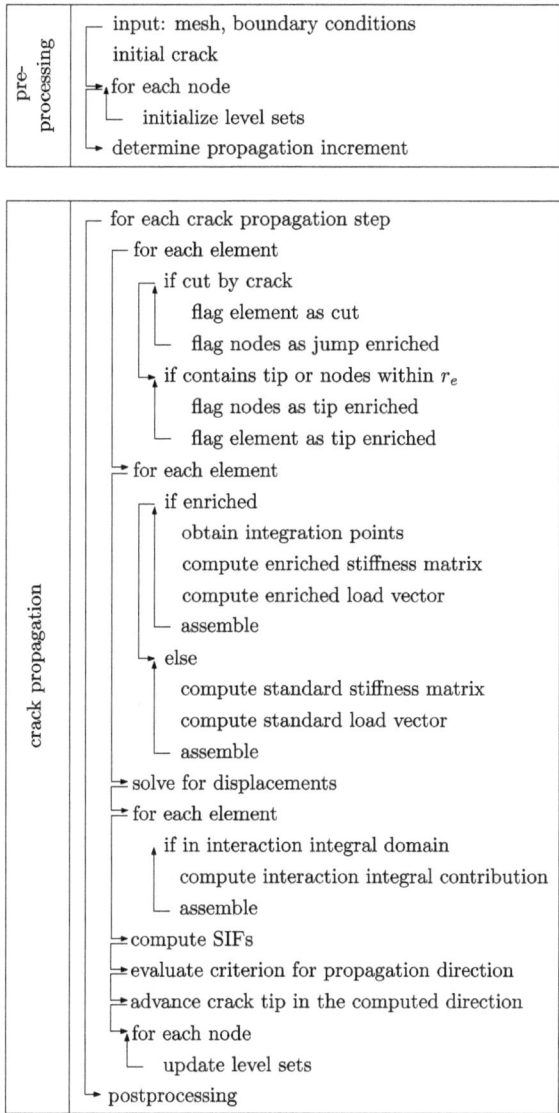

Figure 41. Steps comprising an XFEM/GFEM crack propagation analysis.

7.2. Crack Propagation by SBFEM

The crack propagation process by SBFEM, enhanced via hybrid balanced quadtree polygon meshes, requires the polygon representation of domain features as input, including the crack. The points comprising the polygons constitute the subdivision criterion for the quadtree decomposition. If more than a user-specified number of points fall within a quadrant, this is subdivided. Together with the balancing operation, these steps entail minimal computational effort. The explicit neighbours of each cell do not need to be calculated, but simply the size of its neighbour. This is efficiently achieved by querying the center of each element, offsetting them by the element size in all four cardinal directions, passing them through the tree structure, and finally returning the size of the final cell. Assuming a balanced mesh, all possible element realizations are precomputable. When the domain features, such as the boundary and strong & weak discontinuities do not align with the Cartesian axes, polygon clipping algorithms are required. Although efficient algorithms exist for polygon clipping, the resulting polygonal elements are no longer precomputable and must therefore be calculated individually. In order to construct the stiffness matrix of an SBFEM element, a Hamiltonian eigen-problem must be solved. This entails a real Schur decomposition, sorting of the eigenvalue blocks and subsequent block-diagonalization, as well as the inversion of the matrix $[E^0]$ and the evaluation of a matrix exponential, if quantities of interest inside the SBFEM element need be determined. For smaller elements, commonly employed on quadtree meshes, this additional step when compared to the standard FEM procedure, does not generate a significant computational overhead. Specifically, Ooi et al. [51] report a reduction of computational effort close to 50% on typical analysis domains, when employing precomputable alongside clipped elements. When larger domains are investigated by using a single SBFEM element for the whole domain and hp-refinement is employed, determining the stiffness matrix dominates the computational effort of the analysis. Unfortunately, the stiffness matrix is fully populated, yet symmetric. Hence, this type of analysis is best suited for problems with small boundary to domain ratios. Determining the gSIFs entails post-processing calculations localized to the element containing the crack tip. The singular modes are identified according to Equation (44) and the gSIFs are calculated by evaluating the components of the stress tensor $\sigma^{(s)}$ in crack extension direction (Figure 2b). The crack propagation angle is selected based on a suitable criterion (Equation (13)), while the crack propagation increment Δa is user specified. After definition of the updated crack tip location, the crack path polyline is updated accordingly and provided as input to the meshing phase of the next iteration. The steps to a standard SBFEM analysis are summarized in Figure 42.

7.3. Crack Propagation by PFM

In PFM fracture is not introduced as an explicit or implicit discontinuity in the displacement field. Rather, it is associated with the evolution of a continuous field, i.e., the phase field. The governing equations of the crack propagation problem emerge through the minimization of the total potential energy established in Equation (70), see, e.g., [69]. This gives rise to the coupled system of equilibrium and phase field governing equations established in Equations (71) and (72). The crack is not explicitly represented but derived from the solution of the coupled system as the region where $c = 0$ (typically values of $c < 10e - 3$). Within the setting of an incremental solution procedure, the phase field is updated at each time step and with it the crack topology. Nucleation, merging, branching and arrest of cracks as well as the associated crack propagation increment is a natural byproduct of the phase field evolution. The mechanical/phase field coupling is enforced by introducing a material degradation function that is dependent on the phase field. The evolution of fracture follows through the solution of this coupled strong form. Existing discontinuities may be introduced into the domain by providing initial values to the phase field. Mesh density is contingent on sufficient resolution of the fracture process zone, mandating a highly refined mesh in its vicinity. The combination of length scale and level of mesh refinement interact and affect the estimation of the fracture energy hence necessitating the scaling of the critical energy release rate. The numerical solution of the PFM-coupled governing equations is performed using either monolithic or staggered solvers. Monolithic solvers are typically based on the Newton-Raphson solution procedure and have been proven to provide accurate fracture paths. However, they have been shown to suffer from poor convergence due to the non-convex nature of the underlying energy functional [277]. Yet, the accuracy provided by monolithic solvers renders them a favourable solution, especially in the case of dynamic fracture problems and several attempts have been suggested in the literature to improve the robustness of monolithic procedures (see, e.g., [279–281]). In staggered methods the displacement and phase field equations are decoupled and solved separately within each load increment. In principle, a staggered algorithm for coupled field problems is based on freezing one field variable at a constant value, solving for the other until convergence is achieved. The staggered approach (also known as alternate minimization approach) provides better convergence rates than the monolithic due to the convexity of the energy functional

(Equation (70)) with respect to the two unknown fields $\{\mathbf{u}, \phi\}$ separately. However, its accuracy is dependent on the incremental step unless stagger iterations are performed; these however increase the computational burden of the analysis. Very recent developments aim towards providing more robust staggered solvers, (see, e.g., [282]). The steps to a typical PFM solution procedure with a staggered solution scheme are summarized in Figure 43.

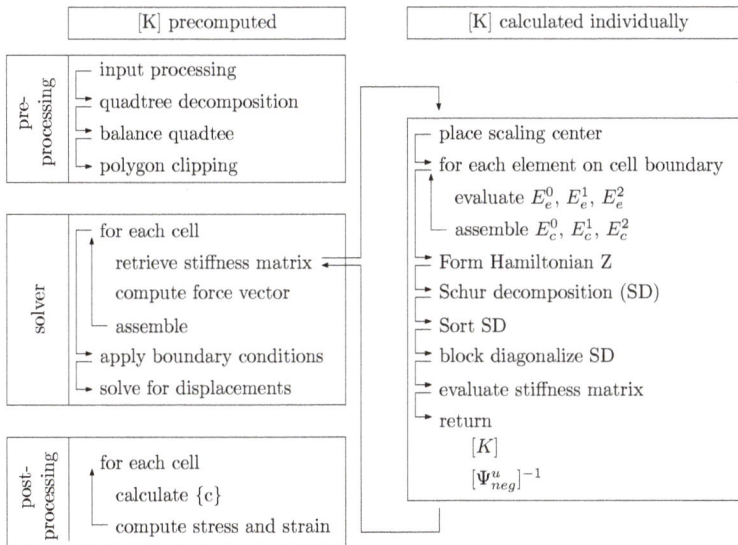

Figure 42. Steps comprising SBFEM analysis.

7.4. Contrasting Discrete and PFM Crack Representation Approaches

The merits of each method within the LEFM setting are discussed by contrasting key features and analysis steps.

For the discrete methods, the representation of the crack is typically available in explicit form. Crack propagation analysis yields a polyline description of the crack topology. Since SBFEM employs polygon clipping, it does not require further information. XFEM, if chosen to employ an implicit enrichment representation, models the crack additionally by associated level sets. Crack path extraction is not necessary, since it is already given as a polyline. A crack consisting of a one-segment polyline is usually provided as input. For the PFM, the crack is represented by a scalar phase-field, with the phase-field variable directly embedded into the constitutive equations. The crack is represented as the region of fully degraded material with $c = 0$. Hence, no explicit crack representation is required during the analysis, albeit readily available in post-processing, if required. Initial defects are introduced in the system by specifying sets of points with corresponding phase field values.

Meshing requirements for analysis by XFEM are largely decoupled due to the level set representation, yet substituted by more involved numerical integration procedures. This permits the use of a constant mesh during crack propagation analysis. This is contrary to analysis by SBFEM, where the initial quadtree decomposition, i.e., the mesh, is updated during each step incrementally. Discontinuities introduced by polygon clipping result in double nodes, such that the nDOF of the system increase gradually as the analysis proceeds. Furthermore, in select cases, clipping can result in non star-convex elements, which the method cannot treat. Delaunay triangulation of the element is required in such instances. Furthermore, due to clipping, elements with poor aspect ratios, in the conventional FEM sense, may arise. Empirically, this does not seem to be as severe an issue manifesting itself in erroneous numerical integration results, when employing SBFEM. In order to adequately represent the fracture process zone, the PFM requires a highly refined mesh in the regions of expected crack propagation as well as at the crack tip, rendering the phase-field method computationally expensive for solving large-scale problems, especially when compared to discrete fracture approaches. It is now accepted that an element size $h \leq l_0$ is required to accurately resolve the crack path. However, this computational burden is effectively addressed using parallel solvers, adaptive mesh refinement [229,283], multiscale computation techniques [284] also within a local/global solution context [285].

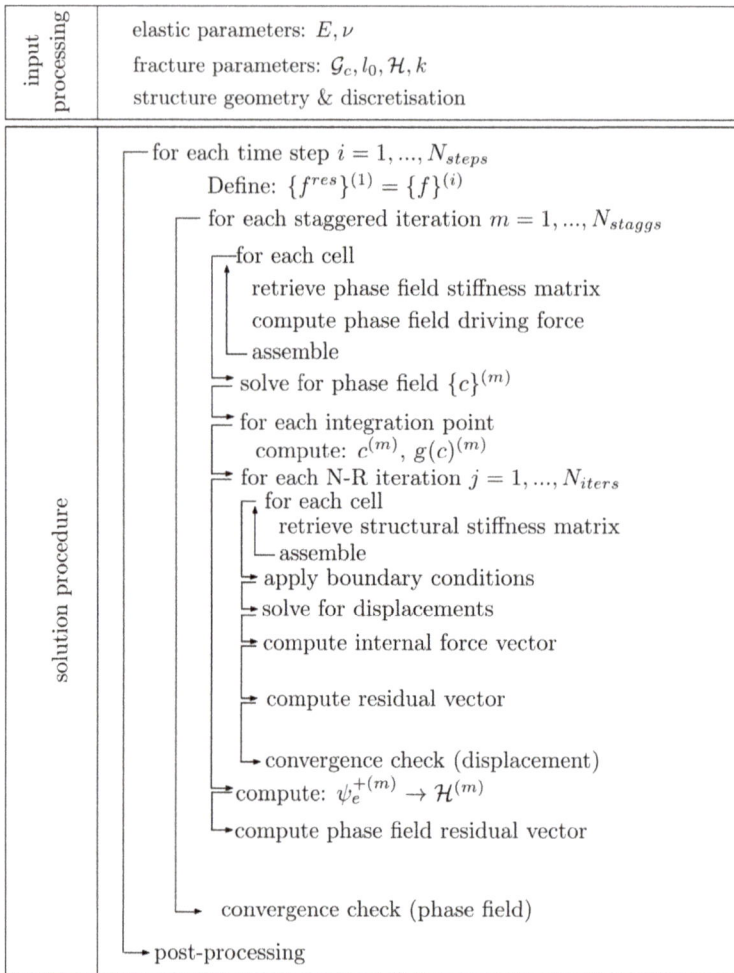

input processing

elastic parameters: E, ν

fracture parameters: $\mathcal{G}_c, l_0, \mathcal{H}, k$

structure geometry & discretisation

solution procedure

─ for each time step $i = 1, ..., N_{steps}$

Define: $\{f^{res}\}^{(1)} = \{f\}^{(i)}$

─ for each staggered iteration $m = 1, ..., N_{staggs}$

─for each cell

retrieve phase field stiffness matrix

compute phase field driving force

└ assemble

⇢ solve for phase field $\{c\}^{(m)}$

⇢ for each integration point

compute: $c^{(m)}, g(c)^{(m)}$

⇢ for each N-R iteration $j = 1, ..., N_{iters}$

─ for each cell

retrieve structural stiffness matrix

└ assemble

⇢ apply boundary conditions

⇢ solve for displacements

⇢ compute internal force vector

⇢ compute residual vector

↳ convergence check (displacement)

⇢compute: $\psi_e^{+(m)} \rightarrow \mathcal{H}^{(m)}$

↳compute phase field residual vector

↳ convergence check (phase field)

↳ post-processing

Figure 43. Steps comprising Phase field analysis.

The methods further differ in the hyper-parameters that ought to be specified by the analyst. XFEM requires the specification of crack tip enrichment type and radius, as well as the region where the interaction integral is to be calculated. Special care must be taken to exclude blending elements from the calculation of the SIFs, which may affect final results. SBFEM similarly requires the analyst to specify the homogenization region about the crack tip. With the exception of the cohesive phase field model, in the PFM implementations discussed in Section 5.2 the specification of the length scale regulates the response, imposing guidelines on mesh discretization and scaling of the critical fracture energy. As further discussed in Section 5.2, early efforts to treat this regulatory effect by introducing stress-based crack-driving forces have been reported in [256] whereas, and most notably, Wu and Nguyen [251] provided length-scale insensitive formulations that also preserve Γ-convergence.

The solution process for both XFEM and SBFEM involves a single elastic solution step. The PFM, as previously described in Section 7.3, comprises either monolithic or staggered approaches within an iterative solution scheme. In the quasi-static regime, displacement or generalised control solution procedures are typically employed. This however necessitates that either Equations (71) and (72) must be solved with very small time-increments (typically 10^{-5}–10^{-6}), or stagger iterations must be performed between both equations to ensure energy convergence. Often both of these options lead to high computational cost.

Therefore, the corresponding load-deflection curve follows from the solution at every time step. In such quasi-static analyses, displacement controlled analysis automatically yields the load-deflection curve along with the softening branch. The discrete crack methods derive the load deflection curve in back-calculation. To this end, an arbitrary loading, e.g., force- or displacement-based, is applied. The resulting equivalent SIF is compared to the critical stress intensity factor. Hence, a scaling factor is derived for the loads and displacements at which crack propagation is initiated. This implies that recovery of the linear branch is a one-step process. Recovering an explicit linear elastic branch with the PFM requires either a linear phase field approximation as in Section 5.2.3 or cubic degradation functions [80]. Absence of these approaches will yield deviations from the linear elastic behaviour contingent on the evolution of material degradation in the process zone. Since the overall system stiffness is underestimated, the associated displacements are overestimated accordingly.

In the PFM a crack is never explicitly propagated, but associated with the evolution of the phase field that emerges from the solution of the phase field governing equation. This is driven by the definition of the crack driving force as discussed in Section 5.2. Depending on the PFM formulation employed, the crack driving force can be established on the basis of either energy or limit-stress criteria. The discrete crack methods, within the LEFM framework, require the calculation of the crack propagation angle and some crack propagation increment. Examples of the later are either user specified or provided by Paris' equation. The crack is assumed to propagate in a straight line, originating from the crack tip determined in the previous analysis step. Hence, the history variables required are none other than the polyline for SBFEM, while XFEM propagates the associated level sets as well. PFM require updating the scalar phase field and specific realization of the PFM require further history variables to impose the crack-irreversibility condition, preventing the crack from healing during cyclic loading.

The fact that the solution of the phase field governing equations emerge from an energy minimization problem, opposite to discrete fracture approaches, enables the resolution of crack initiation without the requirement for a crack path to be defined a priori. Furthermore, crack nucleation, growth and coalescence happen automatically; this results in a robust method with enormous flexibility to model complex cracking patterns including the simulation of curvillinear cracks, crack merging, and crack branching without the need for ad-hoc crack tracking methods. Finally, the method is naturally extended to 3D [229], considering also the case of fracture under multi-physics scenaria, e.g., temperature induced fracture [256,286], hydraulic fracturing [81,82,84,287,288], and diffusion [289]. These advantages, render the phase-field approach a robust crack prediction method. Compared to discrete fracture approaches, the variational structure upon which the phase field theory emerges, equips it with significant capabilities for modelling diverse and complex fracture problems in a unified and consistent manner.

The major advantage of the diffuse crack methods and the PFM specifically lies in their generality. Extending the discrete crack methods to exhibit similar capabilities involves significant algorithmic changes, as these codes are custom and not readily extendable to further types of analysis. Furthermore, extension to 3D problems is not straightforward, in addition, the definition of crack propagation increment in 3D is difficult to specify. Furthermore, judging if a crack arrests or the method simply does not permit continuation across obstacles, requires expert knowledge.

Author Contributions: The idea for this review emerged from discussions held between E.C. and S.P.T.; A.E. was responsible for the SBFEM source code implementation and simulations, the discussion on the merits and drawbacks of the examined methods, and the consistent presentation of the results; K.A. was responsible for the XFEM source code implementation and simulations; U.P., E.K., and S.P.T. were responsible for the phase field source code implementation and simulations. E.C., S.P.T., and I.A.A., advised on the simulation methods, the manuscript organization, and contributed in the discussion of the results. The first three authors contributed equally to paper writing.

Funding: This research was performed under the auspices of the Swiss National Science Foundation (SNSF), Grant # 200021_153379, A Multiscale Hysteretic XFEM Scheme for the Analysis of Composite Structures. Konstantinos Agathos would like to acknowledge funding received from the European Union's Horizon 2020 research and innovation programme under the Marie Sklodowska-Curie grant agreement No. 795917 "SiMAero, Simulation-Driven and On-line Condition Monitoring with Applications to Aerospace". Udit Pillai, Savvas Triantafyllou, and Ian Aschroft would like to acknowledge funding received from the European Union's Horizon 2020 research and innovation programme under the Marie Skłodowska-Curie SAFE-FLY project, grant agreement No. 721455.

Acknowledgments: The authors are grateful to the University of Nottingham for access to its High Performance Computing facility and extend their gratitude to the group of Song from UNSW and in particular Albert Saputra for his guidance on SBFEM implementation.

Conflicts of Interest: The authors declare no conflict of interest. The funders had no role in the design of the study; in the collection, analyses, or interpretation of data; in the writing of the manuscript, or in the decision to publish the results.

References

1. Zheng, J.; Liu, P. Elasto-plastic stress analysis and burst strength evaluation of Al-carbon fiber/epoxy composite cylindrical laminates. *Comput. Mater. Sci.* **2008**, *42*, 453–461. [CrossRef]
2. Xu, P.; Zheng, J.; Liu, P. Finite element analysis of burst pressure of composite hydrogen storage vessels. *Mater. Des.* **2009**, *30*, 2295–2301. [CrossRef]
3. Liu, P.; Zheng, J. Recent developments on damage modeling and finite element analysis for composite laminates: A review. *Mater. Des.* **2010**, *31*, 3825–3834. [CrossRef]
4. Ravi-Chandar, K.; Knauss, W. An experimental investigation into dynamic fracture: III. On steady-state crack propagation and crack branching. *Int. J. Fract.* **1984**, *26*, 141–154. [CrossRef]
5. Ravi-Chandar, K. Dynamic fracture of nominally brittle materials. *Int. J. Fract.* **1998**, *90*, 83–102. [CrossRef]
6. Anderson, T.L. *Fracture Mechanics: Fundamentals and Applications*; CRC Press: Boca Raton, FL, USA, 2017.
7. Murakami, S. *Continuum Damage Mechanics: A Continuum Mechanics Approach to the Analysis of Damage and Fracture*; Springer Science & Business Media: Berlin, Germany, 2012; Volume 185,.
8. Bittencourt, T.; Wawrzynek, P.; Ingraffea, A.; Sousa, J. Quasi-automatic simulation of crack propagation for 2D LEFM problems. *Eng. Fract. Mech.* **1996**, *55*, 321–334. [CrossRef]
9. Bouchard, P.; Bay, F.; Chastel, Y. Numerical modelling of crack propagation: Automatic remeshing and comparison of different criteria. *Comput. Methods Appl. Mech. Eng.* **2003**, *192*, 3887–3908. [CrossRef]
10. Azócar, D.; Elgueta, M.; Rivara, M.C. Automatic LEFM crack propagation method based on local Lepp–Delaunay mesh refinement. *Adv. Eng. Softw.* **2010**, *41*, 111–119. [CrossRef]
11. Kirk, B.S.; Peterson, J.W.; Stogner, R.H.; Carey, G.F. libMesh: A C++ library for parallel adaptive mesh refinement/coarsening simulations. *Eng. Comput.* **2006**, *22*, 237–254. [CrossRef]
12. Geuzaine, C.; Remacle, J.F. Gmsh: A 3-D finite element mesh generator with built-in pre- and post-processing facilities. *Int. J. Numer. Methods Eng.* **2009**, *79*, 1309–1331. [CrossRef]
13. Barsoum, R. On the use of isoparametric finite elements in linear fracture mechanics. *Int. J. Numer. Methods Eng.* **1976**, *10*, 25–37. [CrossRef]
14. Moran, B.; Shih, C. Crack tip and associated domain integrals from momentum and energy balance. *Eng. Fract. Mech.* **1987**, *27*, 615–642. [CrossRef]
15. Gosz, M.; Moran, B. An interaction energy integral method for computation of mixed-mode stress intensity factors along non-planar crack fronts in three dimensions. *Eng. Fract. Mech.* **2002**, *69*, 299–319. [CrossRef]
16. Courtin, S.; Gardin, C.; Bézine, G.; Ben Hadj Hamouda, H. Advantages of the J-integral approach for calculating stress intensity factors when using the commercial finite element software ABAQUS. *Eng. Fract. Mech.* **2005**, *72*, 2174–2185. [CrossRef]

17. Kim, J.H.; Paulino, G.H. The interaction integral for fracture of orthotropic functionally graded materials: Evaluation of stress intensity factors. *Int. J. Solids Struct.* **2003**, *40*, 3967–4001. [CrossRef]

18. Rybicki, E.; Kanninen, M. A finite element calculation of stress intensity factors by a modified crack closure integral. *Eng. Fract. Mech.* **1977**, *9*, 931–938. [CrossRef]

19. Raju, I. Calculation of strain-energy release rates with higher order and singular finite elements. *Eng. Fract. Mech.* **1987**, *28*, 251–274. [CrossRef]

20. Krueger, R. Virtual crack closure technique: History, approach, and applications. *Appl. Mech. Rev.* **2004**, *57*, 109. [CrossRef]

21. Karihaloo, B.; Xiao, Q. Accurate determination of the coefficients of elastic crack tip asymptotic field by a hybrid crack element with p-adaptivity. *Eng. Fract. Mech.* **2001**, *68*, 1609–1630. [CrossRef]

22. Karihaloo, B.L.; Xiao, Q.Z. Asymptotic fields at the tip of a cohesive crack. *Int. J. Fract.* **2008**, *150*, 55–74. [CrossRef]

23. Wang, Y.; Cerigato, C.; Waisman, H.; Benvenuti, E. XFEM with high-order material-dependent enrichment functions for stress intensity factors calculation of interface cracks using Irwin's crack closure integral. *Eng. Fract. Mech.* **2017**, *178*, 148–168. [CrossRef]

24. Belytschko, T.; Lu, Y.; Gu, L. Element-free Galerkin methods. *Int. J. Numer. Methods Eng.* **1994**, *37*, 229–256. [CrossRef]

25. Belytschko, T.; Gu, L.; Lu, Y. Fracture and crack growth by element free Galerkin methods. *Model. Simul. Mater. Sci. Eng.* **1994**, *2*, 519. [CrossRef]

26. Lu, Y.; Belytschko, T.; Gu, L. A new implementation of the element free Galerkin method. *Comput. Methods Appl. Mech. Eng.* **1994**, *113*, 397–414. [CrossRef]

27. Nguyen, V.P.; Rabczuk, T.; Bordas, S.; Duflot, M. Meshless methods: A review and computer implementation aspects. *Math. Comput. Simul.* **2008**, *79*, 763–813. [CrossRef]

28. Sulsky, D.; Chen, Z.; Schreyer, H.L. A particle method for history-dependent materials. *Comput. Methods Appl. Mech. Eng.* **1994**, *118*, 179–196. [CrossRef]

29. Cottet, G.H.; Raviart, P.A. On particle-in-cell methods for the Vlasov-Poisson equations. *Transp. Theory Stat. Phys.* **1986**, *15*, 1–31. [CrossRef]

30. Nairn, J.A. Material point method calculations with explicit cracks. *Comput. Model. Eng. Sci.* **2003**, *4*, 649–664.

31. Moutsanidis, G.; Kamensky, D.; Zhang, D.Z.; Bazilevs, Y.; Long, C.C. Modeling strong discontinuities in the material point method using a single velocity field. *Comput. Methods Appl. Mech. Eng.* **2019**, *345*, 584–601. [CrossRef]

32. Kakouris, E.; Triantafyllou, S.P. Phase-field material point method for brittle fracture. *Int. J. Numer. Methods Eng.* **2017**, *112*, 1750–1776. [CrossRef]

33. Kakouris, E.; Triantafyllou, S. Material point method for crack propagation in anisotropic media: A phase field approach. *Arch. Appl. Mech.* **2018**, *88*, 287–316. [CrossRef]

34. Maschke, H.G.; Kuna, M. A review of boundary and finite element methods in fracture mechanics. *Theor. Appl. Fract. Mech.* **1985**, *4*, 181–189. [CrossRef]

35. Rokhlin, V. Rapid solution of integral equations of classical potential theory. *J. Comput. Phys.* **1985**, *60*, 187–207. [CrossRef]

36. Hackbusch, W. A Sparse Matrix Arithmetic Based on $\Cal H$ -Matrices. Part I: Introduction to ${\Cal H}$ -Matrices. *Computing* **1999**, *62*, 89–108. [CrossRef]

37. Hughes, T.; Cottrell, J.; Bazilevs, Y. Isogeometric analysis: CAD, finite elements, NURBS, exact geometry and mesh refinement. *Comput. Methods Appl. Mech. Eng.* **2005**, *194*, 4135–4195. [CrossRef]

38. Nguyen, V.P.; Anitescu, C.; Bordas, S.P.; Rabczuk, T. Isogeometric analysis: An overview and computer implementation aspects. *Math. Comput. Simul.* **2015**, *117*, 89–116. [CrossRef]

39. Li, K.; Qian, X. Isogeometric analysis and shape optimization via boundary integral. *Comput.-Aided Des.* **2011**, *43*, 1427–1437. [CrossRef]

40. Simpson, R.N.; Bordas, S.; Trevelyan, J.; Rabczuk, T. A two-dimensional isogeometric boundary element method for elastostatic analysis. *Comput. Methods Appl. Mech. Eng.* **2012**, *209*, 87–100. [CrossRef]

41. Scott, M.; Simpson, R.; Evans, J.; Lipton, S.; Bordas, S.; Hughes, T.; Sederberg, T. Isogeometric boundary element analysis using unstructured T-splines. *Comput. Methods Appl. Mech. Eng.* **2013**, *254*, 197–221. [CrossRef]

42. Nguyen, T.T.; Yvonnet, J.; Zhu, Q.Z.; Bornert, M.; Chateau, C. A phase-field method for computational modeling of interfacial damage interacting with crack propagation in realistic microstructures obtained by microtomography. *Comput. Methods Appl. Mech. Eng.* **2016**, *312*, 567–595. [CrossRef]

43. Peng, X.; Atroshchenko, E.; Kerfriden, P.; Bordas, S. Isogeometric boundary element methods for three dimensional static fracture and fatigue crack growth. *Comput. Methods Appl. Mech. Eng.* **2017**, *316*, 151–185. [CrossRef]

44. Moës, N.; Dolbow, J.; Belytschko, T. A finite element method for crack growth without remeshing. *Int. J. Numer. Methods Eng.* **1999**, *46*, 131–150. [CrossRef]

45. Strouboulis, T.; Babuška, I.; Copps, K. The design and analysis of the generalized finite element method. *Comput. Methods Appl. Mech. Eng.* **2000**, *181*, 43–69. [CrossRef]

46. Babuška, I.; Caloz, G.; Osborn, J. Special finite element methods for a class of second order elliptic problems with rough coefficients. *SIAM J. Numer. Anal.* **1994**, *31*, 945–981. [CrossRef]

47. Babuška, I.; Melenk, J. The partition of unity method. *Int. J. Numer. Methods Eng.* **1996**, *40*, 727–758. [CrossRef]

48. Melenk, J.; Babuška, I. The partition of unity finite element method: Basic theory and applications. *Comput. Methods Appl. Mech. Eng.* **1996**, *139*, 289–314. [CrossRef]

49. Wolf, J.P.; Song, C. Consistent infinitesimal finite-element cell method: In-plane motion. *Comput. Methods Appl. Mech. Eng.* **1995**, *123*, 355–370. [CrossRef]

50. Song, C.; Tin-Loi, F.; Gao, W. A definition and evaluation procedure of generalized stress intensity factors at cracks and multi-material wedges. *Eng. Fract. Mech.* **2010**, *77*, 2316–2336. [CrossRef]

51. Ooi, E.; Man, H.; Natarajan, S.; Song, C. Adaptation of quadtree meshes in the scaled boundary finite element method for crack propagation modelling. *Eng. Fract. Mech.* **2015**, *144*, 101–117. [CrossRef]

52. Barenblatt, G.I. The Mathematical Theory of Equilibrium Cracks in Brittle Fracture. *Adv. Appl. Mech.* **1962**, *7*, 55–129.

53. Dugdale, D.S. Yielding of Steel Sheets Containing Slits. *J. Mech. Phys. Solids* **1960**, *8*, 100–104. [CrossRef]

54. Xu, X.P.; Needleman, A. Numerical simulations of fast crack growth in brittle solids. *J. Mech. Phys. Solids* **1994**, *42*, 1397–1434. [CrossRef]

55. Chen, Z.; Bunger, A.P.; Zhang, X.; Jeffrey, R.G. Cohesive zone finite element-based modeling of hydraulic fractures. *Acta Mech. Solida Sin.* **2009**, *22*, 443–452. [CrossRef]

56. Salen, A.L.; Aliabadi, M.H. Crack growth analysis in concrete using boundary element method. *Eng. Fract. Mech.* **1995**, *51*, 533–545.

57. Nguyen, T.C.; Bui, H.H.; Nguyen, P.V.; Nguyen, G.D. A Conceptual Approach to Modelling Rock Fracture using the Smoothed Particle Hydrodynamics and Cohesive Cracks. In Proceedings of the ISRM Regional Symposium-EUROCK 2015, Salzburg, Austria, 7–10 October 2015; International Society for Rock Mechanics and Rock Engineering: Lisbon, Portugal, 2015.

58. Klein, P.A.; Foulk, J.W.; Chen, E.P.; Wimmer, S.A.; Gao, H.J. Physics-based modeling of brittle fracture: Cohesive formulations and the application of meshfree methods. *Theor. Appl. Fract. Mech.* **2001**, *37*, 99–166. [CrossRef]

59. Soparat, P.; Nanakorn, P. Analysis of Cohesive Crack Growth by the Element-Free Galerkin Method. *J. Mech.* **2008**, *24*, 45–54. [CrossRef]

60. Remmers, J.J.C.; de Borst, R.; Needleman, A. A cohesive segments method for the simulation of crack growth. *Comput. Mech.* **2003**, *31*, 69–77. [CrossRef]

61. Remmers, J.J.C.; de Borst, R.; Needleman, A. The simulation of dynamic crack propagation using the cohesive segments method. *J. Mech. Phys. Solids* **2008**, *56*, 70–92. [CrossRef]

62. Rabczuk, T.; Zi, G. A Meshfree Method based on the Local Partition of Unity for Cohesive Cracks. *Comput. Mech.* **2007**, *39*, 743–760. [CrossRef]

63. Barbieri, E.; Meo, M. A Meshless Cohesive Segments Method for Crack Initiation and Propagation in Composites. *Appl. Compos. Mater.* **2011**, *18*, 45–63. [CrossRef]

64. Msekh, M.A.; Sargado, J.M.; Jamshidian, M.; Areias, P.M.; Rabczuk, T. Abaqus implementation of phase-field model for brittle fracture. *Comput. Mater. Sci.* **2015**, *96*, 472–484. [CrossRef]

65. Rashid, Y. Ultimate strength analysis of prestressed concrete pressure vessels. *Nucl. Eng. Des.* **1968**, *7*, 334–344. [CrossRef]

66. Peerlings, R.D.; De Borst, R.; Brekelmans, W.D.; De Vree, J. Gradient enhanced damage for quasi-brittle materials. *Int. J. Numer. Methods Eng.* **1996**, *39*, 3391–3403. [CrossRef]

67. Simone, A.; Wells, G.N.; Sluys, L.J. From continuous to discontinuous failure in a gradient-enhanced continuum damage model. *Comput. Methods Appl. Mech. Eng.* **2003**, *192*, 4581–4607. [CrossRef]

68. Moës, N.; Stolz, C.; Bernard, P.E.; Chevaugeon, N. A level set based model for damage growth: The thick level set approach. *Int. J. Numer. Methods Eng.* **2011**, *86*, 358–380. [CrossRef]

69. Bourdin, B.; Francfort, G.A.; Marigo, J.J. The variational approach to fracture. *J. Elast.* **2008**, *91*, 5–148. [CrossRef]

70. de Borst, R.; Verhoosel, C.V. Gradient damage vs phase-field approaches for fracture: Similarities and differences. *Comput. Methods Appl. Mech. Eng.* **2016**, *312*, 78–94. [CrossRef]

71. Mandal, T.K.; Nguyen, V.P.; Heidarpour, A. Phase field and gradient enhanced damage models for quasi-brittle failure: A numerical comparative study. *Eng. Fract. Mech.* **2019**, *207*, 48–67. [CrossRef]

72. Cazes, F.; Moës, N. Comparison of a phase-field model and of a thick level set model for brittle and quasi-brittle fracture. *Int. J. Numer. Methods Eng.* **2015**, *103*, 114–143. [CrossRef]

73. Francfort, G.A.; Marigo, J.J. Revisiting brittle fracture as an energy minimization problem. *J. Mech. Phys. Solids* **1998**, *46*, 1319–1342. [CrossRef]

74. Ambrosio, L.; Tortorelli, V.M. Approximation of functional depending on jumps by elliptic functional via Γ-convergence. *Commun. Pure Appl. Math.* **1990**, *43*, 999–1036. [CrossRef]

75. Aldakheel, F.; Hudobivnik, B.; Hussein, A.; Wriggers, P. Phase-field modeling of brittle fracture using an efficient virtual element scheme. *Comput. Methods Appl. Mech. Eng.* **2018**, *341*, 443–466. [CrossRef]

76. Moutsanidis, G.; Kamensky, D.; Chen, J.; Bazilevs, Y. Hyperbolic phase field modeling of brittle fracture: Part II-immersed IGA-RKPM coupling for air-blast-structure interaction. *J. Mech. Phys. Solids* **2018**, *121*, 114–132. [CrossRef]

77. Alessi, R.; Vidoli, S.; De Lorenzis, L. A phenomenological approach to fatigue with a variational phase-field model: The one-dimensional case. *Eng. Fract. Mech.* **2018**, *190*, 53–73. [CrossRef]

78. Wu, J.; Wang, D.; Lin, Z.; Qi, D. An efficient gradient smoothing meshfree formulation for the fourth-order phase field modeling of brittle fracture. *Comput. Part. Mech.* **2019**. [CrossRef]

79. Ambati, M.; Kruse, R.; De Lorenzis, L. A phase-field model for ductile fracture at finite strains and its experimental verification. *Comput. Mech.* **2016**, *57*, 149–167. [CrossRef]

80. Borden, M.J.; Hughes, T.J.; Landis, C.M.; Anvari, A.; Lee, I.J. A phase-field formulation for fracture in ductile materials: Finite deformation balance law derivation, plastic degradation, and stress triaxiality effects. *Comput. Methods Appl. Mech. Eng.* **2016**, *312*, 130–166. [CrossRef]

81. Wilson, Z.A.; Landis, C.M. Phase-field modeling of hydraulic fracture. *J. Mech. Phys. Solids* **2016**, *96*, 264–290. [CrossRef]

82. Miehe, C.; Mauthe, S. Phase field modeling of fracture in multi-physics problems. Part III. Crack driving forces in hydro-poro-elasticity and hydraulic fracturing of fluid-saturated porous media. *Comput. Methods Appl. Mech. Eng.* **2016**, *304*, 619–655. [CrossRef]

83. Heider, Y.; Markert, B. A phase-field modeling approach of hydraulic fracture in saturated porous media. *Mech. Res. Commun.* **2017**, *80*, 38–46. [CrossRef]

84. Ehlers, W.; Luo, C. A phase-field approach embedded in the Theory of Porous Media for the description of dynamic hydraulic fracturing. *Comput. Methods Appl. Mech. Eng.* **2017**, *315*, 348–368. [CrossRef]

85. Pillai, U.; Heider, Y.; Markert, B. A diffusive dynamic brittle fracture model for heterogeneous solids and porous materials with implementation using a user-element subroutine. *Comput. Mater. Sci.* **2018**, *153*, 36–47. [CrossRef]

86. Griffith, A. The phenomena of rupture and flow in solids. *Philos. Trans. R. Soc. Lond. Ser. A Contain. Pap. Math. Phys. Character* **1920**, *221*, 163–198. [CrossRef]

87. Erdogan, F.; Sih, G. On the crack extension in plates under plane loading and transverse shear. *J. Basic Eng.* **1963**, *85*, 519–525. [CrossRef]

88. Nuismer, R. An energy release rate criterion for mixed mode fracture. *Int. J. Fract.* **1975**, *11*, 245–250. [CrossRef]

89. Sih, G. Strain-energy-density factor applied to mixed mode crack problems. *Int. J. Fract.* **1974**, *10*, 305–321. [CrossRef]

90. Abdelaziz, Y.; Hamouine, A. A survey of the extended finite element. *Comput. Struct.* **2008**, *86*, 1141–1151. [CrossRef]
91. Belytschko, T.; Gracie, R.; Ventura, G. A review of extended/generalized finite element methods for material modeling. *Model. Simul. Mater. Sci. Eng.* **2009**, *17*, 043001. [CrossRef]
92. Fries, T.; Belytschko, T. The extended/generalized finite element method: An overview of the method and its applications. *Int. J. Numer. Methods Eng.* **2010**, *84*, 253–304. [CrossRef]
93. Sukumar, N.; Dolbow, J.; Moës, N. Extended finite element method in computational fracture mechanics: A retrospective examination. *Int. J. Fract.* **2015**, *196*, 189–206. [CrossRef]
94. Zhang, Q.; Banerjee, U.; Babuška, I. Higher order stable generalized finite element method. *Numer. Math.* **2014**, *128*, 1–29. [CrossRef]
95. Griebel, M.; Schweitzer, M. A particle-partition of unity method part VII: Adaptivity. In *Meshfree Methods for Partial Differential Equations III*; Springer: Berlin, Germany, 2007; pp. 121–147.
96. Hong, W.; Lee, P. Mesh based construction of flat-top partition of unity functions. *Appl. Math. Comput.* **2013**, *219*, 8687–8704. [CrossRef]
97. Belytschko, T.; Black, T. Elastic crack growth in finite elements with minimal remeshing. *Int. J. Numer. Methods Eng.* **1999**, *620*, 601–620. [CrossRef]
98. Hansbo, A.; Hansbo, P. A finite element method for the simulation of strong and weak discontinuities in solid mechanics. *Comput. Methods Appl. Mech. Eng.* **2004**, *193*, 3523–3540. [CrossRef]
99. Mariani, S.; Perego, U. Extended finite element method for quasi-brittle fracture. *Int. J. Numer. Methods Eng.* **2003**, *58*, 103–126. [CrossRef]
100. Cheng, K.; Fries, T. Higher-order XFEM for curved strong and weak discontinuities. *Int. J. Numer. Methods Eng.* **2010**, *82*, 564–590. [CrossRef]
101. Agathos, K.; Chatzi, E.; Bordas, S. A unified enrichment approach addressing blending and conditioning issues in enriched finite elements. *Comput. Methods Appl. Mech. Eng.* **2019**, *349*, 673–700. [CrossRef]
102. Duarte, C.; Reno, L.; Simone, A. A high-order generalized FEM for through-the-thickness branched cracks. *Int. J. Numer. Methods Eng.* **2007**, *72*, 325–351. [CrossRef]
103. Daux, C.; Moës, N.; Dolbow, J.; Sukumar, N.; Belytschko, T. Arbitrary branched and intersecting cracks with the extended finite element method. *Int. J. Numer. Methods Eng.* **2000**, *48*, 1741–1760. [CrossRef]
104. Stazi, F.; Budyn, E.; Chessa, J.; Belytschko, T. An extended finite element method with higher-order elements for curved cracks. *Comput. Mech.* **2003**, *31*, 38–48. [CrossRef]
105. Laborde, P.; Pommier, J.; Renard, Y.; Salaün, M. High-order extended finite element method for cracked domains. *Int. J. Numer. Methods Eng.* **2005**, *64*, 354–381. [CrossRef]
106. Béchet, E.; Minnebo, H.; Moës, N.; Burgardt, B. Improved implementation and robustness study of the X-FEM for stress analysis around cracks. *Int. J. Numer. Methods Eng.* **2005**, *64*, 1033–1056. [CrossRef]
107. Duarte, C.; Babuška, I.; Oden, J. Generalized finite element methods for three-dimensional structural mechanics problems. *Comput. Struct.* **2000**, *77*, 215–232. [CrossRef]
108. Xiao, Q.; Karihaloo, B. Direct evaluation of accurate coefficients of the linear elastic crack tip asymptotic field. *Fatigue Fract. Eng. Mater. Struct.* **2003**, *26*, 719–729. [CrossRef]
109. Liu, X.; Xiao, Q.; Karihaloo, B. XFEM for direct evaluation of mixed mode SIFs in homogeneous and bi-materials. *Int. J. Numer. Methods Eng.* **2004**, *59*, 1103–1118. [CrossRef]
110. Zamani, A.; Gracie, R.; Eslami, M. Cohesive and non-cohesive fracture by higher-order enrichment of XFEM. *Int. J. Numer. Methods Eng.* **2012**, *90*, 452–483. [CrossRef]
111. Gupta, V.; Duarte, C.; Babuška, I.; Banerjee, U. A stable and optimally convergent generalized FEM (SGFEM) for linear elastic fracture mechanics. *Comput. Methods Appl. Mech. Eng.* **2013**, *266*, 23–39. [CrossRef]
112. Gupta, V.; Duarte, C.; Babuška, I.; Banerjee, U. Stable GFEM (SGFEM): Improved conditioning and accuracy of GFEM/XFEM for three-dimensional fracture mechanics. *Comput. Methods Appl. Mech. Eng.* **2015**, *289*, 355–386. [CrossRef]
113. Nicaise, S.; Renard, Y.; Chahine, E. Optimal convergence analysis for the extended finite element method. *Int. J. Numer. Methods Eng.* **2011**, *86*, 528–548. [CrossRef]
114. Chevaugeon, N.; Moës, N.; Minnebo, H. Improved crack tip enrichment functions and integration for crack modeling using the extended finite element method. *J. Multiscale Comput. Eng.* **2013**, *11*, 597–631. [CrossRef]
115. Zi, G.; Belytschko, T. New crack-tip elements for XFEM and applications to cohesive cracks. *Int. J. Numer. Methods Eng.* **2003**, *57*, 2221–2240. [CrossRef]

116. Babuška, I.; Banerjee, U. Stable generalized finite element method (SGFEM). *Comput. Methods Appl. Mech. Eng.* **2012**, *201*, 91–111. [CrossRef]

117. Chessa, J.; Wang, H.; Belytschko, T. On the construction of blending elements for local partition of unity enriched finite elements. *Int. J. Numer. Methods Eng.* **2003**, *57*, 1015–1038. [CrossRef]

118. Gracie, R.; Wang, H.; Belytschko, T. Blending in the extended finite element method by discontinuous Galerkin and assumed strain methods. *Int. J. Numer. Methods Eng.* **2008**, *74*, 1645–1669. [CrossRef]

119. Tarancón, J.; Vercher, A.; Giner, E.; Fuenmayor, F. Enhanced blending elements for XFEM applied to linear elastic fracture mechanics. *Int. J. Numer. Methods Eng.* **2009**, *77*, 126–148. [CrossRef]

120. Chahine, E.; Laborde, P.; Renard, Y. A non-conformal eXtended Finite Element approach: Integral matching Xfem. *Appl. Numer. Math.* **2011**, *61*, 322–343. [CrossRef]

121. Agathos, K.; Chatzi, E.; Bordas, S. Stable 3D extended finite elements with higher order enrichment for accurate non planar fracture. *Comput. Methods Appl. Mech. Eng.* **2016**, *306*, 19–46. [CrossRef]

122. Fries, T. A corrected XFEM approximation without problems in blending elements. *Int. J. Numer. Methods Eng.* **2008**, *75*, 503–532. [CrossRef]

123. Chahine, E.; Laborde, P. Crack tip enrichment in the XFEM using a cutoff function. *Int. J. Numer. Methods Eng.* **2008**, *75*, 629–646. [CrossRef]

124. Ventura, G.; Gracie, R.; Belytschko, T. Fast integration and weight function blending in the extended finite element method. *Int. J. Numer. Methods Eng.* **2009**, *77*, 1–29. [CrossRef]

125. Menk, A.; Bordas, S. A robust preconditioning technique for the extended finite element method. *Int. J. Numer. Methods Eng.* **2011**, *85*, 1609–1632. [CrossRef]

126. Lang, C.; Makhija, D.; Doostan, A.; Maute, K. A simple and efficient preconditioning scheme for heaviside enriched XFEM. *Comput. Mech.* **2014**, *54*, 1357–1374. [CrossRef]

127. Loehnert, S. A stabilization technique for the regularization of nearly singular extended finite elements. *Comput. Mech.* **2014**, *54*, 523–533. [CrossRef]

128. Ventura, G.; Tesei, C. Stabilized X-FEM for Heaviside and nonlinear enrichments. In *Advances in Discretization Methods*; Springer: Berlin, Germany, 2016; pp. 209–228.

129. Agathos, K.; Bordas, S.; Chatzi, E. Improving the conditioning of XFEM/GFEM for fracture mechanics problems through enrichment quasi-orthogonalization. *Comput. Methods Appl. Mech. Eng.* **2019**, *346*, 1051–1073. [CrossRef]

130. Agathos, K.; Chatzi, E.; Bordas, S.; Talaslidis, D. A well-conditioned and optimally convergent XFEM for 3D linear elastic fracture. *Int. J. Numer. Methods Eng.* **2016**, *105*, 643–677. [CrossRef]

131. Duarte, C.; Hamzeh, O.; Liszka, T.; Tworzydlo, W. A generalized finite element method for the simulation of three-dimensional dynamic crack propagation. *Comput. Methods Appl. Mech. Eng.* **2001**, *190*, 2227–2262. [CrossRef]

132. Sukumar, N.; Moës, N.; Moran, B.; Belytschko, T. Extended finite element method for three-dimensional crack modelling. *Int. J. Numer. Methods Eng.* **2000**, *48*, 1549–1570. [CrossRef]

133. Osher, S.; Sethian, J. Fronts propagating with curvature-dependent speed: Algorithms based on Hamilton-Jacobi formulations. *J. Comput. Phys.* **1988**, *79*, 12–49. [CrossRef]

134. Sethian, J. *Level Set Methods and Fast Marching Methods: Evolving Interfaces in Computational Geometry, Fluid Mechanics, Computer Vision, and Materials Science*; Cambridge University Press: Cambridge, UK, 1999; Volume 3.

135. Stolarska, M.; Chopp, D.; Moës, N.; Belytschko, T. Modelling crack growth by level sets in the extended finite element method. *Int. J. Numer. Methods Eng.* **2001**, *51*, 943–960. [CrossRef]

136. Moës, N.; Gravouil, A.; Belytschko, T. Non-planar 3D crack growth by the extended finite element and level sets-Part I: Mechanical model. *Int. J. Numer. Methods Eng.* **2002**, *53*, 2549–2568. [CrossRef]

137. Gravouil, A.; Moës, N.; Belytschko, T. Non-planar 3D crack growth by the extended finite element and level sets-Part II: Level set update. *Int. J. Numer. Methods Eng.* **2002**, *53*, 2569–2586. [CrossRef]

138. Sukumar, N.; Chopp, D.; Béchet, E.; Moës, N. Three-dimensional non-planar crack growth by a coupled extended finite element and fast marching method. *Int. J. Numer. Methods Eng.* **2008**, *76*, 727–748. [CrossRef]

139. Duflot, M. A study of the representation of cracks with level sets. *Int. J. Numer. Methods Eng.* **2007**, *70*, 1261–1302. [CrossRef]

140. Elguedj, T.; de Saint Maurice, R.; Combescure, A.; Faucher, V.; Prabel, B. Extended finite element modeling of 3D dynamic crack growth under impact loading. *Finite Elem. Anal. Des.* **2018**, *151*, 1–17. [CrossRef]

141. Fries, T.; Baydoun, M. Crack propagation with the extended finite element method and a hybrid explicit-implicit crack description. *Int. J. Numer. Methods Eng.* **2012**, *89*, 1527–1558. [CrossRef]

142. Ventura, G.; Budyn, E.; Belytschko, T. Vector level sets for description of propagating cracks in finite elements. *Int. J. Numer. Methods Eng.* **2003**, *58*, 1571–1592. [CrossRef]

143. Agathos, K.; Ventura, G.; Chatzi, E.; Bordas, S. Stable 3D XFEM/vector level sets for non-planar 3D crack propagation and comparison of enrichment schemes. *Int. J. Numer. Methods Eng.* **2018**, *113*, 252–276. [CrossRef]

144. Agathos, K.; Ventura, G.; Chatzi, E.; Bordas, S. Well Conditioned Extended Finite Elements and Vector Level Sets for Three-Dimensional Crack Propagation. In *Geometrically Unfitted Finite Element Methods and Applications*; Springer: Berlin, Germany, 2017; pp. 307–329.

145. Sadeghirad, A.; Chopp, D.; Ren, X.; Fang, E.; Lua, J. A novel hybrid approach for level set characterization and tracking of non-planar 3D cracks in the extended finite element method. *Eng. Fract. Mech.* **2016**, *160*, 1–14. [CrossRef]

146. Fries, T.; Omerović, S.; Schöllhammer, D.; Steidl, J. Higher-order meshing of implicit geometries—Part I: Integration and interpolation in cut elements. *Comput. Methods Appl. Mech. Eng.* **2017**, *313*, 759–784. [CrossRef]

147. Paul, B.; Ndeffo, M.; Massin, P.; Moës, N. An integration technique for 3D curved cracks and branched discontinuities within the extended Finite Element Method. *Finite Elem. Anal. Des.* **2017**, *123*, 19–50. [CrossRef]

148. Ventura, G. On the elimination of quadrature subcells for discontinuous functions in the eXtended Finite-Element Method. *Int. J. Numer. Methods Eng.* **2006**, *66*, 761–795. [CrossRef]

149. Ventura, G.; Benvenuti, E. Equivalent polynomials for quadrature in Heaviside function enriched elements. *Int. J. Numer. Methods Eng.* **2015**, *102*, 688–710. [CrossRef]

150. Natarajan, S.; Mahapatra, D.; Bordas, S. Integrating strong and weak discontinuities without integration subcells and example applications in an XFEM/GFEM framework. *Int. J. Numer. Methods Eng.* **2010**, *83*, 269–294. [CrossRef]

151. Mousavi, S.; Sukumar, N. Generalized Gaussian quadrature rules for discontinuities and crack singularities in the extended finite element method. *Comput. Methods Appl. Mech. Eng.* **2010**, *199*, 3237–3249. [CrossRef]

152. Loehnert, S.; Mueller-Hoeppe, D.; Wriggers, P. 3D corrected XFEM approach and extension to finite deformation theory. *Int. J. Numer. Methods Eng.* **2011**, *86*, 431–452. [CrossRef]

153. Minnebo, H. Three-dimensional integration strategies of singular functions introduced by the XFEM in the LEFM. *Int. J. Numer. Methods Eng.* **2012**, *92*, 1117–1138. [CrossRef]

154. González-Albuixech, V.; Giner, E.; Tarancon, J.; Fuenmayor, F.; Gravouil, A. Convergence of domain integrals for stress intensity factor extraction in 2-D curved cracks problems with the extended finite element method. *Int. J. Numer. Methods Eng.* **2013**, *94*, 740–757. [CrossRef]

155. González-Albuixech, V.; Giner, E.; Tarancón, J.; Fuenmayor, F.; Gravouil, A. Domain integral formulation for 3-D curved and non-planar cracks with the extended finite element method. *Comput. Methods Appl. Mech. Eng.* **2013**, *264*, 129–144. [CrossRef]

156. Lan, M.; Waisman, H.; Harari, I. A direct analytical method to extract mixed-mode components of strain energy release rates from Irwin's integral using extended finite element method. *Int. J. Numer. Methods Eng.* **2013**, *95*, 1033–1052. [CrossRef]

157. Lan, M.; Waisman, H.; Harari, I. A High-order extended finite element method for extraction of mixed-mode strain energy release rates in arbitrary crack settings based on Irwin's integral. *Int. J. Numer. Methods Eng.* **2013**, *96*, 787–812. [CrossRef]

158. Song, G.; Waisman, H.; Lan, M.; Harari, I. Extraction of stress intensity factors from Irwin's integral using high-order XFEM on triangular meshes. *Int. J. Numer. Methods Eng.* **2015**, *102*, 528–550. [CrossRef]

159. Wang, Y.; Waisman, H. An arc-length method for controlled cohesive crack propagation using high-order XFEM and Irwin's crack closure integral. *Eng. Fract. Mech.* **2018**, *199*, 235–256. [CrossRef]

160. Schätzer, M.; Fries, T.P. Stress Intensity Factors Through Crack Opening Displacements in the XFEM. In *Advances in Discretization Methods*; Springer: Berlin, Germany, 2016; pp. 143–164.

161. Sukumar, N.; Prévost, J. Modeling quasi-static crack growth with the extended finite element method Part I: Computer implementation. *Int. J. Solids Struct.* **2003**, *40*, 7513–7537. [CrossRef]

162. Huang, R.; Sukumar, N.; Prévost, J. Modeling quasi-static crack growth with the extended finite element method Part II: Numerical applications. *Int. J. Solids Struct.* **2003**, *40*, 7539–7552. [CrossRef]

163. Budyn, E.; Zi, G.; Moës, N.; Belytschko, T. A method for multiple crack growth in brittle materials without remeshing. *Int. J. Numer. Methods Eng.* **2004**, *61*, 1741–1770. [CrossRef]

164. Bordas, S.; Moran, B. Enriched finite elements and level sets for damage tolerance assessment of complex structures. *Eng. Fract. Mech.* **2006**, *73*, 1176–1201. [CrossRef]

165. Lecampion, B. An extended finite element method for hydraulic fracture problems. *Commun. Numer. Methods Eng.* **2009**, *25*, 121–133. [CrossRef]

166. Sutula, D.; Bordas, S. Minimum energy multiple crack propagation. Part III: XFEM computer implementation and applications. *Eng. Fract. Mech.* **2018**, *191*, 257–276.. [CrossRef]

167. Bordas, S.; Nguyen, P.; Dunant, C.; Guidoum, A.; Nguyen-Dang, H. An extended finite element library. *Int. J. Numer. Methods Eng.* **2007**, *71*, 703–732. [CrossRef]

168. Malekan, M.; Silva, L.; Barros, F.; Pitangueira, R.; Penna, S. Two-dimensional fracture modeling with the generalized/extended finite element method: An object-oriented programming approach. *Adv. Eng. Softw.* **2018**, *115*, 168–193. [CrossRef]

169. Sutula, D.; Bordas, S. Minimum energy multiple crack propagation. Part II: Discrete Solution with XFEM. *Eng. Fract. Mech.* **2018**, *191*, 225–256. [CrossRef]

170. Sutula, D.; Kerfriden, P.; van Dam, T.; Bordas, S. Minimum energy multiple crack propagation. Part I: Theory and state of the art review. *Eng. Fract. Mech.* **2018**, *191*, 205–224. [CrossRef]

171. Belytschko, T.; Chen, H.; Xu, J.; Zi, G. Dynamic crack propagation based on loss of hyperbolicity and a new discontinuous enrichment. *Int. J. Numer. Methods Eng.* **2003**, *58*, 1873–1905. [CrossRef]

172. Réthoré, J.; Gravouil, A.; Combescure, A. An energy-conserving scheme for dynamic crack growth using the extended finite element method. *Int. J. Numer. Methods Eng.* **2005**, *63*, 631–659. [CrossRef]

173. Menouillard, T.; Rethore, J.; Combescure, A.; Bung, H. Efficient explicit time stepping for the eXtended Finite Element Method (X-FEM). *Int. J. Numer. Methods Eng.* **2006**, *68*, 911–939. [CrossRef]

174. Asadpoure, A.; Mohammadi, S.; Vafai, A. Modeling crack in orthotropic media using a coupled finite element and partition of unity methods. *Finite Elem. Anal. Des.* **2006**, *42*, 1165–1175. [CrossRef]

175. Legrain, G.; Moes, N.; Verron, E. Stress analysis around crack tips in finite strain problems using the extended finite element method. *Int. J. Numer. Methods Eng.* **2005**, *63*, 290–314. [CrossRef]

176. Prabel, B.; Combescure, A.; Gravouil, A.; Marie, S. Level set X-FEM non-matching meshes: Application to dynamic crack propagation in elastic–plastic media. *Int. J. Numer. Methods Eng.* **2007**, *69*, 1553–1569. [CrossRef]

177. Moës, N.; Belytschko, T. Extended finite element method for cohesive crack growth. *Eng. Fract. Mech.* **2002**, *69*, 813–833. [CrossRef]

178. Kausel, E. Thin-layer method: Formulation in the time domain. *Int. J. Numer. Methods Eng.* **1994**, *37*, 927–941. [CrossRef]

179. Kita, E.; Kamiya, N. Trefftz method: An overview. *Adv. Eng. Softw.* **1995**, *24*, 3–12. [CrossRef]

180. Patera, A.T. A spectral element method for fluid dynamics: Laminar flow in a channel expansion. *J. Comput. Phys.* **1984**, *54*, 468–488. [CrossRef]

181. Nelson, R.; Dong, S.; Kalra, R. Vibrations and waves in laminated orthotropic circular cylinders. *J. Sound Vib.* **1971**, *18*, 429–444. [CrossRef]

182. Silvester, P.; Lowther, D.; Carpenter, C.; Wyatt, E. Exterior finite elements for 2-dimensional field problems with open boundaries. *Proc. Inst. Electr. Eng.* **1977**, *124*, 1267. [CrossRef]

183. Dasgupta, G. A Finite Element Formulation for Unbounded Homogeneous Continua. *J. Appl. Mech.* **1982**, *49*, 136–140. [CrossRef]

184. Wolf, J.P.; Song, C. Dynamic-stiffness matrix in time domain of unbounded medium by infinitesimal finite element cell method. *Earthq. Eng. Struct. Dyn.* **1994**, *23*, 1181–1198. [CrossRef]

185. Wolf, J.P.; Song, C. *Finite-Element Modelling of Unbounded Media*; Wiley: Chichester, UK ; New York, NY, USA, 1996.

186. Wolf, J.P. *The Scaled Boundary Finite Element Method*; Wiley: Chichester, UK; Hoboken, NJ, USA, 2003.

187. Deeks, A.J.; Wolf, J.P. A virtual work derivation of the scaled boundary finite-element method for elastostatics. *Comput. Mech.* **2002**, *28*, 489–504. [CrossRef]

188. Chidgzey, S.R.; Deeks, A.J. Determination of coefficients of crack tip asymptotic fields using the scaled boundary finite element method. *Eng. Fract. Mech.* **2005**, *72*, 2019–2036. [CrossRef]

189. Song, C. Evaluation of power-logarithmic singularities, T-stresses and higher order terms of in-plane singular stress fields at cracks and multi-material corners. *Eng. Fract. Mech.* **2005**, *72*, 1498–1530. [CrossRef]

190. Song, C. A super-element for crack analysis in the time domain. *Int. J. Numer. Methods Eng.* **2004**, *61*, 1332–1357. [CrossRef]

191. Müller, A.; Wenck, J.; Goswami, S.; Lindemann, J.; Hohe, J.; Becker, W. The boundary finite element method for predicting directions of cracks emerging from notches at bimaterial junctions. *Eng. Fract. Mech.* **2005**, *72*, 373–386. [CrossRef]

192. Lindemann, J.; Becker, W. Free-Edge Stresses around Holes in Laminates by the Boundary Finite-Element Method. *Mech. Compos. Mater.* **2002**, *38*, 407–416. [CrossRef]

193. Yang, Z. Fully automatic modelling of mixed-mode crack propagation using scaled boundary finite element method. *Eng. Fract. Mech.* **2006**, *73*, 1711–1731. [CrossRef]

194. Yang, Z.; Deeks, A. Fully-automatic modelling of cohesive crack growth using a finite element–scaled boundary finite element coupled method. *Eng. Fract. Mech.* **2007**, *74*, 2547–2573. [CrossRef]

195. Yang, Z.J.; Deeks, A.J. Modelling cohesive crack growth using a two-step finite element-scaled boundary finite element coupled method. *Int. J. Fract.* **2007**, *143*, 333–354. [CrossRef]

196. Ooi, E.; Yang, Z. Modelling multiple cohesive crack propagation using a finite element–scaled boundary finite element coupled method. *Eng. Anal. Bound. Elem.* **2009**, *33*, 915–929. [CrossRef]

197. Ooi, E.; Yang, Z. Efficient prediction of deterministic size effects using the scaled boundary finite element method. *Eng. Fract. Mech.* **2010**, *77*, 985–1000. [CrossRef]

198. Zhu, C.; Lin, G.; Li, J. Modelling cohesive crack growth in concrete beams using scaled boundary finite element method based on super-element remeshing technique. *Comput. Struct.* **2013**, *121*, 76–86. [CrossRef]

199. Ooi, E.T.; Yang, Z.J. Modelling dynamic crack propagation using the scaled boundary finite element method. *Int. J. Numer. Methods Eng.* **2011**, *88*, 329–349. [CrossRef]

200. Ooi, E.T.; Yang, Z.J.; Guo, Z.Y. Dynamic cohesive crack propagation modelling using the scaled boundary finite element method: Dynamic cohesive crack propagation modelling. *Fatigue Fract. Eng. Mater. Struct.* **2012**, *35*, 786–800. [CrossRef]

201. Ooi, E.T.; Yang, Z.J. Modelling crack propagation in reinforced concrete using a hybrid finite element–scaled boundary finite element method. *Eng. Fract. Mech.* **2011**, *78*, 252–273. [CrossRef]

202. Ooi, E.T.; Song, C.; Tin-Loi, F.; Yang, Z. Polygon scaled boundary finite elements for crack propagation modelling: Scaled boundary polygon FINITE elements for crack propagation. *Int. J. Numer. Methods Eng.* **2012**, *91*, 319–342. [CrossRef]

203. Talischi, C.; Paulino, G.H.; Pereira, A.; Menezes, I.F.M. PolyMesher: A general-purpose mesh generator for polygonal elements written in Matlab. *Struct. Multidiscip. Optim.* **2012**, *45*, 309–328. [CrossRef]

204. Chiong, I.; Ooi, E.T.; Song, C.; Tin-Loi, F. Scaled boundary polygons with application to fracture analysis of functionally graded materials: Scaled boundary polygons for functionally graded materials. *Int. J. Numer. Methods Eng.* **2014**, *98*, 562–589. [CrossRef]

205. Chen, X.; Luo, T.; Ooi, E.; Ooi, E.; Song, C. A quadtree-polygon-based scaled boundary finite element method for crack propagation modeling in functionally graded materials. *Theor. Appl. Fract. Mech.* **2018**, *94*, 120–133. [CrossRef]

206. Zhang, Z.; Dissanayake, D.; Saputra, A.; Wu, D.; Song, C. Three-dimensional damage analysis by the scaled boundary finite element method. *Comput. Struct.* **2018**, *206*, 1–17. [CrossRef]

207. Zhang, Z.; Liu, Y.; Dissanayake, D.D.; Saputra, A.A.; Song, C. Nonlocal damage modelling by the scaled boundary finite element method. *Eng. Anal. Bound. Elem.* **2019**, *99*, 29–45. [CrossRef]

208. Lin, G.; Zhang, Y.; Hu, Z.; Zhong, H. Scaled boundary isogeometric analysis for 2D elastostatics. *Sci. China Phys. Mech. Astron.* **2014**, *57*, 286–300. [CrossRef]

209. Natarajan, S.; Wang, J.; Song, C.; Birk, C. Isogeometric analysis enhanced by the scaled boundary finite element method. *Comput. Methods Appl. Mech. Eng.* **2015**, *283*, 733–762. [CrossRef]

210. Auricchio, F.; Calabrò, F.; Hughes, T.; Reali, A.; Sangalli, G. A simple algorithm for obtaining nearly optimal quadrature rules for NURBS-based isogeometric analysis. *Comput. Methods Appl. Mech. Eng.* **2012**, *249–252*, 15–27. [CrossRef]

211. Cottrell, J.A.; Hughes, T.J.R.; Bazilevs, Y. *Isogeometric Analysis: Toward Integration of CAD and FEA*; Wiley: Chichester, UK; Hoboken, NJ, USA, 2009; OCLC: 441875062.

212. Song, C. *The Scaled Boundary Finite Element Method: Introduction to Theory and Implementation*; John Wiley & Sons: Hoboken, NJ, USA, 2018.

213. Song, C.; Ooi, E.T.; Natarajan, S. A review of the scaled boundary finite element method for two-dimensional linear elastic fracture mechanics. *Eng. Fract. Mech.* **2018**, *187*, 45–73. [CrossRef]

214. Song, C.; Wolf, J.P. The scaled boundary finite-element method—alias consistent infinitesimal finite-element cell method—For elastodynamics. *Comput. Methods Appl. Mech. Eng.* **1997**, *147*, 329–355. [CrossRef]

215. Hu, Z.; Lin, G.; Wang, Y.; Liu, J. A Hamiltonian-based derivation of Scaled Boundary Finite Element Method for elasticity problems. *IOP Conf. Ser. Mater. Sci. Eng.* **2010**, *10*, 012213. [CrossRef]

216. Song, C. A matrix function solution for the scaled boundary finite-element equation in statics. *Comput. Methods Appl. Mech. Eng.* **2004**, *193*, 2325–2356. [CrossRef]

217. Egger, A.W.; Chatzi, E.N.; Triantafyllou, S.P. An enhanced scaled boundary finite element method for linear elastic fracture. *Arch. Appl. Mech.* **2017**, *87*, 1667–1706. [CrossRef]

218. Zienkiewicz, O.C.; Zhu, J.Z. The superconvergent patch recovery anda posteriori error estimates. Part 1: The recovery technique. *Int. J. Numer. Methods Eng.* **1992**, *33*, 1331–1364. [CrossRef]

219. Zienkiewicz, O.C.; Zhu, J.Z. The superconvergent patch recovery anda posteriori error estimates. Part 2: Error estimates and adaptivity. *Int. J. Numer. Methods Eng.* **1992**, *33*, 1365–1382. [CrossRef]

220. Deeks, A.J.; Wolf, J.P. Stress recovery and error estimation for the scaled boundary finite-element method. *Int. J. Numer. Methods Eng.* **2002**, *54*, 557–583. [CrossRef]

221. Ooi, E.; Shi, M.; Song, C.; Tin-Loi, F.; Yang, Z. Dynamic crack propagation simulation with scaled boundary polygon elements and automatic remeshing technique. *Eng. Fract. Mech.* **2013**, *106*, 1–21. [CrossRef]

222. Wachspress, E.L. A Rational Basis for Function Approximation. *IMA J. Appl. Math.* **1971**, *8*, 57–68. [CrossRef]

223. Sutton, O.J. The virtual element method in 50 lines of MATLAB. *Numer. Algorithms* **2017**, *75*, 1141–1159. [CrossRef]

224. Ooi, E.T.; Natarajan, S.; Song, C.; Ooi, E.H. Crack propagation modelling in concrete using the scaled boundary finite element method with hybrid polygon-quadtree meshes. *Int. J. Fract.* **2017**, *203*, 135–157. [CrossRef]

225. Ooi, E.T.; Natarajan, S.; Song, C.; Ooi, E.H. Dynamic fracture simulations using the scaled boundary finite element method on hybrid polygon–quadtree meshes. *Int. J. Impact Eng.* **2016**, *90*, 154–164. [CrossRef]

226. Saputra, A.; Talebi, H.; Tran, D.; Birk, C.; Song, C. Automatic image-based stress analysis by the scaled boundary finite element method: Automatic image-based stress analysis by the scaled boundary fem. *Int. J. Numer. Methods Eng.* **2017**, *109*, 697–738. [CrossRef]

227. Liu, G.; Li, Q.; Msekh, M.A.; Zuo, Z. Abaqus implementation of monolithic and staggered schemes for quasi-static and dynamic fracture phase-field model. *Comput. Mater. Sci.* **2016**, *121*, 35–47. [CrossRef]

228. Miehe, C.; Welschinger, F.; Hofacker, M. Thermodynamically consistent phase-field models of fracture: Variational principles and multi-field FE implementations. *Int. J. Numer. Methods Eng.* **2010**, *83*, 1273–1311. [CrossRef]

229. Borden, M.J.; Verhoosel, C.V.; Scott, M.A.; Hughes, T.J.; Landis, C.M. A phase-field description of dynamic brittle fracture. *Comput. Methods Appl. Mech. Eng.* **2012**, *217*, 77–95. [CrossRef]

230. Gültekin, O.; Dal, H.; Holzapfel, G.A. A phase-field approach to model fracture of arterial walls: Theory and finite element analysis. *Comput. Methods Appl. Mech. Eng.* **2016**, *312*, 542–566. [CrossRef]

231. Schlüter, A.; Willenbücher, A.; Kuhn, C.; Müller, R. Phase field approximation of dynamic brittle fracture. *Comput. Mech.* **2014**, *54*, 1141–1161. [CrossRef]

232. Miehe, C.; Hofacker, M.; Welschinger, F. A phase field model for rate-independent crack propagation: Robust algorithmic implementation based on operator splits. *Comput. Methods Appl. Mech. Eng.* **2010**, *199*, 2765–2778. [CrossRef]

233. Kuhn, C.; Müller, R. A continuum phase field model for fracture. *Eng. Fract. Mech.* **2010**, *77*, 3625–3634. [CrossRef]

234. Quintanas-Corominas, A.; Reinoso, J.; Casoni, E.; Turon, A.; Mayugo, J. A phase field approach to simulate intralaminar and translaminar fracture in long fiber composite materials. *Compos. Struct.* **2019**. [CrossRef]

235. Natarajan, S.; Annabattula, R.K. Modeling crack propagation in variable stiffness composite laminates using the phase field method. *Compos. Struct.* **2019**, *209*, 424–433.

236. Hansen-Dörr, A.C.; de Borst, R.; Hennig, P.; Kästner, M. Phase-field modelling of interface failure in brittle materials. *Comput. Methods Appl. Mech. Eng.* **2019**, *346*, 25–42. [CrossRef]

237. Smith, M. *ABAQUS/Standard User's Manual, Version 6.9*; Simulia: Johnston, RI, USA, 2009.

238. Li, H.; Zhang, H.; Zheng, Y. A coupling extended multiscale finite element method for dynamic analysis of heterogeneous saturated porous media. *Int. J. Numer. Methods Eng.* **2015**, *104*, 18–47. [CrossRef]

239. Li, B.; Maurini, C. Crack kinking in a variational phase-field model of brittle fracture with strongly anisotropic surface energy. *J. Mech. Phys. Solids* **2019**, *125*, 502–522. [CrossRef]

240. Abinandanan, T.; Haider, F. An extended Cahn-Hilliard model for interfaces with cubic anisotropy. *Philos. Mag. A* **2001**, *81*, 2457–2479. [CrossRef]

241. Torabi, S.; Lowengrub, J. Simulating interfacial anisotropy in thin-film growth using an extended Cahn-Hilliard model. *Phys. Rev. E* **2012**, *85*, 041603. [CrossRef]

242. Shen, R.; Waisman, H.; Guo, L. Fracture of viscoelastic solids modeled with a modified phase field method. *Comput. Methods Appl. Mech. Eng.* **2019**, *346*, 862–890. [CrossRef]

243. Ambati, M.; Gerasimov, T.; De Lorenzis, L. Phase-field modeling of ductile fracture. *Comput. Mech.* **2015**, *55*, 1017–1040. [CrossRef]

244. Kuhn, C.; Noll, T.; Müller, R. On phase field modeling of ductile fracture. *GAMM-Mitteilungen* **2016**, *39*, 35–54. [CrossRef]

245. Ambati, M.; De Lorenzis, L. Phase-field modeling of brittle and ductile fracture in shells with isogeometric NURBS-based solid-shell elements. *Comput. Methods Appl. Mech. Eng.* **2016**, *312*, 351–373. [CrossRef]

246. Kiendl, J.; Ambati, M.; De Lorenzis, L.; Gomez, H.; Reali, A. Phase-field description of brittle fracture in plates and shells. *Comput. Methods Appl. Mech. Eng.* **2016**, *312*, 374–394. [CrossRef]

247. Reinoso, J.; Paggi, M.; Linder, C. Phase field modeling of brittle fracture for enhanced assumed strain shells at large deformations: Formulation and finite element implementation. *Comput. Mech.* **2017**, *59*, 981–1001. [CrossRef]

248. Verhoosel, C.V.; de Borst, R. A phase-field model for cohesive fracture. *Int. J. Numer. Methods Eng.* **2013**, *96*, 43–62. [CrossRef]

249. Vignollet, J.; May, S.; De Borst, R.; Verhoosel, C.V. Phase-field models for brittle and cohesive fracture. *Meccanica* **2014**, *49*, 2587–2601. [CrossRef]

250. Geelen, R.J.; Liu, Y.; Hu, T.; Tupek, M.R.; Dolbow, J.E. A phase-field formulation for dynamic cohesive fracture. *arXiv* **2018**, arXiv:1809.09691.

251. Wu, J.Y.; Nguyen, V.P. A length scale insensitive phase-field damage model for brittle fracture. *J. Mech. Phys. Solids* **2018**, *119*, 20–42. [CrossRef]

252. Lorentz, E. A nonlocal damage model for plain concrete consistent with cohesive fracture. *Int. J. Fract.* **2017**, *207*, 123–159. [CrossRef]

253. Wu, J.Y.; Qiu, J.F.; Nguyen, V.P.; Mandal, T.K.; Zhuang, L.J. Computational modeling of localized failure in solids: XFEM vs PF-CZM. *Comput. Methods Appl. Mech. Eng.* **2019**, *345*, 618–643. [CrossRef]

254. Nguyen, V.P.; Wu, J.Y. Modeling dynamic fracture of solids with a phase-field regularized cohesive zone model. *Comput. Methods Appl. Mech. Eng.* **2018**, *340*, 1000–1022. [CrossRef]

255. Lorentz, E.; Godard, V. Gradient damage models: Toward full-scale computations. *Comput. Methods Appl. Mech. Eng.* **2011**, *200*, 1927–1944. [CrossRef]

256. Miehe, C.; Schaenzel, L.M.; Ulmer, H. Phase field modeling of fracture in multi-physics problems. Part I. Balance of crack surface and failure criteria for brittle crack propagation in thermo-elastic solids. *Comput. Methods Appl. Mech. Eng.* **2015**, *294*, 449–485. [CrossRef]

257. Tanné, E.; Li, T.; Bourdin, B.; Marigo, J.J.; Maurini, C. Crack nucleation in variational phase-field models of brittle fracture. *J. Mech. Phys. Solids* **2018**, *110*, 80–99. [CrossRef]

258. Miehe, C.; Mauthe, S.; Teichtmeister, S. Minimization principles for the coupled problem of Darcy–Biot-type fluid transport in porous media linked to phase field modeling of fracture. *J. Mech. Phys. Solids* **2015**, *82*, 186–217. [CrossRef]

259. Borden, M.J.; Hughes, T.J.; Landis, C.M.; Verhoosel, C.V. A higher-order phase-field model for brittle fracture: Formulation and analysis within the isogeometric analysis framework. *Comput. Methods Appl. Mech. Eng.* **2014**, *273*, 100–118. [CrossRef]

260. Dittmann, M.; Aldakheel, F.; Schulte, J.; Wriggers, P.; Hesch, C. Variational phase-field formulation of non-linear ductile fracture. *Comput. Methods Appl. Mech. Eng.* **2018**, *342*, 71–94. [CrossRef]

261. Franke, M.; Hesch, C.; Dittmann, M. Phase-field approach to fracture for finite-deformation contact problems. *PAMM* **2016**, *16*, 123–124. [CrossRef]

262. Pham, K.; Amor, H.; Marigo, J.J.; Maurini, C. Gradient damage models and their use to approximate brittle fracture. *Int. J. Damage Mech.* **2011**, *20*, 618–652. [CrossRef]

263. Gerasimov, T.; De Lorenzis, L. On penalization in variational phase-field models of brittle fracture. *arXiv* **2018**, arXiv:1811.05334.

264. Amor, H.; Marigo, J.J.; Maurini, C. Regularized formulation of the variational brittle fracture with unilateral contact: Numerical experiments. *J. Mech. Phys. Solids* **2009**, *57*, 1209–1229. [CrossRef]

265. Ambati, M.; Gerasimov, T.; De Lorenzis, L. A review on phase-field models of brittle fracture and a new fast hybrid formulation. *Comput. Mech.* **2015**, *55*, 383–405. [CrossRef]

266. Wu, J.Y. A unified phase-field theory for the mechanics of damage and quasi-brittle failure. *J. Mech. Phys. Solids* **2017**, *103*, 72–99. [CrossRef]

267. Bourdin, B.; Francfort, G.A.; Marigo, J.J. Numerical experiments in revisited brittle fracture. *J. Mech. Phys. Solids* **2000**, *48*, 797–826. [CrossRef]

268. Karma, A.; Kessler, D.A.; Levine, H. Phase-field model of mode III dynamic fracture. *Phys. Rev. Lett.* **2001**, *87*, 045501. [CrossRef] [PubMed]

269. Kuhn, C.; Schlüter, A.; Müller, R. On degradation functions in phase field fracture models. *Comput. Mater. Sci.* **2015**, *108*, 374–384. [CrossRef]

270. Lorentz, E.; Cuvilliez, S.; Kazymyrenko, K. Modelling large crack propagation: From gradient damage to cohesive zone models. *Int. J. Fract.* **2012**, *178*, 85–95. [CrossRef]

271. Alessi, R.; Marigo, J.J.; Vidoli, S. Gradient damage models coupled with plasticity: Variational formulation and main properties. *Mech. Mater.* **2015**, *80*, 351–367. [CrossRef]

272. Bellettini, G.; Coscia, A. Discrete approximation of a free discontinuity problem. *Numer. Funct. Anal. Optim.* **1994**, *15*, 201–224. [CrossRef]

273. Hillerborg, A.; Modéer, M.; Petersson, P.E. Analysis of crack formation and crack growth in concrete by means of fracture mechanics and finite elements. *Cem. Concr. Res.* **1976**, *6*, 773–781. [CrossRef]

274. Pham, K.; Ravi-Chandar, K.; Landis, C. Experimental validation of a phase-field model for fracture. *Int. J. Fract.* **2017**, *205*, 83–101. [CrossRef]

275. Shao, Y.; Duan, Q.; Qiu, S. Adaptive consistent element-free Galerkin method for phase-field model of brittle fracture. *Comput. Mech.* **2019**. [CrossRef]

276. Chowdhury, M.S.; Song, C.; Gao, W. Highly accurate solutions and Padé approximants of the stress intensity factors and T-stress for standard specimens. *Eng. Fract. Mech.* **2015**, *144*, 46–67. [CrossRef]

277. Wu, J.Y.; Nguyen, V.P.; Nguyen, C.T.; Sutula, D.; Bordas, S.; Sinaie, S. Phase field modeling of fracture. *Adv. Appl. Mech. Multi-Scale Theory Comput.* **2018**, *53*, in press.

278. Winkler, B.J. *Traglastuntersuchungen von Unbewehrten und Bewehrten Betonstrukturen auf der Grundlage eines Objektiven Werkstoffgesetzes Für Beton*; Innsbruck University Press: Innsbruck, Austria, 2001.

279. Gerasimov, T.; De Lorenzis, L. A line search assisted monolithic approach for phase-field computing of brittle fracture. *Comput. Methods Appl. Mech. Eng.* **2016**, *312*, 276–303. [CrossRef]

280. Heister, T.; Wheeler, M.F.; Wick, T. A primal-dual active set method and predictor-corrector mesh adaptivity for computing fracture propagation using a phase-field approach. *Comput. Methods Appl. Mech. Eng.* **2015**, *290*, 466–495. [CrossRef]

281. Singh, N.; Verhoosel, C.; De Borst, R.; Van Brummelen, E. A fracture-controlled path-following technique for phase-field modeling of brittle fracture. *Finite Elem. Anal. Des.* **2016**, *113*, 14–29. [CrossRef]

282. Brun, M.K.; Wick, T.; Berre, I.; Nordbotten, J.M.; Radu, F.A. An iterative staggered scheme for phase field brittle fracture propagation with stabilizing parameters. *arXiv* **2019**, arXiv:1903.08717.

283. Nagaraja, S.; Elhaddad, M.; Ambati, M.; Kollmannsberger, S.; De Lorenzis, L.; Rank, E. Phase-field modeling of brittle fracture with multi-level hp-FEM and the finite cell method. *Comput. Mech.* **2017**, *63*, 1283–1300. [CrossRef]

284. Patil, R.; Mishra, B.; Singh, I. An adaptive multiscale phase field method for brittle fracture. *Comput. Methods Appl. Mech. Eng.* **2018**, *329*, 254–288. [CrossRef]

285. Gerasimov, T.; Noii, N.; Allix, O.; De Lorenzis, L. A non-intrusive global/local approach applied to phase-field modeling of brittle fracture. *Adv. Model. Simul. Eng. Sci.* **2018**, *5*, 14. [CrossRef] [PubMed]

286. Kuhn, C.; Müller, R. Phase field simulation of thermomechanical fracture. *Proc. Appl. Math. Mech.* **2009**, *9*, 191–192. [CrossRef]
287. Nguyen, V.P.; Lian, H.; Rabczuk, T.; Bordas, S. Modelling hydraulic fractures in porous media using flow cohesive interface elements. *Eng. Geol.* **2017**, *225*, 68–82. [CrossRef]
288. Zhou, S.; Zhuang, X.; Rabczuk, T. Phase-field modeling of fluid-driven dynamic cracking in porous media. *Comput. Methods Appl. Mech. Eng.* **2019**, *350*, 169–198. [CrossRef]
289. Wu, T.; Lorenzis, L.D. A phase-field approach to fracture coupled with diffusion. *Comput. Methods Appl. Mech. Eng.* **2016**, *312*, 196–223. [CrossRef]

MDPI

St. Alban-Anlage 66

4052 Basel

Switzerland

Tel. +41 61 683 77 34

Fax +41 61 302 89 18

www.mdpi.com

Applied Sciences Editorial Office

E-mail: applsci@mdpi.com

www.mdpi.com/journal/applsci

www.ingramcontent.com/pod-product-compliance
Lightning Source LLC
Chambersburg PA
CBHW051707210326
41597CB00032B/5398